T0133269

Aquatic and Standing Water Plants of the Central Midwest

Books in the Aquatic and Standing Water Plants of the Central Midwest Series by Robert H. Mohlenbrock

Cyperaceae: Sedges

Filicineae, Gymnospermae, and Other Monocots, Excluding Cyperaceae: Ferns, Conifers, and Other Monocots, Excluding Sedges

Acanthaceae to Myricaceae: Water Willows to Wax Myrtles

Nelumbonaceae to Vitaceae: Water Lotuses to Grapes

Other Southern Illinois University Press Books by Robert H. Mohlenbrock

Guide to the Vascular Flora of Illinois, revised and enlarged edition

Distribution of Illinois Vascular Plants, with Douglas M. Ladd

A Flora of Southern Illinois, with John W. Voigt

In the Illustrated Flora of Illinois Series

Ferns, 2nd edition

Flowering Plants: Basswoods to Spurges

Flowering Plants: Flowering Rush to Rushes

Flowering Plants: Hollies to Loasas

Flowering Plants: Lilies to Orchids

Flowering Plants: Magnolias to Pitcher Plants

Flowering Plants: Nightshades to Mistletoe

Flowering Plants: Pokeweeds, Four-o'clocks, Carpetweeds, Cacti, Purslanes, Goosefoots, Pigweeds, and Pinks

Flowering Plants: Smartweeds to Hazelnuts

Flowering Plants: Willows to Mustards

Grasses: Bromus to Paspalum, 2nd edition

Grasses: Panicum to Danthonia, 2nd edition

Sedges: Carex

Sedges: Cyperus to Scleria, 2nd edition

Aquatic and Standing Water Plants of the Central Midwest

Nelumbonaceae to Vitaceae

Water Lotuses to Grapes

Robert H. Mohlenbrock

Southern
Illinois
University Press

*Carbondale and
Edwardsville*

Printed in the United States of America
13 12 11 10 4 3 2 1

Cover Illustration by Mark W. Mohlenbrock
Library of Congress Cataloging-in-Publication Data
Mohlenbrock, Robert H., 1931–
Nelumbonaceae to Vitaceae : water lotuses to grapes / Robert H. Mohlenbrock.
p. cm. — (Aquatic and standing water plants of the central Midwest)
Includes bibliographical references and index.
ISBN-13: 978-0-8093-2894-9 (alk. paper)
ISBN-10: 0-8093-2894-1 (alk. paper)
1. Angiosperms—Middle West—Identification. 2. Angiosperms—Middle West—Pictorial works. I. Title. II. Series: Mohlenbrock, Robert H., 1931– Aquatic and standing water plants of the central Midwest
QK128.M645 2010
581.977—dc22 2009025824

Printed on recycled paper. ♻

The paper used in this publication meets the minimum requirements of American National Standard for Information Sciences—Permanence of Paper for Printed Library Materials, ANSI Z39.48-1992. ∞

This book is dedicated to Grenville Lucas, former director of Kew Gardens and former chairman of the Species Survival Commission of the International Union for the Conservation of Nature, who selected me to chair the North American Plant Group of the Species Survival Commission, a position I held for fourteen years.

Contents

Illustrations

Series Preface

The purpose of the four books in the Aquatic and Standing Water Plants of the Central Midwest series is to provide illustrated guides to the plants of the central Midwest that may live in standing or running water at least three months a year, though a particular species may not necessarily live in standing or running water during a given year. The states covered by these guides include Iowa, Illinois, Indiana, Ohio, Kansas, Kentucky, Missouri, and Nebraska, except for the Cumberland Mountain region of eastern Kentucky, which is in a different biological province. Since 1990, I have taught weeklong wetland plant identification courses in all of these states on several occasions.

The most difficult task has been to decide what plants to include and what plants to exclude from these books. Three groups of plants are within the guidelines of the manuals. One group includes those aquatic plants that spend their entire life with their vegetative parts either completely submerged or at least floating on the water's surface.

This group includes obvious submerged aquatics such as *Ceratophyllum*, the Najadaceae, the Potamogetonaceae, *Elodea, Cabomba, Brasenia, Nymphaea,* some species of *Ranunculus, Utricularia,* and a few others.

Plants in a second group are called emergents. These plants typically are rooted under water, with their vegetative parts standing above the water surface. Many of these plants can live for a long period of time, even their entire life, out of the water. Included in this group are *Sagittaria, Alisma, Peltandra, Pontederia, Saururus, Justicia,* and several others.

The most difficult group of plants that I had to consider is made up of those wetland plants that live most or all of their lives out of the water, but which on occasion can live at least three months in water. I concluded that I would include within these books only those species that I personally have observed in standing water during the year, or which have been reported in the literature as living in water. In this last group, for example, I have included *Poa annua*, since Yatskievych, in his *Steyermark's Flora of Missouri* (1999), indicates that this species may occur in standing water, even though I have not observed this myself.

In these books, I have included most plants that live in lakes, ponds, rivers, and streams, as well as plants that live in marshes, bogs, fens, wet meadows, sedge meadows, wet prairies, swampy woods, and temporary depressions in woods, on cliffs, and in barrens.

Swink and Wilhelm (1999) consider a marsh as a transition between aquatic communities and drier communities, or in large flats which are regularly inundated by shallow surface water for much of the growing season. Cattails are often frequent in marshes. Trees are generally uncommon in a typical marsh.

Wet meadows, as treated here, are similar to marshes, but without pools of water. Woody plants are generally absent, and cattails do not dominate and are often not present.

Sedge meadows are similar, except that the overwhelming majority of species is composed of members of the sedge family (Cyperaceae). Wet prairies are also similar, but the dominant vegetation consists of grasses. Ladd (1995) recognizes dry, moist, and wet prairies, and only the last of these is included in these manuals.

Fens are wetlands where the underlying groundwater is rich with calcium or magnesium carbonates. Fens may or may not have woody plants present. Swink and Wilhelm (1999) also recognize marly fens that occur on open prairie slopes and hillside fens that are wooded seeps on steep bluffs.

Bogs are habitats typified by acidic, usually organic, substrates. However, bogs that have been influenced by carbonate-rich water have been called alkaline bogs by Swink and Wilhelm. Where minerotrophic water is insignificant, acid bogs develop. Floating sedge mats are acidic and often develop in sand flats or basins that rise and fall with the water table.

Swampy woods are forested wetlands in poorly drained flats or basins. In areas around Lake Michigan, swamps occur in wet sandy flats and on the moraine in wet depressions and in large flats behind the high dunes. In the southern part of the central Midwest, swampy woods may have standing water throughout the year.

Occasional depressions that fill with water harbor species that are included in these books. These depressions may occur in various kinds of woodlands, on exposed sandstone blufftops, and, in Kentucky, even on rocky barrens. Seeps may occur on the faces of cliffs, and the plants that live in this constantly wet habitat are also included in these books.

It is likely that I failed to include a few plants that should have been included, but that I had not observed myself.

The nomenclature that I have used in these manuals reflects my own opinion as to what I believe the scientific names should be. If these names differ from those used by the United States Fish and Wildlife Service, I have indicated this. A partial list of synonymy is included for each species, particularly accounting for synonyms that have been in use for several decades.

After the description of each plant, I have indicated the habitats in which the plant may be found, followed by the states in which the plant occurs. I have indicated the U.S. Fish and Wildlife Service wetland designation for each species for the states in which each occurs. In 1988, the National Wetland Inventory Section of the United States Fish and Wildlife Service attempted to give a wetland designation for every plant occurring in the wild in the United States. The states covered by these aquatic manuals occur in three regions of the Fish and Wildlife Service. Kentucky and Ohio are in region 1; Illinois, Indiana, Iowa, and Missouri are in region 3; and Kansas and Nebraska are in region 5. Definitions of the Fish and Wildlife Service wetland categories are:

OBL (Obligate Wetland). Occur almost always under natural conditions in wetlands, at least 99% of the time.

FACW (Facultative Wetland). Usually occur in wetlands 67–99% of the time, but occasionally found in non-wetlands.

FAC (Facultative). Equally likely to occur in wetlands or non-wetlands 34–66% of the time.

FACU (Facultative Upland). Usually occur in non-wetlands 67–99% of the time, but occasionally found in wetlands.

UPL (Upland). Occur in uplands at least 99% of the time, but under natural conditions not found in wetlands.

NI (Not Indicated). Due to insufficient information.

A plus or minus sign (+ or –) may appear after FACW, FAC, and FACU. The plus means leaning toward a wetter condition; the minus means leaning toward a drier condition.

Although the Fish and Wildlife Service made changes to the wetland status of several species in an updated version in 1997, this later list has never been approved by the Congress of the United States.

Following this is one or more common names currently employed in the central Midwest. A brief discussion of distinguishing characteristics and nomenclatural notes is often included. Illustrations accompany each species, showing the diagnostic characteristics. In some of the illustrations, a gap in the stem signifies that a portion of the stem has been omitted due to space limitations.

The sequence of families in these aquatic manuals is as follows:

1. Azollaceae
2. Blechnaceae
3. Equisetaceae
4. Isoetaceae
5. Lycopodiaceae
6. Marsileaceae
7. Onocleaceae
8. Osmundaceae
9. Thelypteridaceae
10. Pinaceae
11. Taxodiaceae
12. Acoraceae
13. Alismataceae
14. Araceae
15. Butomaceae
16. Cyperaceae
17. Eriocaulaceae
18. Hydrocharitaceae
19. Iridaceae
20. Juncaceae
21. Juncaginaceae
22. Lemnaceae
23. Maranthaceae
24. Najadaceae
25. Orchidaceae
26. Poaceae
27. Pontederiaceae
28. Potamogetonaceae
29. Ruppiaceae
30. Scheuchzeriaceae
31. Sparganiaceae
32. Typhaceae
33. Xyridaceae
34. Zannichelliaceae
35. Acanthaceae
36. Aceraceae
37. Amaranthaceae
38. Anacardiaceae
39. Apiaceae
40. Apocynaceae
41. Aquifoliaceae
42. Araliaceae
43. Aristolochiaceae
44. Asclepiadaceae
45. Asteraceae
46. Balsaminaceae
47. Betulaceae
48. Bignoniaceae
49. Boraginaceae
50. Brassicaceae
51. Cabombaceae
52. Caesalpiniaceae
53. Callitrichaceae
54. Campanulaceae
55. Caprifoliaceae
56. Caryophyllaceae
57. Ceratophyllaceae
58. Convolvulaceae
59. Cornaceae
60. Corylaceae
61. Cuscutaceae
62. Droseraceae
63. Elatinaceae
64. Ericaceae
65. Escalloniaceae
66. Euphorbiaceae
67. Fabaceae
68. Fagaceae
69. Gentianaceae
70. Grossulariaceae
71. Haloragidaceae
72. Hamamelidaceae
73. Hippuridaceae
74. Hydrophyllaceae
75. Hypericaceae
76. Juglandaceae
77. Lamiaceae
78. Lauraceae
79. Leitneriaceae
80. Lentibulariaceae
81. Limnanthaceae

82. Linaceae
83. Lythraceae
84. Malvaceae
85. Melastomaceae
86. Menyanthaceae
87. Molluginaceae
88. Myricaceae
89. Nelumbonaceae
90. Nymphaeaceae
91. Nyssaceae
92. Oleaceae
93. Onagraceae
94. Parnassiaceae
95. Plantaginaceae
96. Platanaceae
97. Podostemaceae
98. Polemoniaceae
99. Polygalaceae
100. Polygonaceae
101. Primulaceae
102. Ranunculaceae
103. Rhamnaceae
104. Rosaceae
105. Rubiaceae
106. Salicaceae
107. Sarraceniaceae
108. Saururaceae
109. Saxifragaceae
110. Scrophulariaceae
111. Solanaceae
112. Styracaceae
113. Ulmaceae
114. Urticaceae
115. Valerianaceae
116. Verbenaceae
117. Violaceae
118. Vitaceae

This volume consists of families 89 to 118. Volume 3 has the remaining dicots, from families 35–88. The first volume in the series was devoted exclusively to the Cyperaceae, family 16. The second volume included the ferns, gymnosperms, and monocots, families 1–15 and 17–34.

Aquatic and Standing Water Plants
of the Central Midwest

The Central Midwest

Descriptions and Illustrations

General Key to Groups of Aquatic and Wetland Dicots

1. Leaves pitcherlike .... 107. Sarraceniaceae
1. Leaves not pitcherlike, or absent.
  2. Plants without chlorophyll .... 61. Cuscutaceae*
  2. Plants with chlorophyll.
    3. Leaves reduced to scales .... *Bartonia* in 69. Gentianaceae*
    3. Leaves not reduced to scales.
      4. Plants aquatic, living in water the entire year .... Group 1
      4. Plants terrestrial, at least part of the year.
        5. Plants with latex .... Group 2
        5. Plants without latex.
          6. Some part of plant prickly or spiny .... Group 3
          6. Plants not prickly nor spiny.
            7. Plants climbing or twining .... Group 4
            7. Plants erect, ascending, prostrate, or trailing, not climbing nor twining.
              8. Plants woody (excluding woody vines that are in Group 4).
                9. Leaves opposite or whorled .... Group 5
                9. Leaves alternate.
                  10. Leaves simple, entire, toothed, or lobed .... Group 6
                  10. Leaves compound .... Group 7
              8. Plants herbaceous.
                11. Leaves all basal or, at most, with one cauline leaf .... Group 8
                11. Leaves cauline (basal leaves may also be present).
                  12. Leaves, or some of them, whorled .... Group 9
                  12. Leaves opposite or alternate.
                    13. At least some of the leaves opposite.
                      14. Leaves simple.
                        15. Leaves entire .... Group 10
                        15. Leaves toothed or lobed .... Group 11
                      14. Leaves compound .... Group 12
                    13. All leaves alternate.
                      16. Leaves simple.
                        17. Leaves entire .... Group 13
                        17. Leaves toothed or lobed .... Group 14
                      16. Leaves compound.
                        18. Leaves trifoliolate or ternate .... Group 15
                        18. Leaves pinnately or palmately compound.
                          19. Leaves pinnately compound .... Group 16
                          19. Leaves palmately compound .... Group 17

Group 1. Plants aquatic, living in standing water the entire year.

1. Plants attached to rocks or wood in flowing water .... 97. Podostemaceae
1. Plants not attached to rocks or wood in flowing water.
  2. Some of the leaves deeply divided or compound.
    3. Flowers borne together in a solitary, yellow head, each flower sharing a common receptacle .... *Megalodonta* in 45. Asteraceae*

* Family names marked with an asterisk are included in volume 3.

3. Flowers not crowded, each with its own receptacle.
4. Leaves trifoliolate .... *Menyanthes* in 86. Menyanthaceae*
4. Leaves 5- to 9-parted or pinnatisect.
5. Leaves 5- to 9-parted .... *Nasturtium* in 50. Brassicaceae*
5. Leaves pinnatisect.
6. Underwater structures bearing small bladders; flowers zygomorphic .... 80. Lentibulariaceae*
6. Underwater structures not bearing bladders; flowers actinomorphic.
7. Some or all the flowers unisexual.
8. Calyx (or involucre) 8- to 12-cleft; stamens 10–20; ovary unlobed .... 57. Ceratophyllaceae*
8. Calyx 3- or 4-cleft; stamens 3–8; ovary 2- to 4-lobed .... 71. Haloragidaceae*
7. All flowers perfect.
9. Pistils 2–several, free, or pistil 1 but deeply 2- to 4-lobed.
10. Sepals 3; petals 3; stamens 3 or 6 .... 51. Cabombaceae*
10. Sepals 4–5; petals 0, 4, or 5; stamens 3, 4, 8, or numerous.
11. Sepals 5; petals 5; stamens numerous .... *Ranunculus* in 102. Ranunculaceae
11. Sepals 4; petals 0 or 4; stamens 3, 4, or 8 .... 71. Haloragidaceae*
9. Pistil 1, unlobed.
12. Sepals 5; petals 5, united at base; stamens 5 .... *Hottonia* in 101. Primulaceae
12. Sepals 4; petals 4, free; stamens 6 .... *Neobeckia* in 50. Brassicaceae*
2. Leaves simple, none deeply divided.
13. Leaves with conspicuous sheathing stipules (ocreae) at base .... 100. Polygonaceae
13. Leaves without sheathing stipules at base.
14. Flowers unisexual; petals absent.
15. Calyx absent; fruit heart-shaped .... 53. Callitrichaceae*
15. Sepals 3; fruit ellipsoid .... *Proserpinaca* in 71. Haloragidaceae*
14. Flowers perfect; perianth, or at least the calyx, present.
16. Leaves peltate.
17. Sepals, petals, and stamens each 5 .... *Hydrocotyle* in 39. Apiaceae*
17. Sepals, petals, and stamens some number other than 5.
18. Sepals 3–4; petals 3–4; stamens 12–18; leaves up to 6 cm across, covered with a gelatinous material .... *Brasenia* in 51. Cabombaceae*
18. Sepals and petals indistinguishable, together totaling more than 8 segments; stamens more than 20; leaves more than 6 cm across, not gelatinous .... 89. Nelumbonaceae
16. Leaves not peltate.
19. Leaves opposite or whorled.
20. Stamen 1; leaves linear .... 73. Hippuridaceae*
20. Stamens 3–5; leaves not linear.
21. Sepals and petals usually 3 each .... *Elatine* in 63. Elatinaceae*
21. Sepals and petals (if present) usually 4 or 5 each.
22. Corolla absent .... *Didiplis* in 83. Lythraceae*
22. Corolla present.
23. Stamens 2; leaves lanceolate or ovate .... *Justicia* in 35. Acanthaceae*
23. Stamens 4; leaves more or less orbicular to oblong .... *Bacopa* in 110. Scrophulariaceae
19. Leaves alternate or basal.
24. Petals absent (sepals appearing petaloid in *Caltha*).

25. Flowers white; pistil 1 .... 108. Saururaceae
25. Flowers yellow; pistils 4–12 .... *Caltha* in 102. Ranunculaceae
24. Petals present (what appear to be petals in *Caltha* are actually sepals).
26. Petals united .... *Nymphoides* in 86. Menyanthaceae*
26. Petals free.
27. Flowers borne in umbels; stamens 5 .... *Hydrocotyle* in 39. Apiaceae*
27. Flowers not borne in umbels; stamens 8–numerous.
28. Petals 4–6; stamens 8–11 .... *Ludwigia* in 93. Onagraceae
28. Petals more than 6; stamens more than 12 .... 90. Nymphaeaceae

Group 2. Plants terrestrial, at least part of the year; plants with latex.

1. Many ray flowers crowded into heads, each sharing the same receptacle .... *Prenanthes* in 45. Asteraceae*
1. Flowers not in heads and not sharing the same receptacle.
2. Leaves opposite.
3. Pistils 2; fruit in pairs; stems often pink or purple .... 40. Apocynaceae*
3. Pistil 1; fruit solitary; stems not pink or purple .... 44. Asclepiadaceae*
2. Leaves alternate.
4. Leaves toothed .... 54. Campanulaceae*
4. Leaves entire.
5. Plants scabrous; leaves up to 5 mm wide .... 54. Campanulaceae*
5. Plants not scabrous; leaves more than 5 mm wide.
6. Pistils 2; fruit in pairs; stems often pink or purple .... 40. Apocynaceae*
6. Pistil 1; fruit solitary; stems not pink or purple .... 44. Asclepiadaceae*

Group 3. Plants terrestrial, at least part of the year; some part of plant prickly or spiny.

1. Plants woody.
2. Leaves compound.
3. Leaves twice-compound .... *Gleditsia* in 52. Caesalpiniaceae*
3. Leaves once-compound or trifoliolate.
4. Leaves trifoliolate.
5. Base of petiole dilated .... *Rosa* in 104. Rosaceae
5. Base of petiole not dilated .... *Rubus* in 104. Rosaceae
4. Leaves once-pinnate, with 5 or more leaflets.
6. Leaflets more than 5.
7. Stamens more than 10; shrubs .... 104. Rosaceae
7. Stamens 5; trees .... *Gleditsia* in 52. Caesalpiniaceae*
6. Leaflets 5.
8. Leaves pinnately compound .... *Rosa* in 104. Rosaceae
8. Leaves palmately compound .... *Rubus* in 104. Rosaceae
2. Leaves simple.
9. Stamens 15 or more .... 104. Rosaceae
9. Stamens 5.
10. Woody vines .... *Solanum* in 111. Solanaceae
10. Shrubs or trees.
11. Leaves usually lobed .... 70. Grossulariaceae*
11. Leaves not lobed .... 103. Rhamnaceae
1. Plants herbaceous.
12. Flowers crowded into heads, each flower sharing the same receptacle .... *Cirsium* in 45. Asteraceae*

12. Flowers not crowded into heads and not sharing the same receptacle.
13. Petals absent; leaves with a sheathing stipule .... 100. Polygonaceae
13. Petals present; leaves without a sheathing stipule .... 74. Hydrophyllaceae*

Group 4. Plants terrestrial, at least part of the year; plants climbing or twining.

1. Plants woody.
2. Leaves simple.
3. Leaves opposite .... *Lonicera* in 55. Caprifoliaceae*
3. Leaves alternate.
4. Some of the leaves lobed.
5. Plants without tendrils .... 111. Solanaceae
5. Plants with tendrils.. 118. Vitaceae
4. None of the leaves lobed.
8. Leaves toothed .... 118. Vitaceae
8. Leaves entire .... 43. Aristolochiaceae*
2. Leaves compound.
9. Leaves opposite..48. Bignoniaceae*
9. Leaves alternate.
10. Leaves bipinnate or tripinnate .... 118. Vitaceae
10. Leaves once-pinnate or trifoliolate.
11. Leaves pinnate, with 5 or more leaflets ....67. Fabaceae*
11. Leaves trifoliolate .... 38. Anacardiaceae*
1. Plants herbaceous.
12. Leaves opposite.
13. Leaves compound .... *Clematis* in 102. Ranunculaceae
13. Leaves simple.
14. Plants with latex; leaves entire.................. *Trachelospermum* in 40. Apocynaceae*
14. Plants without latex; leaves toothed .... *Mikania* in 45. Asteraceae*
12. Leaves alternate.
15. Leaves compound ....67. Fabaceae*
15. Leaves simple, sometimes with a pair of tiny lobes at base.
16. Tendrils present; stamens 8................................. *Brunnichia* in 48. Polygonaceae
16. Tendrils absent; stamens 5.
17. Some part of plant prickly............................... . *Tracaulon* in 48. Polygonaceae
17. Plants not prickly.
18. Leaves entire... 58. Convolvulaceae*
18. Some leaves with a pair of basal lobes............................. 111. Solanaceae

Group 5. Plants terrestrial, at least part of the year; plants woody (excluding vines); leaves opposite or whorled.

1. Leaves simple.
2. Some or all the leaves whorled.
3. Flowers white; petals 4, united; stamens 4; ovary inferior; flowers and fruits in globose clusters .... *Cephalanthus* in 105. Rubiaceae
3. Flowers pink; petals 5, free; stamens 10; ovary superior; flowers and fruits in axillary whorls.. *Decodon* in 83. Lythraceae*
2. All leaves opposite.
4. Leaves entire.
5. Sepals 5; petals 5 .... 55. Caprifoliaceae*
5. Sepals 4; petals 4.
6. Flowers in globose heads, white .... *Cephalanthus* in 105. Rubiaceae
6. Flowers not in globose heads, yellow or white.

7. Petals united at base, yellow; stamens 2 .... *Forestiera* in 92. Oleaceae
7. Petals free, white; stamens 4 .... 59. Cornaceae*
4. Leaves toothed or lobed.
8. Leaves lobed.
9. Petals small and inconspicuous or absent; fruit a samara............... 36. Aceraceae*
9. Petals white, conspicuous; fruit fleshy.................. *Viburnum* in 55. Caprifoliaceae*
8. Leaves toothed.
10. Sepals 4; petals 0; stamens 2 .... *Forestiera* in 92. Oleaceae
10. Sepals 5; petals 5; stamens 5 .... 55. Caprifoliaceae*
1. Leaves compound.
11. Stamens 2; sepals 4 .... *Fraxinus* in 92. Oleaceae
11. Stamens 5; sepals 5.
12. Petals white; fruit a berry .... *Sambucus* in 55. Caprifoliaceae*
12. Petals yellow or green or absent; fruit a samara................................. 36. Aceraceae*

Group 6. Plants terrestrial, at least part of the year; plants woody (excluding vines); leaves alternate, simple, entire, toothed, or lobed.

1. Leaves entire.
2. Fruit an acorn.. 68. Fagaceae*
2. Fruit not an acorn.
3. Petals absent.
4. Sepals present.
5. Sepals 6; stamens 9; plants aromatic... 78. Lauraceae*
5. Sepals 5; stamens 5 or 10; plants not aromatic.
6. Leaves 3-veined from the base; stamens 5...................... .*Celtis* in 113. Ulmaceae
6. Leaves not 3-veined from the base; stamens 10 .... 91. Nyssaceae
4. Sepals absent.
7. Leaves glandular, aromatic... 88. Myricaceae*
7. Leaves eglandular, not aromatic.
8. Fruit drupelike; leaves 10–15 cm long.................................... 79. Leitneriaceae*
8. Fruit a capsule; leaves up to 8 cm long .... *Salix* in 106. Salicaceae
3. Petals present.
9. Petals free.
10. Ovary superior; hypanthium absent.
11. Sepals absent; petals yellow .... *Nemopanthus* in 41. Aquifoliaceae*
11. Sepals present; petals usually not yellow .... 64. Ericaceae*
10. Ovary inferior; hypanthium present.
12. Stamens numerous.. 104. Rosaceae
12. Stamens 4, 5, or 10 .... 53. Ericaceae*
9. Petals united .... 53. Ericaceae*
1. Leaves toothed or lobed.
13. Leaves toothed.
14. Stems spiny... *Crataegus* in 104. Rosaceae
14. Stems not spiny.
15. Flowers unisexual.
16. Petals present .... 41. Aquifoliaceae*
16. Petals absent.
17. Leaves aromatic, glandular .... 88. Myricaceae*
17. Leaves not aromatic, not glandular.
18. Fruit an acorn... 68. Fagaceae*
18. Fruit not an acorn.
19. Sepals 5; fruit fleshy or a samara............................ 113. Ulmaceae

19. Sepals absent (at least in the staminate flowers); fruit dry, but not a samara.
20. Carpels 2; seeds with a tuft of hairs .... 106. Salicaceae
20. Carpel 1; seeds without a tuft of hairs.
21. Staminate flowers without a calyx; pistillate flowers not in catkins .... 60. Corylaceae*
21. Staminate flowers with a calyx; pistillate flowers in catkins ....47. Betulaceae*
15. Flowers perfect.
22. Petals absent ....103. Rhamnaceae
22. Petals present.
23. Petals 4 ....64. Ericaceae*
23. Petals 5.
24. Petals united.
25. Pubescence stellate...112. Styracaceae
25. Pubescence not stellate, or absent ....64. Ericaceae*
24. Petals free.
26. Stamens 5.
27. Flowers white, in racemes .... 65. Escalloniaceae*
27. Flowers not white, not in racemes................103. Rhamnaceae
26. Stamens 15 or more ....104. Rosaceae
13. Leaves lobed.
28. Leaves star-shaped...72. Hamamelidaceae*
28. Leaves not star-shaped.
29. Flowers unisexual.
30. Petals present; fruit dry, globose .... 96. Platanaceae
30. Petals absent; fruit an acorn..68. Fagaceae*
29. Flowers perfect.
31. Stamens 5 ....70. Grossulariaceae*
31. Stamens 10 or more.. 104. Rosaceae

Group 7. Plants woody (at least part of the year); leaves alternate, compound.

1. Leaves palmately compound, with at least 5 leaflets.
2. Plants prickly; flowers white ....*Rubus* in 104. Rosaceae
2. Plants not prickly; flowers yellow ....*Pentaphylloides* in 104. Rosaceae
1. Leaves pinnately compound.
3. Plants prickly or spiny.
4. Leaves bipinnately or tripinnately compound ....52. Caesalpiniaceae*
4. Leaves once pinnately compound .... 104. Rosaceae
3. Plants neither prickly nor spiny.
5. Flowers, or some of them, unisexual; stamen 1 ....76. Juglandaceae*
5. Flowers perfect.
6. Flowers zygomorphic; stamens 9 or 10...67. Fabaceae*
6. Flowers actinomorphic; stamens 15 or more.. 104. Rosaceae

Group 8. Plants terrestrial, at least part of the year; herbaceous; leaves all basal.

1. Leaves compound.
2. Leaves ternately decompound.
3. Sepals 5; petals 5; stamens 5...42. Araliaceae*
3. Sepals 4 or more; petals absent; stamens numerous....................... 102. Ranunculaceae
2. Leaves trifoliolate ....*Coptis* in 102. Ranunculaceae
1. Leaves simple.

4. Leaves or some of them lobed.
  5. Leaves peltate .... *Hydrocotyle* in 39. Apiaceae*
  5. Leaves not peltate.
    6. Flowers in umbels; plants usually aquatic........ 39. *Hydrocotyle* in 39. Apiaceae*
    6. Flowers not in umbels; plants rarely aquatic........ 109. Saxifragaceae
4. Leaves not lobed.
  7. Leaves toothed.
    8. Petals united, at least at base........ 101. Primulaceae
    8. Petals free.
      9. Stamens numerous........ *Dalibarda* in 104. Rosaceae
      9. Stamens 5........ 117. Violaceae
  7. Leaves entire.
    10. Leaves with sticky hairs........ 62. Droseraceae*
    10. Leaves without sticky hairs.
      11. Sepals 4; petals 4.
        12. Petals white or yellow, free; stamens 6; ovary superior........ 50. Brassicaceae*
        12. Petals translucent, united; stamens 2–4; ovary inferior........ 95. Plantaginaceae
      11. Sepals or calyx lobes 5; petals or corolla lobes 5.
        13. Leaves linear; sepals spurred; pistils numerous........ *Myosurus* in 102. Ranuculaceae
        13. Leaves not linear; sepals not spurred; pistil 1.
          14. Petals united, at least at base........ 101. Primulaceae
          14. Petals free........ 94. Parnassiaceae

Group 9. Plants terrestrial, at least part of the year; herbaceous; leaves whorled.

1. Flowers crowded in heads, each sharing the same receptacle........ 45. Asteraceae*
1. Flowers not crowded in heads, at least not sharing the same receptacle.
  2. Leaves compound........ *Hottonia* in 101. Primulaceae
  2. Leaves simple.
    3. Leaves toothed........ *Veronicastrum* in 110. Scrophulariaceae
    3. Leaves entire.
      4. Flowers zygomorphic; petals 3; stamens 8........ 99. Polygalaceae
      4. Flowers actinomorphic; petals 4–5 (rarely 3 in some species of *Galium*); stamens 3, 4, 5, or 10.
        5. Petals 4........ 105. Rubiaceae
        5. Petals 3 or 5.
          6. Petals free; ovary superior........ 56. Caryophyllaceae*
          6. Petals united; ovary inferior........ 105. Rubiaceae

Group 10. Plants terrestrial, at least part of the year; herbaceous; leaves, or some of them, opposite, simple, entire.

1. Flowers crowded into heads, each sharing the same receptacle........ 45. Asteraceae*
1. Flowers not crowded into heads, at least not sharing the same receptacle.
  2. Latex present.
    3. Pistils 2; fruits borne in pairs........ 40. Apocynaceae*
    3. Pistil 1; fruits borne singly........ 44. Asclepiadaceae*
  2. Latex absent.
    4. Petals absent.
      5. Stamen 1; dwarf plant; fruits heart-shaped........ 53. Callitrichaceae*
      5. Stamens 3–5, 8 or 10 or 20.
        6. Stamens 8 or 10 or 20.

7. Plants lying on the ground .... *Glinus* in 87. Molluginaceae*
7. Plants ascending to erect .... 56. Caryophyllaceae*
6. Stamens 3–5.
8. Sepals 4; stamens 4.
9. Plants prostrate.
10. Ovary inferior .... *Ludwigia* in 93. Onagraceae
10. Ovary superior .... *Didiplis* in 83. Lythraceae*
9. Plants erect .... 83. Lythraceae*
8. Sepals 3 or 5; stamens 3 or 5.
11. Stipules present .... 56. Caryophyllaceae*
11. Stipules absent .... 37. Amaranthaceae*
4. Petals present.
12. Petals free.
13. Stamens 8–numerous.
14. Stamens numerous; leaves often punctate .... 75. Hypericaceae*
14. Stamens 8–12.
15. Stamens 8 .... 85. Melastomaceae*
15. Stamens 9–12.
16. Stamens 9; flowers maroon or pink .... 75. Hypericaceae*
16. Stamens 10–12.; flowers purple or white.
17. Stipules present .... 56. Caryophyllaceae*
17. Stipules absent .... 83. Lythraceae*
13. Stamens 4–5.
18. Sepals 4; petals 4 .... *Rotala* in 83. Lythraceae*
18. Sepals 5; petals 5.
19. Stipules present; flowers white .... 56. Caryophyllaceae*
19. Stipules absent; flowers yellow .... 82. Linaceae*
12. Petals united.
20. Corolla 4-lobed.
21. Ovary superior .... 77. Lamiaceae*
21. Ovary inferior .... 105. Rubiaceae
20. Corolla 5-lobed, or petals 5.
22. Stamens 5; flowers actinomorphic.
23. Corolla rotate .... 101. Primulaceae
23. Corolla tubular.
24. Ovary 1-locular .... 69. Gentianaceae*
24. Ovary 3-locular .... 98. Polemoniaceae
22. Stamens 2 or 4; flowers zygomorphic.
25. Ovary 4-parted; fruit separating into 4 nutlets .... 77. Lamiaceae*
25. Ovary not 4-parted; fruit not separating into 4 nutlets.
26. Flowers nearly actinomorphic .... 35. Acanthaceae*
26. Flowers zygomorphic .... 110. Scrophulariaceae

Group 11. Plants terrestrial, at least part of the year; herbaceous; leaves opposite, simple, or lobed.

1. Flowers crowded into heads, each sharing the same receptacle .... 45. Asteraceae*
1. Flowers not crowded into heads, at least not sharing the same receptacle.
2. Leaves toothed.
3. Petals absent .... 114. Urticaceae
3. Petals present.
4. Petals 4; sepals 4 .... 77. Lamiaceae*
4. Petals 5; sepals 5, or absent.

5. Petals free; leaves glandular-tooted; flowers actinomorphic .... *Bergia* in 63. Elatinaceae*
5. Petals united; leaves not glandular-toothed; flowers zygomorphic.
6. Calyx absent; ovary inferior .... 115. Valerianaceae
6. Calyx present; ovary superior.
7. Ovary 4-parted.
8. Ovary deeply 4-lobed, the style arising from between the lobes .... 77. Lamiaceae*
8. Ovary shallowly 4-lobed or, if more deeply lobed, the style arising from the top of the ovary .... 116. Verbenaceae
7. Ovary single, not 4-parted .... 116. Verbenaceae
2. Leaves lobed.
9. Ovary 4-parted; fruit separating into 4 nutlets .... 77. Lamiaceae*
9. Ovary not 4-parted; fruit solitary .... 110. Scrophulariaceae

Group 12. Plants terrestrial, at least part of the year; herbaceous; leaves opposite, compound.

1. Flowers crowded into heads, each sharing the same receptacle .... 45. Asteraceae*
1. Flowers not crowded into heads, at least not sharing the same receptacle .... 115. Valerianaceae

Group 13. Plants terrestrial, at least part of the year; herbaceous; leaves alternate, simple, entire.

1. Flowers crowded into heads, each sharing the same receptacle .... 45. Asteraceae*
1. Flowers not crowded into heads, at least not sharing the same receptacle.
2. Latex present.
3. Flowers zygomorphic; corolla split nearly to base .... 4. Campanulaceae*
3. Flowers actinomorphic; corolla not split nearly to base.
4. Pistils 2; fruit borne in pairs; flowers deep blue .... *Amsonia* in 40. Apocynaceae*
4. Pistils1; fruit not borne in pairs; flowers not deep blue .... 54. Campanulaceae*
2. Latex absent.
5. Flowers unisexual.
6. Perianth and bracts scarious .... 37. Amaranthaceae*
6. Perianth and bracts not scarious .... 66. Euphorbiaceae*
5. Flowers perfect.
7. Petals absent.
8. Flowers zygomorphic; stamens 6 .... 43. Aristolochiaceae*
8. Flowers actinomorphic; stamens 5 or 8, less commonly 6.
9. Leaves with sheaths at base, not cordate .... 100. Polygonaceae
9. Leaves without sheaths at base, cordate .... 108. Saururaceae
7. Petals present.
10. Flowers zygomorphic; stamens 8; petals 3 .... 99. Polygalaceae
10. Flowers actinomorphic; stamens 4, 5, 8, or 10; petals absent or other than 3.
11. Petals united.
12. Ovary distinctly 4-lobed .... 49. Boraginaceae*
12. Ovary not distinctly 4-lobed.
13. Flowers in a scorpioid cyme .... *Heliotropium* in 49. Boraginaceae*
13. Flowers not in a scorpioid cyme.
14. Ovary inferior .... *Samolus* in 101. Primulaceae
14. Ovary superior.
15. Stigmas 3 .... 98. Polemoniaceae
15. Stigma 1 .... 101. Primulaceae

11. Petals free.
16. Petals 6 .... 83. Lythraceae*
16. Petals 4 or 5.
17. Petals 5.
18. Stamens united into a tube.......... 82. Linaceae*
18. Stamens free .... *Ludwigia* in 93. Onagraceae
17. Petals 4.
19. Stamens 6; ovary superior.......... 50. Brassicaceae*
19. Stamens 8; ovary inferior .... 93. Onagraceae

Group 14. Plants terrestrial, at least for part of the year; herbaceous; leaves simple, alternate, toothed or lobed.

1. Flowers crowded into heads, each sharing the same receptacle.......... 45. Asteraceae*
1. Flowers not in heads, at least not sharing the same receptacle.
2. Leaves toothed.
3. Petals absent.
4. Sepals 5–several, petaloid.
5. Leaves cordate; stamens numerous; pistils 4–12 ...... *Caltha* in 102. Ranunculaceae
5. Leaves not cordate; stamens 10; pistils 5.......... *Penthorum* in 109. Saxifragaceae
4. Sepals 3, not petaloid, green.......... *Proserpinaca* in 71. Haloragidaceae*
3. Petals present.
6. Flowers zygomorphic.
7. Petals free, one of them spurred .... 46. Balsaminaceae*
7. Petals united, not spurred.
8. Corolla tube split; stamens 5 .... *Lobelia* in 54. Campanulaceae*
8. Corolla tube not split; stamens 2 or 4 .... 110. Scrophulariaceae
6. Flowers actinomorphic.
9. Stamens and pistils attached to a central column; stamens numerous .... 84. Malvaceae*
9. Stamens and pistils not attached to a central column; stamens 5, 6, or 8.
10. Ovary inferior; stamens 8 .... 93. Onagraceae
10. Ovary superior; stamens 5 or 6.
11. Sepals 4; petals 4, free; stamens 6 .... 50. Brassicaceae*
11. Sepals 5; petals 5; stamens 5.......... *Eryngium* in 39. Apiaceae*
2. Leaves lobed.
12. Petals absent; sepals 3; stamens 3 .... *Proserpinaca* in 71. Haloragidaceae*
12. Petals present; sepals 4 or more; stamens 2 or 4 or more.
13. Petals free.
14. Stamens and pistils borne on a central column.......... 84. Malvaceae*
14. Stamens and pistils not borne on a central column.
15. Plants trailing; flowers borne in a small purple head .... *Eryngium* in 39. Apiaceae*
15. Plants upright; flowers not borne in a head, yellow or white .... 50. Brassicaceae*
13. Petals united.
16. Flowers zygomorphic; stamens 4 .... 110. Scrophulariaceae
16. Flowers actinomorphic; stamens 5 .... 111. Solanaceae

Group 15. Plants terrestrial; herbaceous; leaves alternate, trifoliolate or ternately compound.

1. Leaves trifoliolate.
2. Flowers zygomorpic; fruit a legume .... 67. Fabaceae*

2. Flowers actinomorphic; fruit not a legume.
3. Stamens 10 or more .... 104. Rosaceae
3. Stamens 5.
4. Petals free; flowers in umbels; ovary inferior .... 39. Apiaceae*
4. Petals united; flowers not in umbels; ovary superior .... *Menyanthes* in 86. Menyanthaceae*
1. Leaves ternately compound.
5. Stamens and pistils numerous; ovary superior .... 102. Ranunculaceae
5. Stamens 5; pistil 1, ovary inferior .... 39. Apiaceae*

Group 16. Plants terrestrial, at least part of the year; herbaceous; leaves alternate, pinnately compound.

1. Flowers crowded into heads, each sharing the same receptacle .... 45. Asteraceae*
1. Flowers not crowded into heads, at least not sharing the same receptacle.
2. Leaves bipinnately or tripinnately compound.
3. Petals absent; ovary or ovaries superior .... 102. Ranunculaceae
3. Petals present; ovary inferior .... 39. Apiaceae*
2. Leaves once pinnately compound.
4. Petals absent; stamens 4 .... *Sanguisorba* in 104. Rosaceae
4. Petals present; stamens some number other than 4.
5. Petals free.
6. Sepals 3; petals 3 .... 81. Limnanthaceae*
6. Sepals 4 or more; petals 4 or more.
7. Sepals 4; petals 4.
8. Stamens 6; ovary superior .... 50. Brassicaceae*
8. Stamens 8; ovary inferior .... 93. Onagraceae
7. Sepals 5 (–7); petals 5 (–7).
9. Stamens usually more than 10 (sometimes as few as 5 in *Agrimonia*); pistils 2–numerous .... 104. Rosaceae
9. Stamens 5, pistil 1 .... 39. Apiaceae*
5. Petals united .... 111. Solanaceae

Group 17. Plants terrestrial, at least part of the year; herbaceous; leaves alternate, palmately compound.

1. Stamens 10 or more; pistils numerous .... 104. Rosaceae
1. Stamens 5; pistil 1 .... *Cynosciadium* in 39. Apiaceae*

89. NELUMBONACEAE—WATER LOTUS FAMILY

Only the following genus comprises the family.

1. **Nelumbo** Adans.—Water Lotus

Large aquatic herbs; leaves peltate, entire; flower solitary, large, perfect, actinomorphic; sepals and petals spirally arranged imperceptibly into one another, the outermost green, the remainder creamy yellow to pink; stamens numerous; pistils numerous, sunken in an enlarged, spongy, top-shaped receptacle; fruit a hard nut sunken in a woody, top-shaped receptacle.

Two species comprise this genus, separated by the color of their flowers. *Nelumbo* differs from all other aquatic genera by its circular, peltate leaves that are not cleft. Some botanists include this family in the Nymphaeaceae.

1. Flowers creamy yellow.. 1. *N. lutea*
1. Flowers pink ....2. *N. nucifera*

1. Nelumbo lutea (Willd.) Pers. 1:92. 1805. Fig. 1.
Nelumbo pentapetala Walt. Fl. Carol. 155. 1788, *nomen illeg.*
Nelumbium luteum Willd. Sp. Pl. 2:1259. 1799.

Large aquatic herbs with stout rhizomes and tubers; leaves circular, peltate, entire, not cleft, up to 70 cm in diameter, glabrous, some floating on the water, others elevated above the water on stout petioles; flower solitary, up to 25 cm across, elevated above the water on a peduncle up to 1 m long; perianth parts numerous, the outer green, the inner creamy yellow, up to 5 cm long; stamens numerous; receptacle flat, spongy, top-shaped, up to 10 cm across, each cavity containing a nut-like, globose fruit up to 1 cm in diameter. June–September.

Rivers, streams, lakes, ponds.

IA, IL, IN, KS, KY, MO, NE, OH (OBL).

Yellow water lotus; yellow water chinquapin.

The tubers and seeds of this species are edible. The seeds, protected by a hard seed coat, may germinate after many years. This species differs from the following non-native species by its creamy yellow flowers.

1. *Nelumbo lutea* (Yellow water lotus). a. Leaf. b. Flower. c. Stamen. d. Fruit. e. Seed.

2. **Nelumbo nucifera** Gaertn. Fruct. & Sem. 1:73. 1788. Not illustrated.

Large aquatic herbs with stout rhizomes and tubers; leaves circular, peltate, entire, not cleft, up to 80 cm in diameter, glabrous, some floating on the water, others elevated above the water on stout petioles; flower solitary, up to 25 cm across, elevated above the water on a peduncle up to 1 m long; perianth parts numerous, the outer green, the inner pink, up to 5 cm long; stamens numerous; receptacle flat, spongy, top-shaped, up to 10 cm across, each cavity containing a nut-like, globose fruit up to 1 cm in diameter. July–September.

Ponds.

MO (OBL).

Oriental sacred lotus; pink water lotus.

This handsome species rarely escapes from cultivation. The seed may remain viable for several hundred years. It is distinguished by its pink flowers.

90. NYMPHAEACEAE—WATER LILY FAMILY

Aquatic perennial herbs from rhizomes; leaves simple, often leathery, cordate at the base or with a narrow cleft, borne on long petioles; flower solitary, perfect, actinomorphic, on a stout peduncle; sepals 4–6, green or petaloid, free from each other; petals 8-numerous, free from each other; stamens numerous; carpels 5 or more, the ovaries superior to inferior, with numerous ovules.

Five genera and about fifty species comprise the Nymphaeaceae. As treated here, the family contains two genera and four species in the central Midwest.

1. Flowers yellow; sepals 5–6; leaves oval, not peltate .... 1. *Nuphar*
1. Flowers white or pinkish; sepals 4; leaves orbicular, peltate .... 2. *Nymphaea*

1. **Nuphar** Sm.—Spatterdock

Aquatic perennial herbs with stout rhizomes; leaves floating, or more commonly elevated above the water, leathery, a little longer than broad, cordate at base; flower solitary, elevated above the water, perfect, actinomorphic; outer sepals 5–6, concave, green (at least on the outside), inner sepals like the outer ones, but yellow; petals numerous, small, almost stamen-like, yellow; stamens numerous, attached to the receptacle beneath the ovary; stigmas disk-like, with up to 25 rays; fruit not surrounded by the remains of the petals.

Approximately fifteen species found in the north temperate regions of the world comprise this genus, although several additional species have been proposed due to the variability of some of the species.

Two species are recognized from the central Midwest.

1. Petioles terete or more or less flattened; sepals green .... 1. *N. advena*
1. Petioles conspicuously flattened; sepals red-tinged .... 2. *N. variegatum*

1. **Nuphar advena** (Soland.) R. Br. in Ait. Hort. Kew., ed. 2, 3:295. 1811. Fig. 2.
Nymphaea advena Soland. in Ait. Hort. Kew. 2:226. 1789.
Nuphar macrophyllum Small, Fl. Graec. Prodr. 1:361. 1809.
Nuphar luteum (L.) Sibth. & Smith ssp. *macrophyllum* (Small) Beal, Journ. Elisha Mitchell Sci. Soc. 72:332. 1956.

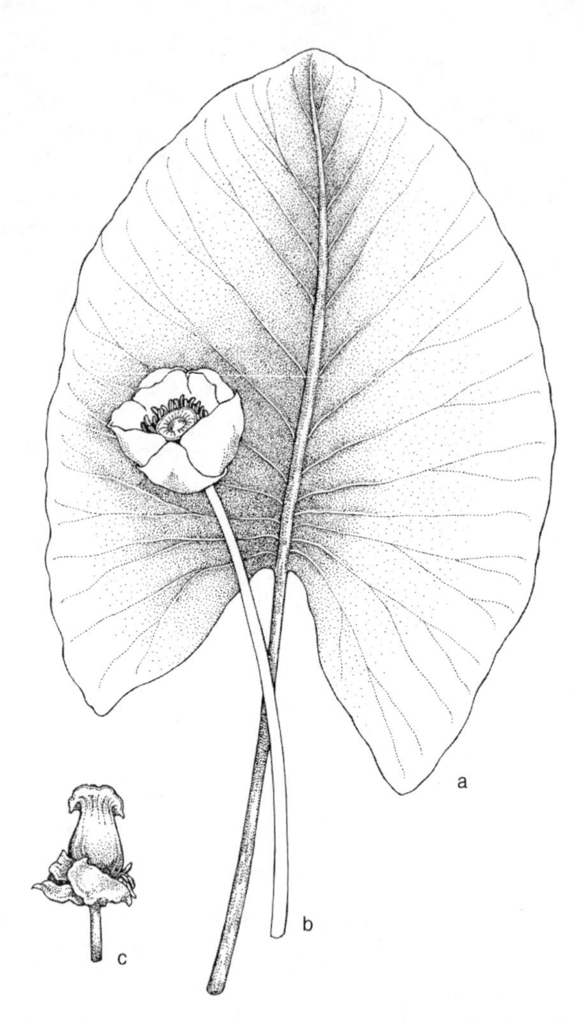

2. *Nuphar advena* (Spatterdock).
a. Leaf.
b. Flower.
c. Flower, with perianth removed.

Aquatic perennial herbs from stout rhizomes; leaves elevated above the water or floating on the surface, ovate or at least longer than broad, subacute to obtuse at the apex, cordate at the base, to 40 cm long, nearly as wide, leathery, glabrous, on terete or somewhat flattened petioles; flower solitary, 2–4 cm high, elevated above the water on a stout peduncle; outer sepals 5–6, ovate, concave, green on the outside, green or yellowish on the inside; inner sepals yellow, about the same size as the outer ones; petals numerous, yellow, small, stamenlike; stamens numerous, on short filaments; fruit 2–5 cm high, green, becoming blackish. May–September.

Swamps, ponds, lakes.

IA, IL, IN, KS, KY, MO, NE, OH (OBL).

Spatterdock; cow lily; pond lily.

The leaves are longer than wide, and the flowers are club-shaped, readily distinguishing *Nuphar* from *Nymphaea. Nuphar advena* differs from *N. variegatum* by lacking purple coloration on the inner face of the outer sepals and by having terete petioles.

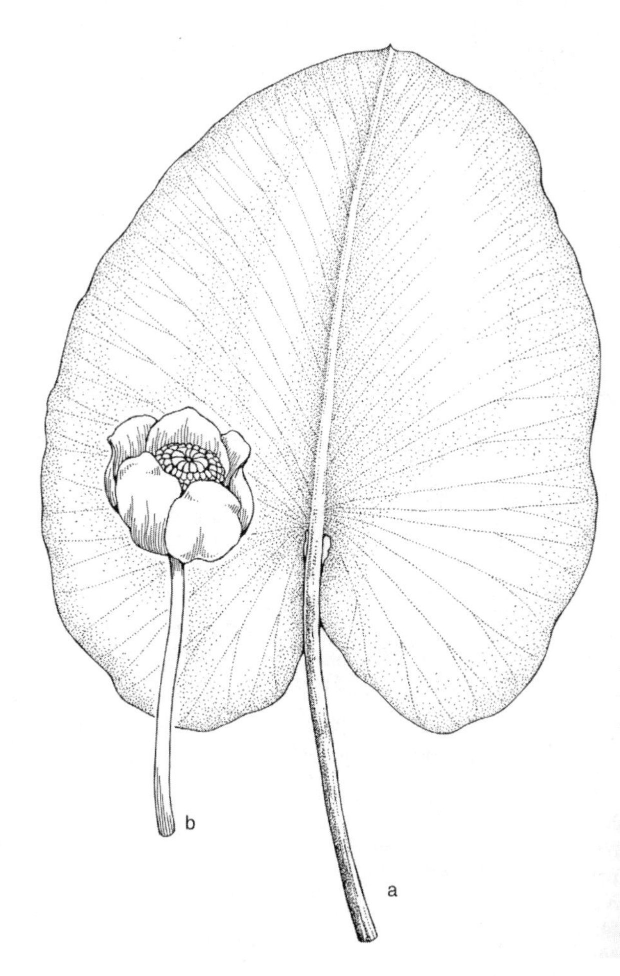

3. *Nuphar variegatum* (Bullhead lily).
a. Leaf.
b. Flower.

2. Nuphar variegatum Engelm. Ann. Rep. State Bot. 19:73. 1809. Fig. 3.
Nuphar luteum (L.) Sibth. & Smith ssp. *variegatum* (Engelm.) Beal, Journ. Elisha Mitchell Sci. Soc. 72:330. 1956.

Aquatic perennial herbs from stout rhizomes; leaves elevated above the water or floating on the surface, ovate or at least longer than broad, subacute to obtuse at the apex, cordate at the base, to 40 cm long, nearly as wide, leathery, glabrous, the petioles flat on the upper side; flower solitary, 2–4 cm high, elevated above the water on a stout peduncle; outer sepals 5–6, ovate, concave, green on the outside, purplish on the inside; inner sepals yellow, about the same size as the outer ones; petals numerous, yellow, small, stamen-like; stamens numerous, on short filaments; fruit 2–5 cm high, green, becoming blackish. May–September.

Ponds, lakes, swamps.

IA, IL, IN, KS, OH (OBL).

Bullhead lily.

This species is very similar to *N. advena*, differing by its outer sepals purplish on the inner face.

2. **Nymphaea** L.—Water Lily

Aquatic perennials with creeping rhizomes; leaves large, alternate from along the rhizome, long-petiolate, with a deep basal sinus; flowers bisexual, showy, white or pink (in the central Midwest), on stout peduncles; sepals 4, free, green; petals numerous, free, gradually passing into the stamens; stamens numerous, free, borne on the ovary; pistil 1, the ovary superior, 12- to 35-locular, with many ovules, the stigma sessile, several-rayed; fruit a depressed-globose berry, enveloped by an aril, maturing under water.

Nymphaea is comprised of about forty wild species in the temperate and tropical regions of the world, in addition to a number of horticultural forms.

1. Petals subacute at tip; flowers fragrant; seeds 2 mm long .... 1. *N. odorata*
1. Petals rounded at tip; flowers not fragrant; seeds 3–4 mm long .... 2. *N. tuberosa*

1. Nymphaea odorata Ait. Hort. Kew. 2:227. 1789. Fig 4.
Castalia odorata (Ait.) Woodville & Wood in Rees, Cycl. 6, no. 1. 1806.
Nymphaea odorata Ait. var. *gigantea* Tricker, Water Garden 186. 1897.

Aquatic perennials from a stout, forking rhizome; leaves floating or ascending, orbicular, to 30 (–60) cm in diameter, the basal sinus narrow, green above, usually purple beneath, usually pubescent beneath, on flat, glabrous or puberulent, purplish petioles; flowers showy, fragrant, 5–20 cm across, white, rarely pink, long-pedunculate; sepals 4, free, green or sometimes purple on the back, ovate, subacute to obtuse at the apex, to 10 cm long, up to one-third as broad; petals numerous, free, usually white, subacute at the apex, to 3 cm long; seeds ellipsoid to oblongoid, about 2 mm long. June–September.

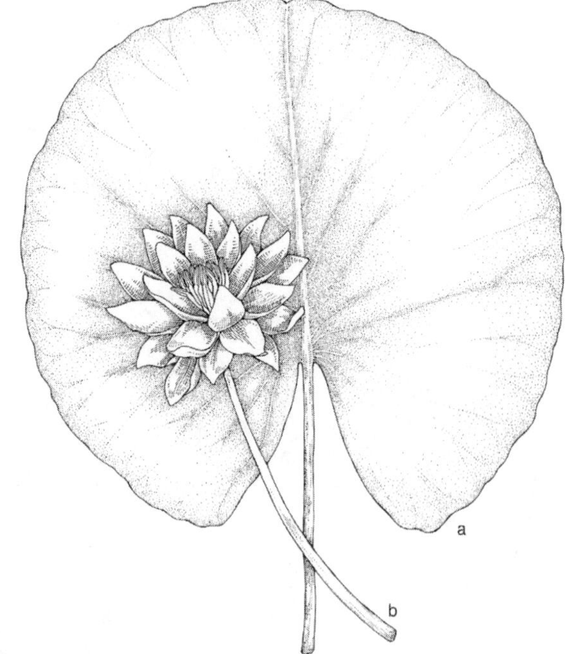

4. *Nymphaea odorata* (Fragrant water lily).
a. Leaf
b. Flower.

Lakes and ponds.

IA, IL, IN, KS, KY, MO, NE, OH (OBL).

Fragrant water lily.

There is some question among botanists whether this species is specifically distinct from *N. tuberosa*. After examining much material of the two, I have decided to recognize both as separate species.

Some botanists recognize var. *gigantea*, a robust taxon with slightly larger flowers and broader leaves that tend to turn up around the edges. At this time, I prefer not to recognize this variety.

2. **Nymphaea tuberosa** Paine, Cat. Pl. Oneida Co. N.Y. 184. 1865. Fig. 5.
Castalia tuberosa (Paine) Greene, Bull. Torrey Club 15:84. 1888.

Aquatic perennials from a stout, forking rhizome; leaves floating, orbicular, to 40 cm across, the basal sinus narrow, green above and below, glabrous or slightly pubescent below, on stout petioles green and with brown stripes above; flowers showy, not fragrant, to 25 cm across, white, long-pedunculate; sepals 4, free, green, narrowly ovate, subacute to obtuse at the apex, to 10 cm long, up to one-third as broad; petals numerous, free, white, rounded at the apex, to 3 cm long; seeds globose-ovoid, 3–4 mm long. June–September.

Lakes and ponds.

IA, IL, IN, KS, MO, NE, OH (OBL).

White water lily.

The tuberous rootstocks are an important source of food for beavers, muskrats, and other wildlife.

5. *Nymphaea tuberosa* (White water lily).
a. Leaf.
b. Flower.

91. NYSSACEAE—NYSSA FAMILY

Trees; leaves simple, alternate, entire or sometimes with a few irregular teeth, without stipules; flowers unisexual, the plants dioecious; staminate flowers numerous, with a 5-cleft calyx and 5 small, free petals, or petals absent, and usually 10 stamens attached to a disk; pistillate flowers 1–8 in a cluster, with no calyx, 5 small, free petals, or petals absent, 5 or 10 usually sterile stamens, and an inferior ovary; fruit a drupe.

One genus and six species in North America and southeast Asia comprise this family. This family is sometimes included in the Cornaceae.

1. **Nyssa** L.—Black Gum

Trees or shrubs; leaves alternate, simple; flowers greenish, unisexual, small, the staminate numerous, with 5 minute sepals, 0 or 5 minute petals, and 5–15 stamens, the pistillate 1–several in a cluster, bracteate, with a 5-lobed calyx, 0 or 5 minute petals, several aborted stamens, and an inferior, 1-celled ovary; fruit a drupe.

1. Most of the petioles 3 cm long or longer; pistillate flower borne singly; fruit 2–3 cm long...... .... 1. *N. aquatica*
1. None of the petioles over 2.5 cm long; pistillate flowers (1–) 2 or more in a cluster; fruit up to 1.5 cm long.
    2. Flowers (1–) 2 (–3) in a cluster; largest leaves usually less than 6 cm long, none of them with irregular teeth........ 2. *N. biflora*
    2. Flowers and fruits (2–) 3–5 in a cluster; largest leaves usually more than 6 cm long, some of them often with irregular teeth .... 3. *N. sylvatica*

1. Nyssa aquatica L. Sp. Pl. 1058. 1743. Fig. 6.

Large trees up to 30 m tall, with a trunk diameter up to 1.3 m and a spreading crown with numerous branchlets; bark light gray to brown, broken into thin scales; twigs stout, more or less angular, gray or brown, glabrous, the buds rounded, glabrous, 2–3 mm long; leaves alternate, simple, oblong to ovate, acute to somewhat acuminate at the apex, rounded or occasionally subcordate at the base, entire or with a few irregularly arranged coarse teeth, dark green, glabrous or puberulent on the upper surface, paler and softly pubescent on the lower surface, to 20 cm long, to 12 cm wide, on stout petioles to 2 cm

6. *Nyssa aquatica* (Tupelo gum). Leaves and fruits.

long, pubescent; staminate and pistillate flowers borne on separate trees, appearing as the leaves begin to unfold, greenish, the staminate several in spherical clusters, the pistillate solitary on long pedicels arising from the axils of the leaves; drupes oblongoid to oval, dark purple with pale speckles, to 2–3 cm long, 1-seeded. April–May.

Swamps, low woods.

IL, KY, MO (OBL).

Tupelo gum; swamp tupelo.

This species often grows in standing water where it develops a buttressed base. It is distinguished from the other species of *Nyssa* by its larger, coarsely toothed leaves and its larger purple oblongoid to oval fruits.

2. **Nyssa biflora** Walt. Fl. Carol. 253. 1788. Fig. 7.
Nyssa sylvatica Marsh var. *biflora* (Walt.) Sarg. Sylva 5:76. 1893.

Medium trees to 20 m tall, with a trunk diameter up to 0.8 m, the crown more or less rounded, some of the branchlets arising at right angles to the main branches; bark dark brown, sometimes broken up into squarrish blocks; twigs brown, glabrous, the buds short-pointed, glabrous, 2–3 mm long; leaves alternate, simple, oblong, obtuse to acute at the apex, tapering to the base, entire, dark green and shiny on the upper surface, usually paler and somewhat pubescent on the lower surface, to 6 cm long, to 3.5 cm wide, on petioles up to 1.5 cm long; staminate and pistillate flowers borne on separate trees, appearing after the leaves begin to unfold, greenish, very small, the staminate several in spherical clusters, the pistillate (1–) 2 (–3) on long pedicels arising from the axils of the leaves; drupes spherical, 2 in a cluster, dark blue, to 8 mm in diameter, 1-seeded. April–May.

Swampy woods.

IL, MO (NI).

Swamp tupelo.

This species is readily recognized by its oblong, often obtuse leaves and its fruits borne two in a cluster. In the southern United States where it is common, is has a designation of OBL.

7. *Nyssa biflora* (Swamp tupelo). Leafy branch (center). Fruits (lower right).

3. **Nyssa sylvatica** Marsh. Arb. Am. 97. 1785. Fig. 8.

Medium to large trees up to 25 m tall, with a trunk diameter up to 1 m, the crown rounded, with many small, pendulous branchlets often arising at right angles to the main branches; bark brown to black, often broken up into squarrish blocks; twigs rather stout, reddish brown, glabrous, sometimes zigzag, the buds short-pointed, yellowish or reddish, glabrous, 2–3 mm long; leaves alternate, simple, obovate, abruptly acuminate at the apex, tapering or occasionally rounded at the base, entire or with an irregularly arranged small tooth, dark green and shiny on the upper surface, paler and usually somewhat pubescent on the lower surface, to 10 cm long, to 6.5 cm wide, on glabrous or sparsely pubescent petioles up to 2 cm long; staminate and pistillate flowers borne on separate trees, appearing after the leaves begin to unfold, greenish, very small, the staminate several in spherical clusters, the pistillate 2–5 on long pedicels arising from the axils of the leaves; drupes spherical to oval, 2–5 in a cluster, dark blue, to 10 mm long, 1-seeded. April–May.

8. *Nyssa sylvatica* (Black gum). Habit, in flower (center). Fruits (left of center).

Dry woods, mesic woods, occasionally in swampy woods.
IL, IN, KS, KY, MO, OH (FAC).
Black gum; sour gum.

Nyssa sylvatica differs from *N. biflora* by the obovate leaves that sometimes bear an irregular tooth. It may be confused with persimmon (*Diospyros virginiana*) but differs by its abruptly acuminate leaves and its twigs that have continuous pith marked by distinct partitions.

92. OLEACEAE—OLIVE FAMILY

Trees or shrubs, dioecious, monoecious, or polygamous; leaves opposite, simple or pinnately compound, without stipules; flowers perfect or unisexual, variously arranged; calyx 4-parted, or absent; corolla 4-parted, or absent; stamens 2 (–6), attached to the corolla if the corolla is present; ovary superior, 2-locular; fruit a samara, capsule, drupe, or berry.

There are about thirty genera and six hundred species in this family. Several genera are of ornamental importance, including *Syringa* (lilac), *Ligustrum* (privet), *Forsythia*, *Osmanthus* (sweet olive), and *Chionanthus* (fringe tree). *Olea*, the olive, is also in this family.

Two genera may have species that occur in wetland habitats in the central Midwest.

1. Leaves simple; fruit a drupe; shrubs .... 1. *Forestiera*
1. Leaves pinnately compound; fruit a samara; trees .... 2. *Fraxinus*

1. **Forestiera** Poir.—Swamp Privet

Shrubs or small trees; leaves opposite, simple; flowers in clusters or small panicles, unisexual, less commonly perfect, the plants dioecious or polygamo-dioecious; calyx 4-parted or nearly absent; corolla absent; stamens 2 (–6); ovary superior; fruit a 1- or 2-seeded drupe.

Forestiera consists of fifteen North American species. Only the following species occurs in the central Midwest.

1. **Forestiera acuminata** (Michx.) Poir. in Lam. Encycl. Suppl. 2:664. 1803. Fig. 9.
Adelia acuminata Michx. Fl. Bor. Am. 2:225. 1803.

Shrubs or small trees to 5 m tall, occasionally with a few stiff short spinelike branchlets, usually pubescent at first, becoming nearly glabrous; leaves opposite, simple, lanceolate to elliptic to narrowly ovate, acute at the apex, tapering to the base, glabrous, sometimes seemingly entire but usually minutely serrulate, to 8 (–10) cm long, to 3 (–5) cm wide, with petioles up to 1.2 cm long; flowers mostly unisexual, the staminate borne in small clusters, the pistillate borne in small panicles before the leaves expand; staminate flowers usually without a perianth, but subtended by 4–5 small bracts, and with 2–6 yellow stamens; pistillate flowers usually with 4 caducous yellow sepals up to 0.2 mm long, without petals, and with a superior ovary, subtended by caducous bracts; drupes ellipsoid to oblongoid, more or less curved, 1.0–1.8 mm long, dark blue but glaucous. February–March.

Swampy woods, floodplain woods, along rivers and streams.
IL, IN, KY, MO, NE (OBL).

Swamp privet.

The leaves of this species resemble those of *Ilex decidua* and *I. verticillata*, but they are opposite in *Forestiera*.

When the flowers bloom before the leaves expand, the plant has a very yellow appearance.

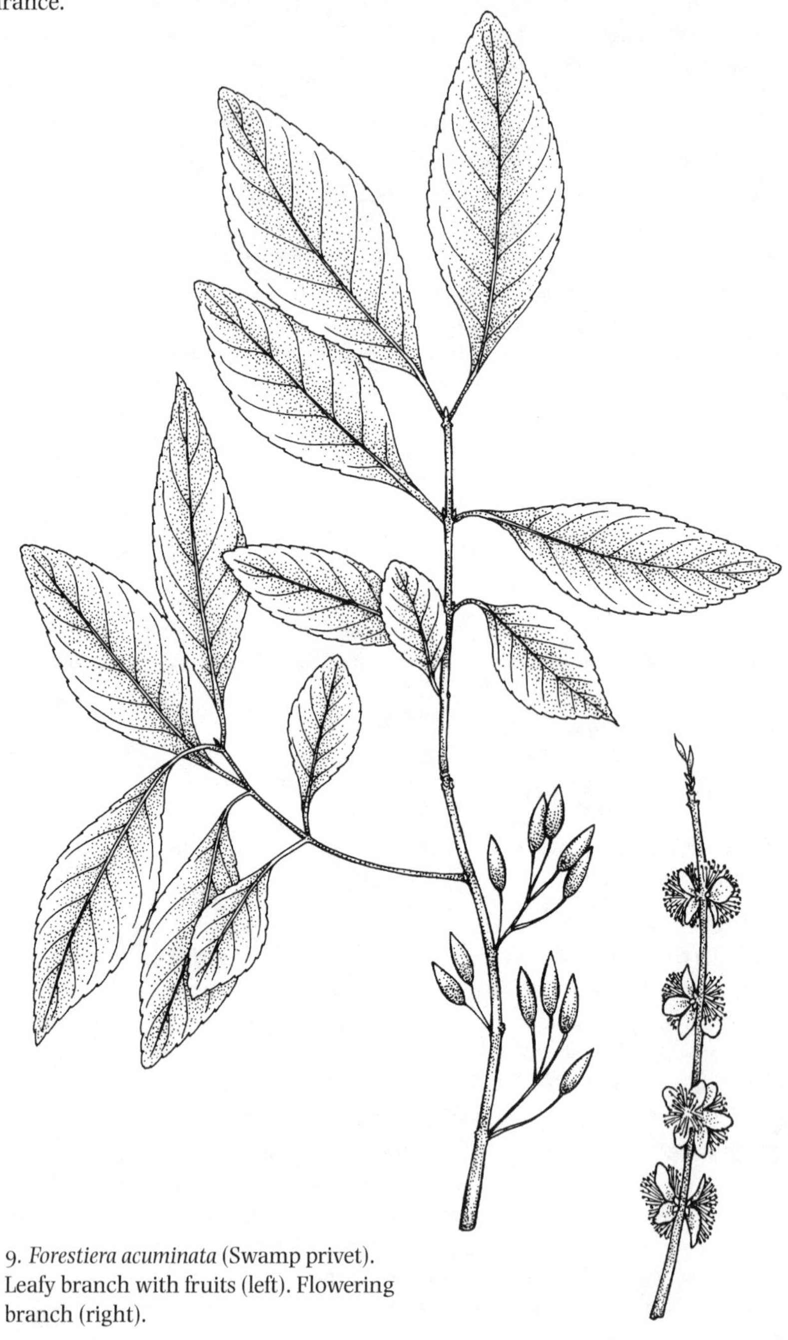

9. *Forestiera acuminata* (Swamp privet). Leafy branch with fruits (left). Flowering branch (right).

2. **Fraxinus** L.—Ash

Trees; leaves opposite, pinnately compound; flowers unisexual or sometimes perfect, appearing before or as the leaves expand, the plants dioecious or polygamous; calyx 4-parted, often caducous, or absent; corolla absent; stamens 2 (–3); ovary superior, sometimes with 2 abortive stamens in the pistillate flowers; samaras 1-seeded, the seed terete or flat.

There are about sixty-five species of *Fraxinus*, all in the Northern Hemisphere. In addition to the species described below, *F. americana*, *F. biltmoreana*, and *F. quadrangulata* occur in the central Midwest.

1. Leaflets sessile (except the terminal one) .... 2. *F. nigra*
1. Leaflets petiolulate.
  2. Leaflets more or less green on both surfaces; wing of samara covering at least ½ of the body.
    3. Twigs, leaflets, and petioles densely velvety-pubescent........................ 3. *F. pennsylvanica*
    3. Twigs, leaflets, and petioles glabrous or nearly so..................................... 1. *F. lanceolata*
  2. Leaflets more or less brownish beneath; wing of samara covering at most ⅓ of the body... .... 4. *F. profunda*

1. **Fraxinus lanceolata** Borkh. Handb. Forst. Bot. 1:826. 1800. Fig. 10.
Fraxinus pennsylvanica Marsh. var. *lanceolata* (Borkh.) Sarg. Silva N. Am. 6:50. 1894.
Fraxinus pennsylvanica Marsh. var. *subintegerrima* (Vahl) Fern. Rhodora 49:159. 1947.

Trees to 25 m tall; twigs glabrous or sparsely pubescent, the upper margin of the leaf scars straight across or slightly concave; leaves opposite, pinnately compound, with 5–9 leaflets; leaflets lanceolate to elliptic to narrowly ovate, acute to acuminate at the apex, tapering or rounded at the short-petiolulate base, green and glabrous on the upper surface, green or slightly paler and glabrous or sparsely pubescent on the lower surface, entire to serrate to serrulate, to 15 cm long, to 5 cm wide, the petiolules up to 4 mm long, narrowly winged; flowers unisexual; calyx cup-shaped, usually persistent at the base of the fruit; corolla absent; stamens 2; ovary superior; samaras narrowly oblanceolate to linear-oblong, 4–5 (–7) cm long, up to 6 (–7) mm wide, the seed terete, the wing reaching more than halfway to the base of the seed. May.

Bottomland forests, swampy woods.

IA, IL, IN, KY, MO, OH (FACW), KS, NE (FACW-).

Green ash.

Young leaves are usually bronze in color. Plants that resemble this species but with velvety-tomentose twigs, petioles, lower leaf surfaces, and leaf rachises are a separate species—*F. pennsylvanica*.

While *F. americana* (white ash) is often similar in appearance to *F. lanceolata*, the leaflets in white ash are distinctly whiter on the lower surface than in green ash. The wing on the samara in white ash does not extend as far down on the seed as it does in green ash. The leaf scars on the twigs of white ash are distinctly concave across the top, while those of green ash are straight across at the top or slightly concave.

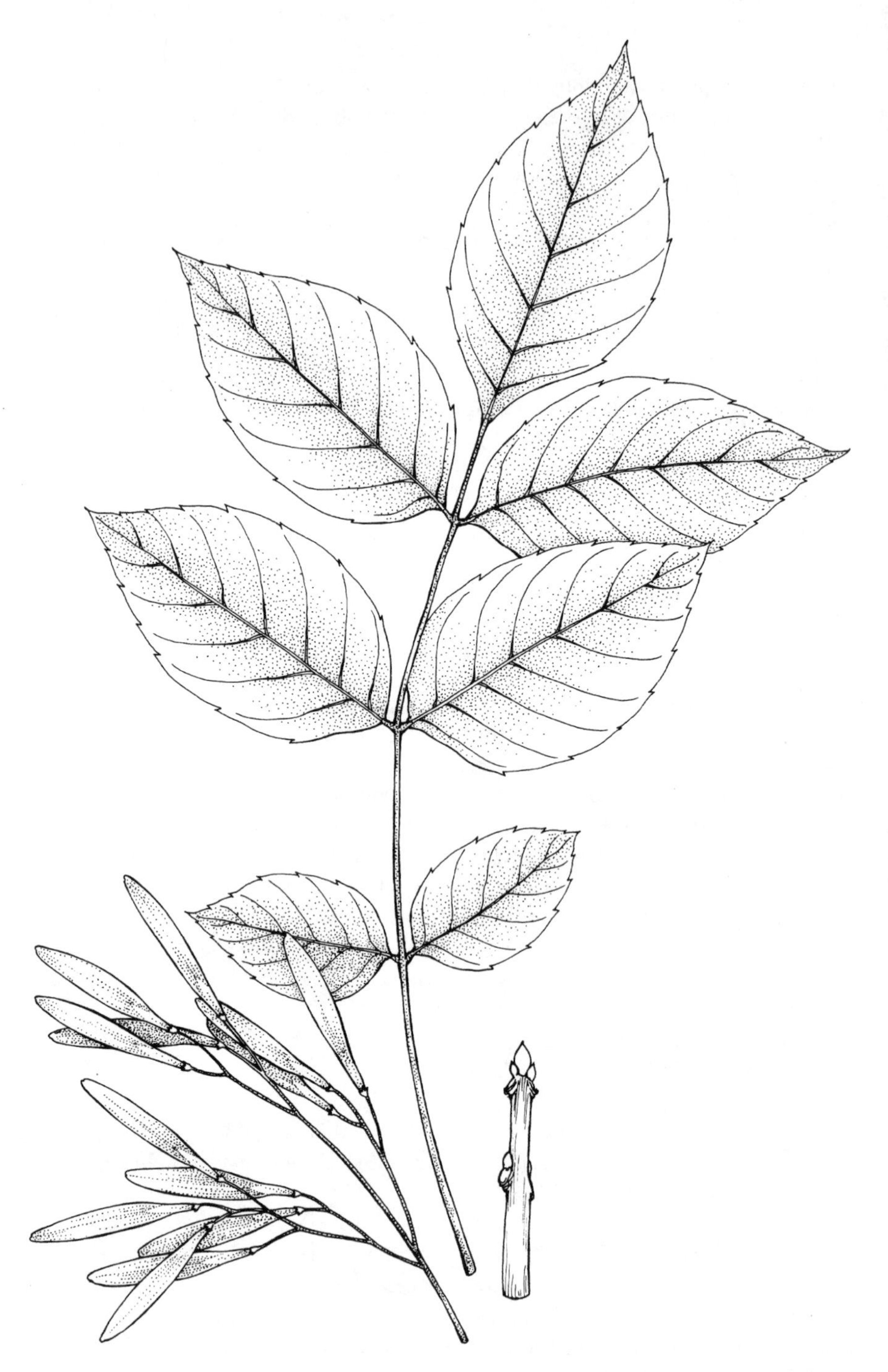

10. *Fraxinus lanceolata* (Green ash). Leaf (center). Cluster of samaras (lower left). Twig (right).

2. **Fraxinus nigra** Marsh. Arb. Am. 51. 1785. Fig. 11.

Trees to 25 m tall; twigs glabrous; upper margin of the leaf scars straight across to concave; leaves opposite, pinnately compound, with 7–11 leaflets; leaflets broadly lanceolate to elliptic, acute to acuminate at the apex, tapering or rounded at the sessile base, green and glabrous on the upper surface, paler and glabrous or somewhat pubescent on the lower surface, serrate, to 15 cm long, to 5 cm wide, the petiolules of the lateral leaflets absent; flowers perfect or unisexual; calyx cup-shaped, caducous, or absent; corolla absent; stamens 2; samaras narrowly oblong to oblong, rounded at both ends, to 4 cm long, to 1 cm wide, the seed flat, the wing often reaching to the base of the seed. May–June.

Swampy woods, bogs, edge of fens, around lakes, sometimes in standing water.

IA, IL, IN (FACW+), OH (FACW).

Black ash.

The distinguishing features of this species are its numerous leaflets that are sessile and by the flat rather than terete seeds.

The leaves, when crushed, have the odor of elderberry (*Sambucus*) leaves.

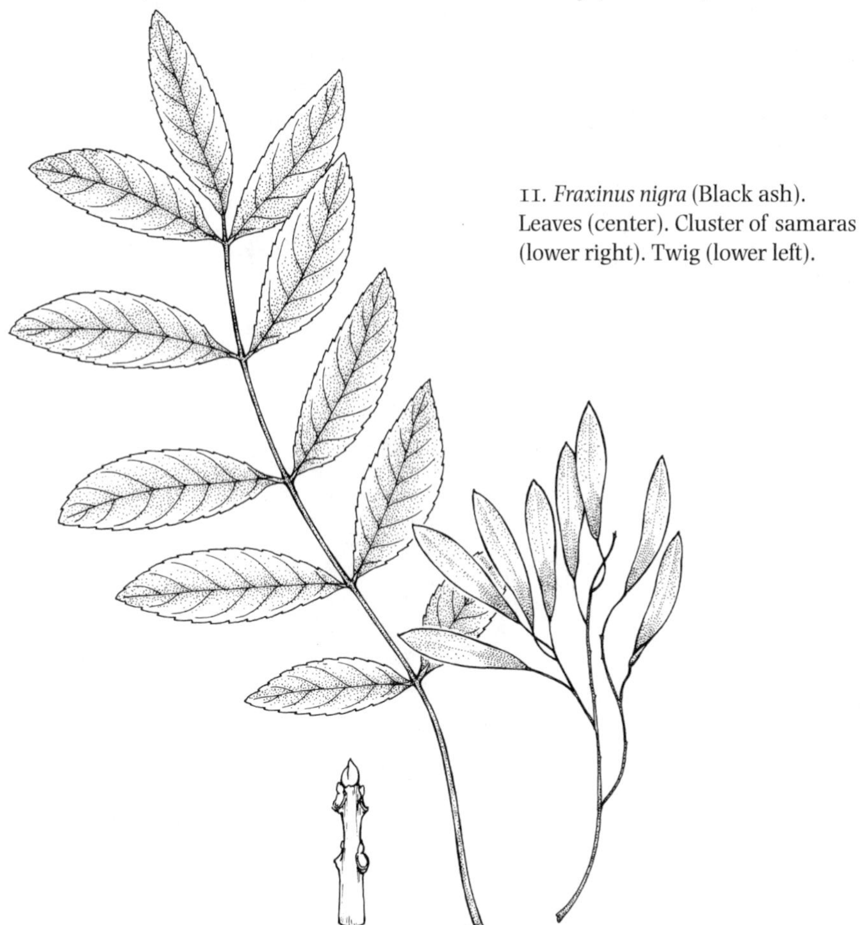

11. *Fraxinus nigra* (Black ash). Leaves (center). Cluster of samaras (lower right). Twig (lower left).

3. **Fraxinus pennsylvanica** Marsh. Arb. Am. 51. 1785. Fig. 12.
Fraxinus pubescens Lam. Encycl. 2:548. 1786.
Fraxinus pennsylvanica Marsh. var. *austinii* Fern. Rhodora 40:452. 1938.

Trees to 25 m tall; twigs velvety-tomentose; upper margin of leaf scars straight across or slightly concave; leaves opposite, pinnately compound, with 5–11 leaflets, with velvety-tomentose rachises; leaflets lanceolate to elliptic to narrowly ovate, acute to acuminate at the apex, tapering or rounded at the short-petiolulate base, green and often pubescent on the upper surface, paler and velvety-tomentose on the lower surface, entire to serrulate to serrate, to 15 cm long, to 5 cm wide, the petiolules up to 4 mm long, narrowly winged; flowers unisexual; calyx cup-shaped, usually persistent at the base of the fruit; corolla absent; stamens 2; samaras narrowly oblanceolate to linear-oblong to spatulate, (3–) 4–5 (–7) cm long, up to 6 (–11) mm wide, the seed terete, the wing reaching more than halfway to the base of the seed. May.

Swampy woods, sometimes in standing water.

IA, IL, IN, KY, MO, OH (FACW), KS, NE (FAC−).

Red ash.

Although this species is frequently merged with the glabrous or nearly so green ash, DNA analysis indicates that red ash should be treated as a distinct species. It is readily recognized by its velvety-pubescent twigs and leaves and its very short (up to 4 mm long) petiolules. Pumpkin ash (*F. profunda*) may be nearly as velvety-pubescent, but its leaflets are on petiolules at least 8 mm long.

Plants with serrate leaflets and shorter, spatulate fruits (3–4 cm long) with wings up to 11 mm wide occur and may be called var. *austinii*.

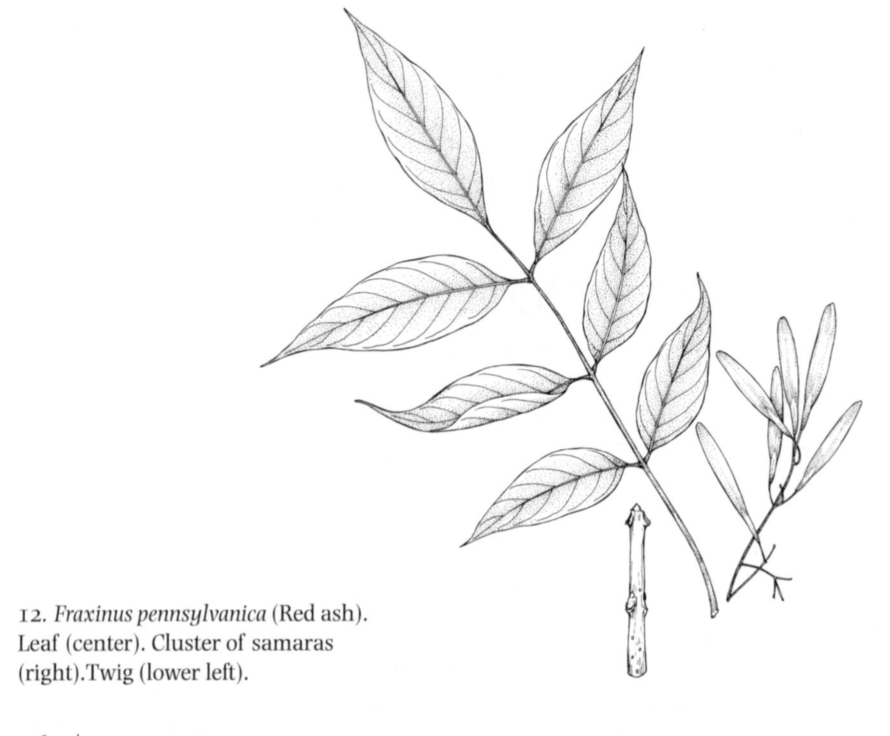

12. *Fraxinus pennsylvanica* (Red ash). Leaf (center). Cluster of samaras (right). Twig (lower left).

4. **Fraxinus profunda** (Bush) Bush in Britt. Man. 725. 1901. Fig. 13.
Fraxinus tomentosa Michx. f. Hist. Arb. For. Am. Sept. 3:112. 1813. *nomen illeg.*
Fraxinus americana L. var. *profunda* Bush, Ann. Rep. Mo. Bot. Gard. Sci. 47. 1894.

Trees to 30 m tall; twigs velvety-tomentose; upper margin of leaf scars straight across or slightly concave; leaves opposite, pinnately compound, with 5–9 leaflets, usually with pubescent rachises; leaflets lanceolate to oblong-lanceolate to ovate-lanceolate, acuminate at the apex, tapering or rounded at the petiolulate base, green and glabrous or pubescent on the upper surface, often reddish on the tomentose or glabrate lower surface, entire or scarcely serrulate, to 20 cm long, to 12 cm wide, the petiolules 8–15 mm long, wingless; flowers unisexual; calyx cup-shaped, 2–5 mm long, persistent at the base of the fruit; corolla absent; stamens 2; samaras spatulate or narrowly oblong, 4.0–7.5 cm long, 6–12 mm wide, the seeds terete, the wing extending to about the middle of the seed. April–May.

Swamps, bottomland forests.

IL, IN, KY, MO, OH (OBL).

Pumpkin ash.

This species often occurs in bald cypress swamps. The base of the trunk is buttressed. Because of its velvety-pubescent leaflets and twigs, this species resembles red ash (*F. pennsylvanica*), but red ash has shorter petiolules (up to 4 mm long). It has been known sometimes as *F. tomentosa*.

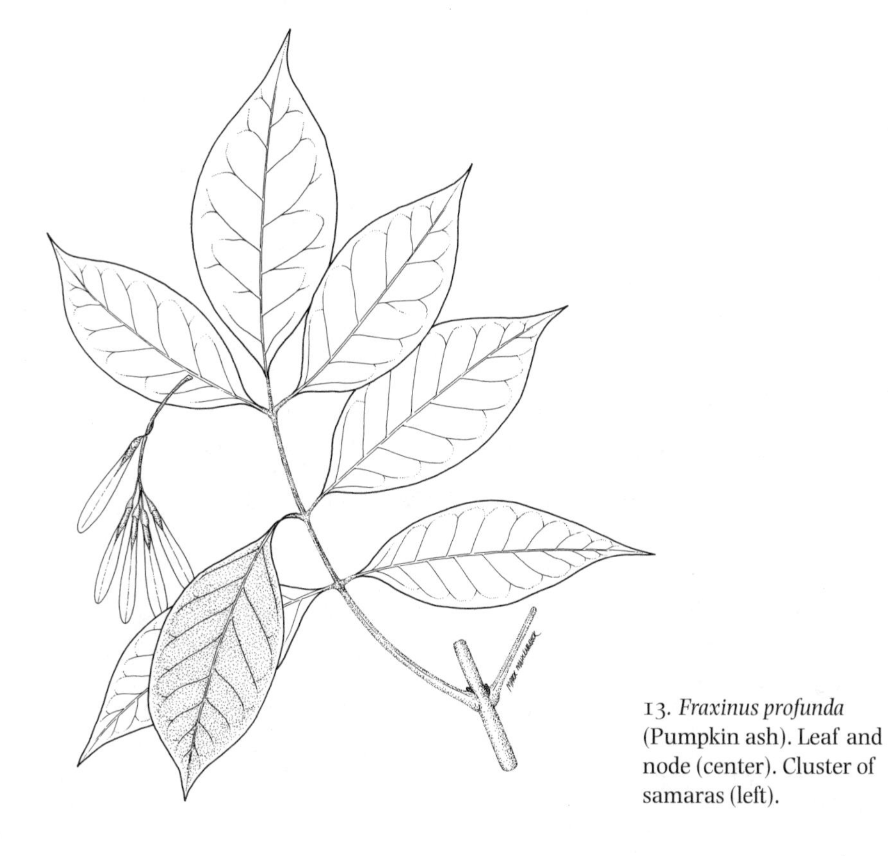

13. *Fraxinus profunda* (Pumpkin ash). Leaf and node (center). Cluster of samaras (left).

93. ONAGRACEAE—EVENING PRIMROSE FAMILY

Annual, biennial, or perennial herbs; leaves simple, alternate or opposite, without stipules; flowers perfect, usually actinomorphic, axillary or in various inflorescence types; sepals usually 4, free; petals usually 4, free, or absent; stamens (4) 8; ovary inferior, 2- to 5-locular, with a hypanthium usually prolonged above the ovary; fruit a capsule or indehiscent and nutlike; seeds without a coma (except in *Chamerion* and *Epilobium*).

This family consists of sixty-four genera and 675 species, many of them in the western United States.

The following genera and species may occur in aquatic or wetland habitats:

1. Leaves opposite.
  2. Leaves entire; seeds not comose .... 4. *Ludwigia*
  2. Leaves serrulate to denticulate; seeds comose.
    3. Floral tube not prolonged beyond the ovary; petals not notched, usually more than 14 mm long .... 1. *Chamerion*
    3. Floral tube prolonged beyond the ovary; petals notched, less than 10 mm long (except in *Epilobium hirsutum*) .... 2. *Epilobium*
1. Leaves alternate.
  4. Petals yellow or greenish; seeds not comose.
    5. Calyx divided to the tip of the ovary, persistent on the fruit .... 4. *Ludwigia*
    5. Calyx not divided to the tip of the ovary, deciduous from the fruit .... 5. *Oenothera*
  4. Petals white or pink; seeds with or without a coma.
    6. Calyx tube not prolonged beyond the ovary; seeds without a coma; plants canescent .... 3. *Gaurella*
    6. Calyx tube prolonged beyond the ovary; seeds with a coma; plants not canescent .... 2. *Epilobium*

1. **Chamerion** Raf.

Perennial herbs, sometimes woody at the base; leaves alternate, simple, entire; inflorescence terminal, consisting of racemes; flowers perfect, somewhat zygomorphic, with a hypanthium not prolonged beyond the ovary; sepals 4, free, caducous; petals 4, free, large, not notched at the tip, pink, white, or purple; stamens 8; stigma 4-lobed; capsules elongated, more or less 4-angled, the seeds comose.

Approximately ten species comprise this genus, which is often included within *Epilobium*. However, the hypanthium in *Epilobium* is extended beyond the ovary, while in *Chamerion*, the hypanthium is not extended beyond the ovary.

1. **Chamerion angustifolium** (L.) Holub, Folia Geobot. Phytotax. 7:86. 1972. Fig. 14.
Epilobium angustifolium L. Sp. Pl. 1:347. 1753.

Perennial herbs from spreading roots; stems erect, branched or unbranched, to 1 m tall, glabrous or sparsely pubescent; leaves alternate, simple, lanceolate, acute at the apex, tapering to the sessile or subsessile base, entire or irregularly denticulate, glabrous, to 25 cm long, to 8 cm wide; flowers in terminal or axillary racemes, actinomorphic; calyx 4-lobed, the lobes oblong-lanceolate, obtuse at the apex, usually purple, 10–15 mm long; petals 4, free, obovate, obtuse at the apex, with a basal claw, bright pink to rose, 14–18 mm long; stamens 8; ovary inferior,

14. *Chamerion angustifolium* (Fireweed). Habit (center). Capsule (right).

4-locular, the stigma 4-lobed; capsules linear, elongate, to 6.5 cm long, short-pubescent, the seeds numerous, fusiform, brown, to 1 mm long, with a light brown or tawny coma. July–September.

Burned areas, dunes, bogs.

IA, IL, IN, NE, OH (FAC).

Fireweed.

The common name is derived from the rapid colonization by this plant after a fire.

Chamerion angustifolium is distinguished by its large, showy flowers and the tawny coma attached to the seed. This species is often placed in the genus *Epilobium*.

2. **Epilobium** L.—Willow Herb

Mostly perennial herbs; leaves alternate or opposite, simple, entire or toothed; flowers in racemes or panicles or solitary in the leaf axils, perfect, usually actinomorphic; sepals 4, free or forming a hypanthium prolonged beyond the ovary; petals 4, free; stamens 8; ovary inferior, the style simple or 4-cleft; fruit an elongated capsule; seeds with a coma.

Epilobium consists of about two hundred species in temperate and boreal parts of the world.

This and *Chamerion* are the only genera in the Onagraceae with comose seeds.

1. Petals 1 cm long or longer; stems softly hirsute; stigma deeply 4-lobed; seeds smooth .... 4. *E. hirsutum*
1. Petals up to 1 cm long; stems variously pubescent but not softly hirsute; stigma entire or slightly notched; seeds papillose.
    2. Leaves entire.

3. Stems and leaves velvety pubescent; most or all the leaves more than 4 mm wide .... 6. *E. strictum*

3. Stems and leaves merely canescent; most or all the leaves less than 4 mm wide .... 5. *E. leptophyllum*

2. Leaves serrulate to denticulate.

4. Inflorescence and capsules glandular-pubescent; seeds short-beaked; coma white.

5. Basal rosette persistent; turions absent; inflorescence branched, not leafy; petals pale pink or white; seeds up to 1.2 mm long .... 1. *E. ciliatum*

5. Basal rosette absent; turions present; inflorescence unbranched, leafy; petals rose-purple; seeds 1.1–1.6 mm long .... 3. *E. glandulosum*

4. Inflorescence and capsules not glandular-pubescent; seeds beakless; coma reddish brown .... 2. *E. coloratum*

1. **Epilobium ciliatum** Raf. Med. Rep. ser. 2, 5:361. 1808. Fig. 15.
Epilobium americanum Hausskn. Oesterr. Bot. Zeit. 29:118. 1879.

Perennial herbs with persistent basal rosettes but without fleshy turions; stems erect, to 1.2 m tall, more or less 4-angled, densely pubescent above, less pubescent below, the pubescence in lines; leaves simple, usually opposite below and alternate above, flat, lanceolate to narrowly ovate, acute at the apex, tapering to the base, serrulate to denticulate, pubescent to nearly glabrous, conspicuously ciliate, not

15. *Epilobium ciliatum* (American willow herb). Habit (center). Capsule (left). Flower (upper right).

rugose, to 12 cm long, to 4.5 cm wide, sessile or on petioles to 4 (–6) mm long; inflorescence branched, not leafy, sometimes becoming paniculate, the flowers crowded in the upper leaf axils; sepals 4, 1.5–5.0 mm long, strigillose; petals 4, pale pink or white, 1.5–5.0 (-8.0) mm long; floral tube 0.5–2.6 mm long; stamens 8; capsules narrow, to 10 cm long, pubescent, with pedicels up to 4 cm long; seeds up to 1.2 mm long, with a white, caducous coma. July–September.

Wet ditches, marshes, along streams.

IA, IL, IN (FACU), KS, NE (OBL), KY, OH (FAC−).

American willow herb.

This species differs from the very similar *E. glandulosum* by having persistent basal rosettes of leaves instead of fleshy turions, by its branched, leafless inflorescence, and by its pale pink to white petals. It differs from *E. coloratum* by its more irregular teeth of the leaves and by its white, caducous coma of the seeds.

Epilobium ciliatum is much more of a wetland species in the western part of our range.

2. **Epilobium coloratum** Biehler, Pl. Nov. Herb. Spreng. 18. 1807. Fig. 16.

Perennial herbs with persistent basal rosettes and slender rhizomes, but not producing fleshy turions; stems erect, branched, to 1 m tall, more or less 4-angled, usually pubescent, at least above, in lines; leaves simple, nearly always opposite, flat, lanceolate to oblong-lanceolate, long-acuminate at the apex, tapering to the base, regularly serrulate, glabrous or pubescent on the veins beneath, minutely ciliate, usually somewhat rugose, to 8 cm long, to 3 cm wide, on petioles up to 4 mm long; inflorescence usually paniculate, with many flowers, the flowers on canescent pedicels about 1 cm long; sepals 4, free, 1.3–3.0 mm long, canescent; petals 4, free, pink or whitish, 3–5 mm long; floral tube 0.3–0.6 mm long; stamens 8; capsules narrow, to 6 cm long, canescent, with pedicels up to 1 (–2) cm long; seeds 1.2–1.7 mm long, with a persistent, cinnamon-colored coma. August–October.

Marshes, along streams, around lakes and ponds, wet ditches.

IA, IL, IN, KS, KY, MO, NE, OH (OBL).

Purple-leaved willow herb.

This species is distinguished by its long-acuminate, serrulate, rugose, opposite leaves and its cinnamon-colored coma of its seeds. Where it occurs with *E. ciliatum*, the two species tend to intergrade. A hybrid between these two species is known as *E. X wisconsinense* Ugent.

3. **Epilobium glandulosum** Lehm. in Hook. Fl. Bor. Am. 1:206. 1833. Fig. 17.
Epilobium adenocaulon Hausskn. Oest. Bot. Zeit. 29:119. 1879.
Epilobium adenocaulon Hausskn. var. *occidentale* Trel. Ann. Rep. Mo. Bot. Gard. 2:95. 1891.
Epilobium occidentale (Trel.) Rydb. Mem. N.Y. Bot. Gard. 1:275. 1900.
Epilobium glandulosum Lehm. var. *adenocaulon* (Hausskn.) Fern. Rhodora 20:35. 1918.
Epilobium ciliatum Raf. ssp. *glandulosum* (Lehm.) Hoch & Raven, Ann. Mo. Bot. Gard. 64:136. 1977.

Perennial herbs without persistent basal rosettes, but with fleshy turions; stems

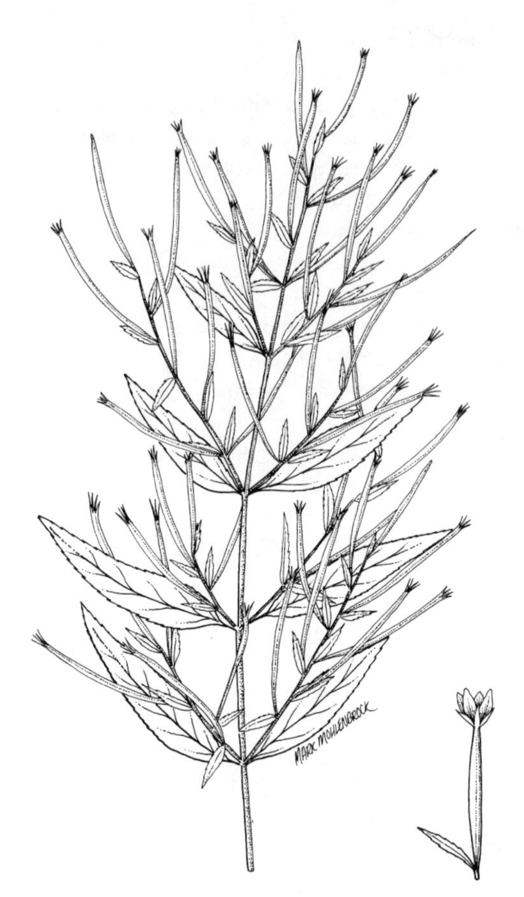

16. *Epilobium coloratum* (Purple-leaved willow herb). Habit (center). Flower and bract (lower right).

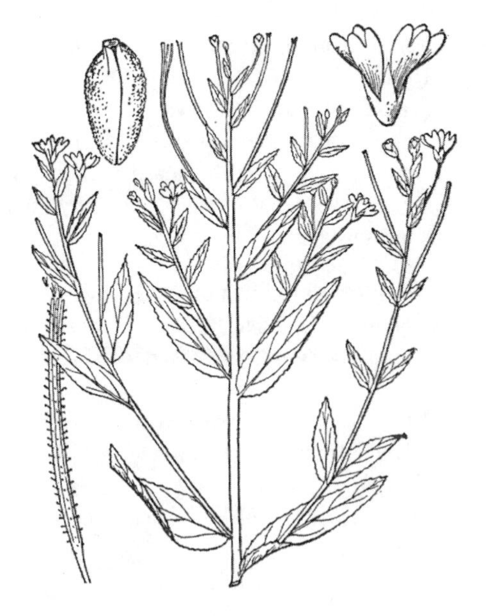

17. *Epilobium glandulosum* (Western water willow). Habit (center). Capsule (lower left). Seed (upper left). Flower (upper right).

erect, to 1.2 m tall, more or less 4-angled, densely pubescent above, less pubescent below, the pubescence in lines; leaves simple, usually opposite below and alternate above, flat, lanceolate to narrowly ovate, acute at the apex, tapering to the base, serrulate to denticulate, pubescent to nearly glabrous, conspicuously ciliate, not rugose, to 12 cm long, to 4.5 cm wide, sessile or on petioles to 4 (–6) mm long; inflorescence unbranched, leafy, sometimes becoming paniculate, the flowers crowded in the upper axils; sepals 4, 1.5–5.0 mm long, strigillose; petals 4, rose-purple 3.5–10.0 mm long; floral tube 0.5–2.6 mm long; stamens 8; capsules narrow, to 10 cm long, pubescent with pedicels up to 4 cm long; seeds 1.1–1.6 mm long, with a white, caducous coma. July–September.

Wet ditches, marshes, along streams.

OH (FAC−). The U.S. Fish and Wildlife Service does not distinguish this species from *E. ciliatum*.

Western water willow.

This species is sometimes considered to be a subspecies of *E. ciliatum*. Plants with broader leaves have sometimes been called *E. adenocaulon* or *E. glandulosum* var. *adenocaulon*.

4. **Epilobium hirsutum** L. Sp. Pl. 1:347–348. 1753. Fig. 18.

Creeping to ascending perennials from thick rhizomes and stolons; stems branched, densely villous to hirsute, to 2 m long; leaves alternate, simple, lanceolate to oblong-lanceolate, acute at the apex, tapering to the sessile and sometimes partly clasping base, softly long-villous, serrate, to 4 cm long, to 2.5 cm wide; flowers in racemes in the axils of the upper leaves; calyx 4-parted, the tube 1.5–2.5 mm long, the lobes apiculate; petals 4, free, notched at the apex, purple, up to 2 cm long; stamens 8; ovary inferior, the style deeply 4-parted; capsules to 5 cm long, with numerous obovoid seeds 1.4–1.5 mm long. July–September.

18. *Epilobium hirsutum* (Hairy herb willow). Habit (center). Flower (right).

Wet disturbed areas, particularly in degraded meadows.

Native to Europe, adventive in IL, IN (FACW+), OH (FACW).

Hairy willow herb.

This species is readily distinguished by its densely villous-hirsute stems and leaves, its thick creeping rhizomes and stolons, and its purple petals.

5. **Epilobium leptophyllum** Raf. Precis Decouv. Somiol. 41. 1814. Fig. 19.

Perennial herbs with very slender stolons and fleshy turions; stems erect, slender, branched or unbranched, terete, to 90 cm tall, densely canescent, the hairs not in lines; leaves opposite, simple, linear to linear-lanceolate, subacute at the apex, tapering to the base, subsessile, sometimes in small fascicles, revolute, entire, densely canescent on both surfaces, to 7 cm long, to 5 (–7) mm wide; inflorescence crowded, with numerous flowers on pedicels 5–12 mm long; sepals 4, free, 2.5–4.5 mm long, canescent; petals 4, free, pale pink to white, 3.5–7.0 mm long, notched at the apex; floral tube 0.8–1.5 mm long; stamens 8; capsules narrow, to 8 cm long, canescent, on pedicels up to 3.5 cm long; seeds 1.5–2.0 mm long, the persistent coma whitish. July–September.

Marshes, bogs, wet meadows, seeps.

IA, IL, IN, KY, MO, OH (OBL), KS, NE (FACW+).

Narrow-leaved willow herb.

This species is distinguished by its narrow revolute opposite leaves that are canescent on both surfaces. The similar appearing *E. strictum* has gray, velvety leaves and a brownish coma.

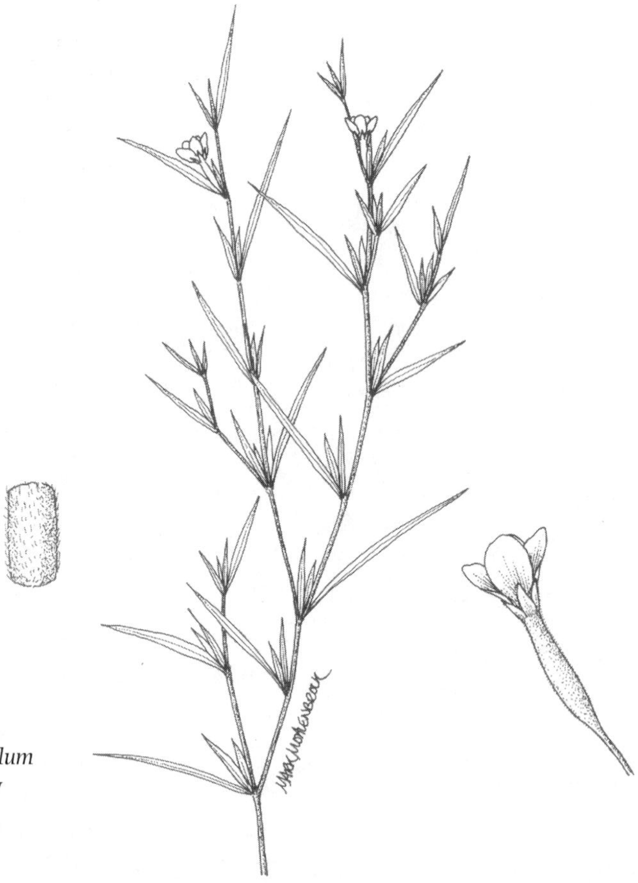

19. *Epilobium leptophyllum* (Narrow-leaved willow herb). Habit (center). Section of stem (left). Flower (right).

6. **Epilobium strictum** Muhl. Cat. 39. 1813. Fig. 20.

Perennial herbs with very slender stolons and fleshy turions; stems erect, slender, branched or unbranched, terete, to 1.2 m tall, grayish velvety; leaves opposite, simple, or the upper ones alternate, linear to linear-lanceolate, acute at the apex, tapering to the base, subsessile, often in fascicles, revolute, entire, grayish velvety on both surfaces, to 4 cm long, to 8 mm wide; inflorescence somewhat crowded, with several flowers, on pedicels up to 1 cm long; sepals 4, free, 4.0–5.5 mm long, notched at the apex; floral tube 1.0–1.8 mm long; stamens 8; capsules narrow, to 8 cm long, grayish velvety, on pedicels up to 2 cm long; seeds 1.7–2.3 mm long, the persistent coma brownish. July–September.

Bogs, wet meadows, swamps.

IA, IL, IN, OH (OBL).

Slender willow herb.

This species differs from the similar *E. leptophyllum* by its gray velvety leaves and its brownish coma.

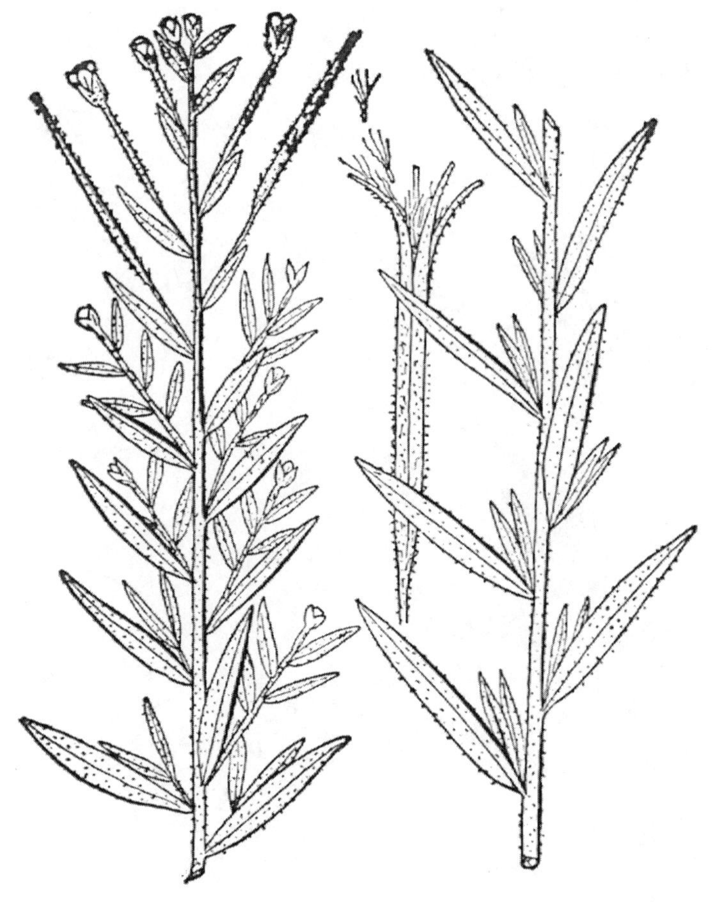

20. *Epilobium strictum* (Slender willow herb). Habit (left). Leafy branch (right). Capsule (center).

3. **Gaurella** Small

Tufted perennials; leaves alternate, simple, entire or denticulate, usually canescent; flowers axillary, sessile, perfect, the hypanthium extended beyond the ovary; sepals 4, free, reflexed; petals 4, free, spotted, white or pink; stamens 8, the filaments unequal in length; stigma 4-lobed; capsules distinctly 4-angled, beaked, the seeds not comose.

Two species comprise this genus. The species of this genus have often been included within *Oenothera*, but *Gaurella* differs by having stamens of unequal lengths and distinctly angular capsules.

1. **Gaurella canescens** (Torr. & Frem.) A. Nels. Bot. Gaz. 47:437. 1909. Fig. 21.
Oenothera canescens Torr. & Frem. Rep. 315. 1845.

Colonial-forming perennials from a thickened caudex and adventitious roots; stems decumbent to ascending, much branched from the base, densely strigillose, usually with fascicles of smaller axillary leaves; flowers in the axils of the leaves, sessile, opening at sunset; sepals 4, green, lanceolate, 8–10 mm long, canescent; petals 4, free, pink or whitish and red-spotted, obovate, 8–12 mm long; stamens 8; ovary inferior; capsules ovoid, 4-angled, beaked, canescent, 7–8 mm long; seeds brown, about 1 mm long. May–July.

21. *Gaurella canescens* (Spotted evening primrose). Habit (center). Capsule (lower left). Cross-section of ovary (above center).

Low areas in depressions, buffalo wallows, ditches, dried ponds.

KS, NE (FAC).

Spotted evening primrose.

The distinguishing features of this species are its sinuate to denticulate leaves and its red-spotted petals.

4. **Ludwigia** L.—Water Primrose, Primrose Willow

Perennial herbs; leaves simple, alternate or opposite, entire, usually with minute caducous stipules; flowers borne in the axils of leaves, usually yellow, with a floral tube; sepals 4–5, green, free, persistent at the top of the fruit; petals 4, 5, or absent, free; stamens 4, 5, 8, or 10; ovary inferior, 4-locular; capsules with many small seeds lacking a coma.

There are about seventy-five species of *Ludwigia*, many of them in the tropics. Species with stamens twice as many as the sepals were placed in the genus *Jussiaea* at one time.

1. Leaves opposite; stems floating in water or creeping on mud.
  2. Capsules with a green vertical stripe on each face .... 7. *L. palustris*
  2. Capsules without a green vertical stripe on each face .... 10. *L. repens*
1. Leaves alternate; stems not floating in water, except for *L. peploides*.
  3. Plants pubescent throughout.
    4. At least the lower part of the stem shaggy-pubescent.
      5. Leaves acute at the apex; only the lower part of the stem shaggy-pubescent; stems erect .... 5. *L. leptocarpa*
      5. Leaves obtuse at the apex; all parts of the stem shaggy-pubescent; stems decumbent to ascending .... 12. *L. uruguayensis*
    4. No part of the stem shaggy-pubescent, but with long-spreading hairs .... 4. *L. hirtella*
  3. Plants glabrous or sparsely pubescent.
    6. Capsules 1.5–2.0 mm long; petals absent .... 6. *L. microcarpa*
    6. Capsules more than 2 mm long; petals present or absent.
      7. Stems floating on water; petioles as long as the blades .... 8. *L. peploides*
      7. Stems not floating on water; petioles shorter than the blades, or absent.
        8. Leaves decurrent; petals usually 5 .... 2. *L. decurrens*
        8. Leaves not decurrent; petals 4, or absent.
          9. Flowers at least 1 cm across, yellow, short-pedicellate .... 1. *L. alternifolia*
          9. Flowers up to 3 mm across, greenish, sessile.
            10. Floral tube 6–10 mm long, about 4 times longer than the calyx lobes .... 3. *L. glandulosa*
            10. Floral tube up to 5 (–7) mm long, at most about 2 times longer than the calyx lobes.
              11. Bracts 2–5 mm long; capsules 4–7 mm long, at most about twice as long as the calyx lobes; plants glabrous .... 9. *L. polycarpa*
              11. Bracts up to 1 mm long; capsules 2.5–3.0 mm long, about as long as the calyx lobes; plants usually sparsely pubescent .... 11. *L. sphaerocarpa*

1. **Ludwigia alternifolia** L. Sp. Pl. 118. 1753. Fig. 22.
Ludwigia alternifolia L. var. *pubescens* Palmer & Steyerm. Ann. Mo. Bot. Gard. 25:772. 1938.

Perennial herbs from a cluster of fleshy roots; stems erect, branched, to 1.2 m tall, sometimes reddish, angled but not square, glabrous to rarely densely pubescent; leaves alternate, simple, lanceolate, acute to acuminate at the apex, tapering to the base, entire, to 10 cm long, to 3.5 cm wide, glabrous to densely pubescent, sessile or on petioles to 10 mm long; flower solitary in the axils of the uppermost leaves, on pedicels up to 5 mm long, the pedicels bearing a pair of bracteoles near the tip; sepals 4, narrowly ovate to ovate, spreading, 7–10 mm long, more than half as wide, green but often becoming reddish; petals 4, free, yellow, 7–10 mm long; stamens 4; capsules 4–6 mm long, nearly as broad, appearing like a small box, wing-angled, glabrous or pubescent, opening by a solitary pore at the top, on pedicels about 5 mm long; seeds numerous, shiny, yellow-brown, up to 0.8 mm long. June–August.

22. *Ludwigia alternifolia* (Seed-box). Habit (center). Capsule with sepals (lower right).

Swamps, around lakes and ponds, along streams, wet ditches.

IA, IL, IN, KS, MO, NE (OBL), KY, OH (FACW+).

Seed-box.

The appropriately named seed-box is square and contains numerous tiny seeds. In southern Illinois and southern Missouri, plants occur that are densely pubescent on the stems, leaves, and capsules. These may be called var. *pubescens.*

2. **Ludwigia decurrens** Walt. Fl. Carol. 89. 1788. Fig. 23.
Jussiaea decurrens (Walt.) DC. Prodr. 3:56. 1828.

Annual herbs from a cluster of fleshy roots; stems erect, branched, 4-angled with wing angles, glabrous, to 2 m tall; leaves alternate, simple, linear-lanceolate to lanceolate, acute at the apex, tapering to the decurrent base, rather thin, entire, glabrous, to 12 (–15) cm long, to 4 cm wide, sessile or subsessile; flowers in the axils of the leaves on pedicels up to 10 mm long; sepals 4, green, up to 10 mm long, persistent on the fruit; petals 4, yellow, up to 10 mm long, about as long as the sepals; stamens 8; capsules obconic, 4-angled, with each angle usually narrowly winged, 1–2 cm long, glabrous; seeds numerous, up to 0.5 mm long. June–October.

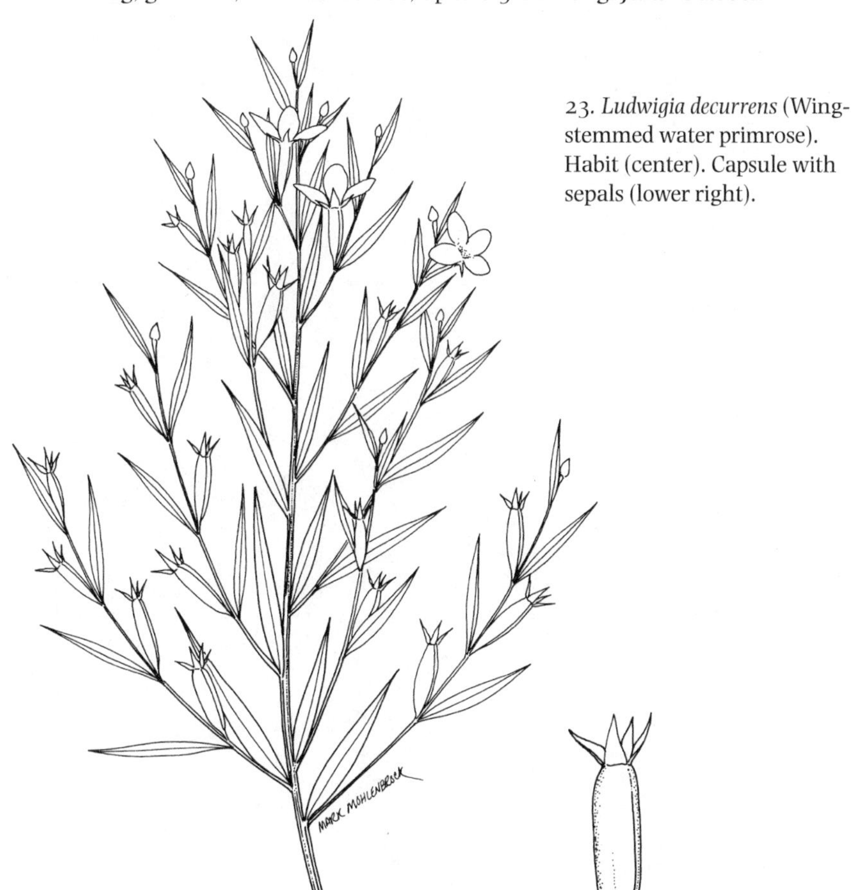

23. *Ludwigia decurrens* (Wing-stemmed water primrose). Habit (center). Capsule with sepals (lower right).

Swamps, wet ditches, often in shallow water.

IL, IN, KY, MO, OH (OBL).

Wing-stemmed water primrose.

This attractive species is readily distinguished by its 4-winged stems due to the base of each leaf, which is decurrent along the stem. Because the stamens are twice the number of the sepals, this plant at one time was segregated into the genus *Jussiaea.*

3. **Ludwigia glandulosa** Walt. Fl. Carol. 88. 1788. Fig. 24.

Perennial herbs with fibrous roots; stems sometimes decumbent at first, soon becoming erect, much branched, to 1.0 (–1.2) m tall, glabrous on strigillose; leaves alternate, simple, lanceolate to elliptic, acute at the apex, tapering to the base, entire, glabrous, to 10 cm long, to 1.5 cm wide, sessile or on short petioles; flower solitary in the axil of the upper leaves, sessile, with bracteoles at the base of the ovary up to 1 mm long; sepals 4, free, green, spreading, persistent on the fruit, 1–2 mm long; petals 0; stamens 4; capsules cylindrical, (4–) 6–10 mm long, up to 2 mm wide, at least four times longer than wide, sessile; seeds numerous. June–September.

Swamps, depressions in wet woods.

IL, IN, KS, KY, MO (OBL).

Cylindric-fruited seed-box.

This species is recognized by its very narrow, cylindrical capsules, its very short sepals, and the absence of petals.

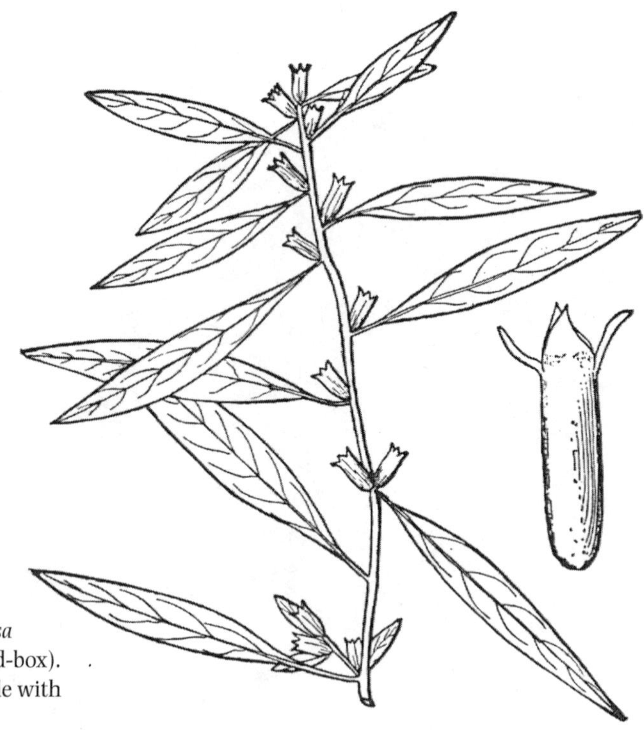

24. *Ludwigia glandulosa* (Cylindric-fruited seed-box). Habit (center). Capsule with sepals (right).

4. **Ludwigia hirtella** Raf. Med. Rep. 2. 5:358. 1808. Fig. 25.

Perennial herbs with tuberous rootstocks; stems erect, branched or unbranched, with long-spreading pubescence, to 75 cm tall; leaves alternate, simple, oblong to ovate-lanceolate, obtuse to subacute at the apex, tapering to or rounded at the sessile base, entire, with spreading pubescence on both surfaces, to 3 cm long, to 1.5 cm wide; flower usually solitary in the axil of the upper leaves, on pubescent pedicels to 15 mm long, subtended by a pair of inconspicuous bracteoles; calyx lobes 4–5, ovate-lanceolate, erect, green, hirsute, 5–8 mm long, reflexed after flowering; petals 4, free, yellow, caducous, oblong to obovate, to 10 mm long, a little longer than the calyx lobes; stamens 4; capsules square but with a rounded base, hirsute, with a single terminal pore, to 4 mm high; seeds numerous. June–September.

Swampy woods.

KY (OBL).

Hairy ludwigia.

This species is distinguished by its hirsute stems, leaves, pedicels, and calyx lobes. *Ludwigia uruguayensis* and *L. leptocarpa*, the only other densely pubescent species of *Ludwigia*, are the other species that have five calyx lobes.

Ludwigia hirtella just enters our range in Kentucky.

25. *Ludwigia hirtella* (Hairy ludwigia). Habit (left). Leafy branch (right). Capsule with sepals (upper right).

5. **Ludwigia leptocarpa** (Nutt.) Hara, Journ. Jap. Bot. 28:292. 1953. Fig. 26.
Jussiaea leptocarpa Nutt. Gen. N. Am. Pl. 1:279–280. 1818.

Annual herbs from fibrous roots; stems erect, branched, softly pubescent, at least below, to 25 cm tall; leaves alternate, simple, lanceolate to narrowly elliptic, acute at the apex, tapering to the base, entire, softly and shaggy-pubescent, sometimes glabrate, to 15 cm long, to 3 cm wide, sessile or on short petioles; flower solitary in the axil of some of the leaves, on pedicels up to 3–15 mm long, subtended by 2 bracteoles; calyx lobes (5–) 6, lanceolate, acuminate, green, 5–8 mm long; petals (5–) 6, yellow, obovate, 5–10 mm long, about as long as the sepals; stamens (10) 12; capsules narrowly cylindric, to 4 (-5) cm long, to 3 mm wide, strongly ribbed, pubescent; seeds 1.0–1.4 mm long, light beige. June–September.

Swamps, marshes, wet ditches, around ponds and lakes.

IL, IN, KY, MO, OH (OBL).

Hairy water primrose.

This *Ludwigia* is densely pubescent throughtout. *Ludwigia uruguayenesis*, which is also shaggy-pubescent, has decumbent or creeping stems and pedicels longer than the floral tube.

6. **Ludwigia microcarpa** Michx. Fl. Bor. Am. 1:88. 1803. Fig. 27.

Perennial herbs with fibrous or slender creeping rootstocks; stems ascending to erect, branched or unbranched, to 40 cm tall, glabrous, forming basal stolons; leaves alternate, simple, oblancolate to nearly orbicular, obtuse to subacute at the apex, tapering to the subsessile base, entire but glandular along the margins, glabrous, to 10 (–12) mm long, to 6 (–8) mm wide; flower solitary in the axils of most of the leaves, sessile or on very short pedicels, subtended by small, linear bracteoles; calyx 4-lobed, the lobes broadly ovate, acute, green, 1.3–1.6 mm long; petals absent; stamens 4; capsules obpyramidal or nearly square, with rounded sides, glabrous, 1.5–2.0 mm long; seeds ellipsoid, 0.4–0.5 mm long, red-brown. July–September.

Along a spring-fed stream.

MO (OBL).

Small-fruited ludwigia.

This diminutive species, which barely enters our range in southern Missouri, is distinguished by its 4 minute sepals, no petals, 4 stamens, and sessile or near sessile squarrish capsules.

7. **Ludwigia palustris** (L.) Ell. Bot. S. Carol. 1:211. 1817. Fig. 28.
Isnardia palustris L. Sp. Pl. 120. 1753.

Perennial herbs, often rooting at the nodes; stems creeping on mud or floating in shallow water, branched or unbranched, glabrous, to 50 cm long; leaves opposite, simple, elliptic to ovate to obovate, acute at the apex, tapering to the base, to 30 mm long, to 20 mm wide, usually much smaller, entire, glabrous, the petioles up to 2.0 (–2.5) cm long; flower solitary in the axils of the leaves, sessile, usually with minute bracteoles at the base of the ovary; calyx lobes 4, green, 1.5–2.0 mm long, persistent at the top of the fruit; petals 0; stamens 4, attached to an elevated disk; capsules oblongoid to obovoid, 4-sided, rounded at the sessile base, 3–5 mm long, 2–3

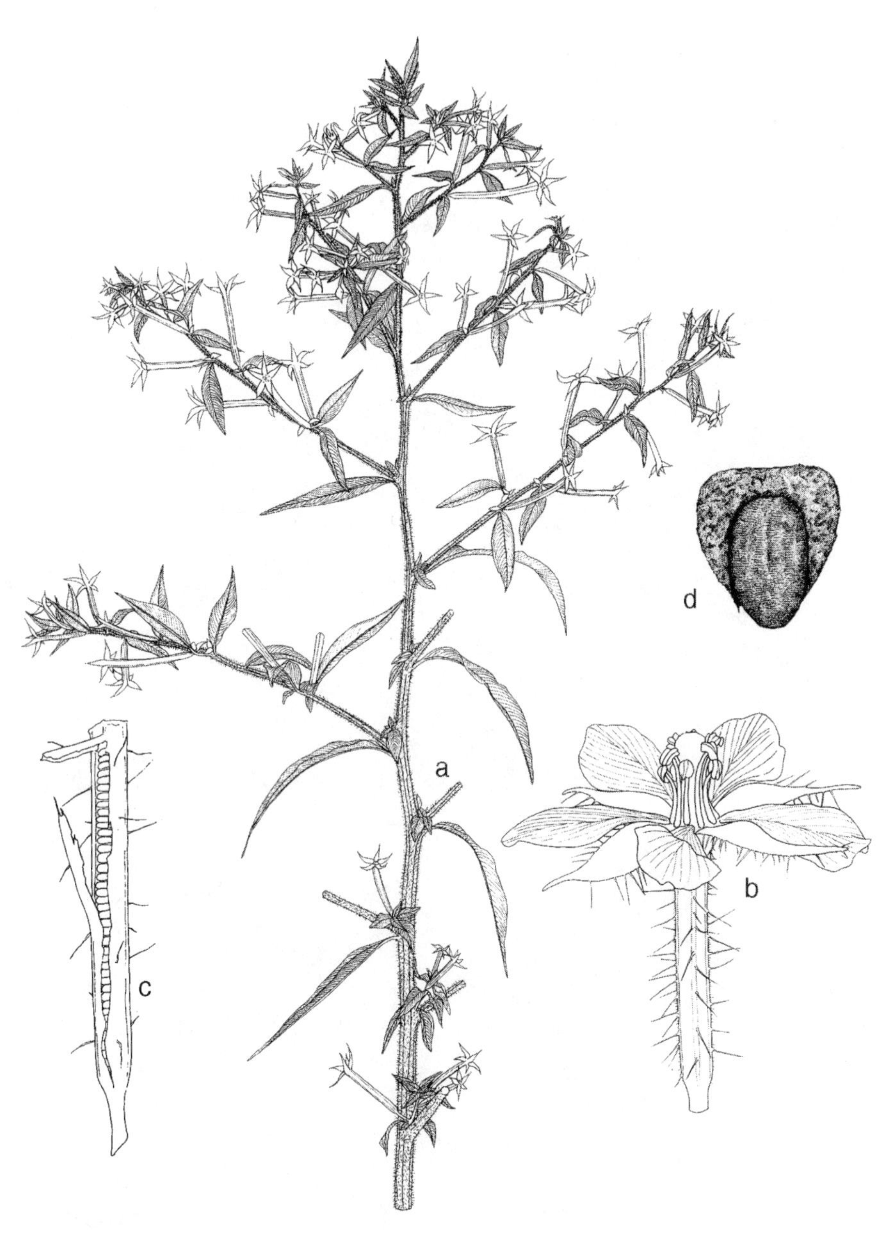

26. *Ludwigia leptocarpa* (Hairy water primrose).

a. Habit.
b. Flower.
c. Capsule.
d. Seed.

27. *Ludwigia microcarpa* (Small-fruited ludwigia). Habit (left). Capsule with sepals (right).

28. *Ludwigia palustris* (Marsh purslane). Habit (center). Capsule with sepals (lower right).

mm across, with a distinct green longitudinal stripe on each face of the capsule; seeds numerous, shiny, white or pale brown, up to 1 mm long. June–October.

Around and in ponds, lakes, streams, wet ditches.

IA, IL, IN, KS, KY, MO, NE, OH (OBL).

Marsh purslane.

This common species crawls on muddy surfaces or floats in shallow water. One must look carefully to observe the obscure, apetalous flowers in the axils of the leaves.

The green stripe on each face of the capsule distinguishes this species from the similar-appearing but rare *L. repens*.

Some botanists distinguish our plants as var. *americana*, differing from the typical Eurasian var. *palustris*, which may be slightly different.

Because of the opposite leaves, its apetalous flowers, and its creeping habit, Linnaeus put this species in *Isnardia* and not in *Ludwigia*, a choice I was tempted to follow.

8. **Ludwigia peploides** (HBK.) Raven, Reinwardtia 6:393. 1963. Fig. 29.
Jussiaea peploides HBK. Nov. Gen. & Sp. Pl. 6:97. 1823.
Jussiaea repens Forsk. var. *glabrescens* Ktze. Rev. Gen. Pl. 1:251. 1891.
Ludwigia peploides (HBK.) Raven ssp. *glabrescens* (Ktze.) Raven, Reinwardtia 6:394. 1963.
Ludwigia peploides (HBK.) Raven var. *glabrescens* (Ktze.) Shinners, Sida 1:386. 1964.

Perennial herbs with fleshy fibrous roots; stems floating in water or sometimes creeping on mud, rooting at the nodes, branched, often reddish, glabrous or sparsely pubescent, to 75 cm long; leaves opposite, simple, oblong to spatulate, obtuse at the apex, tapering to the base, entire, glabrous or somewhat pubescent, on petioles up to 2 (–4) cm long; stipules usually present, up to 1.5 mm long; flower solitary in the axils of the upper leaves, on pedicels up to 6 cm long, with bracteoles up to 1 mm long at the base; calyx lobes 5, green, linear-lanceolate, up to 12 mm long, persistent at the tip of the fruit; petals 5, free, bright yellow, showy, to 2.5 cm long; stamens 10; capsules cylindrical, up to 4 cm long, up to 4 mm wide, strongly nerved, glabrous or somewhat pubescent; seeds numerous, 1.0–1.5 mm long. May–October.

Ponds, lakes, swamps, streams.

IA, IL, IN, KS, KY, MO, NE, OH (OBL)

Yellow water primrose.

This aquatic is readily recognized by its floating or creeping reddish stems, its showy yellow flowers with 5 petals and 10 stamens, and its long petioles, which are often longer than the blades.

Plants in the central Midwest may be glabrous or sparsely pubescent and may be known as var. *glabrescens*.

This species sometimes grows in great profusion, often clogging up the shallow necks of lakes.

Because the stamens are twice the number of the sepals, this species was long regarded in the genus *Jussiaea*.

29. *Ludwigia peploides* (Yellow water primrose). Habit (center). Capsule with sepals (lower left).

9. **Ludwigia polycarpa** Short & Peter, Transyl. Journ. Med. 8:581. 1826. Fig. 30.

30. *Ludwigia polycarpa* (Many-seeded seed-box). Habit (center). Flower (lower right).

Perennial herbs from a stoloniferous base; stems erect or ascending, branched or unbranched, 4-angled, glabrous, to 1 m tall; leaves alternate, simple, narrowly lanceolate to lanceolate, acute at the apex, tapering to the base, entire, glabrous, to 12 cm long, to 4 cm wide, on winged petioles up to 8 mm long; flower solitary in the axils of almost every leaf, sessile, subtended by bracteoles 2–5 mm long; calyx lobes 4, green, 2.5–4.0 mm long, persistent on the tip of the fruit; petals 0 or occasionally 4 and minute and greenish; stamens 4; capsules short-cylindric to turbinate, somewhat 4-angled, 4–7 mm long, 3–6 mm wide, glabrous, sessile; seeds numerous, pale yellow, up to 1 mm long. June–September.

Around ponds and lakes, marshes, wet prairies, along streams, depressions in woods.

IA, IL, IN, KS, KY, MO, NE, OH (OBL).

Many-seeded seed-box.

This species resembles *L. glandulosa*, but this latter species has capsules several times longer than wide.

10. **Ludwigia repens** Forsk. Fl. Am. Sept. 6. 1771. Fig. 31.
Ludwigia natans Ell. Sketch Bot. S. Carol. 1:581. 1821.

Perennial herbs; stems floating in water or creeping on mud, rooting at the nodes, usually branched, glabrous or nearly so, to 50 cm long; leaves opposite, simple, elliptic to obovate, obtuse to subacute at the apex, tapering to the base, entire, glabrous, on petioles up to 2 mm long; flower solitary in the axil of the leaves, sessile or subsessile, subtended by subulate bracteoles; calyx lobes 4, green, 2.5–4.0 mm long, persistent on the tip of the fruit; petals 4, free, yellow, 4–5 mm long, caducous; stamens 4, on an elevated disk; capsules short-cylindric, often squarrish in appearance to turbinate, 4–6 mm long, nearly as wide, without a green stripe on each face; seeds numerous, yellow-brown, 0.5–0.8 mm long. July–September.

Along spring-fed streams.

KS, MO (OBL).

Southern marsh purslane.

This primarily southern species is very similar in appearance to *L. palustris,* but its capsules lack the vertical green stripe on each face. Although *L. repens* has 4 yellow petals, they fall off so early that they are seldom seen.

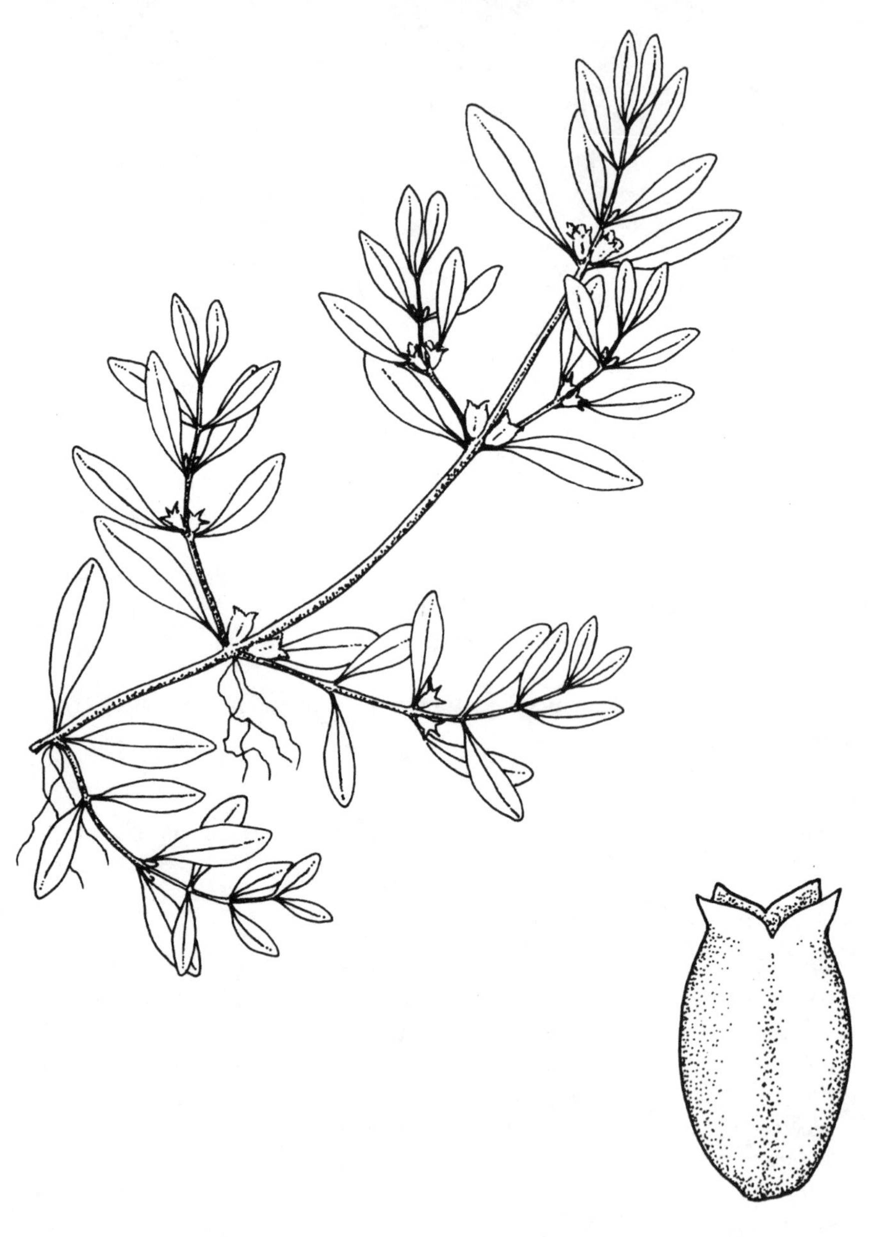

31. *Ludwigia repens* (Southern marsh purslane). Habit (center). Capsule (lower right).

11. **Ludwigia sphaerocarpa** Ell. Bot. S. Carol. 1:213. 1817. Fig. 32.
Ludwigia sphaerocarpa Ell. var. *deamii* Fern. & Grisc. Rhodora 37:174–175. 1935.

Perennial herbs from stoloniferous bases; stems erect, branched, glabrous or sometimes pubescent, to 1 m tall; leaves alternate, simple, linear-lanceolate to oblong, acute at the apex, tapering to the base, entire, glabrous or sometimes pubescent, to 10 cm long, to 5 mm wide, the leaves near the top of the stem much smaller, on petioles up to 2 mm long; flower solitary in the axil of the leaves, subsessile, with or without minute bracts; calyx lobes 4, green, 2.5–4.5 mm long, puberulent, persistent on the tip of the fruit; petals 0, or 4 and minute and greenish; stamens 4; capsules globose or subglobose, usually a little wider than high, pilose, 2.5–3.0 mm high, 3–4 mm wide; seeds numerous, about 1 mm long. July–September.

Swamps, around ponds and lakes, occasionally in standing water.

IL, IN (OBL).

Round-fruited seed-box.

This species differs from all other species of *Ludwigia* by its nearly spherical capsules. In the northern part of the central Midwest, capsules that are perfectly round may be known as var. *deamii*.

32. *Ludwigia sphaerocarpa* (Round-fruited seed-box). Habit (center). Flower, side view (bottom center). Flower, face view (bottom right).

12. **Ludwigia uruguayensis** (Camb.) Hara, Journ. Jap. Bot. 28:294. 1953. Fig. 33.
Jussiaea uruguayensis Camb. Fl. Bras. Merid. 2:264. 1829.

Perennial herbs from many fibrous roots; stems decumbent to ascending, rooting at the lower nodes, branched or unbranched, densely shaggy-pubescent; leaves alternate, simple, extremely variable, ranging from oblanceolate to suborbicular, obtuse at the apex, tapering to the base, sessile, or the lowermost on short, shaggy-pubescent petioles, the veins and margins of the leaves densely shaggy-pubescent; flower solitary in the axil of the upper leaves, on shaggy-pubescent pedicels up to 5 cm long; calyx lobes 5 (–6), subulate to narrowly lanceolate, green, up to 10 mm long, densely pubescent; petals 5 (–6), free, yellow, obovate, up to 2 cm long and often nearly as wide; stamens 10, rarely 12; capsules short-cylindric to obconic, 1.5–2.5 cm long, up to 4 mm in diameter, pubescent; seeds numerous, 1.2–1.5 mm long. July–September.

33. *Ludwigia uruguayensis* (Southern water primrose). Upper part of plant (center). Bract (lower left). Stipules (right). Flower (upper left). Flower with petals removed (left center).

Marshes, along rivers, wet ditches.

Native to South America; KY, MO (OBL).

Southern water primrose.

This *Ludwigia* is distinguished by its very shaggy-pubescent stems and leaves and its bright yellow flowers with 5 petals.

5. **Oenothera** L.—Evening Primrose

Annual, biennial, or perennial herbs; leaves alternate or basal, simple, without stipules; flowers variously arranged, perfect, actinomorphic; sepals 4, reflexed at time of flowering, not persistent on the fruit; petals 4, free, yellow, white, or pink; floral tube prolonged beyond the ovary; stamens 8; ovary inferior; fruit a capsule, or nutlike; seeds without a coma.

About eighty species in the New World comprise this genus.

1. Petals up to 1 cm long; flower buds nodding; anthers to 2.5 mm long .... 2. *O. perennis*
1. Petals 1 cm long or longer; flower buds erect; anthers 4 mm long or longer.
 2. Stems with appressed pubescence; calyx tube 0.5–1.5 cm long.
  3. Capsules and inflorescence with non-glandular hairs; capsules shallowly 4-winged .... 1. *O. fruticosa*
  3. Capsules and inflorescence usually with glandular hairs; capsules strongly 4-winged .. ..4. *O. tetragona*
 2. Stems with spreading pubescence; calyx tube 1.5–2.5 cm long......................3. *O. pilosella*

1. **Oenothera fruticosa** L. Sp. Pl. 1:346–347. 1753. Fig. 34.
Oenothera linearis Michx. Fl. Bor. Am. 1:225. 1803.
Kneiffia fruticosa (L.) Raimann, Nat. Pflanzenf. 3:214. 1893.
Oenothera fruticosa L. var. *linearis* (Michx.) S. Wat. Proc. Am. Acad. Art. 8:584. 1895.

Perennial herbs from fibrous roots; stems erect to ascending, sometimes cespitose, branched or unbranched, glabrous, or strigose, to 1 m tall; leaves simple, alternate, the basal ovate or narrowly ovate and petiolate, the cauline leaves linear to oblong to lanceolate, acute to acuminate at the apex, tapering to the sessile or subsessile base, to 9 cm long, to 4.5 cm wide, often much smaller, entire, glabrous or appressed-pubescent; flowers few to several, terminal or in the upper leaf axils, actinomorphic, perfect, short-pedicellate; calyx lobes 4, 0.5–2.5 mm long, green, long-attenuate, pubescent with appressed or spreading hairs, the calyx tube to 1.5 cm long; petals 4, free, yellow, to 2.8 cm long, obtuse at the tip; stamens 8; ovary inferior; capsules ovoid to clavate, to 1.5 cm long, narrowed to a stipitate base, pubescent with appressed or spreading hairs, with numerous seeds. May–August.

Woods, fields, wet meadows, wet open woods.

IL, IN, MO (FACU+), KY, OH (FAC).

Sundrops.

This species is distinguished by its large, yellow flowers and its ovoid to clavate capsules. The capsules are variously pubescent, which has led to the naming of several varieties and subspecies. Typical *O. fruticosa* has non-glandular, appressed hairs on the capsules, calyx, and hypanthium. Var. *linearis* has spreading, non-glandular hairs. *Oenothera tetragona* is similar but has glandular hairs on the capsules and in the inflorescence.

2. **Oenothera perennis** L. Syst. Nat., ed. 10, 998. 1758. Fig. 35.
Kneiffia perennis (L.) Pennell, Bull. Torrey Club 46:373. 1919.

Perennial herbs from fibrous roots, often with a basal rosette of leaves present; stems erect to ascending, to 75 cm tall, strigose to glabrous; basal leaves simple, oblanceolate, obtuse at the apex, tapering to the base, entire, glabrous or strigose; cauline leaves alternate, simple, linear-lanceolate to elliptic, obtuse to subacute at the apex, tapering to a short-petiolate base, glabrous or strigose, to 5.5 cm long, to 2.2 cm wide, the upper progressively smaller; flowers few in a terminal cluster, nodding, actinomorphic, perfect; calyx lobes 4, 5–8 mm long, green to reddish, the tube 5–10 mm long, glandular-pubescent; petals 4, free, yellow, notched at the summit, to 9 mm long; stamens 8; ovary inferior; capsules obovoid to clavate, to 10 mm long, glandular-pubescent, on a stipe to 4 mm long, with numerous seeds. May–August.

34. *Oenothera fruticosa* (Sundrops). Habit (center). Capsule (lower left).

35. *Oenothera perennis* (Perennial sundrops). Habit (left). Lower part of plant (right). Fruit (bottom center).

Wet meadows, fields, woods.
IA, IL, IN, KY, MO, OH (FAC−).
Perennial sundrops.
This species is similar to *O. fruticosa* and *O. pilosella* but has smaller flowers and shorter capsules.

3. **Oenothera pilosella** Raf. Ann. Nat. 1:15. 1820. Fig. 36.
Kneiffia pilosella (Raf.) Heller, Cat. N. Am. Pl. ed. 2, 8. 1900.

Perennial herb from fibrous roots; stems erect, branched or unbranched, hirsute to villous, at least in the lower part, to 80 cm tall; leaves alternate, simple, lanceolate to lance-ovate, acute at the apex, tapering to the sessile or subsessile base, entire to denticulate, pilose on both surfaces, to 10 cm long, to 3 cm wide; flowers few-several in a terminal cluster or from the axils of the upper leaves, actinomorphic, perfect; calyx lobes 4, green, hirsute, long-attenuate, 2.5–4.0 mm long, the tube up to 2.5 cm long; petals 4, free, bright yellow, to 2.8 cm long, usually notched at the apex; stamens 8; capsules clavate, to 1.5 cm long, with spreading hairs, sessile or short-stipitate, with numerous seeds. June–July.

Wet meadows, wet prairies, woods, fields.
IA, IL, IN, MO (FAC−), KY, OH (FAC).
Prairie sundrops.
This species differs from the similar-appearing *O. fruticosa* by its longer calyx tube.

4. **Oenothera tetragona** Roth, Cat. Bot. 2:39. 1800. Fig. 37.
Oenothera glauca Michx. Fl. Bor. Am. 1:224. 1803.
Kneiffia tetragona (Roth) Pennell, Bull. Torrey Club 46:370. 1919.
Oenothera fruticosa L. ssp. *glauca* (Michx.) Straley, Ann. Mo. Bot. Gard. 64:403. 1977.

Perennial herb from fibrous roots; stems erect, branched or unbranched, villous to glabrous, rarely glaucous, to 1 m tall; leaves alternate, simple, narrowly elliptic to lanceolate to ovate, acute at the apex, tapering to the sessile or subsessile base, entire or sparsely denticulate, glabrous or somewhat pubescent, to 10 cm long, to 2.2 (–4.0) cm wide; flowers few–several in a terminal cluster or from the axils of the upper leaves, actinomorphic, perfect; calyx lobes 4, green, glabrous or with gland-tipped hairs, long-attenuate, 2.5–4.0 mm long, the tube up to 2.5 cm long; petals 4, free, bright yellow, to 3 cm long, sometimes notched at the apex; stamens 8; ovary

36. *Oenothera pilosella* (Prairie sundrops). Habit (center). Flower (lower left).

inferior; capsules oblongoid to ellipsoid, strongly 4-winged, to 1.3 cm long, usually with glandular hairs, sessile or short-stipitate, with numerous seeds. June–July.

Wet meadows, wet prairies, woods, fields.

IA, IL, IN, MO (FAC−), KY, OH (FAC). This species is not distinguished from *O. fruticosa* by the U.S. Fish and Wildlife Service.

Square-fruited sundrops.

Although similar in appearance to *O. fruticosa*, this species has more strongly winged capsules and usually glandular-pubescent stems, leaves, and capsules.

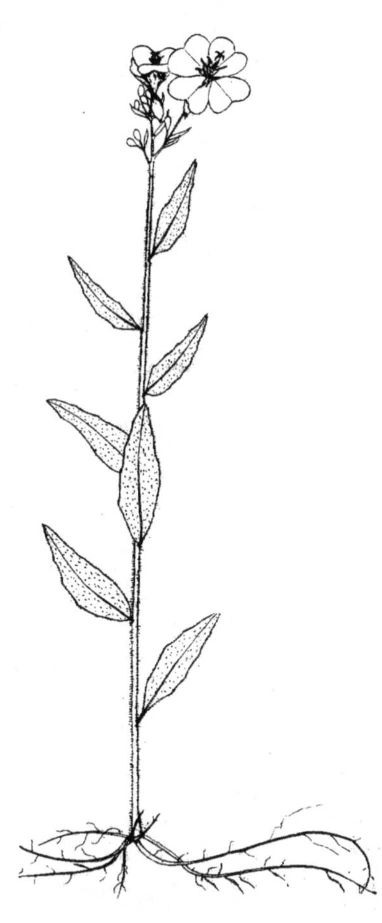

37. *Oenothera tetragona* (Square-fruited sundrops). Habit.

94. PARNASSIACEAE—GRASS-OF-PARNASSUS FAMILY

Only the following genus comprises this family.

1. **Parnassia** L.—Grass-of-Parnassus

Perennial herbs; leaves basal except for one usually present cauline leaf, without stipules; flower solitary, perfect, actinomorphic; sepals 5, united at the base, persistent on the fruit; petals 5, free, conspicuously veined, each with a glandular staminodium at the base; stamens 5; ovary superior or half-inferior, 1-locular, with 4 stigmas; capsules with numerous seeds.

There are approximately fifty species in this genus, all in the Northern Hemisphere.

Many botanists include this genus and several others in the diverse *Saxifragaceae*, but I believe its characters are sufficient enough to merit family status.

1. Larger basal leaves to 5 cm long; petals broadly ovate, to 1.8 cm long; staminodia shorter or equaling the stamens .... 1. *P. glauca*
1. Larger basal leaves 7–10 cm long; petals elliptic to oblong, to 2.2 cm long; staminodia longer than the stamens .... 2. *P. grandifolia*

1. **Parnassia glauca** Raf. Aut. Bot. 42. 1840. Fig. 38.
Parnassia americana Muhl. Cat. 33. 1813, misapplied.

Perennial herbs from short rhizomes; stems erect, to 40 cm tall, glabrous, glaucous; basal leaves several, oblong to broadly ovate, obtuse to subacute at the apex, tapering to the base or rounded or subcordate, coriaceous, glabrous, glaucous, to 5 cm long, often nearly as wide, palmately 5- to 9-veined, petiolate; cauline leaf solitary or absent, similar to the basal leaves but smaller and sessile; sepals 5, united at the base, coriaceous, oblong to ovate, obtuse at the apex, green with a hyaline margin, 3- to 7-nerved, persistent and reflexed on the fruit; petals 5, free, cream-colored to greenish white, to 1.8 cm long, broadly ovate, with 9 conspicuous veins; stamens 5, to 10 mm long, as long as or longer than the 3-lobed, white staminodia; capsules ovoid, to 1 cm long, with many seeds about 1 mm long, July–October.

Bogs, wet meadows, swamps, fens.

IA, IL, IN, OH (OBL).

Glaucous grass-of-Parnassus.

This species is similar to *P. grandifolia* but is smaller in all respects and has staminodia shorter than or equaling the stamens.

38. *Parnassia glauca* (Glaucous grass-of-Parnassus). Habit (center). Staminodia with stamen (right).

2. **Parnassia grandifolia** DC. Prodr. 1:320. 1824. Fig. 39.

Perennial herbs; stems erect, to 40 cm tall, glabrous, somewhat glaucous; basal leaves several, broadly ovate to broadly oblong, obtuse to subacute at the apex, tapering or rounded or subcordate at the base, coriaceous, glabrous, somewhat glaucous, the larger ones 7–10 cm long, nearly as wide, palmately 5- to 9-nerved, petiolate; cauline leaf solitary, sessile or subsessile, similar to the basal leaves but smaller; sepals 5, united at the base, oblong to ovate, coriaceous, oblong to ovate, obtuse at the apex, green with hyaline margins, 3- to 7-nerved, persistent and reflexed on the fruit; petals 5, free, cream-colored, to 2.2 cm long, elliptic to oblong, with 7 or 9 conspicuous veins; stamens 5, to 10 mm long, much shorter than the 3- to 5-lobed, white staminodia; capsules ovoid, to 1 cm long, with many seeds about 1 mm long. August–October.

Wet meadows.

KY, MO (OBL).

Large grass-of-Parnassus.

This primarily southeastern species barely gets into the central Midwest in Kentucky and the Missouri Ozarks. It is larger than *P. glauca* in all respects, and the staminodia are much longer than the stamens.

39. *Parnassia grandiflora* (Large grass-of-Parnassus). Habit (center). Flower, with one petal removed (lower right).

95. PLANTAGINACEAE—PLANTAIN FAMILY

Annual, biennial, or perennial herbs; leaves simple, basal or opposite; flowers in spikes or heads, perfect or unisexual, actinomorphic; sepals 4, united at the base; petals 4, united at the base, dry and scarious; stamens 2 or 4; ovary superior, 2-locular; fruit a capsule, achene, or nutlet.

In addition to *Plantago*, there are two other genera in the family, *Bougeria* and *Littorella*, each monotypic.

1. **Plantago** L.—Plantain

Annual, biennial, or perennial herbs; leaves mostly basal, although a few opposite cauline leaves may be present in some species, without stipules; flowers perfect or rarely unisexual, actinomorphic, borne in spikes or heads; sepals 4, united at the base, persistent; petals 4, united at the base, dry or scarious, with reflexed or spreading lobes; stamens 2 or 4; ovary superior, 2-locular; fruit a capsule.

About two hundred species found worldwide comprise this genus. Only the following occurs in wetlands in the central Midwest.

1. **Plantago cordata** Lam. Tabl. Encycl. 1:338. 1791. Fig. 40.

Perennial herbs from a thickened caudex and fleshy roots; flowering stems leafless, glabrous, hollow, up to 30 cm long; leaves all basal, oval to ovate, subacute to acute at the apex, cordate at the base, glabrous, pinnately veined, to 30 cm long, to 22 cm wide, on long, glabrous petioles; flowers in an interrupted spike up to 30 cm long, each one subtended by a bract about 3 mm long; sepals 4, united at the base, broadly ovate, obtuse at the apex, keeled, 3–4 mm long; petals 4, attached at the base, dry and scarious; stamens 4; capsules globose to ovoid, glabrous, 4–7 mm long and often about as wide, circumscissile about the middle, with 2–4 ellipsoid, dark brown seeds about 3 mm long. July–September.

In flowing streams.

IA, IL, IN, KY, MO, OH (OBL).

Heart-leaved plantain.

This species is found in shallow water of clear streams. Its large, heart-shaped leaves and hollow flowering stems are distinctive.

During the latter half of the twentieth century, this species became so rare that it was listed as a federally threatened species.

96. PLATANACEAE—SYCAMORE FAMILY

Trees with peeling bark; leaves alternate, simple, palmately lobed, the base of the petiole excavated to fit over the bud; flowers unisexual but on the same tree, arranged in globose heads, each flower subtended by bracteoles; sepals 0; petals 0; stamens 3–8; pistils 2–9, free from each other, with several staminodia, the ovary 1-locular; fruit a globose head of narrow nutlets with long hairs attached.

Only the following genus with eight species comprise the family.

1. **Platanus** L.—Sycamore

Characters of the family.

40. *Plantago cordata* (Heart-leaved plantain). Leaf and fruiting spike (center). Fruit (upper right).

1. **Platanus occidentalis** L. Sp. Pl. 999. 1753. Fig. 41.

Large tree to 35 m tall, with a trunk diameter up to 2.6 m, the crown broad but irregular; bark reddish brown when young, quickly breaking into thin, flat scales, falling away in sections to expose large patches of whitish or greenish inner bark; twigs glabrous, light brown, somewhat zigzag, the buds light brown, acute, to 5.5 mm long, completely covered by the hollowed out base of the petiole; leaves alternate, simple, usually wider than high, palmately 3- to 5-lobed, the lobes with large, pointed teeth, the base of the leaf cordate or truncate, up to 18 cm long, up to 22 cm wide, usually larger on vigorous shoots, green and more or less glabrous on the upper surface, paler on the lower surface with floccose pubescence, the petioles up to 12 cm long, pubescent, hollow at base; stipules resembling the leaves but only up to 2 cm long, often persistent; staminate and pistillate flowers borne separately on the same tree, minute, crowded into spherical heads, the staminate with 3–8 stamens, the pistillate with 2–9 pistils and several staminodia; fruit globose, up to 1.8 cm in diameter, on long, pendulous stalks, each head consisting of numerous crowded narrow nutlets with persistent long hairs. May–June.

Bottomland forests, along rivers and streams, around lakes and ponds.

IA, IL, IN, MO (FACW), KS, NE (FAC), KY, OH.(FACW−).

Sycamore.

The large palmately lobed leaves and the brown and gray mottled bark are distinguishing characteristics.

97. PODOSTEMACEAE—RIVERWEED FAMILY

Aquatic annual herbs; leaves and stems scarcely differentiated, appearing thalloid; flowers minute, perfect, borne from within a short, tubular spathe; sepals 0; petals 0; stamens 2–numerous; ovary superior, sessile or stipitate, 2- to 3-locular; styles 2–3; capsules 2- to 3-celled, with numerous minute seeds.

Approximately twenty genera and 180 species comprise this family, almost all of them tropical.

1. **Podostemum** Michx.—Riverweed

Aquatic annual herbs; leaves and stems scarcely differentiated, appearing thalloid; flowers minute, perfect, sessile, borne from a short, tubular spathe; sepals 0; petals 0; stamens 2, with 2 staminodia; ovary superior, 2-locular; styles 2; capsules 2-celled, with numerous minute seeds.

There are approximately twelve species of *Podostemum*, found in various parts of the world.

1. **Podostemum ceratophyllum** Michx, Fl. Bor. Am. 2:165. 1803. Fig. 42.

Aquatic annual herbs with slender roots attached to rocks in running water; stems and leaves scarcely differentiated, stiff, to 25 cm long, the segments linear to filiform, glabrous; flowers few in the axils of the branches, perfect, 1.2–1.5 mm long, protruding from a short, tubular spathe; sepals 0; petals 0; stamens 2, with 2 staminodia; ovary superior, with 2 styles; capsules ovoid to oval, 1.5–1.8 mm long, obtuse at the apex, stipitate, with 8 strong vertical ribs; seeds minute, numerous. July–September.

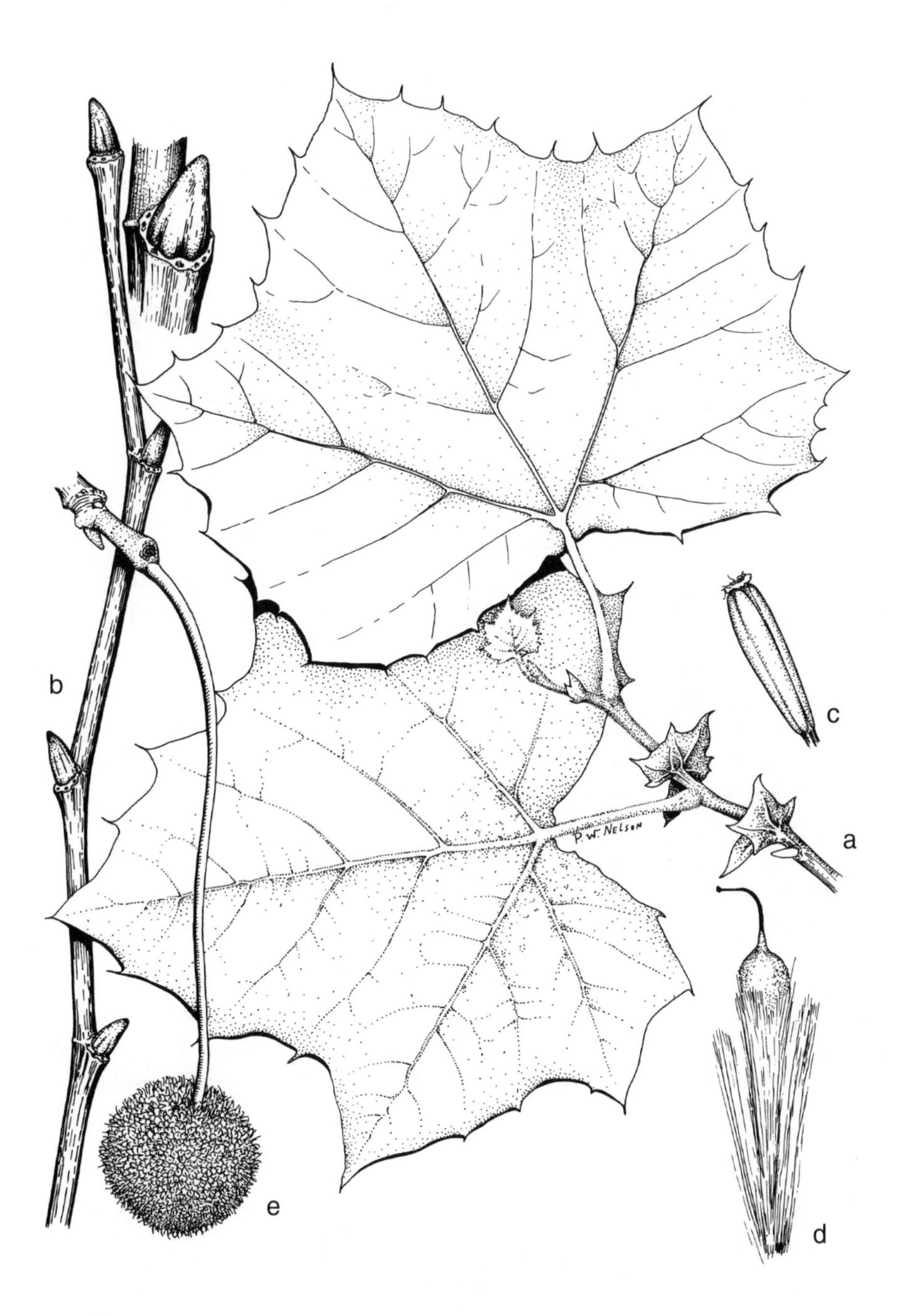

41. *Platanus occidentalis* (Sycamore).

a. Leafy branch, with stipules.
b. Winter twig and buds.
c. Stamen.
d. Ovary.
e. Fruiting head.

Attached to rocks in flowing water.

KY, OH (OBL).

Riverweed.

This slender, branched aquatic grows attached to rocks and stones in shallow running water of streams. The tiny flowers protrude from a short, funnel-shaped spathe.

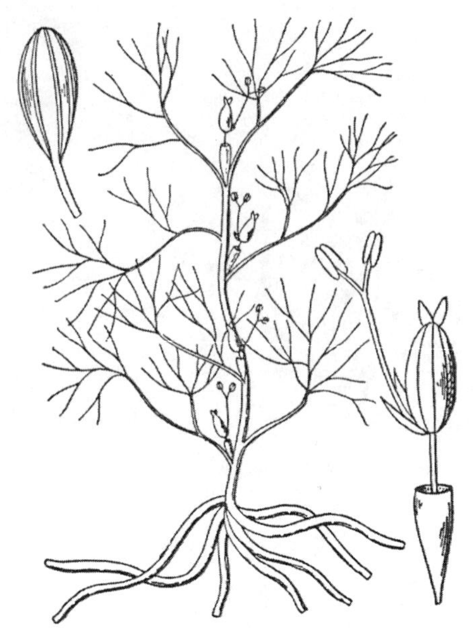

42. *Podostemum ceratophyllum* (Riverweed). Habit (center). Capsule (upper left). Flower (lower right).

98. POLEMONIACEAE—PHLOX FAMILY

Annual or perennial herbs, rarely shrubs or small trees; leaves alternate or opposite, simple or compound, without stipules; inflorescence mostly cymose; flowers perfect, actinomorphic; calyx 4- to 5-lobed; corolla 4- to 5-parted; stamens 5, attached to the corolla tube; disk usually present; pistil 1, the ovary superior, (2–) 3-locular, with 1–several ovules on each axile placenta; stigmas 3; fruit a capsule.

This family is comprised of about thirteen genera and 275 species native to North and South America.

The chief diagnostic characters of the family are the usually trilocular ovary with numerous ovules and the united sepals and united petals.

Three genera occur in wetlands in the central Midwest.

1. Leaves simple, entire or pinnatifid.
   2. Leaves pinnatifid; calyx teeth spinulose .... 1. *Navarretia*
   2. Leaves entire; calyx teeth not spinulose .... 2. *Phlox*
1. Leaves pinnately compound .... 3. *Polemonium*

1. **Navarretia** R.& P.

Annual herbs; leaves alternate, simple but usually pinnatifid; flowers in terminal heads, actinomorphic, bracteate; calyx 4- to 5-lobed, short-tubular below, the lobes sharp-tipped and sometimes with spinulose teeth; corolla 4- to 5-lobed, tubular below, salverform; stamens 5, attached to the corolla; ovary superior, 2- to 3-locular; fruit a capsule; seeds few to several, often mucilaginous.

This genus consists of about thirty species in the western United States and in southern South America.

1. **Navarretia intertexta** (Benth.) Hook. Fl. Bor. Am. 2:75. 1837. Not illustrated.
Aegochloa intertexta Benth. Bot. Reg. 19:pl. 1622. 1833.

Erect annuals; stems 5–10 (–20) cm tall, branched or unbranched, glabrous or with reflexed glandular hairs; leaves alternate, simple, pinnatifid to bipinnatifid, the segments subulate to linear, with pungent, spinose tips, glabrous to glandular-pubescent to pilose, sessile; flowers in dense capitate clusters at the tips of the branches, subtended by bracts similar in shape to the leaves; calyx 4- to 5-lobed, cylindrical, white-pilose, the lobes more or less dentate and often pungent; corolla 4- to 5-lobed, salverform or funnelform, white, the lobes 9–10 mm long; stamens 4 or 5, attached to the corolla tube, exserted; ovary superior, oblong, 2- to 3-locular; capsules indehiscent or tardily dehiscent, with 1–4 (–6) seeds that are mucilaginous when wet. May–August.

Vernal ponds, buffalo wallows, ephemeral wet sites.

IA, NE, OH (not listed for any of these states by the U.S. Fish and Wildlife Service).

Navarretia.

This species, which barely enters our range in western Nebraska, is distinguished by its narrowly divided, spinescent leaves, its dense capitate cluster of white flowers, and its white-pilose, dentate calyx teeth.

2. **Phlox** L.—Phlox

Perennial herbs; leaves opposite, simple, entire; flowers mostly in cymes, actinomorphic, perfect; calyx 5-lobed; corolla salverform, 5-lobed; stamens 5, unequally inserted on the tube of the corolla, usually included; pistil 1, the ovary superior, 3-locular, with 1–4 ovules per locule.

Phlox is a genus of about seventy species, all but one native to North America. Several species are grown as garden ornamentals.

The following may occur in wetland habitats in the central Midwest.

1. Leaves with conspicuous lateral veins and reticulations; calyx teeth subulate .... 4. *P. paniculata*
1. Leaves without conspicuous lateral veins and reticulations; calyx teeth lanceolate.
    2. Flowers in panicles usually longer than broad; stems usually purple-spotted .... 3. *P. maculata*
    2. Flowers in corymbs nearly as broad as long; stems green or purplish, not spotted.
        3. Calyx to 7.5 mm long .... 2. *P. glaberrima*
        3. Calyx 9–12 mm long .... 1. *P. carolina*

1. **Phlox carolina** L. ssp. **angusta** Wherry, Baileya 4:98. 1956. Fig. 43.

Perennial herb; stems erect, simple on branched, green, usually glabrous, to 70 cm tall; leaves simple, opposite, linear to lanceolate, acute to acuminate at the apex, cuneate to somewhat rounded at the base, entire, usually glabrous, up to 10 cm long, up to 2 cm broad, the lateral veins not prominent; inflorescence cymose, usually about as broad as long, the flowers short-pedicellate; calyx 9–12 mm long, green, glabrous, 5-lobed, the lobes subulate-lanceolate, less than half as long as the calyx tube; corolla salverform, to 2.5 cm long, pink, 5-lobed; stamens 5, included; capsule oval, to 5 mm long. May–June.

Open moist or wet habitats.

IL, IN, MO (FACW), KY (FACU).

Carolina phlox.

This species differs from other wetland species of *Phlox* by the leaves without conspicuous lateral veins and the calyx 9–12 mm long. *Phlox glaberrima* is similar but has a calyx up to 7.5 mm long.

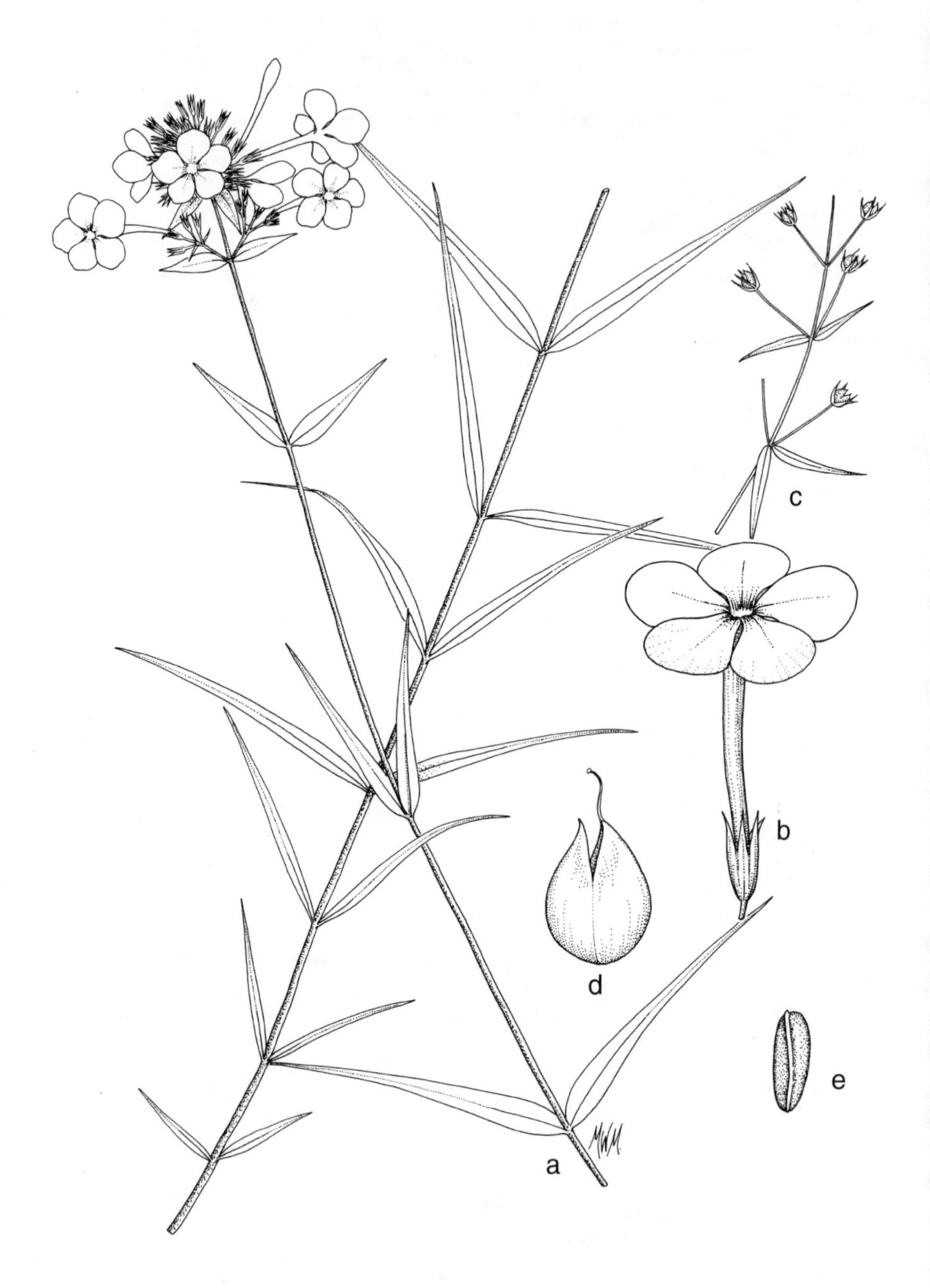

43. *Phlox carolina* (Carolina phlox).

a. Habit.
b. Flower.
c. Fruiting branch.
d. Fruit.
e. Seed.

2. **Phlox glaberrima** L. ssp. **interior** (Wherry) Wherry, Baileya 4:98. 1956. Fig. 44.
Phlox glaberrima L. var. *interior* Wherry, Bartonia 14:19. 1932.

Perennial herb; stems erect, simple, glabrous, to 85 cm tall; leaves linear-lanceolate to lanceolate, acuminate at the apex, cuneate at the base, entire, glabrous, the lateral veins not prominent, to 10 cm long; flowers borne in a flat-topped cyme, on short pedicels; calyx to 7.5 mm long, green, glabrous, 5-lobed, the lobes subulate-lancolate, less than half as long as the calyx tube; corolla salverform, to 2.5 cm long, pink, 5-lobed; stamens 5, included; capsule oval, to 5 mm long. May–August.

Woods, wet prairies.

IL, IN, MO (FACW), KY, OH (FAC).

Smooth phlox.

Our specimens fall under ssp. *interior*, a Midwestern taxon with the calyx shorter than in the typical subspecies. This is a handsome species with showy rose-pink flowers.

Almost all parts of the plant are glabrous.

44. *Phlox glaberrima* (Smooth phlox). a. Habit. b. Calyx.

3. **Phlox maculata** L. Sp. Pl. 152. 1753. Fig. 45.
Phlox suaveolens Ait. Hort. Kew. 1:206. 1789.
Phlox maculata L. var. *purpurea* Michx. Fl. Bor. Am. 1:143. 1803.
Phlox pyramidalis J.E. Smith, Exot. Bot. 2:55. 1804.
Phlox maculata L. ssp. *pyramidalis* (J.E. Smith) Wherry, Castanea 16:100. 1951.

Perennial herb; stems erect, simple or branched, speckled with purple, glabrous or pubescent, to nearly 1 m tall; lower leaves simple, opposite, the lower ones linear to lanceolate, acuminate at the apex, rounded at the base, the upper ones lanceolate to lance-ovate, acuminate at the apex, rounded or subcordate at the base, entire, glabrous or puberulent, the lateral veins not prominent, to 10 cm long; flowers borne in narrow conical to cylindrical to pyramidal panicles, on short pedicels; calyx to 7 mm long, green, glabrous or puberulent, 5-lobed, the lobes lanceolate,

45. *Phlox maculata* (Speckled phlox). a, c. Habit. b, d. Flower.

acute to acuminate, about one-fourth as long as the calyx tube; corolla salverform, to 2.5 cm long, pink or purple, 5-lobed; stamens 5, included; capsules oval, to 1 cm long. May–June.

Wet prairies, near bogs and swamps, in moist woodlands and meadows.

IA, IL, IN, MO (FACW+), KY, OH (FACW).

Speckled phlox; wild sweet William.

This species, which is readily distinguished by its pyramidal inflorescences, may be called spp. *pyramidalis*. Purple-flowered specimens may be called var. *purpurea.*

4. **Phlox paniculata** L. Sp. Pl. 151. 1753. Fig. 46.
Phlox acuminata Pursh, Fl. Am. Sept. 2:730. 1814.
Phlox paniculata L. var. *acuminata* (Pursh) Chapm. Fl. S. U. S. 338. 1860.

Perennial herb; stems erect, simple or branched, green, glabrous or puberulent, to 2 m tall; leaves oblong to lance-ovate, acute to acuminate at the apex, cuneate at the base, or the uppermost leaves subcordate at the base, entire, glabrous or puberulent, the lateral veins prominent, to 15 cm long; flowers in pyramidal panicles, on short pedicels; calyx to 1 cm long, green, glabrous or puberulent, 5-lobed, the lobes subulate, about half as long as the calyx tube; corolla salverform, to 2.5 cm long, pink or purple, 5-lobed; stamens 5, included; capsule oval, to 1 cm long. July–September.

Mesic woods, wet meadows.

IA, IL, IN, KS, KY, MO, NE, OH (FACU).

Garden phlox; sweet William.

This species is one of several plants in Illinois called sweet William.

Phlox paniculata is related to *P. maculata* but differs in the prominent leaf venation and absence of purple markings on the stem.

3. **Polemonium** L.—Greek Valerian

Annual or perennial herbs; leaves alternate, pinnately compound; flowers perfect, actinomorphic, in cymose panicles or thyrses; calyx campanulate, 5-lobed; corolla campanulate to funnelform, 5-lobed; stamens 5, attached to the corolla; ovary superior, 3-locular; fruit a 3-valved capsule.

Polemonium is a genus of nearly fifty species, mostly native to western North America.

Only the following species occurs in the central Midwest.

1. **Polemonium reptans** L. Syst. ed. 10, 2:925. 1759. Fig. 47.
Polemonium reptans L. var. *villosum* E. L. Braun, Rhodora 42:50. 1940.
Polemonium reptans L. f. *villosum* (E. L. Braun) Wherry, Am. Midl. Nat. 27:753. 1942.

Perennial herbs; stems widely spreading, much branched, glabrous or rarely villous, to 45 cm long; lowermost leaves with 11–17 leaflets, the leaflets lanceolate to oblong to ovate, acute at the apex, subcuneate at the sessile or nearly sessile base, glabrous or rarely villous, to 3 cm long, the uppermost leaves with 3–5 leaflets or even simple; flowers in corymbs, to nearly 2 cm across, on pedicels to 1.5 cm long; calyx campanulate, green, glabrous or nearly so, the 5 lobes deltoid to lanceolate, more or less acute; corolla campanulate, blue, 5-lobed; stamens 5, included; capsule ovoid, glabrous, 3-valved, with about 3 seeds. April–June.

Prairies, fens, dry woods, mesic woods.

IA, IL, IN, KY, MO, OH (FAC), KS (FACU).

Jacob's-ladder.

This is one of our more attractive spring wild flowers. Its common habitat is woodlands, but occasional specimens may be found in calcareous fens and prairies.

Specimens with villous stems and leaves are rarely encountered. They may be called var. *villosum*.

46. *Phlox paniculata* (Garden phlox). a. Habit. b. Flower. c. Calyx. d. Fruit.

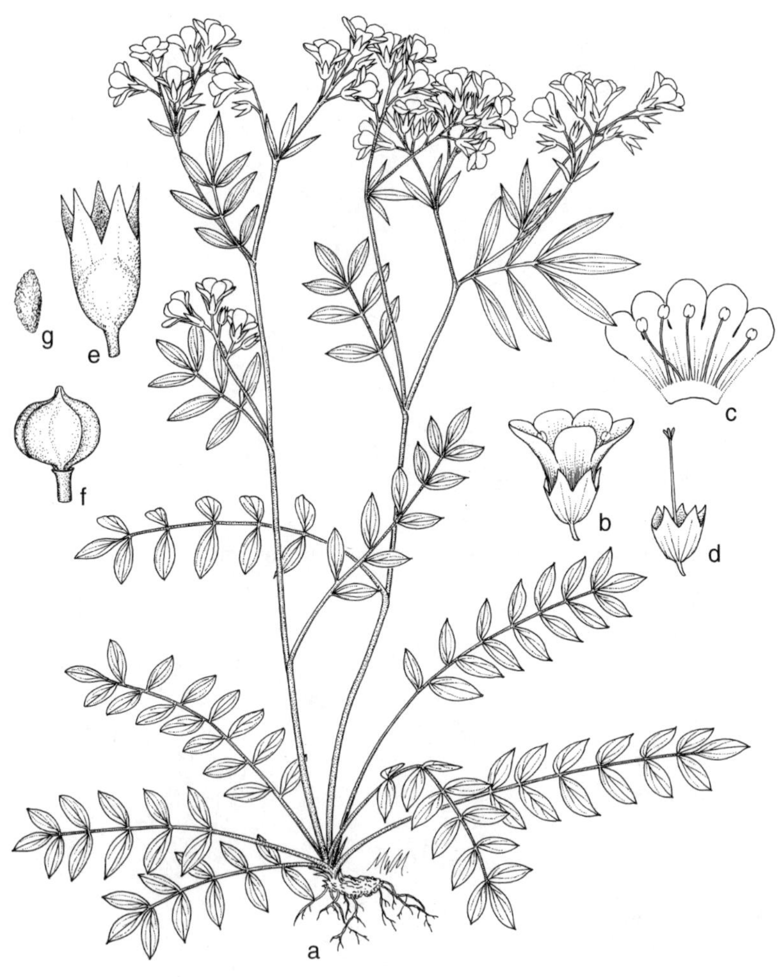

47. *Polemonium reptans* (Jacob's-ladder).
a. Habit.
b. Flower.
c. Corolla opened out.
d. Flower with corolla and stamens removed.
e. Calyx.
f. Capsule.
g. Seed.

99. POLYGALACEAE—MILKWORT FAMILY

Herbaceous or some woody plants; leaves simple, alternate or less commonly opposite or whorled, without stipules; flowers variously arranged, mostly zygomorphic, usually perfect; calyx 5-parted, the segments usually unequal in size and shape; petals 3, united below and attached to the staminal tube; stamens 4, 6, or 8, united into 1 or 2 tubes; ovary superior, 2-locular; capsules 2-locular and 2-seeded; seeds usually arillate.

Approximately twelve genera and eight hundred species are in this family, found throughout the world.

Only the following genus occurs in the central Midwest.

1. **Polygala** L.—Milkwort

Annual or perennial herbs; leaves simple, alternate or occasionally whorled, usually entire, without stipules; flowers in various inflorescence types, zygomorphic, perfect; calyx 5-parted, irregular in size, the uppermost and 2 lower ones small and green, the 2 lateral ones larger and colored like the petals, all of them persistent; petals 3, united at base, the lower one keeled; stamens 6 or 8, united into 1 or 2 tubes, the anthers 1-locular; ovary superior, 2-locular, each with one ovule; fruit a capsule with 2 seeds, usually flattened, the seeds arillate.

The flowers in this genus appear to have 5 petals, but 2 of these are actually sepals.

Approximately six hundred species are in the genus.

The following may be found in wetland habitats.

1. At least the lowermost leaves in whorls .... 1. *P. cruciata*
1. All leaves alternate .... 2. *P. sanguinea*

1. **Polygala cruciata** L. Sp. Pl. 2:706. 1753. Fig. 48.
Polygala cruciata L. var. *aquilonia* Fern. & Schub. Rhodora 50:163–166. 1948.

Annual herbs; stems simple or branched, erect, to 35 (–50) cm tall, glabrous; leaves simple, at least the lowermost in whorls of 3–4, linear to linear-lanceolate to narrowly oblanceolate, to 2 cm long, to 7 mm wide, obtuse to subacute at the apex, tapering to the base, entire, glabrous; flowers in dense cylindrical racemes up to 3.5 (–4.0) cm long, subtended by persistent bracts 2–3 mm long; lateral sepals petal-like, deltoid, cordate, 2.5–5.0 mm long, purple to green to white; seeds more or less rugulose. July–October.

Wet meadows, wet savannas.

IA, IL, IN, KY, OH (FACW+).

Whorled milkwort.

The whorled leaves and often purplish racemes are distinctive for this species.

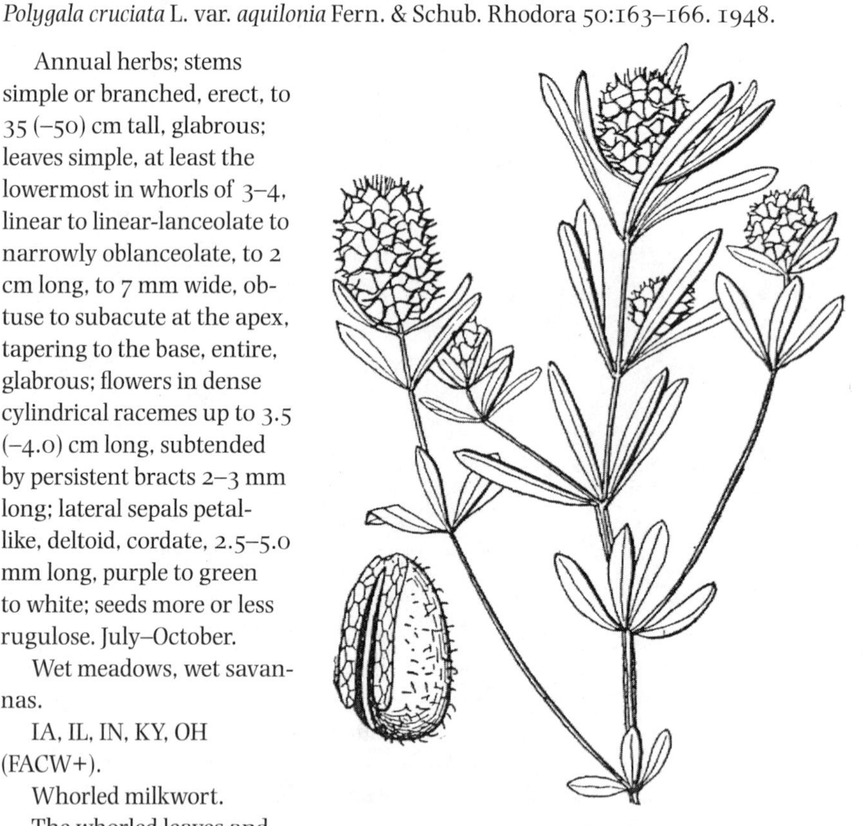

48. *Polygala cruciata* (Whorled milkwort). Habit (center). Seed (lower left).

2. **Polygala sanguinea** L. Sp. Pl. 2:705. 1753. Fig. 49.

Annual herbs; stems simple to much branched, erect, to 40 cm tall, glabrous; leaves simple, alternate, elliptic to narrowly lanceolate, acute at the apex, tapering to the base, entire, glabrous, to 2.5 cm long, to 1 cm wide; flowers crowded into capitate or cylindrical racemes, up to 2 cm long, up to 1.5 cm wide, pink or white or rarely greenish; lateral sepals petallike, broadly rounded, to 6 mm long; seeds pilose. June–October.

Wet meadows, old fields.

IA, IL, IN, KY, MO, OH (FACU), KS, NE (FAC).

Milkwort.

The flower color varies from pink to white to even greenish. In addition to an occasional wetland habitat, this species is common in old fields.

49. *Polygala sanguinea* (Milkwort). Habit (center). Sepal (right).

100. POLYGONACEAE—SMARTWEED FAMILY

Annual or perennial herbs or less commonly woody, some of them climbing; leaves alternate, simple, entire, with a sheath (ocrea) that surrounds the stem; flowers solitary in the axils of the leaves, or variously arranged in spikes, panicles, or corymbs; calyx 4- to 6-parted, usually petaloid, free from each other; petals absent; stamens 3–9; ovary superior, with 2 or 3 styles; achenes lenticular or trigonous.

This family consists of thirty to forty genera and approximately one thousand species found worldwide. Several upland genera and species occur in the central Midwest, in addition to the wetland species described below.

1. Climbing shrubs with tendrils; pedicels winged on one side by the calyx........... 1. *Brunnichia*
1. Prostrate or ascending or erect herbs or, if climbing, then without tendrils; pedicels not winged by the calyx.
    2. Sepals 6; stamens 6..5. *Rumex*
    2. Sepals usually 5; stamens 3–8.
        3. Stems retrorse prickly..6. *Tracaulon*
        3. Stems not retrorse prickly.
            4. Stems twining or trailing, neither erect nor prostrate.................................. 2. *Fallopia*
            4. Stems erect or prostrate, neither twining nor trailing.
                5. Flowers borne in small axillary clusters ....4. *Polygonum*
                5. Flowers borne in terminal and/or axillary spikes or racemes................ 3. *Persicaria*

1. **Brunnichia** Banks—Buckwheat Vine

Woody vines climbing by tendrils; leaves alternate, entire, petiolate; ocreae absent, with only a pubescent ring at each node; inflorescence a series of slender racemes; flowers perfect, zygomorphic, with the pedicel winged on one side; calyx 5-parted, petaloid, winged on one side; petals absent; stamens 8; styles 3; ovary superior, with a solitary ovule; achene enclosed by the persistent calyx and the winged pedicel.

Our species and three from tropical Africa comprise this genus.

Only the following species occurs in the central Midwest.

1. **Brunnichia ovata** (Walt.) Shinners, Sida 3:115. 1967. Fig. 50.
Bajania ovata Walt. Fl. Carol. 247. 1788.
Brunnichia cirrhosa Banks ex Gaertn. Fruct. & Sem. 1:213, pl. 45. 1788.

High-climbing woody vine; stems grooved, glabrous, much branched, up to 2 cm in diameter, with pubescent rings at each node representing the stipules; leaves ovate, abruptly acuminate at the apex, subcordate to truncate at the base, puberulent on the lower surface, up to 15 cm long, up to 7 cm wide; inflorescence a series of racemes, the lowest in the axils of leaves, the uppermost in a panicle; flowers perfect, the pedicels jointed, winged on one side, to 10 mm long; calyx to 4 mm long, greenish, tubular below, 5-lobed above, the lobes oblong, obtuse, winged on one side; petals absent; stamens 8, slightly exserted; styles 3, with bilobed stigmas; fruits up to 3.5 cm long, up to 8 mm wide, brown, the enlarged, coriaceous calyx persistent; achene oblongoid, to 1 cm long, 3-angled. June–August.

Swamp woods.

IL, KY, MO (FACW).

Ladies' ear-drops.

The flowers and fruits are bizarre structures in which a wing extends down one side of the calyx and along one side of the pedicel. The wing persists during fruiting.

The U.S. Fish and Wildlife Service calls this plant *B. cirrhosa*.

50. *Brunnichia ovata* (Ladies' ear-drops).
a. Flowering branch.
b. Flower.
c. Sepal.
d. Pistil.
e. Fruit.

2. **Fallopia** Adans.—Climbing Buckwheat

Perennial vines; stems twining without tendrils, angular, branched; leaves alternate, simple, entire, cordate to truncate at the base; ocreae oblique, without bristles; flowers in racemes from the axils of the leaves; sepals 5, in 2 series, greenish, greatly enlarging during fruiting; petals 0; stamens 8; styles 3; ovary superior; achenes 3-angular, subtended by the outer three broadly winged sepals.

Many botanists have maintained this genus in the very diverse genus *Polygonum*, but I believe its features are distinct enough to merit the rank of genus as I indicated in my Vascular Flora of Illinois as early as 2002.

Two species may occur in wetlands in the central Midwest, although they are more frequently found in drier habitats. There are about twelve species in the genus.

1. Calyx in fruit up to 10 mm long; achenes up to 3 mm long.................................. 1. *F. cristata*
1. Calyx in fruit more than 10 mm long; achenes more than 3 mm long .... 2. *F. scandens*

1. **Fallopia cristata** (Engelm. & Gray) Holub, Folia Geobot. Phytotax. 5:176. 1971. Fig. 51.
Polygonum cristatum Engelm. & Gray, Boston Journ. Nat. Hist. 5:259. 1845.
Tiniaria cristata (Engelm. & Gray) Small, Fl. S.E. U.S. 382. 1903.
Reynoutria scandens (L.) Shinners var. *cristata* (Engelm. & Gray) Shinners, Sida 3:118. 1967.
Bilderdykia cristata (Engelm. & Gray) Greene, Leaflets Bot. Obs. 1:23. 1904.

Perennial; stems twining, angled, branched, minutely roughened, to 4 m long; leaves ovate to deltoid, acute at the apex, cordate to truncate at the base, to 8 (–10) cm long, to 4 (–5) cm wide, glabrous or minutely roughened, the lower on petioles to 2 cm long, the upper becoming sessile or nearly so; ocreae oblique, without bristles; inflorescence of racemes or shortened clusters from the axils of the leaves, the racemes up to 10 cm long; calyx composed of 5 sepals in 2 series, the sepals greenish white, 1.5–2.5 mm long, becoming up to 10 mm long in fruit; stamens 8; styles usually 3; achene 3-angular, black, shiny, smooth, 2–3 mm long, subtended by the outer three broadly winged sepals, the wings entire or incised. July–November.

Woodlands, along streams, on bluffs.

IA, IN, IL, KY, MO, OH (FAC).

Crested false buckwheat.

This species differs from *F. scandens* by its smaller calyx in fruit and by its smaller achenes.

The U.S. Fish and Wildlife Service considers this species to be the same as *F. scandens*.

2. **Fallopia scandens** (L.) Holub, Folia Geobot. Phytotax. 6:176. 1971. Fig. 52.
Polygonum scandens L. Sp. Pl. 364. 1753.
Polygonum scandens L. var. *scandens* (L.) Gray, Man. Bot., ed. 5, 41. 1867.
Tiniaria scandens (L.) Small, Fl. S.E. U.S. 382. 1903.
Reynoutria scandens (L.) Shinners, Sida 3:118. 1967.

Perennial; stems twining, angled, branched, minutely roughened, to 7 m long; leaves ovate, acuminate at the apex, cordate at the base, to 10 (–12) cm long, to 5

(–6) cm wide, roughened along the margins, the lower on petioles to 2 cm long, the upper becoming sessile or nearly so; ocreae oblique, without bristles; inflorescence of racemes from the axils of the leaves, the racemes to 20 cm long; calyx composed of 5 sepals in 2 series, the sepals greenish yellow or greenish white, 1.5–2.5 mm long, becoming 10–15 mm long in fruit; stamens 8; styles 3; achene 3-angular, black, shiny, smooth, 3–5 mm long, subtended by the outer 3 broadly winged sepals, the wings entire or incised. August–October.

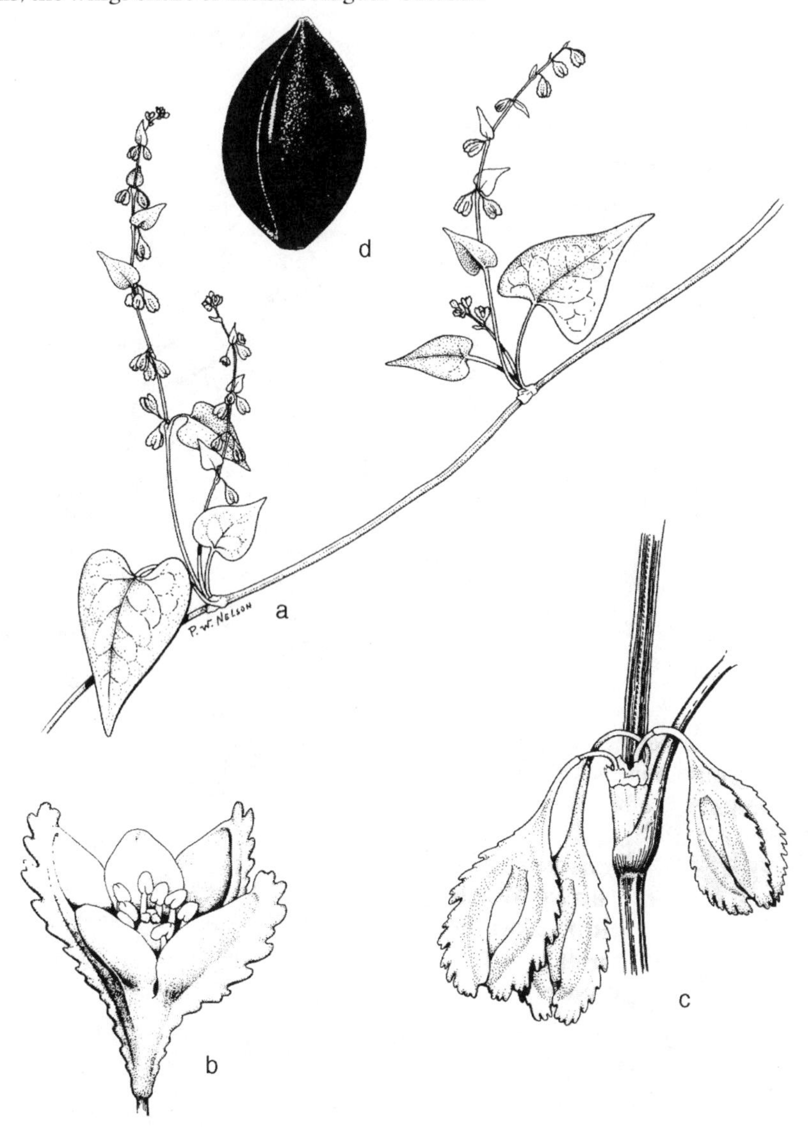

51. *Fallopia cristata* (Crested false buckwheat).

a. Segment of stem, with leaves and fruits.
b. Flower.
c. Ocrea, with fruits.
d. Achene.

Woods and thickets.

IL, IN, KY, MO, OH (FAC), KS, NE (FACU). The U.S. Fish and Wildlife Service calls this species *Polygonum scandens*.

False buckwheat.

False buckwheat is a common vine throughout the central Midwest. It is frequently confused with the climbing *F. cristata*, but the calyx during fruiting and the achenes are much larger in *F. scandens*.

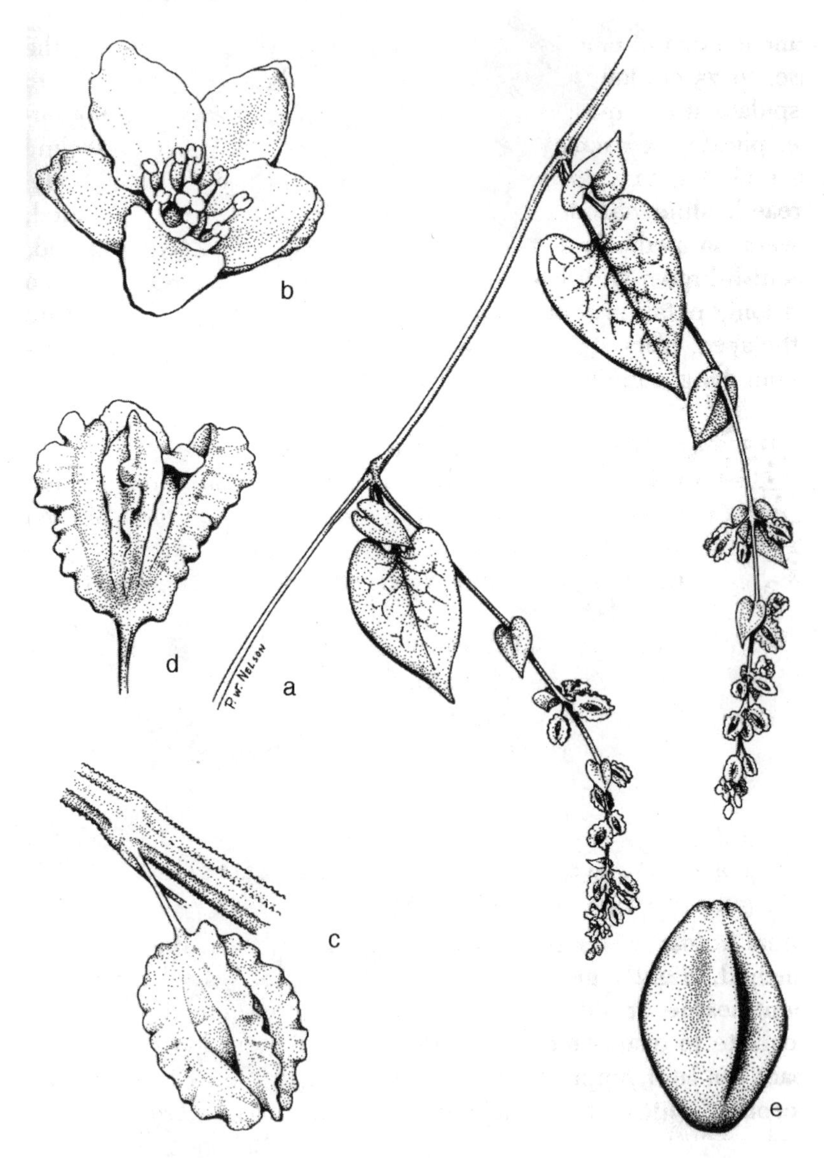

52. *Fallopia scandens* (False buckwheat). a. Segment of stem, with leaves and fruit. b. Flower. c. Node with fruit. d. Fruit. e. Achene.

3. **Persicaria** (Turcz.) Nakai—Smartweed

Annual or perennial herbs (in our area) or shrubs; leaves simple, alternate, entire; ocreae present; flowers in terminal and/or axillary racemes, without foliaceous bracts, perfect, actinomorphic; calyx 3- to 6-parted (usually 5-), petaloid; petals 0; stamens 3–8; ovary superior, 1-locular, with 1 ovule, the styles 2–3; achenes lenticular or trigonous.

Persicaria consists of about 150 species that are found throughout most of the world.

This genus has often been included as a section of *Polygonum* (*sensu lato*). It is my belief that a better treatment of *Polygonum* is to divide it into several smaller genera, based on a combination of characteristics, a view I published in my Vascular Flora of Illinois in 2002. In the central Midwest, wetland species that are often included in *Polygonum* (*sensu lato*) are now placed in *Fallopia, Persicaria, Polygonum,* and *Tracaulon.* This concept of *Polygonum* was prevalent nearly a century ago by such botanists as John K. Small, Edward Greene, Per Axel Rydberg, and Nathaniel Lord Britton.

Most species of *Persicaria* in the central Midwest occupy wetland habitats.

1. Sheaths with bristles.
    2. Perennials with thick rhizomes or stolons; spikes 1–2 per stem.
        3. Flowers white .... 5. *P. glabra*
        3. Flowers pink or red.
            4. Spikes cylindrical, usually more than 4 cm long; peduncles densely pubescent; stipules without a flange .... 4. *P. coccinea*
            4. Spikes conical to ovoid, usually less than 4 cm long; peduncles sparsely pubescent; stipules often with a flange .... 1. *P. amphibia*
    2. Annuals with fibrous roots or perennials with slender rhizomes; spikes more than 2 per stem.
        5. Peduncles with stipitate glands.
            6. Some of the flowers with long-exserted stamens and styles .... 2. *P. bicornis*
            6. None of the flowers with long-exserted stamens and styles .... 3. *P. careyi*
        5. Peduncles without stipitate glands.
            7. Sepals punctate or glandular-dotted.
                8. Achenes roughened, dull .... 6. *P. hydropiper*
                8. Achenes smooth, shiny.
                    9. Sepals glandular-punctate only near base .... 7. *P. hydropiperoides*
                    9. Sepals glandular-punctate throughout.
                        10. Racemes interrupted; leaves up to 2.5 cm wide .... 14. *P. punctata*
                        10. Racemes crowded; leaves 2.0–4.5 cm wide .... 15. *P. robustior*
            7. Sepals neither punctate nor glandular-dotted.
                11. Racemes crowded, except near base; annuals with fibrous roots.
                    12. Flowers bright rose-pink; racemes up to 7 mm thick; leaves without a dark blotch.
                        13. Base of raceme interrupted .... 11. *P. minor*
                        13. Racemes crowded throughout .... 9. *P. longiseta*
                    12. Flowers pink or pale pink; racemes more than 7 mm thick; leaves oftenwith a dark blotch .... 10. *P. maculosa*
                11. Racemes loosely flowered, usually interrupted ; perennials with slender rhizomes.
                    14. Leaves strigose on the upper surface; sepals white .... 16. *P. setacea*
                    14. Leaves glabrous or scabrous on the upper surface; sepals mostly pink or greenish.
                        15. Achene partly exserted; sepals greenish or greenish purple .... 12. *P. opelousana*

15. Achene enclosed by the perianth; sepals rosy or pinkish........................... .... 7. *P. hydropiperoides*

1. Sheaths without bristles.
 16. Perennials with rhizomes or stolons.
  17. Sepals white ....5. *P. glabra*
  17. Sepals red or pink.
   18. Racemes cylindrical, usually more than 4 cm long; peduncles densely pubescent; sheaths without a flange.. 4. *P. coccinea*
   18. Racemes conical or ovoid, usually less than 4 cm long; peduncles sparsely pubescent; sheaths often with a flange.. 1. *P. amphibia*
 16. Annuals with fibrous roots.
  19. Spikes arching or pendulous; achenes up to 2 mm broad; sepals usually white.......... ....8. *P. lapathifolia*
  19. Spikes not arching; achenes broader than 2 mm; sepals pink ....13. *P. pensylvanica*

1. **Persicaria amphibia** (L.) S.F. Gray, Nat. Arr. Brit. Pl. 2:268. 1821. Fig. 53, 54.
Polygonum amphibium L. Sp. Pl. 361. 1753.
Polygonum amphibium L. var. *emersum* Michx. Fl. Bor. Am. 1:240. 1803.
Polygonum amphibium L. var. *natans* Michx. Fl. Bor. Am. 1:240. 1803.
Polygonum natans (Michx.) Eaton, Man. Bot., ed. 3, 400. 1822.
Polygonum amphibium L. var. *aquaticum* Torr. Fl. N. & Mid. U.S. 1:404. 1824.
Polygonum amphibium L. var. *muhlenbergii* Meisn. ex DC. Prodr. 14:116. 1856.
Polygonum hartwrightii Gray, Proc. Am. Acad. 8:294. 1870.
Polygonum amphibium L. var. *stipulaceum* Coleman, Cat. Fl. Pl. S. Penins. Mich. 32.1874.
Polygonum muhlenbergii (Meisn.) S. Wats. Proc. Am. Acad. 14:295. 1879.
Polygonum emersum (Michx.) Britt. Trans. N.Y. Acad. Sci. 8:73. 1889.
Polygonum amphibium L. var. *hartwrightii* (Gray) Bissell, Rhodora 4:104. 1902.
Persicaria hartwrightii (Gray) Greene, Leaflets 1:24. 1904.
Persicaria muhlenbergii (Meisn.) Small ex Rydb. Fl. Colo. 111. 1906.
Polygonum amphibium L. var. *marginatum* f. *hartwrightii* Farw. Papers Mich. Acad. Sci. 1:93. 1923.
Polygonum natans (Michx.) Eaton f. *hartwrghtii* (Gray) Stanford, Rhodora 27:160. 1925.
Polygonum amphibium L. var. *stipulaceum* Coleman f. *hirtuosum* (Farw.) Fern. Rhodora 48:49. 1946.
Polygonum fluitans Eaton f. *hartwrightii* (Gray) G.N. Jones, Fl. Ill. 124. 1950.

Aquatic or terrestrial perennial herbs from rhizomes or stolons; stems floating or trailing or erect, glabrous or pubescent; leaves oblong to oblong-lanceolate, acute to acuminate at the apex, rounded or tapering to the base, entire, sometimes ciliate along the margins, glabrous or with short stiff hairs on the upper surface, to 20 cm long, to 8 cm wide, usually short-petiolate; ocreae strigose, sometimes flared into a collar at the summit, with or without bristles; inflorescence of 1–3 thick cylindric spikes up to 4 cm long, up to 2 cm thick, densely flowered, on glabrous peduncles, the ocreolae glabrous; flowers pedicellate; calyx 5-parted, the sepals rose to pink, 3–5 mm long; petals 0; stamens 5, some included, some exserted; style 2-cleft; achenes lenticular, obovoid to oblongoid, shiny, smooth or somewhat granular, dark brown to black, 2.5–3.0 mm long. June–September.

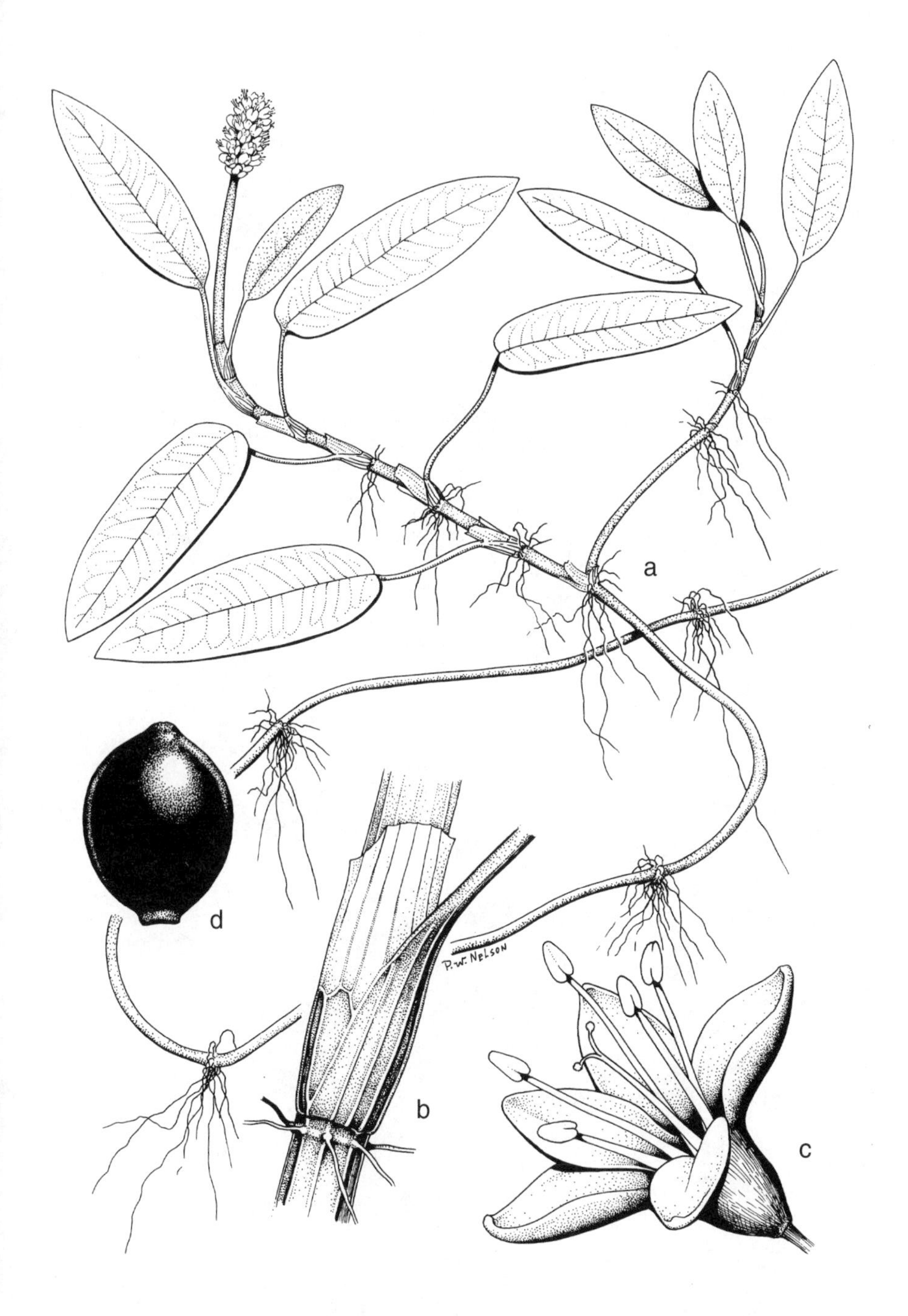

53. *Persicaria amphibia* (Water smartweed). a. Upper part of plant, with inflorescence. b. Stem with ocrea. c. Flower. d. Achene.

54. *Persicaria coccinea* (Scarlet smartweed).

a. Habit, aquatic form.
b. Ocrea.
c. Flowers.
e. Flower.
f. Seed.

Standing water; wet ground.

IA, IL, IN, KS, KY, MO, NE, OH (OBL), as *Polygonum amphibium*.

Water smartweed.

This extremely variable species may occur in standing water or on muddy soil. Leaves of terrestrial plants have short, stiff hairs on the upper surface. In var. *stipulaceum*, the ocreae are flared out at the summit to form a collar.

This species is sometimes merged with *P. coccinea*, which has longer and more slender racemes and pubescent peduncles.

2. **Persicaria bicornis** (Raf.) Nieuwl. Am. Midl. Nat. 3:201. 1914. Fig. 55.
Polygonum bicorne Raf. Fl. Ludov. 29. 1817.
Polygonum longistylum Small, Bull. Torrey Club 21:169. 1894.
Persicaria longistyla (Small) Small, Fl. SE. U.S. 337. 1903.

Annual from slender roots; stems erect, branched or unbranched, glabrous below, glandular-pubescent above, to 1.8 m tall; leaves lanceolate to ovate-lanceolate, to 13 cm long, to 2.3 cm wide, acute to acuminate at the apex, rounded or tapering at the base, entire and often ciliate along the margins, glabrous or nearly so on both surfaces, punctate below, sometimes with a purple blotch, the petioles to 2 cm long; ocreae glabrous, without bristles or with very short bristles; inflorescence a panicle of racemes, the racemes pendulous, to 6 cm long, to 1.8 cm thick, on glandular-pubescent peduncles; flowers pedicellate, the pedicels 1.5–5.0 mm long; calyx 5-parted, the sepals pink, 2.5–4.5 mm long; petals absent; stamens 6–8; style 2- or 3-parted, exceeding the stamens in some flowers, shorter than the stamens in others; achenes mostly 3-angular, ovoid, black, shining, smooth, 2.2–3.0 mm long. July–October.

Wet ground.

IA, IL, KS, MO, NE (FACW+). The U.S. Fish and Wildlife Service does not distinguish this plant from *P. pensylvanica*.

Smartweed.

Although this species has been called *Polygonum longistylum* for years, Rafinesque's *Polygonum bicorne* predates it.

This species is unique in that either the stamens or the styles of a particular flower are long-exserted, a condition known as heterostyly. Plants with bristly ocreae resemble *P. careyi*, but this latter species is not heterostylous.

3. **Persicaria careyi** (Olney) Greene, Leaflets 1:24. 1904. Fig. 56.
Polygonum careyi Olney, Proc. Providence Franklin Soc. 1:29. 1847.

Annual from fibrous roots; stems erect, branched or unbranched, glandular-hairy, to 1.5 m tall; leaves lanceolate, or the uppermost linear, acute at the apex, tapering at the base, rough-hairy, to 25 cm long, to 3 cm wide, somewhat punctate, on short, glandular-hairy petioles or subsessile; ocreae hispid, fringed with bristles; inflorescence of several terminal, pendulous racemes, the racemes to 10 cm long, rather loosely flowered, on glandular-hairy peduncles; flowers pedicellate; calyx 5-parted, usually purple; petals absent; stamens 5 (–8); achenes lenticular, obovoid to ovoid, smooth, shining 1.8–2.5 mm long, 1.5–2.0 mm broad. July–September.

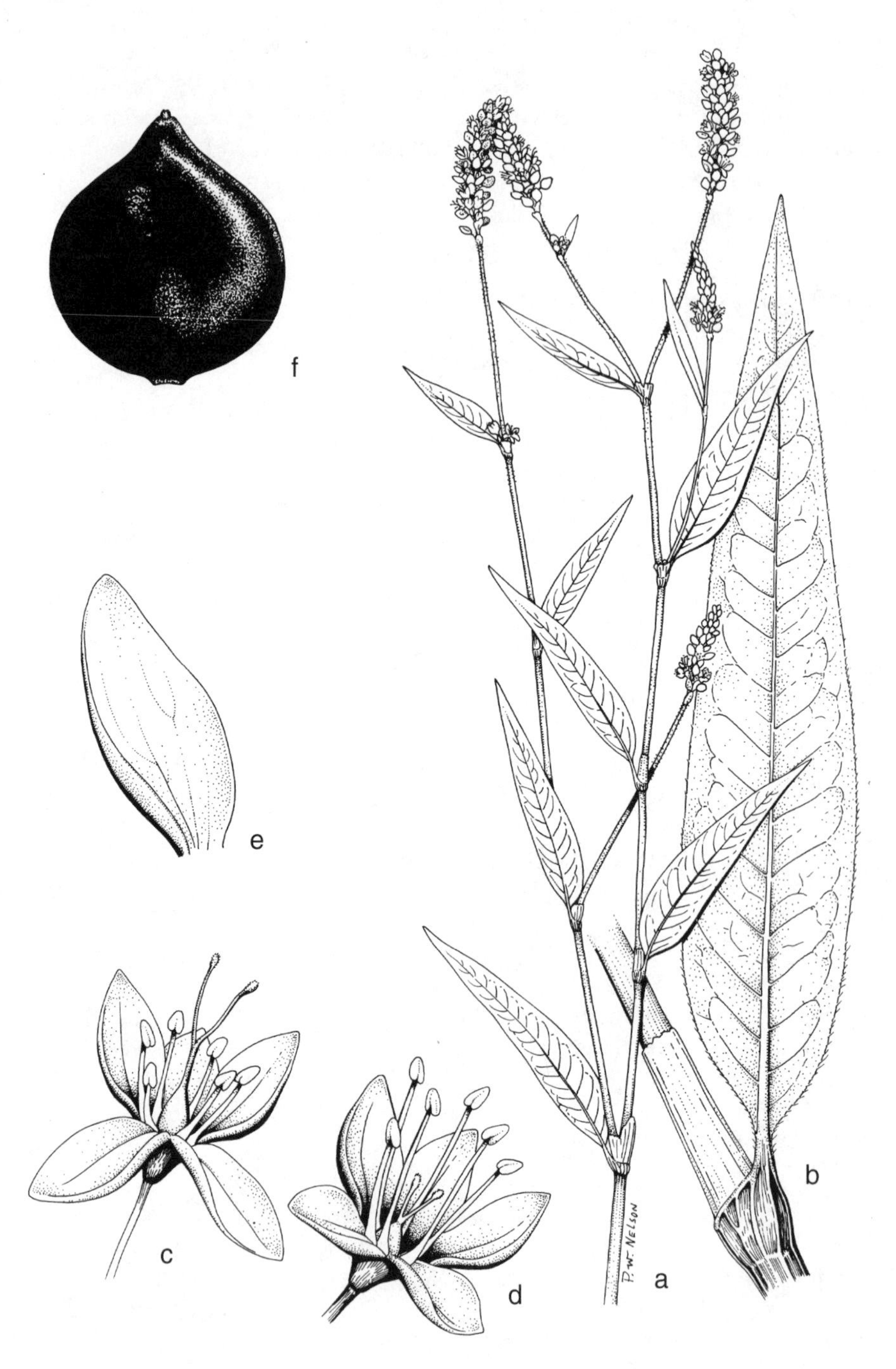

55. *Persicaria bicornis* (Smartweed).

a. Upper part of plant.
b. Ocrea and leaf.
c. Flower with long styles.
d. Flower with short styles.
e. Sepal.
f. Achene.

56. *Persicaria careyi* (Carey's smartweed).

a. Habit.
b. Stem with ocrea and leaf.
c. Upper part of plant, with immature inflorescences.
d. Flower.
e. Sepals.
f. Achene.

Sandy soil.

IL, IN (FACW+), KY, OH, (FACW), as *Polygonum careyi.*

Carey's smartweed.

Carey's smartweed is readily recognized by its glandular-hairy stems and peduncles and its ocreae with bristles. It differs from the glandular-pubescent *P. bicornis* by lacking long-exserted stamens or styles.

4. **Persicaria coccinea** (Muhl.) Greene, Leaflets 1:24. 1904. Fig. 57.
Polygonum coccineum Muhl. ex Willd. Enum. Pl. 1:428. 1809.
Polygonum coccineum Muhl. var. *terrestre* Willd. Enum. Pl. 1:428. 1809.
Polygonum fluitans Eaton, Man. Bot., ed. 6, 274. 1833.
Persicaria pratincola Greene, Leaflets 1:36. 1904.
Persicaria spectabilis Greene, Leaflets 1:37. 1904.
Polygonum coccineum Muhl. f. *terrestre* (Muhl.) Stanford, Rhodora 27:162 1925.
Polygonum coccineum Muhl. var. *pratincola* (Greene) Stanford, Rhodora 27:162. 1925.
Polygonum coccineum Muhl. f. *natans* (Wieg.) Stanford, Rhodora 27:165. 1925.

Aquatic or terrestrial perennial herbs from rhizomes or stolons; stems floating or trailing or erect, glabrous or pubescent; leaves lanceolate to lance-ovate, acute to acuminate at the apex, rounded at the base or cordate in some aquatic specimens, entire, glabrous or with soft hairs, particularly on the upper surface, to 20 cm long, to 8 cm wide, usually short-petiolate; ocreae strigose to glabrous, not flared into a collar at the summit; inflorescence of 1–3 slenderly cylindric spikes 4–18 cm long, up to 1.5 cm thick, on pubescent peduncles, the ocreolae strigose; flowers pedicellate; calyx 5-parted, the sepals rose to pink, 3–5 mm long; petals 0; stamens 5, some included, some exserted; style 2-cleft; achenes lenticular, obovoid to oblongoid, shiny, smooth or somewhat granular, dark brown to black, 2.5–3.0 mm long. June–September

Standing water; wet ground.

IA, IL, IN, KS, KY, MO, NE, OH (OBL), as *Polygonum amphibium.*

Scarlet smartweed.

This species, which may be either aquatic or terrestrial, differs from *P. amphibia* by its longer and more slender spikes and its pubescent peduncles and ocreolae.

Terrestrial plants with shorter spikes up to 8 cm long and with glabrous or strigose stems are the typical variety. Those with spikes up to 18 cm long and with densely pubescent stems may be known as var. *pratincola.* These occur mostly in the western part of the central Midwest. Aquatic plants with cordate leaves may be known as f. *natans.*

5. **Persicaria glabra** (Willd.) M. Gomez, An. Inst. Seg. Ensen. Hab. 2:278. 1896. Fig. 58.
Polygonum glabrum Willd. Sp. Pl. 2:447. 1799.
Polygonum densiflorum Meisn. in Mart. Fl. Bras. 5:13. 1855.
Polygonum portoricense Bertero ex Meisn. in DC. Prodr. 14:121. 1856.
Persicaria densiflora (Meisn.) Moldenke, Torreya 34:7. 1945.

Perennial herbs with rhizomes; stems decumbent to erect, to 1.5 tall, glabrous, branched, punctate; leaves alternate, simple, lanceolate, acute to acuminate at the

57. *Persicaria amphibia* (Water smartweed).

a. Upper part of plant.
b. Ocrea.
c. Flower.
d. Achene.
e. Habit.
f. Ocrea.

apex, tapering at the base, entire, to 30 cm long, to 5 cm wide, glabrous or strigose on the upper surface, glabrous or scabrous on the veins on the lower surface, occasionally punctate, the petioles to 2 cm long; ocreae glabrous, without bristles at the tip; inflorescence a panicle of racemes, the racemes more or less erect, to 10 cm long, to 10 mm thick, densely flowered, uninterrupted, on glabrous, punctate peduncles; flowers pedicellate, the pedicels up to 5 mm long; calyx 5-parted, the sepals white or pink, 3.0–3.5 mm long; petals absent; stamens 5 or 7; style 2- or 3-cleft; achenes biconvex, dark brown to black, smooth, shiny, 2.0–2.2 mm long. August–November.

Swamps, marshes, sometimes in standing water.

IL, IN, KY, MO, OH (OBL), as *Polygonum densiflorum.*

Smooth smartweed.

This species, which lacks bristles at the summit of the ocreae, differs from *P. pensylvanica* by its white flowers and from *P. lapathifolia* by its erect rather than nodding spikes and its biconvex achenes. For many years this species was known as *P. densiflora* or *Polygonum densiflorum.*

58. *Persicaria glabra* (Smooth smartweed). Habit (center). Flower (upper right). Seed (left).

6. **Persicaria hydropiper** (L.) Opiz, Seznam 72. 1852. Fig. 59.
Polygonum hydropiper L. Sp. Pl. 361. 1753.
Polygonum hydropiper L. var. *projectum* Stanford, Rhodora 29:87. 1927.

Annual from fibrous roots; stems prostrate to erect, often purple-red, branched or unbranched, glabrous or nearly so, to 55 cm tall; leaves linear-lanceolate to ovate-lanceolate, acute to acuminate at the apex, tapering to somewhat rounded at the base, entire to undulate along the margins, glabrous or sometimes pubescent on the veins, usually punctate, with a strong peppery taste, to 8 cm long, to 2.5 cm wide, short-petiolate; ocreae glabrous to pubescent, fringed with short bristles;

59. *Persicaria hydropiper* (Water pepper).

a. Upper part of plant.
b. Ocrea.
c. Inflorescence.
d. Achenes enclosed by calyx.
e. Sepal.
f. Achene.

inflorescence of panicled racemes, the racemes pendulous or arching, to 7 cm long, usually interrupted; flowers short-pedicellate; calyx 4-parted, the sepals greenish, often tinged with purple and bordered with white, punctate, 1.5–2.5 mm long, becoming slightly larger during fruiting; petals absent; stamens 4 (–6); style 2- to 3-parted; achenes lenticular to 3-angled, ovoid, dull, striate, dark brown, 2.0–3.5 mm long. May–November.

Wet ground.

IA, IL, IN, KS, KY, MO, NE, OH (OBL), as *Polygonum hydropiper.*

Water pepper; smartweed.

Supposedly native to Europe, Asia, and North America, this species occurs throughout the United States, where some of the populations are undoubtedly adventive. American plants have sometimes been called var. *projectum* to distinguish them from Old World plants.

The common name water pepper is derived from the peppery taste of the leaves and stems, although the white-flowered *P. punctata* has a similar peppery taste.

There is considerable variation in sepal color, stamen number, and achene shape and size.

7. **Persicaria hydropiperoides** (Michx.) Small, Fl. SE. U.S. 378. 1903. Fig. 60.
Polygonum barbatum Walt. Fl. Carol. 131. 1788, non L. (1753).
Polygonum hydropiperoides Michx. Fl. Bor. Am. 1:239. 1803.
Polygonum persooni Engelm. ex Mead, Prairie Farmer 6:119. 1848, *nomen nudum.*

Perennial herb from rhizomes; stems decumbent and rooting at the nodes to erect, branched or unbranched, glabrous or sparingly pubescent, to 1 m tall; leaves linear-lanceolate to oblong-lanceolate, to 25 cm long, to 3.5 cm wide, acute at the apex, tapering at the base, entire and usually ciliate along the margins, punctate, sessile or with petioles up to 2 cm long; ocreae strigose, fringed with bristles, to 10 mm long; inflorescence a panicle of racemes, the racemes loosely flowered, to 8 cm long, to 5 mm wide, erect, on nearly glabrous peduncles; flowers pedicellate, the pedicels 1.0–1.5 mm long; calyx 5-parted, the sepals usually pink to rose to purple, rarely greenish, 2–4 mm long, sometimes punctate near the base; petals absent; stamens usually 8; style 3-parted; achenes 3-angular, ovoid, shining, smooth, brown to black, 1.5–3.0 mm long. June–November.

Wet ground.

IA, IL, IN, KS, KY, MO, NE, OH (OBL), as *Polygonum hydropiperoides.*

Mild water pepper; smartweed.

This species occurs in a variety of wet habitats, from bogs to marshes to shallow standing water.

This species differs from *P. setacea* by its pink to rose to purple calyx and its achenes, which are completely included within the persistent calyx. It is also similar in appearance to *P. hydropiper*, but it lacks the peppery taste of the leaves.

8. **Persicaria lapathifolia** (L.) S.F. Gray, Nat. Arr. Brit. Pl. 2:270. 1821. Fig. 61.
Polygonum lapathifolium L. Sp. Pl. 360. 1753.
Polygonum incarnatum L. var. *incarnatum* (Ell.) Wats. in Gray, Man. Bot., ed. 6, 440. 1890.
Polygonum lapathifolium L. var. *nodosum* Small, Mem. Torrey Club 5:140. 1894.

60. *Persicaria hydropiperoides* (Mild water pepper).

a. Upper part of plant.
b. Ocrea.
c. Inflorescence.
d. Ocrea with three developing fruits.
e. Sepals.
f. Achenes.

61. *Persicaria lapathifolia* (Pale smartweed).

a. Upper part of plant.
b. Ocrea.
c, d. Flower enclosing the achenes.
e. Sepal.
f. Achene.

Annual from slender roots; stems ascending to erect, branched or unbranched, glabrous except sometimes for the presence of sessile glands, to 2.5 m tall; leaves lanceolate, acute to acuminate at the apex, tapering at the base, entire and ciliate along the margins, usually glabrous, sometimes with sessile glands on the lower surface, inconspicuously punctate, to 25 cm long, to 5 cm wide, short-petiolate; ocreae usually glabrous, without bristles; inflorescence a panicle of racemes, the racemes arching to pendulous, densely flowered, to 6.5 cm long, to 8 mm thick, the peduncles glabrous but occasionally with sessile glands; flowers short-pedicellate; calyx 5-parted, the sepals white, often with a greenish tinge, 2.0–3.5 mm long; petals absent; stamens usually 6; style 2-parted; achenes lenticular, oblongoid to ovoid, dark brown to black, shining, reticulate, 1.7–2.3 mm long. June–November.

Wet ground.

IA, IL, IN, KY, MO, OH (FACW+), KS, NE (OBL), as *Polygonum lapathifolium.*

Pale smartweed; nodding smartweed.

This very common species of moist soil is highly variable. Many of the variations have received varietal or formal epithets, but none seems worthy of recognition because of the great inconstancy of characters. Similar plants with erect racemes are probably *P. densiflora.*

The achenes are an excellent food source for wildlife.

9. **Persicaria longiseta** (DeBruyn) Kitagawa, Rep. Inst. Sci. Res. Manchoukuo 1:322. 1937. Fig. 62.
Polygonum longisetum DeBruyn in Miq. Pl. Jungh. 3:307. 1854.
Persicaria cespitosa (Blume) Nakai, Bot. Mag. 48:779. 1934, misapplied.
Polygonum cespitosum Blume var. *longisetum* (DeBruyn) Stewart, Cont. Gray Herb. 88:67. 1950.

Annual from fibrous roots; stems prostrate to tardily ascending, much branched, glabrous or rarely puberulent, to 75 cm long; leaves lanceolate to elliptic to ovate, acute to acuminate at the apex, tapering at the subsessile base, glabrous above, puberulent on the veins below, entire and ciliate along the margins, to 10 cm long, to 3 cm wide, the petioles to 6 mm long; ocreae strigose, fringed with long bristles to 12 mm long; inflorescence solitary or a panicle of racemes, the racemes densely flowered, uninterrupted, 4 (–6) cm long, to 7 mm thick, on glabrous peduncles; flowers pedicellate, the pedicels 1–2 mm long; calyx 5-parted, the sepals rose, 2.0–2.5 mm long; petals absent; stamens usually 5; style 3-parted; achenes 3-angular, shining, smooth, black, 1.6–2.5 mm long. May–November.

Disturbed soil, particularly in wet areas, floodplain forests, streambanks.

Native of southeastern Asia; adventive in the northeastern United States; IA, IL, IN, MO (UPL), KY, OH (FACU−), KS, NE (not listed by the U.S. Fish and Wildlife Service), as *Polygonum cespitosum.*

Bristly lady's-thumb.

This species has short but showy racemes and very long bristles on the ocreae. It generally creeps along the ground. The bristles are often longer than the body of the ocreae.

The similar *P. cespitosa* of Asia has shorter ocreal bristles.

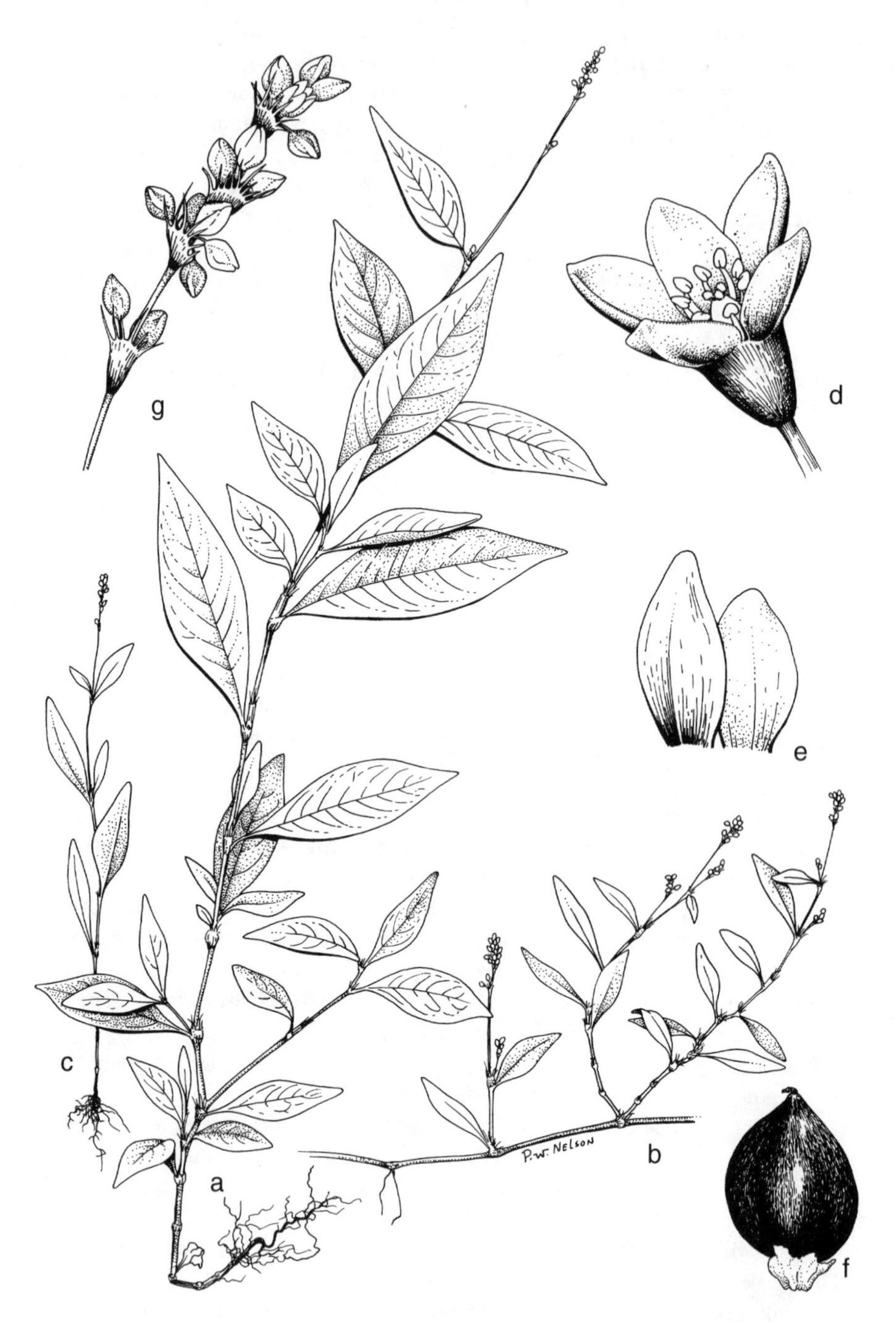

62. *Persicaria longiseta* (Bristly lady's thumb).

a, b. Habit.
c. Young plant.
d. Flower.
e. Sepals.
f. Achene.
g. Inflorescence.

10. **Persicaria maculosa** Gray, Nat. Arr. Brit. 2:269. 1821. Fig. 63.
Polygonum persicaria L. Sp. Pl. 361. 1753.
Polygonum vulgaris Webb & Moq. Hist. Nat. Iles Can. 3:219. 1831.
Persicaria persicaria (L.) Small, Fl. SE. U.S. 378. 1903.

Annual from fibrous roots; stems decumbent to ascending to erect, branched or unbranched, glabrous or sparingly pubescent, to 75 cm tall; leaves lanceolate to linear-lanceolate, acute at the apex, tapering at the base, entire and often ciliate along the margins, with a dark red-purple blotch on the upper surface, glabrous or puberulent, not punctate, to 15 cm long, to 3 cm wide, sessile or with petioles up to 8 mm long; ocreae glabrous or more uncommonly strigose, fringed with bristles up to 5 mm long; inflorescence of solitary or panicled racemes, the racemes to 4.5 cm long, to 1.2 cm thick, densely flowered, not interrupted, usually on glabrous peduncles, with a few cleistogamous flowers usually hidden within some of the ocreolae; flowers short-pedicellate, the pedicels 1.0–2.5 mm long; calyx (4–) 5-parted, the sepals pink to deep red, sometimes greenish, 2.0–3.5 mm long, slightly enlarging during fruiting; petals absent; stamens (4–) 6 (–8); style 2- to 3-parted; achenes lenticular, ovoid, shining, smooth, brownish to black, 1.5–2.7 mm long. May–November.

Wet, disturbed soil.

Native of Europe; adventive throughout the United States and Canada; IA, IL, IN, KY, MO, OH (FACW), KS, NE (OBL), as *Polygonum persicaria.*

Lady's-thumb.

The lady's-thumb is characterized by its short, thickened racemes and the dark red-purple blotch on the upper surface of nearly all of the leaves.

In addition to the obvious flowers that comprise the racemes, there are cleistogamous flowers hidden in and enclosed by some of the ocreolae.

The flowers of this species range from pink to deep red, and may be greenish or greenish-tinged.

11. **Persicaria minor** (Huds.) Opiz, Seznam 72. 1852. Not illustrated.
Polygonum minus Huds. Fl. Angl. 148. 1762.

Annual herbs from fibrous roots; stems decumbent to ascending, branched or unbranched, glabrous or commonly scabrous, to 35 cm tall; leaves lanceolate to lance-ovate, acute to acuminate at the apex, tapering at the base, to 8 cm long, to 1.5 cm wide, entire but scabrous on the margins, glabrous above, glabrous or strigose on the veins below, without a dark red-purple blotch, sessile or on petioles to 2 mm long; ocreae glabrous or strigose, with a fringe of bristles to 4 mm long; inflorescence of terminal and axillary racemes, the racemes up to 5 cm long, up to 2.5 cm thick, interrupted near the base; flowers pedicellate, the pedicels 0.5–1.0 mm long; calyx 5-parted, the sepals pink, 2.5–3.0 mm long; stamens 5–6; style 2- to 3-parted; achenes biconvex or 3-angled, dark brown to black, smooth, shiny, 1.5–2.5 mm long. July–October.

Wet disturbed ground.

Native to Europe: adventive in IN (not listed by the U.S. Fish and Wildlife Service for the central Midwest).

Smartweed.

Although this small species has generally been included within *P. maculosa*, its leaves do not have the dark red-purple blotch.

63. *Persicaria maculosa* (Lady's thumb).

a. Habit (silhouette).
b. Habit.
c. Stem with ocrea and leaf.
d. Flower.
e. Fruit enclosed by sepals.
f. Achene.

12. **Persicaria opelousana** (Riddell) Small, Fl. SE. U.S. 378. 1903. Fig. 64.
Polygonum opelousanum Riddell ex Small, Bull. Torrey Club 19:354. 1892.

Perennial herb from rhizomes; stems decumbent and rooting at the nodes to erect, branched or unbranched, glabrous or nearly so, to 1 m tall; leaves linear-lanceolate to lanceolate, acute to acuminate at the apex, tapering at the subsessile or sessile base, entire and usually ciliate along the margins, to 12 cm long, to 2 cm wide; ocreae strigose, fringed with bristles to 10 mm long; inflorescence a panicle of

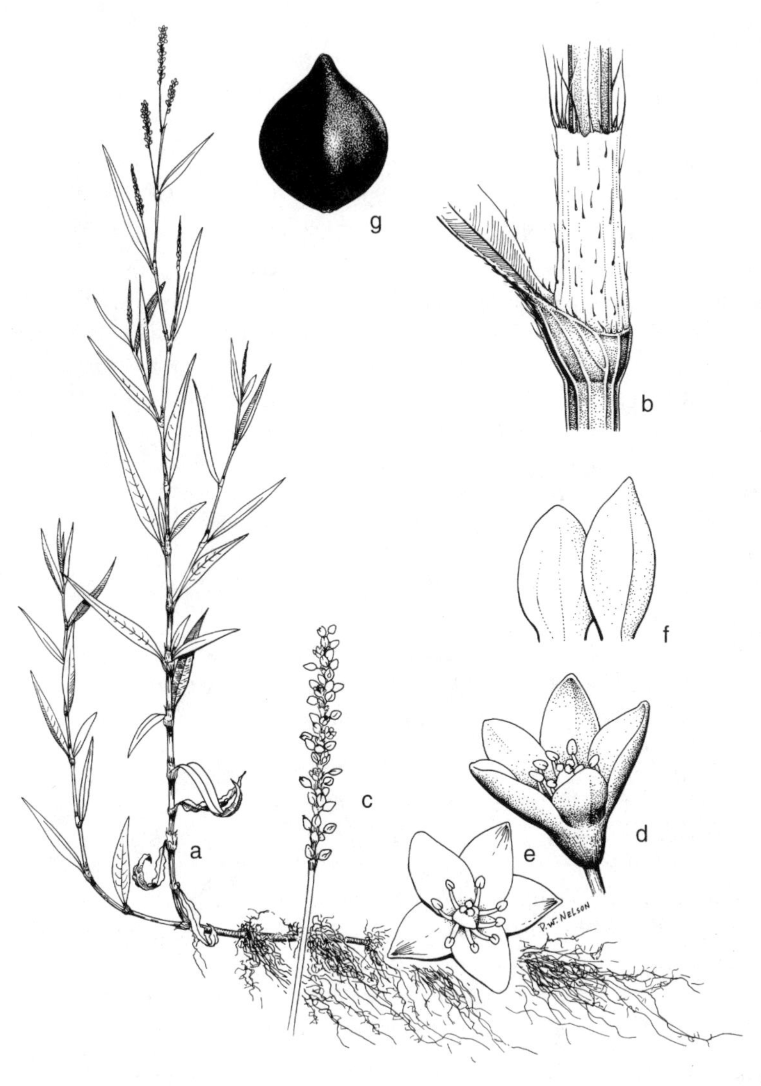

64. *Persicaria opelousana* (Water pepper).

a. Habit.
b. Ocrea.
c. Inflorescence.
d. Flower, side view.
e. Flower, face view.
f. Sepals.
g. Achene.

racemes, the racemes erect, to 6 cm long, loosely flowered; flowers pedicellate; calyx 5-parted, the sepals greenish white to greenish purple, scaly, 2.0–2.5 mm long; petals absent; stamens usually 8; style 3-parted; achenes 3-angular, ovoid to obovoid, shining, smooth, black, 1.5–2.5 mm long, partly exserted from the persistent calyx. July–October.

Wet ground.

IL, MO, OH (OBL), KS (not listed), as *Polygonum opelousanum.*

Water pepper.

This species is distinguished by its nearly glabrous leaves, greenish sepals, and partly exserted achenes.

13. **Persicaria pensylvanica** (L.) Small, Fl. SE. U.S. 377. 1903. Fig. 65.
Polygonum pensylvanicum L. Sp. Pl. 362. 1753.
Polygonum pensylvanicum L. var. *laevigatum* Fern. Rhodora 19:78. 1917.
Polygonum pensylvanicum L. var. *durum* Stanford, Rhodora 27:178. 1925.

Annual from slender roots; stems erect, branched or unbranched, glabrous or pubescent, sometimes glandular, to 2 m tall; leaves lanceolate to oval-lanceolate, acute to acuminate at the apex, tapering at the base, entire and ciliate along the margins, glabrous to strigose, sometimes glandular, to 25 cm long, to 8 cm wide, short-petiolate; ocreae glabrous, without bristles; inflorescence a panicle of racemes, the racemes erect, to 5 cm long, to 1.5 cm thick, densely flowered, on glabrous to pubescent to glandular-stipitate peduncles; flowers short-pedicellate; calyx 5-parted, the sepals pink to white, 2.0–3.5 mm long; petals absent; stamens (6–) 8; style 2-parted; achenes lenticular, dark brown to black, ovoid to orbicular, dull to sublustrous, smooth, 2.5–3.5 mm long. May–October.

Wet ground.

IA, IL, IN, KS, MO, NE (FACW+), KY, OH (FACW), as *Polygonum pensylvanicum.*

Pinkweed; common smartweed.

Typical var. *pensylvanica* has glabrous leaves and peduncles with spreading, gland-tipped hairs. Variety *durum* has glabrous leaves and peduncles without gland-tipped hairs. Variety *laevigatum* has strigose leaves.

14. **Persicaria punctata** (Ell.) Small, Fl. SE. U.S. 379. 1903. Fig. 66.
Polygonum punctatum Ell. Bot. S.C. & Ga. 1:455. 1817.
Polygonum acre HBK. Nov. Gen. 2:179. 1817, *non* Lam. (1788).
Polygonum acre HBK. α *confertiflorum* Meisn. ex DC. Prodr. 14:108. 1856.
Polygonum acre HBK. var. *leptostachyum* Meisn. ex DC. 14:108. 1856.
Polygonum punctatum Ell. var. *leptostachyum* (Meisn.) Small, Bull. Torrey Club 19:356. 1892.
Polygonum punctatum Ell. var. *typicum* Fassett, Brittonia 6:371. 1948.
Polygonum punctatum Ell. var. *confertiflorum* (Meisn.) Fassett, Brittonia 6:377. 1948.

Annual or perennial herbs with elongated rootstocks; stems prostrate at the base and rooting at the nodes to ascending, green to purple-red, branched or unbranched, glabrous or nearly so, to 1.2 m tall; leaves linear-lanceolate to lanceolate to elliptic, acute to acuminate at the apex, tapering at the base, to 15 cm long, to 2.5

65. *Persicaria pensylvanica* (Pinkweed).

a. Habit.
b. Ocrea.
c. Flowers.

d. Sepal.
e. Achene.
f. Inflorescence.

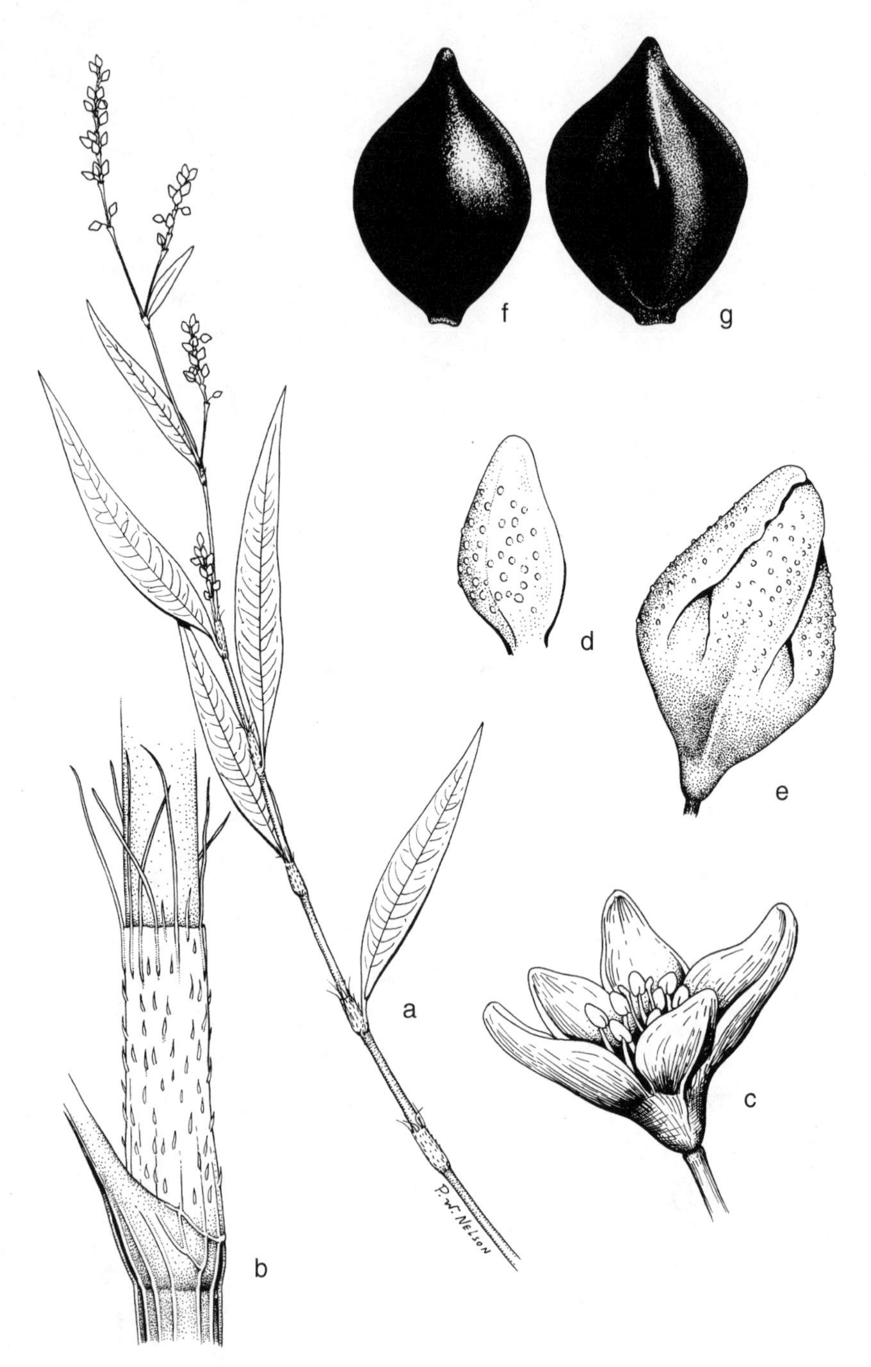

66. *Persicaria punctata* (Dotted smartweed).

a. Habit.
b. Ocrea.
c. Flower.
d. Sepal.
e. Sepals enclosing achene.
f. Lenticular achene.
g. Trigonous achene.

cm wide, entire, essentially glabrous except for the midrib, usually punctate, to 15 cm long, sessile or on petioles to 1 cm long; ocreae glabrous to strigose, fringed with bristles to 11 mm long; inflorescence of panicled racemes, the racemes slender, erect to arched, to 20 cm long, interrupted; flowers short, pedicellate; calyx 5-parted, the sepals greenish or greenish white or white, punctate, 3.0–3.5 mm long; petals absent; stamens 6–8; style 2- to 3-parted, achenes lenticular to 3-angular, oblongoid, shiny, smooth, 2.2–3.2 mm long. July–November.

Wet soil of woods and fields.

IA, IL, IN, KS, KY, MO, NE, OH (OBL), as *Polygonum punctatum*.

Dotted smartweed.

This common and variable species is an inhabitant of wet soil. It occurs in floodplain woods as well as in damp fields and moist, disturbed areas. The leaves have a sharp peppery taste when eaten.

Some botanists divide this species into a typical variety and var. *leptostachya*. The typical variety is a prostrate perennial that roots at the nodes and has larger dark green leaves and racemes without remote fascicles of flowers at the base. On the other hand, var. *leptostachya* is an ascending annual with smaller, pale green leaves and racemes bearing remote flowers all the way to the base.

Waterfowl and other birds use the achenes of this species in their diet, and deer browse on the herbage.

15. **Persicaria robustior** (Small) E. P. Bickn. Bull. Torrey Club 36:455. 1909. Fig. 67.
Polygonum coarctatum Willd. var. *majus* Meisn. Prod. Syst. 14:101. 1856.
Polygonum punctatum Ell. var. *robustior* Small, Bull. Torrey Club 21:477. 1894.
Persicaria punctata (Ell.) Small var. *robustior* (Small) Small, Fl. S. E. U. S. 379. 1903.
Polygonum robustius (Small) Fern. Rhodora 23:147. 1921.
Polygonum punctatum Ell. var. *majus* (Meisn.) Fassett, Brittonia 6:373–374. 1949.

Perennial herbs with rhizomes and often stolons; stems decumbent, becoming erect to ascending, to 2 m tall, green or reddish, branched, glabrous, punctate; leaves lanceolate to elliptic-lanceolate, acute to acuminate at the apex, tapering at the base, to 20 cm long, to 4.5 cm wide, entire although strigose on the margin, glabrous except for scabrous veins on the lower surface, punctate, the petioles to 2 cm long; ocreae strigose, punctate, fringed with bristles to 12 mm long; inflorescence of terminal and axillary panicled racemes, the racemes rather stout, erect, to 80 cm long, not interrupted; flowers pedicellate, the pedicels to 5 mm long; calyx 5-parted, usually white, punctate, 3–4 mm long; petals absent; stamens 6–8; style 3-parted; achenes 3-angled, oblongoid, shiny, smooth, 2.5–3.5 mm long. July–October.

Along streams, often in standing water.

IL, IN, MO (NI), KY, OH (OBL), as *Polygonum punctatum*.

Stout smartweed.

This species is often considered to be a more robust form of *P. punctata*. It differs from *P. punctata* by its uninterrupted racemes and larger leaves. The leaves usually have a sharp peppery taste.

67. *Persicaria robustior* (Stout smartweed). Habit (center). Shoot (upper left). Flower (lower left). Seed (lower right).

16. **Persicaria setacea** (Baldw.) Small, Fl. SE. U.S. 379. 1903. Fig. 68.
Polygonum setaceum Baldw. in Ell. Sketch S. Carol. 1:455. 1817.
Polygonum setaceum Baldw. var. *interjectum* Fern. Rhodora 40:414. 1938.
Polygonum hydropiperoides Michx. var. *setaceum* (Baldw.) Gl. Phytologia 4:23. 1952.

Perennial herb from rhizomes; stems erect, branched or unbranched, glabrous or pubescent at the nodes, to 1.5 m tall; leaves lanceolate to oblanceolate, to 15 cm long, to 3.5 cm wide, acute to acuminate at the apex, tapering at the subsessile base, strigose to glabrous above and below, entire and sometimes ciliate along the margins, the petioles hirsute, to 5 mm long; ocreae sparsely strigose, fringed with bristles to 12 mm long; inflorescence solitary or a panicle of racemes, the racemes loosely flowered, to 8 cm long, to 8 mm thick, erect, on nearly glabrous peduncles; flowers pedicellate, the pedicels 1–3 mm long; calyx 5-parted, the sepals mostly white or greenish, 2–3 mm long; petals absent; stamens usually 5; style 3-parted; achenes 3-angular, oblongoid, shining, smooth, brown or black, 2–3 mm long. July–October.

Wet ground.

IL, IN, KY, MO, OH (OBL), as *Polygonum setaceum.*

Slender smartweed.

The complex of taxa that includes *P. setacea, P. hydropiperoides, P. robustior,* and *P. opelousana* is poorly understood. Some botanists consider *P. setacea* and *P. opelousana* to be varieties of *P. hydropiperoides* since there is some intergradation.

Polygonum setacea can generally be distinguished from *P. hydropiperoides* and *P. opelousana* by its strigose leaves and its white flowers.

68. *Persicaria setacea* (Slender smartweed).

a, b. Habit.
c. Inflorescence.
d. Flower.
e. Sepal.
f. Achene.

3. **Polygonum** L.—Knotweed

Annual or perennial herbs; leaves alternate, simple, entire, jointed on a very short petiole adnate to the short sheath of the stipules, with ocreae; flowers in axillary fascicles, subtended by foliaceous bracts, perfect, actinomorphic; calyx 5-parted, united at base; petals 0; stamens 3–8 (–9); ovary superior; styles 3; achenes trigonous.

In this work, *Polygonum* contains those species that have flowers in axillary fascicles and leaves jointed on a very short petiole adnate to the short sheath of the stipules. All but the following are upland species in the central Midwest.

1. **Polygonum ramosissimum** Michx. Fl. Bor. Am. 1:237. 1803. Fig. 69.

Annual from slender roots; stems ascending to erect, branched, striate, glabrous, to 1 m tall; leaves linear to lanceolate, acute at the apex, tapering at the base, yellow-green, entire, glabrous, to 5 cm long, to 1 cm broad, short-petiolate; ocreae becoming shredded with age; inflorescence a cluster of 1–3 flowers from the axils of the upper ocreolae; flowers short-pedicellate; calyx 5- (6-) parted, yellowish to yellow-green, 2.5–3.5 mm long; petals absent; stamens (5–) 6; achene 3-angular, ovoid, dark brown, shining, smooth, 3–4 mm long, generally included within the persistent calyx. July–October.

Sandy soil.

IA, IL, IN, MO (FAC−), KS,KY, MO, NE, OH (FAC).

Knotweed.

This is a much branched, bushy species that is found primarily in sandy soils.

There is some variation in fruit size and shape, but all specimens examined had fruits never exceeding 4 mm in length.

5. **Rumex** L.—Dock; Sorrel

Annual or perennial herbs; leaves alternate, entire or undulate, with sheathing stipules; inflorescence with verticillate flowers in paniculate racemes; flowers small, perfect, less commonly unisexual, on jointed pedicels; sepals 6, green to reddish, the outer 3 smaller, the inner 3 becoming enlarged in fruit; petals absent; stamens 6; styles 3; ovary superior, with 1 ovule; fruit an achene with the 3 inner sepals enlarged and attached, 1, 2, or all 3 of the sepals with a tubercle.

About 150 species native to both Old and New World temperate regions make up this genus.

As the fruit develops, the three innermost sepals enlarge, becoming veiny, and forming wings on the achene. At this stage, the sepals are referred to as valves. Sometimes, along the midvein of 1 or more of these valves, a tubercle, known as a grain, may be formed. The valves may also develop small spinelike projections along their margins.

1. Valves with spinulose bristles or conspicuously dentate.
    2. Valves with conspicuously dentate margins; leaves very crispate .... 8. *R. stenophyllus*
    2. Fruiting sepals with spinulose bristles; leaves flat or somewhat crispate.
        3. Stems hollow; tubercle of valves long and slender; bristles of valves longer than width of the sepals .... 5. *R. fueginus*
        3. Stems firm; tubercle of valves only slightly longer than broad; bristles of valves not longer than width of the sepals .... 6. *R. obtusifolius*

69. *Polygonum ramosissimum* (Knotweed).

a. Habit (silhouette).
b. Habit.
c. Node.
d. Flower.
e. Sepal.
f. Achene.

1. Valves entire or only minutely toothed or erose.
  4. None of the valves with a tubercle .... 7. *R. occidentalis*
  4. Valves with 1–3 tubercles.
    5. Only 1 of the 3 valves with a tubercle.
      6. Leaves wavy along the margins... 4. *R. crispus*
      6. Leaves flat ....1. *R. altissimus*
    5. Each of the valves with a tubercle.
      7. Leaves with conspicuous wavy margins (crispate)................................... 4. *R. crispus*
      7. Leaves flat, entire to crenulate.
        8. Leaves crenulate; lateral veins of leaves forming right angles with the vertical veins.. 2. *R. brittanica*
        8. Leaves entire; lateral veins of leaves ascending.
          9. Pedicels of fruits 2–5 times longer than the calyx..................... 10. *R. verticillatus*
          9. Pedicels of fruits shorter than to about twice as long as the calyx.
            10. Leaves narrowly lanceolate, never more than 3 cm wide .... 9. *R. triangulivalvis*
            10. Leaves broadly lanceolate, at least some of them more than 3 cm wide.
              11. Inflorescence crowded, without bracts ....1. *R. altissimus*
              11. Inflorescence strongly interrupted, with numerous bracts........................ .... 3. *R. conglomeratus*

1. **Rumex altissimus** Wood, Class-book 477. 1847. Fig. 70.

Perennial herb from a thickened root; stems erect, branched or unbranched, glabrous, pale green, to 1.5 m tall; leaves glabrous, pale green, not crispate along the margin, the lower leaves oblong-lanceolate to ovate-lanceolate, acute at the apex, tapering or rounded at the base, to 15 cm long, to 5.5 cm wide, the petioles usually shorter than the blades, the upper leaves lanceolate, acute to acuminate at the apex, tapering at the base, to 10 cm long, to 3 cm wide, the petioles much shorter than the blades; inflorescence paniculate, with crowded, usually bractless verticils of flowers; flowers perfect or unisexual, on pedicels to 5 mm long, the pedicels jointed near the base; valves orbicular to ovate, 4–6 mm long, entire along the margins; tubercle usually 1, less commonly 3, ovoid, 2–3 mm long; achenes dark reddish brown, 2.0–2.8 mm long, smooth, shiny, angular. April–May.

Swamps; wet, disturbed soils.

IA, IL, IN, KY, MO, OH (FACW−), KS, NE (FAC).

Pale dock; smooth dock.

This pale green, smooth species is a common inhabitant of moist soil.

The presence of usually a single tubercle per fruit and the flat, pale leaves serve to distinguish this species. Occasional fruits with three tubercles may be found.

2. **Rumex britannica** L. Sp. Pl. 1:334. 1753. Fig. 71.
Rumex orbiculatus Gray, Man. Bot., ed. 5, 420. 1867.

Perennial herb from a thickened root; stems erect, branched or unbranched, glabrous, to 2 m tall; leaves glabrous, crenulate along the margin, the lateral veins at right angles to the main vein, the lower leaves oblong-lanceolate, subacute to acute at the apex, tapering at the base, to 30 cm long, to 15 cm wide, the petioles usually as long as the blades, the upper leaves lanceolate, acute at the apex, tapering at the base, to 15 cm long, to 6 cm wide, the petioles shorter than the blades; inflorescence a

70. *Rumex altissimus* (Pale dock).

a. Habit.
b. Staminate flower.
c. Pistillate flower.
d. Sepal, pistillate flower.
e. Fruiting branch.
f. Fruit.

panicle of racemes, the racemes crowded, with few leaflike bracts; flowers perfect, on pedicels to 1 cm long, the pedicels obscurely jointed near the base; valves orbicular to ovate, 5–8 mm long, veiny, usually denticulate along the margins; tubercles 3, lanceoloid, 2–4 mm long; achenes brown, 5–8 mm long, smooth, shiny, angular. April–June.

Along streams, in marshes and ditches, sometimes in standing water.

IA, IL, IN, NE, OH (OBL), as *Rumex orbiculatus.*

Great water dock.

This coarse, wetland species has flat, crenulate leaves, denticulate valves, and three tubercles per fruit. In the past, this species has been known as *Rumex orbiculatus*.

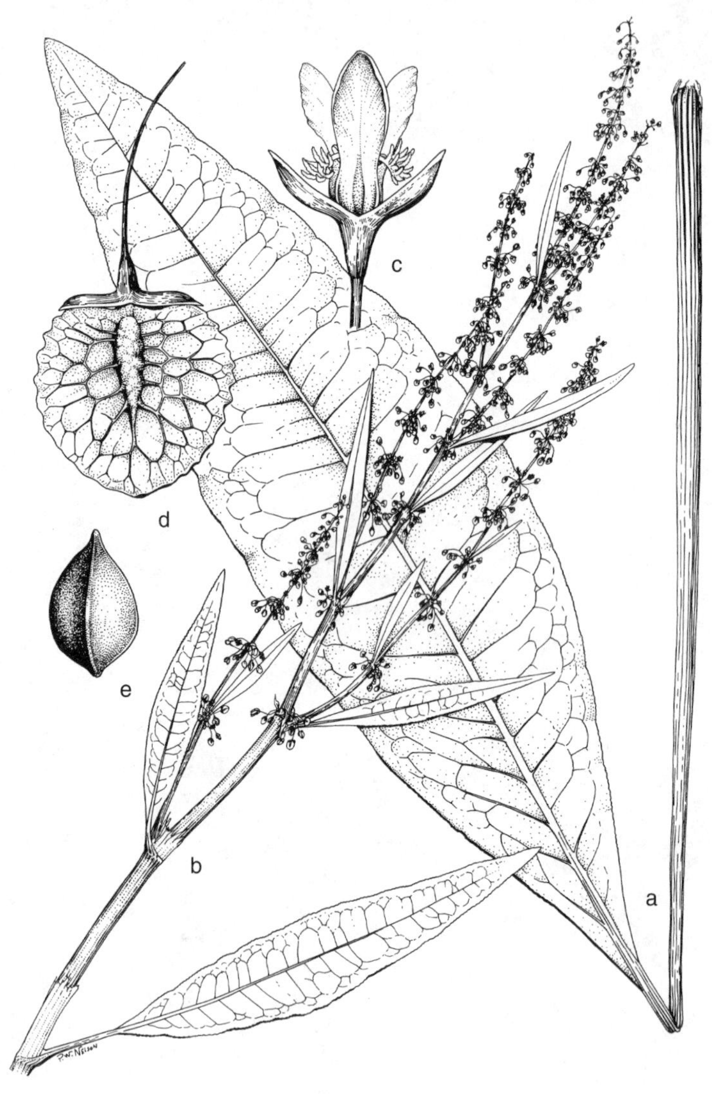

71. *Rumex britannica* (Great water dock).

a. Lower leaf.
b. Upper part of plant with flowers.
c. Flower. d. Fruit. e. Achene.

3. **Rumex conglomeratus** Murray, Prodr. Stirp. Gott. 52. 1770. Fig. 72.

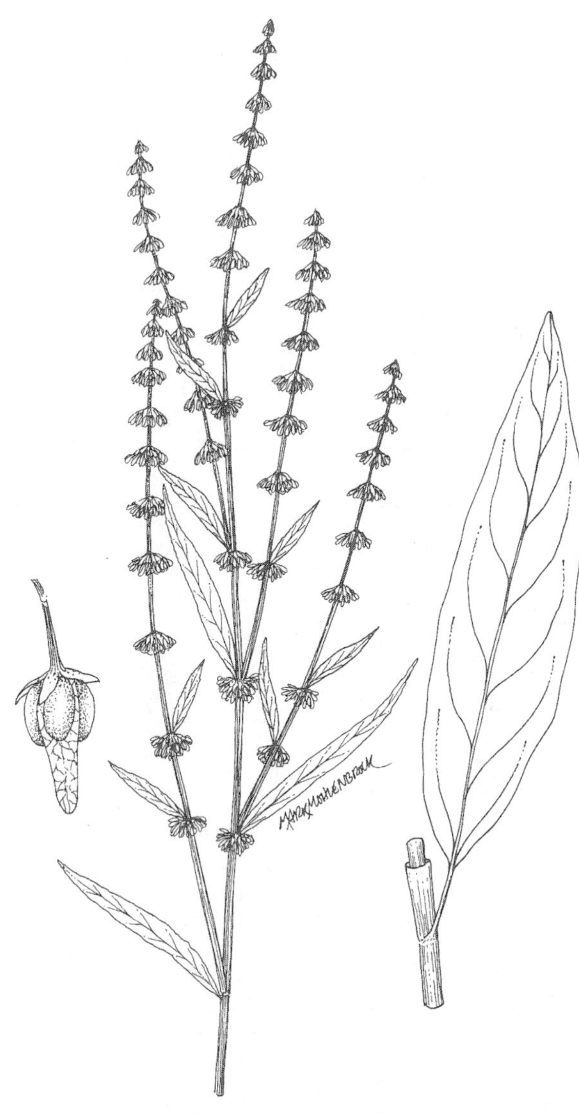

72. *Rumex conglomeratus* (Clustered green dock). Habit (center). Fruit (lower left). Leaf and ocrea (right).

Perennial herbs from a thickened root; stems erect, branched, glabrous, to 1 m tall; leaves oblong-lanceolate to lanceolate, obtuse to acute at the apex, tapering, rounded, or subcordate at the base, to 30 cm long, to 6 cm wide, glabrous, entire, the petioles about ¼–⅓ the length of the blade; inflorescence paniculate, terminal, interrupted, with occasional leaflike bracts; flowers perfect of unisexual, on pedicels up to 4 mm long, the pedicels distinctly swollen at the joint; valves oblong, 2–3 mm long, veiny, entire along the margin; tubercles 3, usually unequal in size, narrowly ellipsoid; achenes dark reddish brown, 1.5–2.0 mm long, smooth, shiny, angular. June–September.

Marshes, wet woods, wet meadows, wet ditches.

Native to Eurasia and Africa; adventive in IL, IN (FACW), OH (FAC).

Clustered green dock.

In the central Midwest, this species most nearly resembles *Rumex cristatus* (not a wetland species), but the perianth parts in *R. conglomeratus* are oblong and the valves are entire, while the perianth parts of *R. cristatus* are deltoid and the valves are irregularly toothed.

4. **Rumex crispus** L. Sp. Pl. 335. 1753. Figs. 73, 74.
Rumex elongatus Guss. Pl. Rar. Neap. 150. 1826.

Perennial herb from a thickened root; stems erect, usually unbranched, glabrous, to 1.5 m tall; leaves glabrous, strongly crispate along the margin, the lower leaves oblong to lance-oblong, acute to subacute at the apex, rounded to subcordate at the base, to 30 cm long, to 6 cm wide, the petioles often nearly as long as the blades, the upper leaves lanceolate, acute at the apex, usually subcordate at the base, to 12 cm long, to 3 cm wide, the petioles shorter than the blades; inflorescence an open panicle, with rather crowded verticils of flowers interspersed with several linear, leaflike bracts; flowers perfect or unisexual, on recurved pedicels to 1 cm

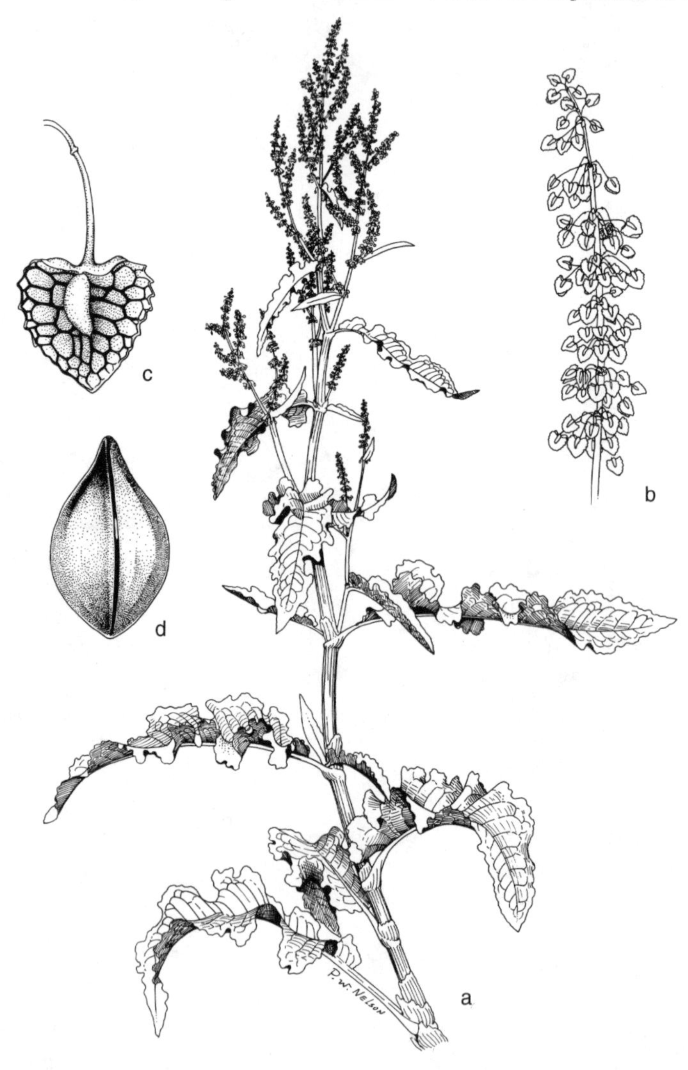

73. *Rumex crispus* (Curly dock). a. Habit. b. Fruiting branch. c. Fruit. d. Achene.

long, the pedicels jointed near the base; valves orbicular to broadly ovate, 4–6 mm long, entire or minutely toothed along the margins; tubercles usually 3, rarely 1, ovoid, 2–3 mm long; achene dark brown, 1.3–1.5 mm long, smooth, shiny, angular. April–June.

Fields, damp disturbed soils, waste ground.

Native of Europe; IA, IL, IN, MO (FAC+), KS, NE (FACW), KY, OH (FACU).

Curly dock.

Specimens with rather slender, elongated inflorescences have been referred to *R. elongatus*, but these two taxa are generally considered to be the same species. Plants with the leaves crispate and the lower leaves more than 6 cm wide are *R. obtusifolius*.

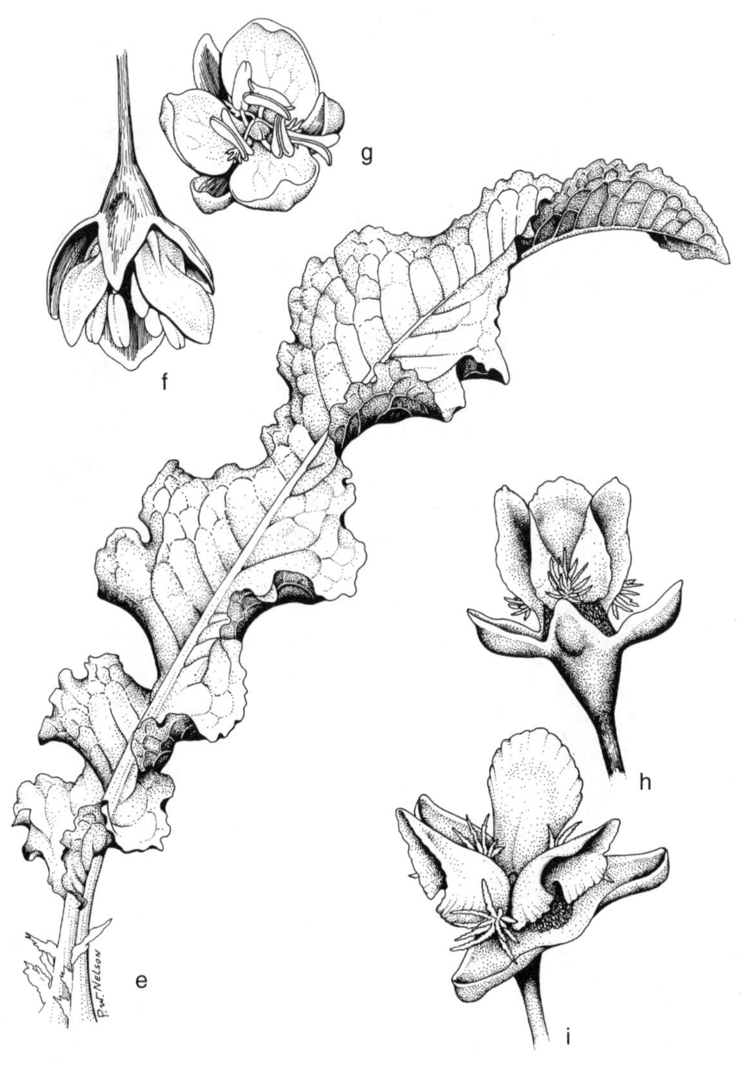

74. *Rumex crispus* (Curly dock). e. Leaf. f, g. Perfect flowers. h, i. Pistillate flowers.

5. **Rumex fueginus** Phil. Anal. Univ. Chile 91:493. 1895. Fig. 75.
Rumex maritimus L. var. *fueginus* (Phil.) Dusén, Sv. Exped. Magell. 3, no. 5:194. 1900.

Annual from fibrous roots; stems ascending to erect, hollow, unbranched or much branched, usually puberulent, to 75 cm long; leaves narrowly lanceolate, acute at the apex, subcordate to truncate to tapering at the base, more or less crispate along the margins, rather thick and fleshy, to 25 cm long, puberulent on the lower surface, on petioles up to 6 cm long; inflorescence a large, much-branched panicle with interrupted verticils, the inflorescence sometimes extending nearly to

75. *Rumex fueginus* (Golden dock). a. Habit. b. Flower. c. Fruit. d. Achene.

the base of the plant and further interrupted by leaves; flowers perfect, on pedicels 4–5 mm long, the pedicels jointed near the base; valves ovate to deltate, 2–3 mm long, with 1–3 spinulose bristles on each margin, the bristles 2–4 mm long; tubercles 3, lanceoloid; achenes reddish brown, 1.3–1.5 mm long, smooth, shiny, angular. May–July.

Sandy shores and muddy banks.

IA, IL, MO (OBL), KS, NE, OH (FACW).

Golden dock.

This species is distinguished by its spinulose valves and its hollow stems. It is found primarily along major rivers and in saline areas around industries.

6. **Rumex obtusifolius** L. Sp. Pl. 335. 1753. Figs. 76, 77.

Perennial herb from a thickened rootstock; stems erect, branched or unbranched, glabrous but rough to the touch, to 1.0 (–1.5) m tall; leaves puberulent on the veins, the lower leaves oblong to ovate, obtuse to acute at the apex, cordate to rounded at the base, more or less crispate along the margin, to 50 cm long, to 30 cm wide, the petioles about as long as the blades, the upper leaves lanceolate to lance-oblong, acute at the apex, tapering or rounded at the base, to 12 cm long, to 6 cm wide, the petioles shorter than the blades; inflorescence an open panicle with interrupted, leafy-bracted verticils; flowers perfect or unisexual, on pedicels 4–7 mm long, the pedicels jointed below the middle; valves deltate to ovate, 3–5 mm long, veiny, with several prominent marginal spiny teeth; tubercle 1, oblongoid; achenes reddish brown, 1.0–1.5 mm long, smooth, shiny, angular. April–June.

Fields, along streams, disturbed areas.

IA, IL, IN, MO (FACW), KS, NE (FAC), KY, OH (FACU−).

Bitter dock; broad-leaved curly dock.

This native of Europe differs from other species of *Rumex* by its spinulose valves and the presence of only one tubercle per fruit. Because of the usually crispate leaves, this species is often misidentified as *R. crispus*, but this latter species has much narrower basal leaves.

The veins of the lower leaves are often red.

One of the most common habitats for this species is in disturbed floodplain woods. It also may be found in cultivated fields.

7. **Rumex occidentalis** S. Wats. Proc. Am. Acad. Art. 12:253. 1877. Fig. 78.
Rumex aquaticus L. ssp. *occidentalis* (S. Wats.) Hulten, Kongl. Svenska Vet. Handl. 13:132. 1971.

Perennial herbs from a thickened rootstock; stems erect, branched, glabrous, to 1.2 m tall; leaves lance-ovate to lance-oblong, acute at the apex, truncate or rounded or cordate at the base, to 35 cm long, to 12 cm wide, entire or sometimes undulate, not crispate, glabrous, the petioles 1/4–1/3 the length of the blades; inflorescence a narrow panicle of racemes, the racemes densely crowded or interrupted; flowers perfect, whorled, on filiform pedicels up to 12 mm long, jointed but the joints not swollen; valves ovate to orbicular, to 10 mm long, nearly as wide, entire or slightly erose; tubercles absent; achenes reddish brown, 3.0–4.5 mm long. June–September.

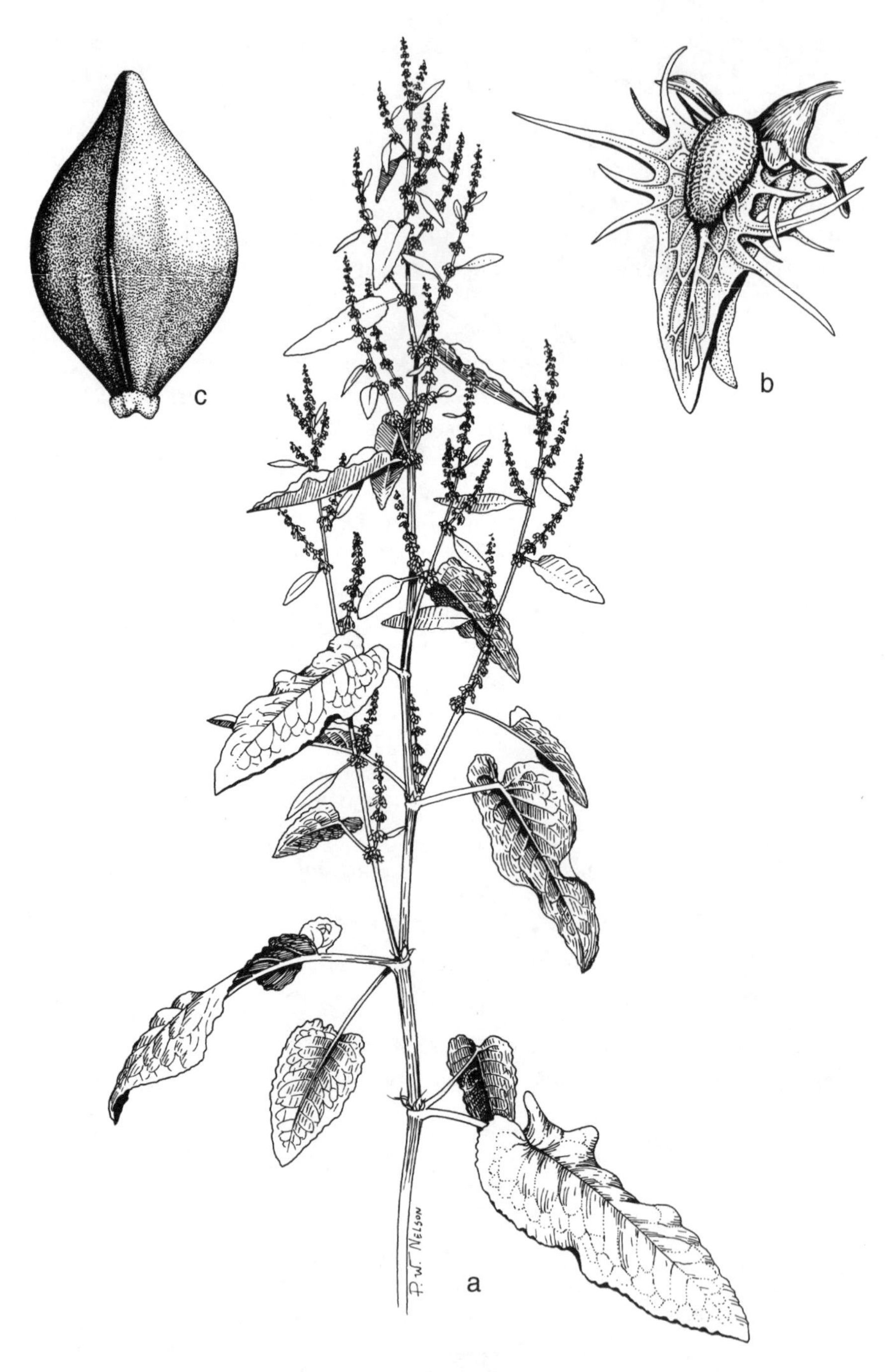

76. *Rumex obtusifolius* (Bitter dock). a. Habit. b. Fruit. c. Achene.

77. *Rumex obtusifolius* (Bitter dock).

d. Lower leaf.
e, f. Pistillate flowers, side view.
g. Pistillate flower, face view.

h. Perfect flower.

Wet meadows, bogs, marshes, along rivers and streams.

IA (OBL).

Western dock.

This species, primarily of the western United States, is distinguished by its non-swollen joints of the pedicels, its often cordate leaves, and its ovate to orbicular valves without tubercles.

78. *Rumex occidentalis* (Western dock). Habit (center). Fruit (lower left).

8. **Rumex stenophyllus** Ledeb. Fl. Altaica 2:58. 1830. Fig. 79.
Rumex crispus L. var. *dentatus* Schur, Enum. Pl. Transsilv. 580. 1866.

Perennial herbs from a thickened rootstock; stems erect, branched, glabrous, to nearly 1 m tall; leaves narrowly lanceolate to oblong-lanceolate, obtuse to acute at the apex, tapering or truncate at the base, to 25 cm long, to 7 cm wide, entire or denticulate, crispate, rarely flat, glabrous above, obscurely papillose on the veins below, the petioles one-half to nearly as long as the blades; inflorescence sometimes occupying as much as half the length of the plant, terminal, densely flowered except near the base; flowers whorled, on filiform pedicels up to 8 mm long, the pedicels with swollen joints; valves orbicular to deltate, dentate, 3.5–5.0 mm long, about as wide; tubercles 3, more or less equal in size; achenes reddish brown to dark brown, 2–3 mm long, 1.0–1.5 mm wide. May–September.

Swamps, marshes, wet meadows, disturbed areas.

Native to Eurasia; IA, IL, MO (FACW), KS, NE (FACW+).

Dentate-fruited curly dock.

The crispate leaves are reminiscent of *R. crispus*, but *R. stenophyllus* has conspicuously dentate valves.

79. *Rumex stenophyllus* (Dentate-fruited curly dock). Branchlet with leaves and flowers (left). Branchlet with fruits (right). Fruit with tubercle (below).

9. **Rumex triangulivalvis** (Danser) Rech. f. Rep. Sp. Nov. 40:297. 1936. Fig. 80.
Rumex mexicanus Meisn. in DC. Prodr. 14:45. 1856, misapplied.
Rumex salicifolius Hook. var. *triangulivalvis* Danser, Nederl. Kruidk. Archief. 1925:415. 1926.

Perennial herb from a thickened root; stem erect to ascending, branched or unbranched, glabrous, to 1 m tall; leaves glabrous, flat, the lower leaves oblong to oblong-lanceolate, acute at the apex, tapering at the base, to 20 cm long, to 3 cm

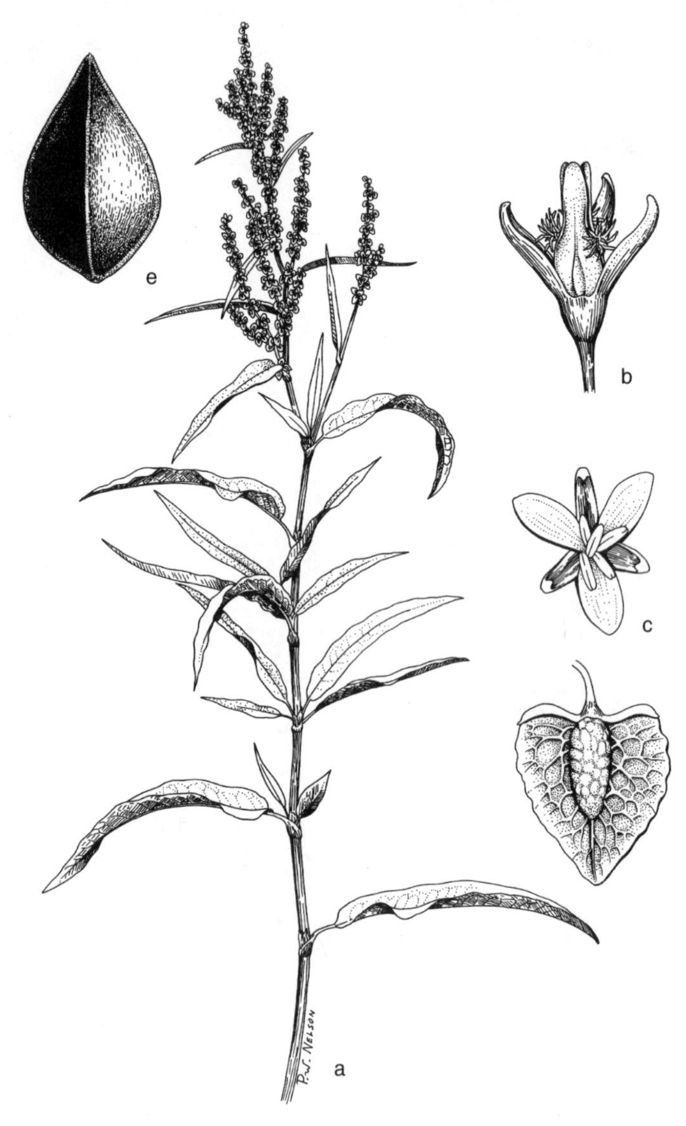

80. *Rumex triangulivalvis* (Narrow-leaved dock).
a. Habit.
b. Pistillate flower.
c. Staminate flower.
d. Fruit.
e. Achene.

wide, the petioles usually as long as or longer than the blades, the upper leaves lanceolate, acute or acuminate at the apex, tapering at the base, to 15 cm long, to 3 cm wide, the petioles usually shorter than the blades; inflorescence paniculate, densely flowered, with few leaflike bracts; flowers perfect or unisexual, on pedicels up to 4 mm long, the pedicels jointed near the base; valves deltate, 3–6 mm long, veiny, entire to undulate along the margins; tubercles 3, ellipsoid to lanceoloid, 2–4 mm long; achene dark red, 1.5–2.2 mm long, smooth, shiny, angular. April–May.

Moist soil.

IA, IL, IN, MO (FACW), KY, OH (FACU), KS, NE (not listed).

Dock.

This very narrow-leaved species occurs in a wide variety of moist soils. It has sharply triangular valves, each with a tubercle.

10. **Rumex verticillatus** L. Sp. Pl. 334. 1753. Fig. 81.

Perennial herb from a thick, deep root; stems erect, unbranched, glabrous, to 1.5 m tall; leaves glabrous, flat, the lower leaves oblong to broadly lanceolate, obtuse to acute at the apex, tapering at the base, to 30 cm long, to 8 cm wide, the petioles usually as long as or longer than the blades, the upper leaves lanceolate to linear-lanceolate, acute at the apex, tapering at the base, to 12 cm long, to 3 cm wide, the petioles usually shorter than the blades; inflorescence paniculate, with ascending racemes, the verticils interrupted, with few leaflike bracts; flowers perfect, on reflexed pedicels to 1.5 cm long, the pedicels jointed near the base, more than twice as long as the valves; valves deltate to ovate, 3–5 mm long, somewhat veiny, entire along the margins; tubercles 3, lanceoloid, 2.0–3.5 mm long; achene reddish brown, 1.2–1.8 mm long, smooth, shiny, angular. April–June.

Swamps, marshes, wet woods.

IA, IL, IN, KS, KY, MO, NE, OH (OBL).

Swamp dock.

This species is recognized by its strongly interrupted verticils of flowers and its elongated pedicels that are more than twice as long as the valves.

Swamp dock frequently grows in standing water.

6. **Tracaulon** Raf.—Tear Thumb

Sprawling annual or perennial herbs; stem 4-angled, reflexed-prickly; leaves alternate, simple, entire, sagittate, or hastately lobed; ocreae oblique, fringed at the summit; inflorescence of short terminal and/or axillary heads or racemes on long peduncles; calyx 5-parted, united at base, pink or white or greenish; petals absent; stamens 6 or 8; styles 2 or 3; achenes lenticular or trigonous.

Members of this genus are often placed in the broad view of *Polygonum*. Three species are in the genus, two of them in the central Midwest.

1. Leaves hastate; achenes lenticular .... 1. *T. arifolium*
1. Leaves sagittate; achenes trigonous.. 2. *T. sagittatum*

81. *Rumex verticillatus* (Swamp dock).

a. Habit.
b. Flower.

c. Fruit.
d. Achene.

1. **Tracaulon arifolium** (L.) Raf. Fl. Tell. 3:13. 1836. Fig. 82.
Polygonum arifolium L. Sp. Pl. 364. 1753.
Polygonum arifolium L. var. *lentiforme* Fern. & Grisc. Rhodora 37:167. 1935.
Polygonum arifolium L. var. *pubescens* (Keller) Fern. Rhodora 48:53. 1946.

Sprawling perennial; stems decumbent, often reclining on other vegetation, 4-angled, reflexed-prickly, to 2 m long; leaves hastate, acuminate to acute at the apex,

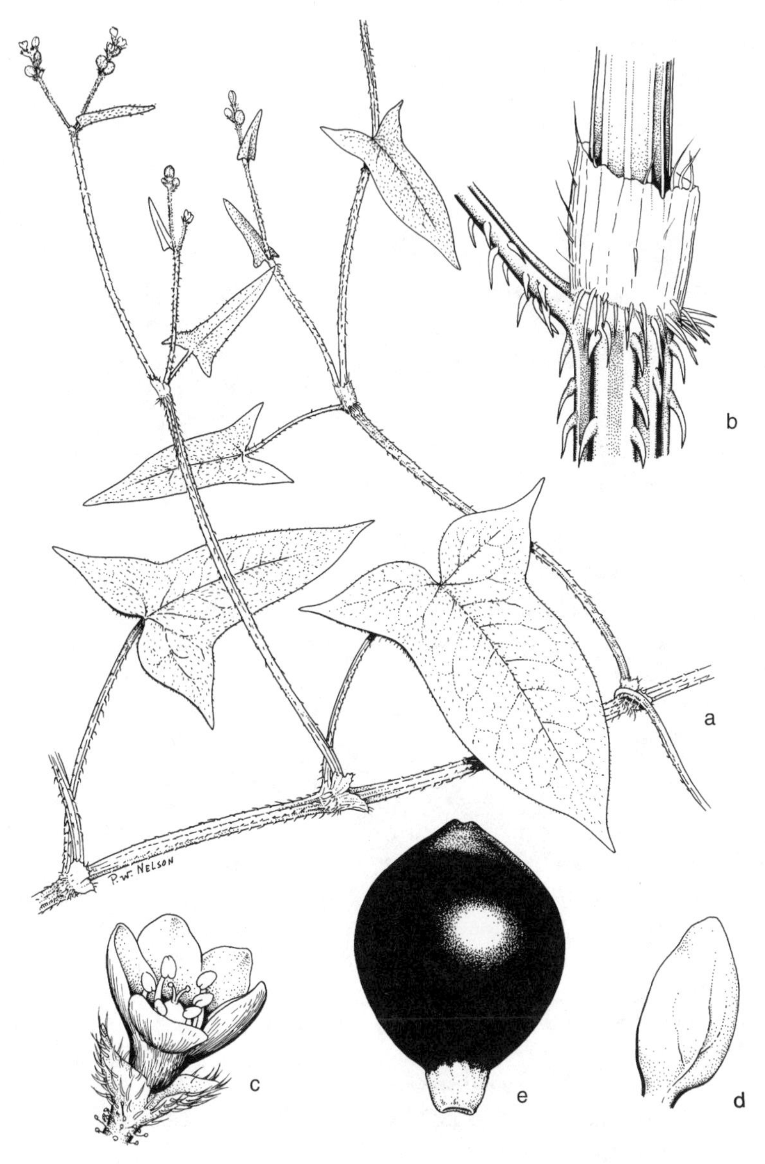

82. *Tracaulon arifolium* (Halberd-leaved tear thumb). a. Stem with leaves and flowers. b. Ocrea. c. Flower. d. Sepal. e. Achene.

truncate at the base, the two hastate lobes acute to acuminate, the leaf to 10 (–20) cm long, to 7 (–15) cm wide, stellate-pubescent, at least below, prickly on the main veins below, the lower leaves on prickly petioles, the upper short-petiolate to subsessile; ocreae oblique, fringed at the apex, bristly near the base; inflorescence of short terminal and/or axillary heads or racemes, on long peduncles; calyx 5-parted, pink to greenish; stamens 6; styles 2-parted; achenes lenticular, obovoid, dark brown, smooth, shining, 3.0–3.5 mm long. June–October.

Wet ground.

IL, IN, KY, MO, OH (OBL), as *Polygonum arifolium.*

Halberd-leaved tear thumb.

The halberd-shaped leaves and the lenticular achenes are the distinguishing features of this species.

2. **Tracaulon sagittatum** (L.) Small, Fl. S.E. U.S. 3981. 1903. Fig. 83.
Polgonum sagittatum L. Sp.Pl. 262, 1753.

Weak, sprawling annual; stems decumbent, often reclining on other vegetation, 4-angled, reflexed-prickly, to 2 m long; leaves lanceolate to elliptic to oblong, obtuse to acute at the apex, sagittate at the base, roughened along the margins, to 10 cm long, to 2.5 cm wide, short-prickly on the main vein below, the lower leaves on short-prickly petioles, the upper leaves sessile or nearly so; ocreae oblique, without apical bristles but with some short prickles near the base; inflorescence of dense, shortened racemes, on long peduncles; calyx 5-parted, pink to white to greenish; stamens 8; styles 3-parted; achene 3-angular, dark brown to black, smooth, shining, 2–3 mm long. June–October.

Swampy ground, marshy ground, burned bogs.

IA, IL, IN, KS, KY, MO, NE, OH (OBL), as *Polygonum sagittatum.*

Tear thumb.

Tear thumb often forms dense entanglements as it climbs and leans over existing vegetation. It differs from the similar *T. arifolium* by its sagittate leaves and its trigonous achenes.

101. PRIMULACEAE—PRIMROSE FAMILY

Herbs; leaves alternate, opposite, whorled, or basal, simple (deeply dissected in *Hottonia*); flowers actinomorphic, perfect, variously arranged; calyx united, (4-) 5- (9-) parted, persistent; corolla united, (4-) 5- (9-) parted; stamens usually as many as the lobes of the corolla and attached to the corolla; ovary superior (subinferior in *Samolus*), 1-locular, with free-central placentation, with 2–several ovules; fruit a capsule or pyxis.

This family is comprised of nearly thirty genera and 800 species, primarily in the north temperate regions of the world.

Members of the Primulaceae are diverse in their appearance, with some of the tiniest herbs in the Midwest found in this family. On the other hand, some species of *Lysimachia* and *Steironema* may reach a height of nearly one meter and almost are shrublike.

The Primulaceae is readily identified by its united, generally 5-parted corolla, its unilocular ovary, and its free-central placentation.

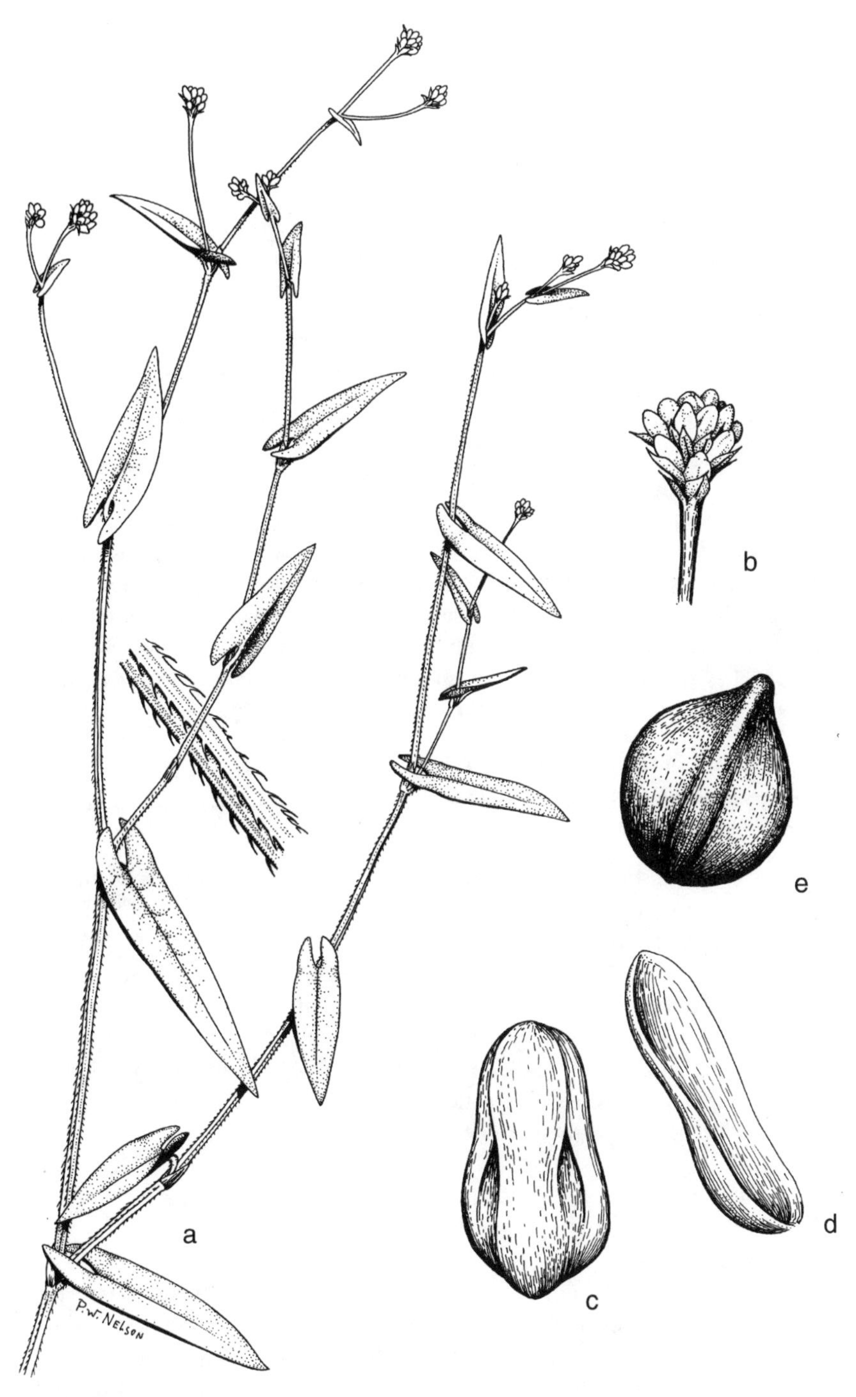

83. *Tracaulon sagittatum* (Tear thumb).

a. Habit.
b. Inflorescence.
c. Sepal.
d. Sepal.
e. Achene.

1. Leaves deeply dissected into threadlike divisions .... 2. *Hottonia*
1. Leaves entire or toothed.
  2. Leaves, or most of them, alternate.
    3. Ovary superior; flowers generally 4-merous; pedicels absent or less than 5 mm long; flower solitary .... 1. *Anagallis*
    3. Ovary inferior; flowers 5-merous; pedicels 1–2 cm long; flowers in racemes .... 5. *Samolus*
  2. Leaves, or most of them, opposite or whorled (occasional alternate leaves may be found in *Trientalis*, which also always has some whorled leaves, *Lysimachia quadrifolia*, and *Lysimachia terrestris*).
    4. Leaves epunctate; corolla lobes erose and apiculate; staminodia present ... 6. *Steironema*
    4. Leaves punctate; corolla lobes entire; staminodia absent.
      5. Flowers in dense axillary racemes; corolla lobes linear .... 4. *Naumbergia*
      5. Flowers in terminal panicles or racemes or, if axillary, solitary on in panicles; corolla lobes lanceolate to orbicular (rarely linear in *L. X commixta*) .... 3. *Lysimachia*

1. **Anagallis** L.—Pimpernel

Annual or perennial herbs from fibrous roots; leaves alternate or opposite, simple, entire; flowers axillary, solitary, perfect, actinomorphic; calyx 4- to 5-parted, sometimes persistent on the fruit; corolla short-tubular, rotate, 4- to 5-lobed, sometimes adhering to the fruit; stamens 4 or 5, inserted on the corolla; ovary superior, with numerous ovules; fruit a pyxis, with numerous minute seeds.

About thirty species found in many parts of the world comprise this genus. It includes species sometimes placed in the genus *Centunculus*.

1. **Anagallis minima** (L.) Krause, Deutschl. Fl. ed. 2, 9:251. 1902. Fig. 84.
Centunculus minimus L. Sp. Pl. 116. 1753.
Centunculus lanceolatus Michx. Fl. Bor. Am. 1:93. 1803.

Low-growing annual from fibrous roots, with simple or branched, slender, glabrous stems to 8 (–15) cm tall; leaves alternate, or the lowermost opposite, spatulate to elliptic to narrowly obovate, obtuse to subacute at the apex, tapering at the base, entire, glabrous, sessile or on glabrous petioles up to 2 mm long; flower solitary in the axils, sessile or on pedicels less than 1 mm long; calyx deeply 4-lobed, the lobes linear to linear-lanceolate, acute to acuminate, glabrous, 1–2 mm long; corolla urceolate-rotate, pink, 0.5–1.5 mm long, the 4 lobes divided nearly to the middle; stamens 4, included; capsule globose, membranaceous, glabrous, 1.5–2.0 mm in diameter, with many very tiny, angular seeds. April–August.

Moist soil.

IL, IN, MO (FACU−), KS, NE (OBL), KY, OH (FACW), as *Centunculus minimus*.

Chaffweed.

The chaffweed is one of the most obscure plants in the flora. Its small stature, together with its tiny, pink flowers, perhaps has caused this species to be overlooked in the past. Plants only about one centimeter tall have been seen in flower.

There is a tendency for the leaves near the base of the stem to be opposite, while those above are invariably alternate.

For many years, this plant was placed in the genus *Centunculus*.

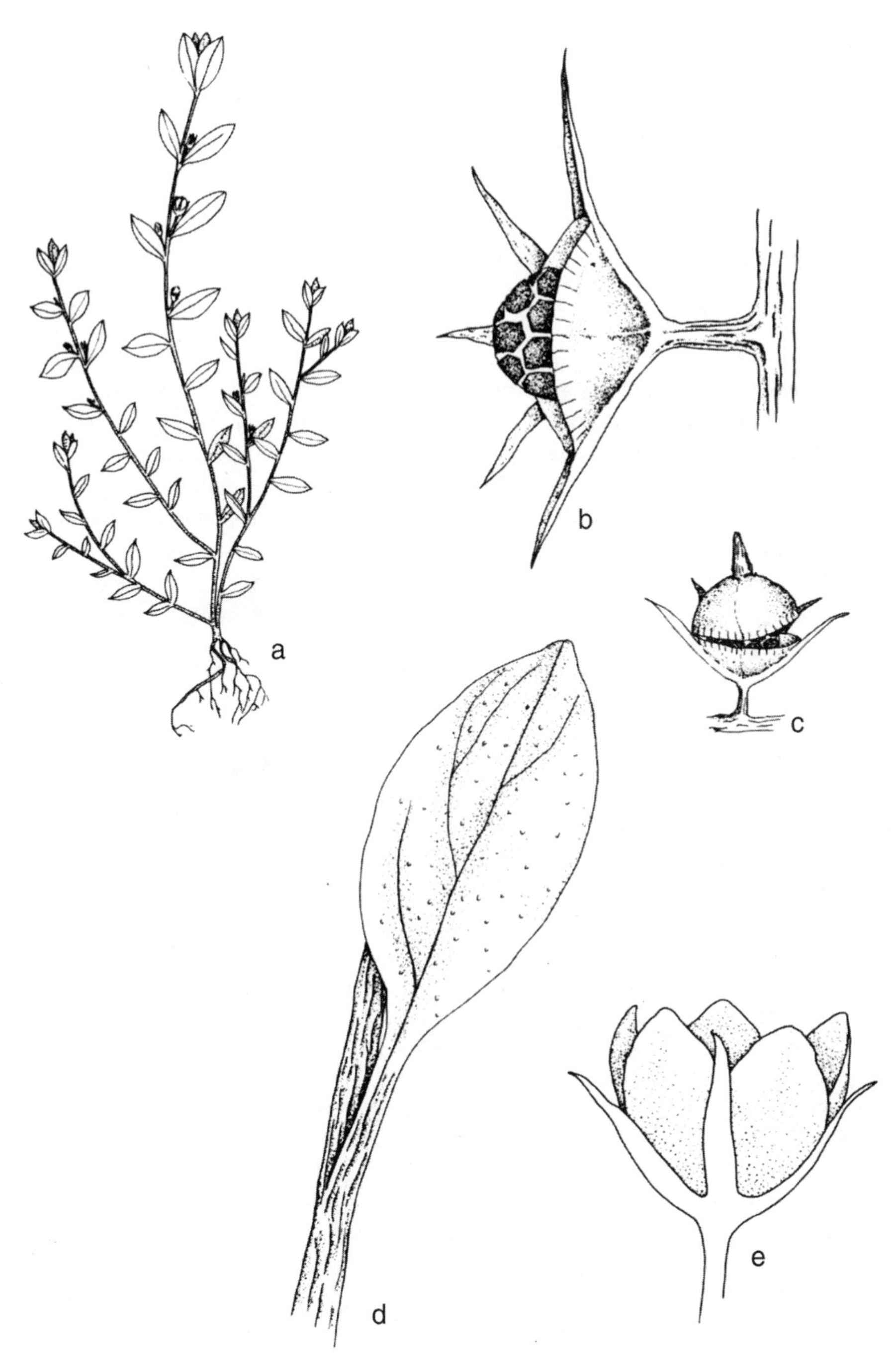

84. *Anagallis minima* (Chaffweed).

a. Habit.
b. Fruit with seeds.
c. Fruit.
d. Leaf.
e. Flower.

2. **Hottonia** L.—Featherfoil

Aquatic herbs; leaves submersed, deeply pinnatifid, crowded near the base; peduncles emersed, inflated, constricted at the nodes; flowers borne in racemes, verticillate; calyx 5-cleft nearly to base; corolla short-tubular, with 5 lobes; stamens 5, inserted on the corolla tube; ovary superior, with numerous ovules; capsule 5-valved, with many angular seeds.

In addition to our species, *H. palustris* occurs in Eurasia.

The often floating habit of this aquatic, along with the inflated and constricted peduncles, sets this genus well apart from all other members of the Primulaceae.

1. **Hottonia inflata** Ell. Bot. S.C. & Ga. 1:231. 1817. Fig. 85.

Aquatic herb either floating in water or rooting in mud; leaves submersed, crowded, divided almost all the way to the rachis into linear or filiform glabrous divisions, the total leaf up to 6 cm long; peduncles several, arising together from above the leaves, partly emersed, inflated but constricted at the joints, to 10 cm long, green, glabrous, each joint giving rise to whorls of 5–10 bracteate flowers; pedicels to 2 cm long, glabrous, ascending; bracts linear, to 3 cm long, glabrous; calyx divided nearly to the base into 5 linear, obtuse to acute, glabrous lobes, up to 10 mm long; corolla white, to 5 mm long, short-tubular, with 5 obtuse to subacute lobes; stamens 5, included; capsules globose, up to 3 mm in diameter, glabrous. May–August.

Swamps, wet ditches;

IL, IN, KY, MO, OH (OBL).

Featherfoil.

This strange-looking aquatic is unlike anything else in the Primulaceae. The inflated structures that arise from the cluster of pinnatifid leaves and stand out of the water are the peduncles. Where each peduncle is constricted, there arises a whorl of 5–10 flowers, each subtended by a narrow bract.

3. **Lysimachia** L.—Yellow Loosestrife

Perennial herbs from rhizomes, stolons, or basal offshoots; leaves alternate, opposite, or verticillate, entire, often punctate; flowers solitary in the upper leaf axils or in terminal racemes, perfect; calyx deeply 5-parted; corolla shallowly campanulate to rotate, mostly 5-parted, the lobes convolute in bud; fertile stamens mostly 5, attached to near the base of the corolla, the filaments free or united at base; staminodia absent; ovary superior, 1-locular; fruit a capsule, with few to many seeds.

Lysimachia consists of about 75 species distributed mostly in temperate regions of the Northern Hemisphere.

Ray (1956), who has studied the New World species of *Lysimachia*, divided the genus into five subgenera.

I believe these subgenera worthy of generic status. The following paragraphs summarize the characteristics of the four genera with representatives in the central Midwest that sometimes are included in *Lysimachia*, together with the names of the species.

Steironema. (Subgenus *Seleucia.*) Corolla lobes in bud each enclosing a stamen; corolla yellow; staminodia present. *S. ciliata, S. lanceolata, S. hybrida, S. radicans, S. quadriflora.*

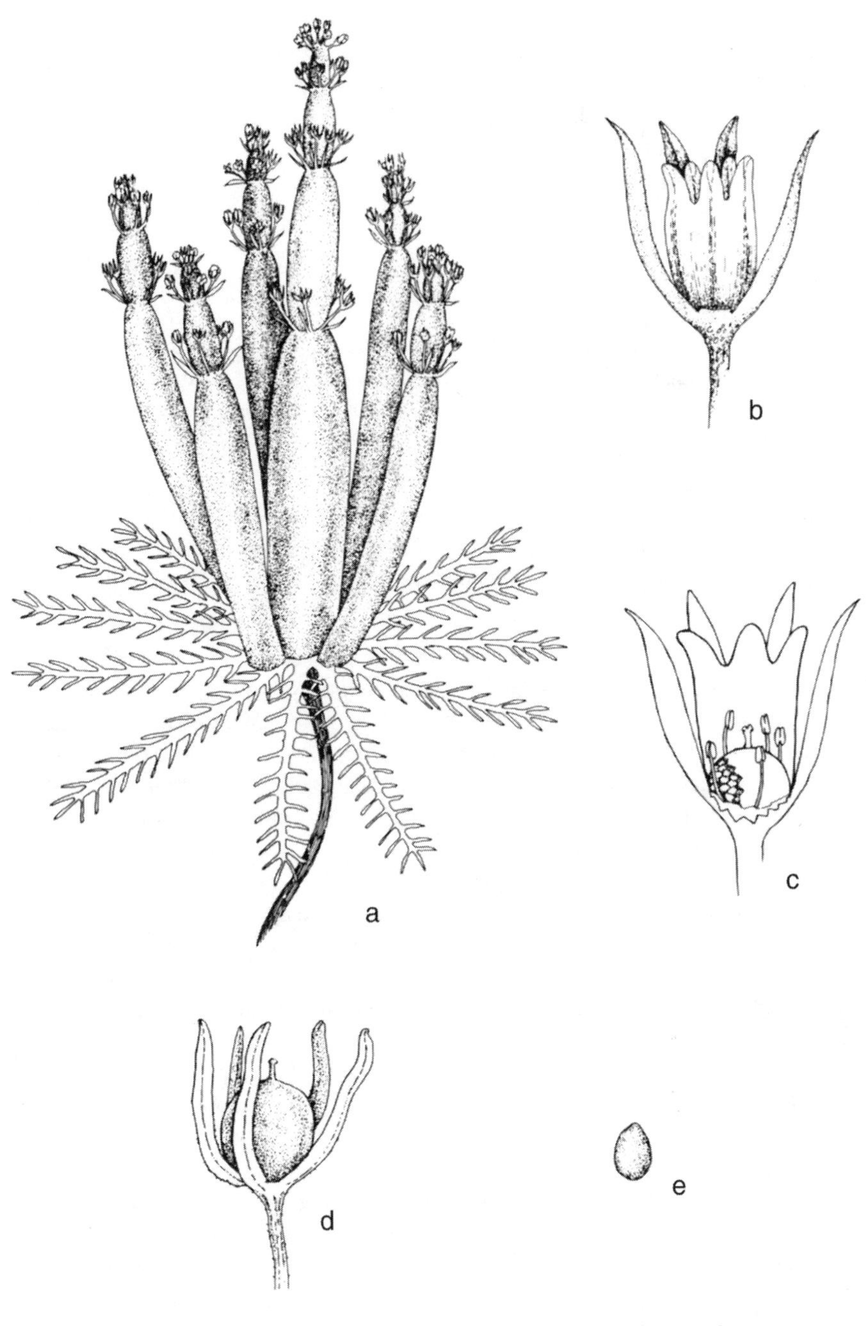

85. *Hottonia inflata* (Featherfoil).

a. Habit.
b. Flower.
c. Flower, longitudinal view.
d. Capsule.
e. Seed.

Lysimachia. (Subgenus *Lysimachia.*) Corolla lobes in bud imbricate; corolla yellow; staminodia absent; flowers axillary, or in terminal racemes. *L. nummularia, L. punctata, L. vulgaris, L. quadrifolia, L. terrestris, L. X commixta, L. X producta,* and *L. fraseri.*

Naumbergia. (Subgenus *Naumbergia.*) Corolla lobes in bud imbricate; corolla yellow; staminodia absent; flowers in axillary racemes. *N. thyrsiflora.*

Subgenus *Palladia.* Corolla lobes in bud imbricate; corolla white; staminodia absent; flowers in terminal, spikelike racemes. *Lysimachia clethroides.* This group does not occur in wetlands.

1. Some or all the leaves in whorls of 3–7.
  2. Stems glabrous; flowers in panicles....................2. *L. fraseri*
  2. Stems usually puberulent; flowers in leafy bracted racemes ....4. *L. X producta*
1. Leaves opposite, or less commonly alternate.
  3. Stems creeping, rooting at the nodes; corolla lobes 10–15 mm long. ....3. *L. nummularia*
  3. Stems ascending to erect, not rooting at the nodes; corolla lobes less than 10 mm long.
    4. Most or all the leaves (except the lowermost scalelike ones) opposite; style 3–4 mm long; capsule 2.8–3.5 mm in diameter....................5. *L. terrestris*
    4. Most or all the leaves (except the lowermost scalelike ones) alternate; style 5–6 mm long; capsule about 2 mm in diameter....................1. *L. X commixta*

1. **Lysimachia X commixta** Fern. Rhodora 52199. 1950. Fig. 86.
Lysimachia terrestris X *thyrsiflora* Fern & Wieg. Rhodora 12:141. 1910.

Perennial herbs from rhizomes; stems erect, simple or branched, to about 1 m tall, often black-dotted, glabrous; leaves subopposite or alternate, the lowermost scalelike, the middle and upper linear-lanceolate to elliptic, acute to acuminate at the apex, tapering to the nearly sessile base, entire, to 12 cm long, up to one-third as wide, glabrous, punctate, green above, pale beneath; inflorescence composed of a terminal raceme and two or more elongated lateral racemes; bracts linear-subulate, to 5 mm long; pedicels to 9 mm long, glabrous or puberulent, spreading to ascending; calyx usually 5-cleft nearly to the base, the lobes narrowly lanceolate, acute, glabrous, punctate, to 4 mm long; corolla yellow, rotate-funnelform, deeply 5-cleft, the lobes narrowly elliptic, obtuse, black-dotted as well as often blotched with red, glandular-pubescent, 5–8 mm long, about one-fourth as wide; stamens 5, the filaments united at the base into a nearly glabrous tube up to 1 mm long; capsules globose, up to 2 mm in diameter, with few ovoid seeds 1.0–1.5 mm long. June–September.

Swamps and marshes.

IL. The U.S. and Fish and Wildlife Service does not list this hybrid, but both of its parents are OBL.

Loosestrife.

Fernald first reported this hybrid of *L. terrestris* and *L. thyrsiflora* (=*Naumbergia thyrsiflora*) in 1910 from a specimen collected in Maine.

I am restricting the concept of *L.* X *commixta* in this study to include those plants with a terminal raceme and two or more well developed lateral racemes. Some specimens with one or two short lateral racemes near the base of the terminal raceme I am retaining in *L. terrestris.*

2. **Lysimachia fraseri** Duby in DC. Prod. 8:65. 1838. Fig. 87.

Herbaceous perennial from spreading rhizomes; stems erect, simple or branched, stipitate-glandular, to 1.5 m tall; leaves in whorls of 3–5, lanceolate to lance-elliptic, to 15 cm long, to 6 cm wide, acute at the apex, cuneate to rounded at the base, entire, glandular-hairy above and below, on glandular-puberulent petioles up to 10 mm long; lowermost leaves reduced; flowers in a leafy terminal panicle, the branches of the inflorescence glandular-puberulent; calyx 5-lobed, the lobes lanceolate, acute to acuminate, glandular, to 5 mm long, their margins purple; corolla yellow, rotate, 5-cleft, the lobes obovate to elliptic, to 8 mm long, to 4 mm broad; stamens 5, the filaments united for about half to one-third their length, glandular-

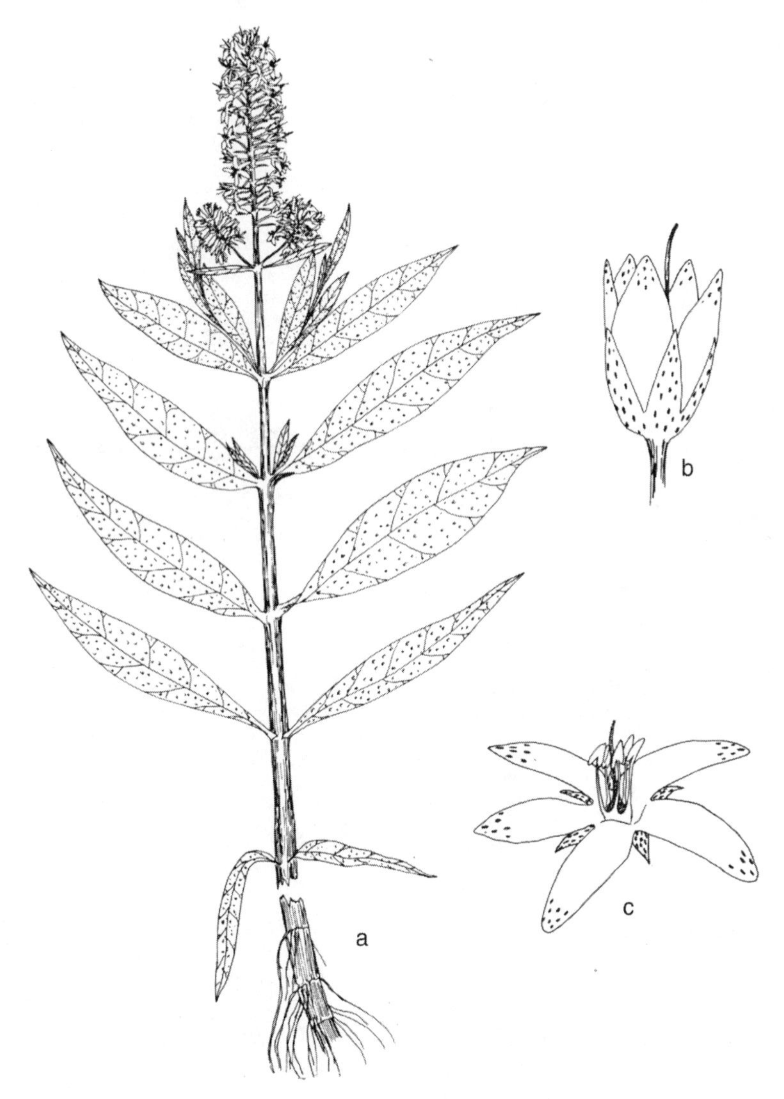

86. *Lysimachia X commixta* (Loosestrife). a. Habit. b, c. Flowers.

puberulent; capsules subglobose, 3–5 mm in diameter, with several seeds 1.5–2.3 mm long. June–September.

Wet soil, particularly along streams.

IL, KY (NI).

Fraser's loosestrife.

The differences between *L. fraseri* and *L. vulgaris* are not great, with the significant differences being in the shape of the corolla and the size of the seeds. Like *L. vulgaris, L. fraseri* is covered by a dense, glandular pubescence. *Lysimachia vulgaris*, a non-native species, does not occur in wetlands.

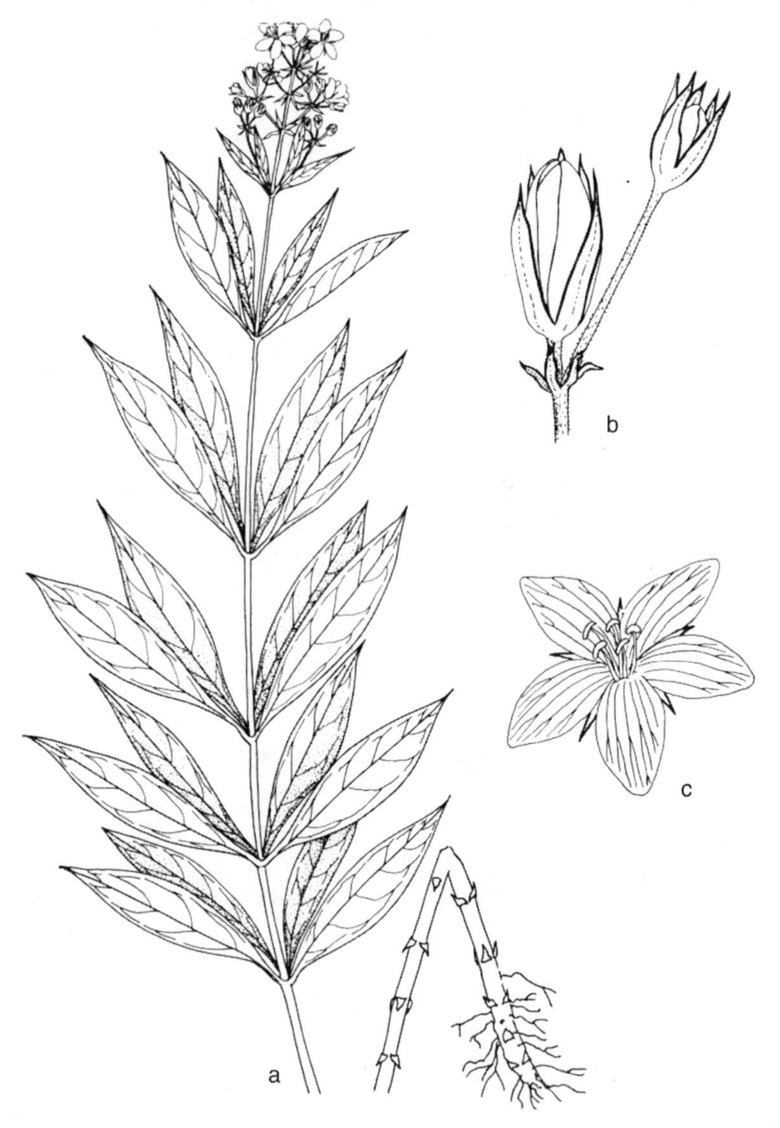

87. *Lysimachia fraseri* (Fraser's loosestrife). a. Habit. b. Capsules. c. Flower.

3. **Lysimachia nummularia** L. Sp. Pl. 148. 1753. Fig. 88.

Herbaceous perennial with fibrous roots; stems creeping, branched, glabrous, up to 60 cm long, sometimes rooting at the nodes; leaves opposite, suborbicular, obtuse at the apex, rounded at the base, to 3.5 cm across, entire, glabrous, reddish-punctate, on glabrous petioles up to 4 mm long; flower solitary from the medial axils, on a glabrous pedicel up to 5 mm long; calyx 5-cleft nearly to the base, the lobes ovate, acute to acuminate, glabrous, 5–8 mm long, more than half as wide; corolla yellow, open-campanulate, the deeply 5-cleft lobes obovate to oblanceolate, 1.0–1.5 cm long, about half as wide, ciliate, punctate or streaked; stamens 5, the filaments united at the base, glandular, unequal, the longest ones to 5 mm long; capsules apparently not formed in the United States. May–August.

Moist, shady areas.

Native of Europe; IA, IL, IN, MO (FACW+), KS, KY, NE, OH (OBL).

Moneywort.

Moneywort is a rather handsome ground cover that sometimes occurs in great abundance in moist, shaded areas.

It is readily distinguished by its creeping habit and suborbicular leaves.

Fruiting specimens are unknown from the United States.

88. *Lysimachia nummularia* (Moneywort). a. Habit. b. Flower.

4. **Lysimachia X producta** (Gray) Fern. Rhodora 1:134. 1899. Not illustrated.

Perennial herbs from rhizomes; stems erect, branched or unbranched, glabrous, to 1.2 m tall; leaves opposite or with a few or all in whorls of 3–5, lanceolate to oblong-lanceolate, acute to acuminate at the apex, tapering to the sessile or short-petiolate base, punctate, to 3.5 cm long, to 1.5 cm wide; flowers perfect, actinomorphic, borne in leafy bracted racemes, the pedicels slender, up to 3 cm long; calyx 5-parted, the lobes united near base, green, lanceolate to ovate-lanceolate, acute to acuminate at the apex, 3–5 mm long; corolla yellow, rotate, 1–2 cm across, deeply 5-cleft, the lobes ovate-oblong to ovate-lanceolate; stamens 5, the filaments united at the base; staminodia absent; capsules subglobose, half to nearly as long as the calyx lobes. June–August.

Wet soil.

IL, IN (NI), KY, OH (FAC).

Hybrid yellow loosestrife.

This plant is a hybrid between *L. terrestris* and *L. quadrifolia*, differing by its leafy bracted racemes. Some or all of the leaves may be whorled.

5. **Lysimachia terrestris** (L.) BSP. Prelim. Cat. N. Y. 34. 1888. Fig. 89.
Viscum terrestre L. Sp. Pl. 1023. 1753.
Lysimachia stricta Ait. Hort. Kew. 1:199. 1789.

Perennial herbs from rhizomes; stems erect, simple or branched, to nearly 1 m tall, glabrous; leaves opposite or nearly so, the lowermost scalelike, the middle and upper elliptic to lanceolate, acute to acuminate at the apex, tapering to the nearly sessile base, to 9 cm long, to 2 cm wide, glabrous, punctate, green above, glaucous below, entire; inflorescence a terminal raceme, rarely also with one or two small lateral racemes at the base of the terminal raceme; bracts linear-subulate, to 6 (–10) mm long; pedicels to 1.6 cm long, glabrous, spreading to ascending; calyx 5-cleft nearly to the base, the lobes lanceolate, acute to acuminate, glabrous, black-dotted or streaked, 2–4 mm long; corolla yellow, rotate, deeply 5-cleft, the lobes elliptic to oblong, more or less obtuse, black-streaked as well as yellow-glandular, 4–6 mm long, less than half as wide; stamens 5, the filaments united at the base into a yellow-glandular tube up to 1 mm long; staminodia absent; capsules subglobose, to 3–5 mm in diameter, with few nearly trigonous seeds about 1.3 mm long. June–August.

Swamps; bogs; open marshes.

IA, IL, IN, KY, OH (OBL).

Swamp candles; swamp loosestrife.

Specimens that have only the terminal racemes are readily distinguished from other members of the genus. There is a tendency for short lateral racemes to be formed near the base of the terminal raceme. Specimens with these small lateral racemes occasionally are referred to as *L.* X *commixta*, the reputed hybrid between *L. terrestris* and *L. thyrsiflora* (=*Naumbergia thyrsiflora*). In this work, however, I am considering those plants with merely one or two short lateral racemes next to the terminal raceme merely variants of *L. terrestris*. *Lysimachia* X *commixta* is here defined to include those plants with several elongated lateral racemes, as well as with a more funnelform corolla and a nearly glabrous staminal tube.

There occasionally will be found in *L. terrestris* moniliform, axillary bulblets in place of fertile flowers. It was a specimen such as this that Linnaeus had in hand when he described this species as *Viscum terrestre*, which he thought to be a parasitic member of the mistletoe family!

4. **Naumbergia** Moench

Herbaceous perennial; leaves simple, opposite or occasionally whorled, entire, punctate; flowers actinomorphic, perfect, in dense axillary racemes; calyx 5-parted, the lobes united near the base; corolla 5-parted, yellow, united at the base; stamens 5, with several staminodia; ovary superior; capsules few-seeded.

There is only one species in this genus. It is often placed in the genus *Lysimachia*.

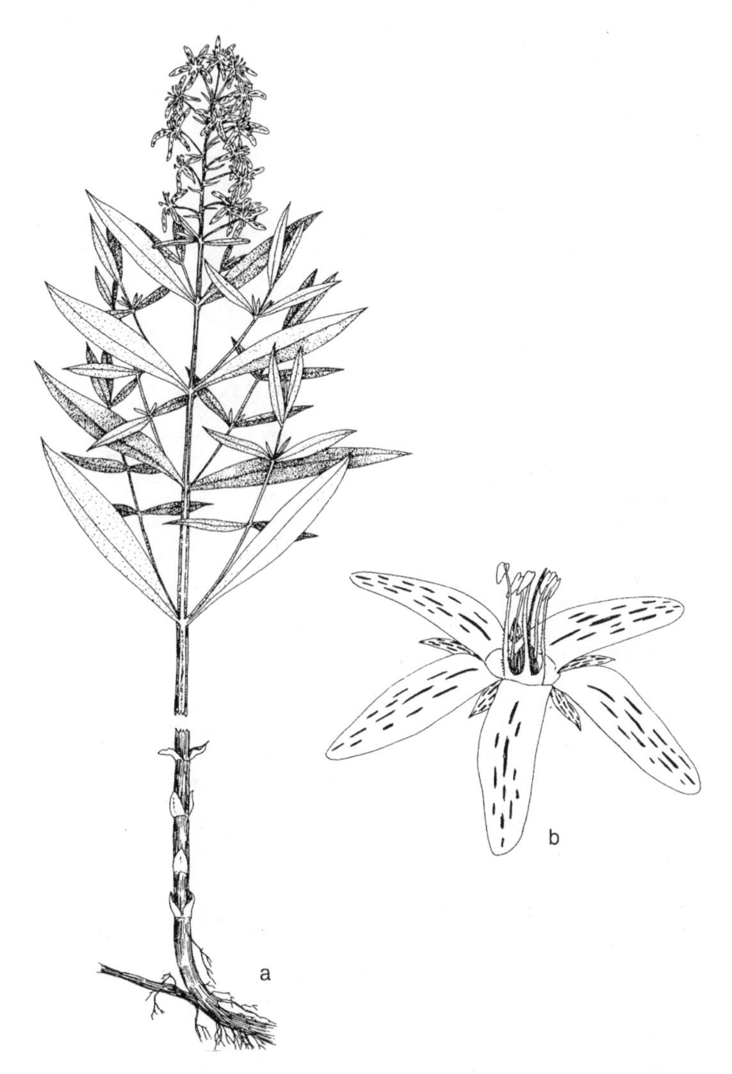

89. *Lysimachia terrestris* (Swamp candles). a. Habit. b. Flower.

1. **Naumbergia thyrsiflora** (L.) Reichenb. Fl. Germ. Exc. 410. 1831. Fig. 90.
Lysimachia thyrsiflora L. Sp. Pl. 147. 1753.

90. *Naumbergia thyrsiflora* (Tufted loosestrife). Habit. Flower (upper right).

Herbaceous perennial from rhizomes; stems erect, simple, angular above, glandular-punctate, often puberulent in the upper half, otherwise glabrous, to 75 cm tall; leaves opposite or occasionally whorled, lanceolate to elliptic, acute to acuminate at the apex, tapering to the nearly sessile base, to 15 cm long, to 6 cm wide, entire or sinuate, black-punctate, glabrous above, glabrous or sparsely villous beneath, the lowermost leaves reduced to small, broadly ovate scales; flowers in dense racemes to 4 cm long, from the axils of the middle leaves, on glabrous or villous peduncles to 3 cm long; calyx 5- to 7-cleft nearly to the base, the lobes linear-subulate, attenuate, black-punctate, to 3 mm long, to 1 mm wide, glabrous; corolla yellow or cream, black-punctate, short-funnelform, the 5–7 lobes cleft nearly to the base, the lobes linear, obtuse to subacute, to 4 mm long, 1–2 mm wide; stamens 5–7, the filaments barely united at the base, glabrous, 4–8 mm long, exserted; capsule subglobose, 2–3 mm in diameter, black-punctate, with few, triangular seeds about 1 mm long. May–July.

Swamps, bogs, bottomlands forests.

IA, IL, IN, KS, MO, NE, OH (OBL), as *Lysimachia terrestris.*

Tufted loosestrife.

This species is quite distinct from species of *Lysimachia* and *Steironema* by its dense, axillary racemes of small flowers. It is the only member of *Naumbergia*.

In addition to the characters of the inflorescence and flowers, this species is further recognized by the dense black punctations on its stems, leaves, calyx, corolla, and capsules.

5. **Samolus** L.—Brookweed

Perennial, usually glabrous, herbs from fibrous roots; leaves cauline and also often basal, simple, entire; flowers in terminal panicles or racemes, sometimes bracteate; calyx 5-lobed, its tube adnate to the ovary; corolla campanulate, 5-lobed; stamens 5, inserted on the corolla, sometimes alternating with 5 staminodia; ovary subinferior or superior, with numerous ovules; fruit a capsule, splitting near the tip into 5 valves.

Samolus is a genus of about ten species, distributed throughout most of the temperate and subtropical regions of the world.

1. Leaves decurrent at base, glaucous; pedicels glandular-stipitate; corolla 4–6 mm across; capsules 3–4 mm in diameter........1. *S. ebracteatus*
1. Leaves not decurrent at base, not glaucous; pedicels eglandular; corolla to 2 mm across; capsules up to 2 mm in diameter........2. *S. parviflorus*

1. **Samolus ebracteatus** Kunth, Nov. Gen. Sp. 2:223. 1817. Fig. 91.
Samolus cuneatus Small, Bull. Torrey Club 24:491. 1897.
Samolus ebracteatus Kunth var. *cuneatus* (Small) Henrickson, Southw. Nat. 29:311. 1983.

Fleshy perennial herb; stems decumbent to ascending, sparingly branched or unbranched, glabrous, to 60 cm tall; leaves opposite, simple, oblanceolate to obovate to spatulate, rounded but sometimes mucronate at the apex, decurrent into broad wings at the base, entire, to 15 cm long, to 6 cm wide, glabrous, glaucous; flowers in racemes up to 3 cm long, on long, glandular-hairy peduncles, with pedicels glandular-stipitate, 1–3 cm long; calyx campanulate, the 5 lobes longer than the tube, green or purple-tinged; corolla white, 4–6 mm across, the 5 lobes as long as or shorter than the tube; stamens 5, without staminodia; ovary superior; capsules more or less globose, 3–4 mm in diameter, with many seeds. May–October.

Marshes, along rivers and streams.

KS (NI).

Water pimpernel.

This primarily southern species has flowers nearly twice as wide as the flowers in *S. parviflorus*. The decurrent, winged leaf bases are distinctive.

2. **Samolus parviflorus** Raf. Am. Month. Mag. 2:176. Jan., 1818. Fig. 92.
Samolus floribundus HBK. Nov. Gen. 2:224. Feb., 1818.
Samolus valerandi L. var. *americanus* Gray, Man. Bot. ed. 2, 274. 1856.
Samolus valerandi L. var. *floribundus* (HBK.) R. Knuth in Engl. Pflanzenr. 237:338. 1905.

Perennial herb from fibrous roots; stems ascending to erect, usually branched, slightly angular, glabrous, to 30 (–40) cm tall; leaves basal and cauline, membranaceous, obovate to elliptic, obtuse to subacute at the apex, narrowed at the base, glabrous, sometimes pellucid-punctate, to 5 cm long, to 3 cm broad, on glabrous petioles 1–2 cm long; inflorescence a raceme or panicle to 10 cm long; pedicels slender, to 1 (–2) cm long, usually bearing a linear bract about midway, the bract up to 1.5 mm long; calyx urceolate, to 2 mm long, with acute lobes up to 0.5 mm long, the lobes becoming cartilaginous and reflexed in fruit; corolla white, 5-lobed, to 2.5

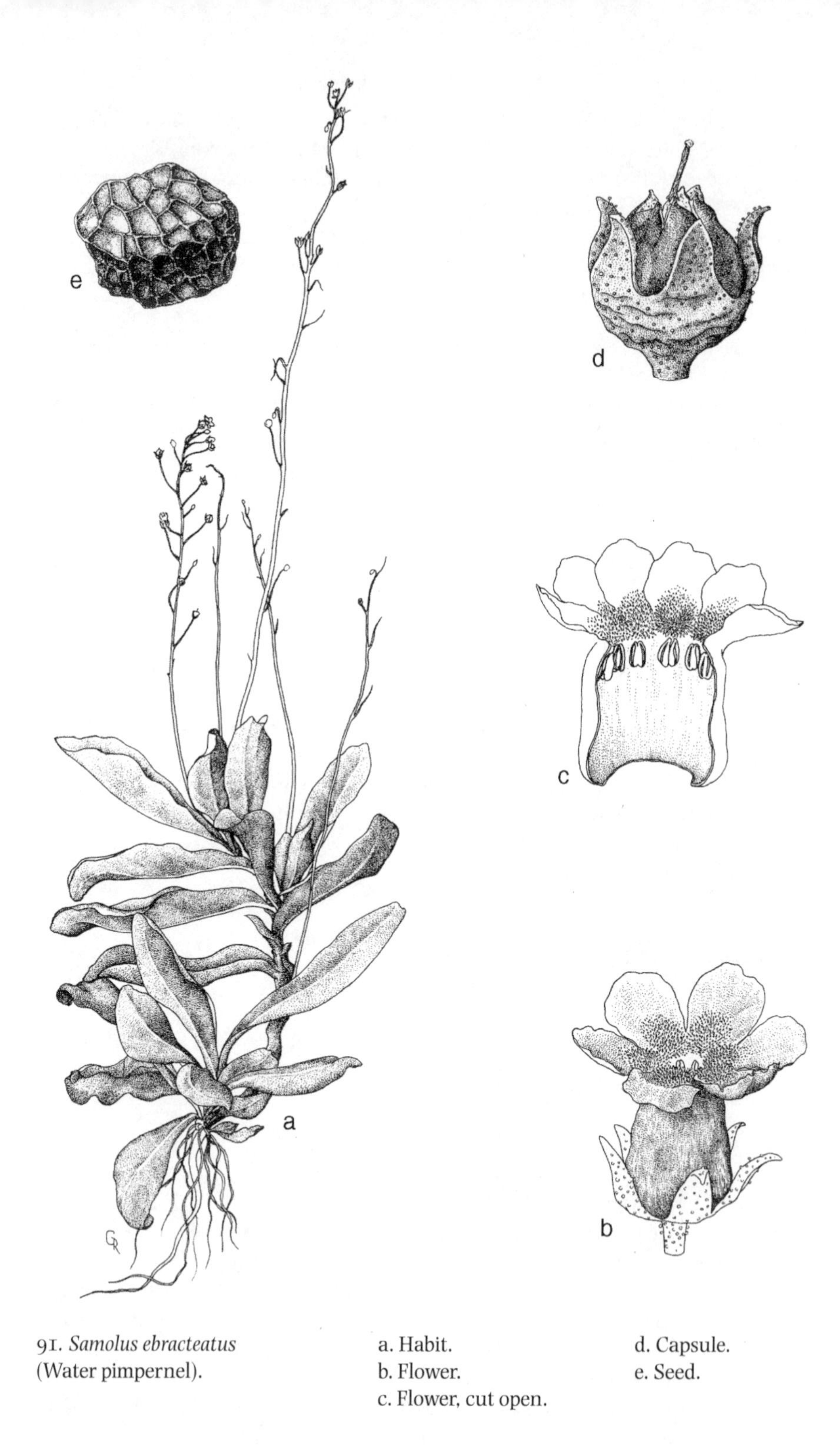

91. *Samolus ebracteatus* (Water pimpernel).

a. Habit.
b. Flower.
c. Flower, cut open.
d. Capsule.
e. Seed.

mm long; stamens included; capsules globose, glabrous, to 2 mm in diameter, the valves spreading after dehiscence; seeds numerous. May–September.

Moist soil.

IL, IN, KY, MO, NE, OH (OBL).

Brookweed.

Brookweed is a fairly common species of moist, usually shaded soil. It differs from all other members of the Primulaceae in the central Midwest by its subinferior ovary.

Great variation is found in the stature of the plants. Slender, delicate specimens less than 6 cm tall have been found in flower, as have robust plants 30 cm tall. In general, the larger specimens tend to be found in open areas, while the dwarfed ones often are heavily sheltered by overhanging cliffs or a dense canopy.

Flowering occurs from mid-May through most of September. The small, rounded fruits shed their many seeds from late June through October.

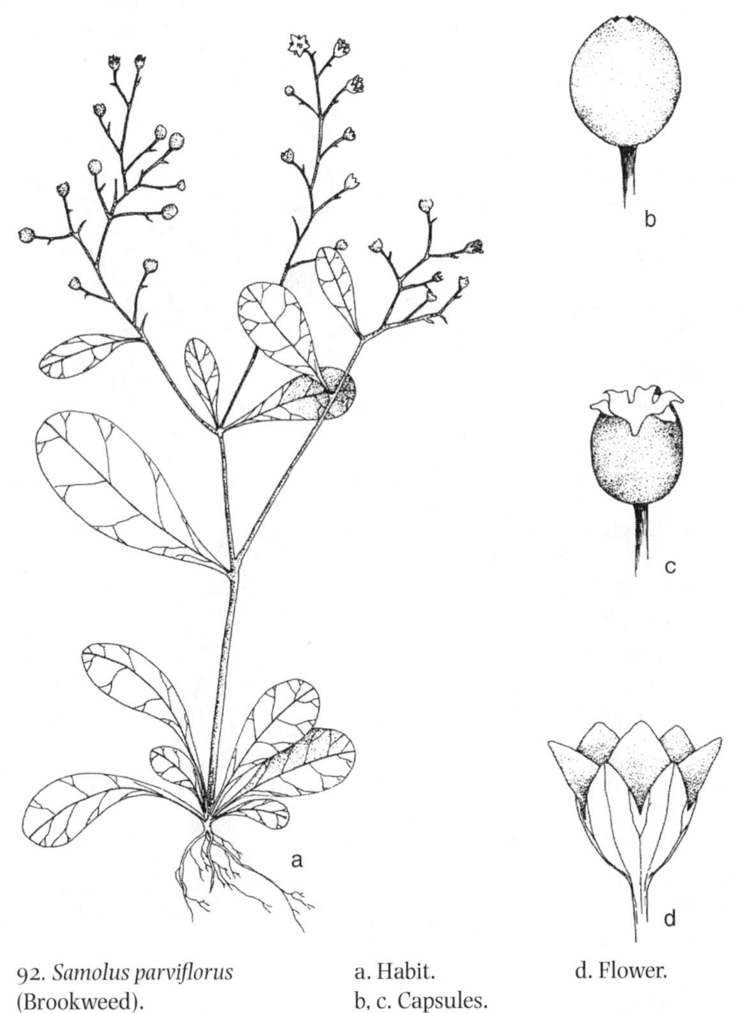

92. *Samolus parviflorus* (Brookweed). a. Habit. b, c. Capsules. d. Flower.

6. **Steironema** Raf.—Yellow Loosestrife

1. Lateral nerves of blades obscure; blades firm .... 4. *S. quadriflorum*
1. Lateral nerves of blades conspicuous; blades thin.
   2. Plants decumbent, rooting at the nodes .... 5. *S. radicans*
   2. Plants erect, not rooting at the nodes.
      3. Leaves ovate to ovate-lanceolate, rounded or subcordate at the base; petioles strongly ciliate.. 1. *S. ciliatum*
      3. Leaves narrowly lanceolate, elliptic, or linear, tapering at the base (rarely sometimes rounded); petioles not usually ciliate.
         4. Basal rosettes developing from slender rhizomes; leaves pale beneath.. 3. *S. lanceolatum*
         4. Basal rosettes not developing from slender rhizomes; leaves green beneath.. 2. *S. hybridum*

1. **Steironema ciliatum** (L.) Raf. Am. Gen. Sc. Phys. 7:193. 1820. Fig. 93.
Lysimachia ciliata L. Sp. Pl. 147. 1753.

Herbaceous perennial from few, slender rhizomes; stems erect, sometimes branched, usually angular above, glabrous except for minute glandular pubescence at some of the nodes, to 1 m tall; cauline leaves opposite, ovate to ovate-lanceolate, to 15 cm long, to 6.5 cm wide, acute to acuminate at the apex, subcordate to rounded to truncate at the base, entire except for cilia on the margins, otherwise glabrous, green above, paler below, the petioles winged, strongly ciliate, to 5 cm long; flower solitary in the upper axils, on minutely glandular-puberulent pedicels to 5 cm long; calyx 5-cleft nearly to the base, the lobes lanceolate, acuminate, minutely glandular, 4–8 mm long, less than half as wide; corolla yellow, rotate, the lobes free nearly to the base, obovate, to 12 mm long, to 9 mm wide, with yellow glands on the inner

93. *Steironema ciliatum* (Fringed yellow loosestrife). Habit. Flower (lower left).

surface; stamens 5, with the filaments minutely puberulent, 2–3 mm long; capsules subglobose, 3.5–5.5 mm in diameter, with several triangular seeds about 2 mm long. June–August.

Moist woods, bottomlands, thickets.

IA, IL, IN, KS, KY, MO, NE, OH (FACW), as *Lysimachia ciliata.*

Fringed yellow loosestrife.

The fringed yellow loosestrife is perhaps the most common species of *Steironema* in the central Midwest. Although it may sometimes be confused with *S. lanceolata*, it is distinguished by its broader leaves and its rounded, subcordate, or truncate leaf bases.

The presence of conspicuous cilia on the petioles is often not sufficient enough to distinguish *S. ciliata* since both *S. lanceolata* and *S. hybrida* usually have ciliate petioles. In these latter two species, however, the cilia are generally confined to near the base of the petiole.

This species is often placed in the genus *Lysimachia.*

2. **Steironema hybridum** (Michx.) Raf. ex Jackson. Index Kew. 2:985. 1895. Fig. 94.
Lysimachia hybrida Michx. Fl. Bor. Am. 1:126. 1803.
Lysimachia lanceolata Walt. var. *hybrida* (Michx.) Gray, Man. Bot., ed. 1, 283. 1848.
Steironema lanceolatum (Walt.) Gray var. *hybridum* (Michx.) Gray, Proc. Am. Acad. 12:63. 1876.
Lysimachia lanceolata Walt. ssp. *hybrida* (Michx.) J.D. Ray, Ill. Biol. Mon. 24:39. 1956.

Herbaceous perennial without slender rhizomes; stems erect or rarely reclining, simple or branched, usually angular above, nearly glabrous, to 1 m tall; basal leaves rosulate, absent at flowering time, oblong to ovate, glabrous or nearly so, on ciliate petioles longer than the blades; cauline leaves opposite, linear to lanceolate to elliptic, to 10 cm long, to 3 cm wide, acute to acuminate at the apex, more or less tapering to the base, entire, glabrous except for some cilia near base, green above and below, sessile or on ciliate petioles to 3.5 cm long; flowers mostly solitary from the upper axils, on minutely glandular-hairy pedicels to 4 cm long; calyx 5-cleft nearly to the base, the lobes lanceolate, acuminate, glabrous, 4–7 mm long, less than half as wide; corolla yellow, rotate, the lobes free nearly to the base, obovate to suborbicular, to 9 mm long, to 7 mm wide, with yellow glands on the inner surface; stamens 5, with the filaments usually minutely glandular, 2–3 mm long; capsules subglobose, 3.5–4.5 mm in diameter, with several triangular seeds 1.2–1.8 mm long. June–September.

Wet soil, often in swampy woods.

IA, IL, IN, KS, KY, MO, NE (OBL), as *Lysimachia hybrida.*

Yellow loosestrife.

Steironema hybridum is considered to be a distinct species, differing from the similar *S. lanceolatum* by the lack of slender rhizomes and by the leaves, which are green on both surfaces. The basal rosette of leaves usually withers by flowering time in *S. hybridum*, apparently due to its frequent inundation in early spring. Cilia on the petioles are usually confined to the lower part of the petioles.

There is a tendency in some specimens of *S. hybridum* to assume a reclining position.

This species is often place in the genus *Lysimachia.*

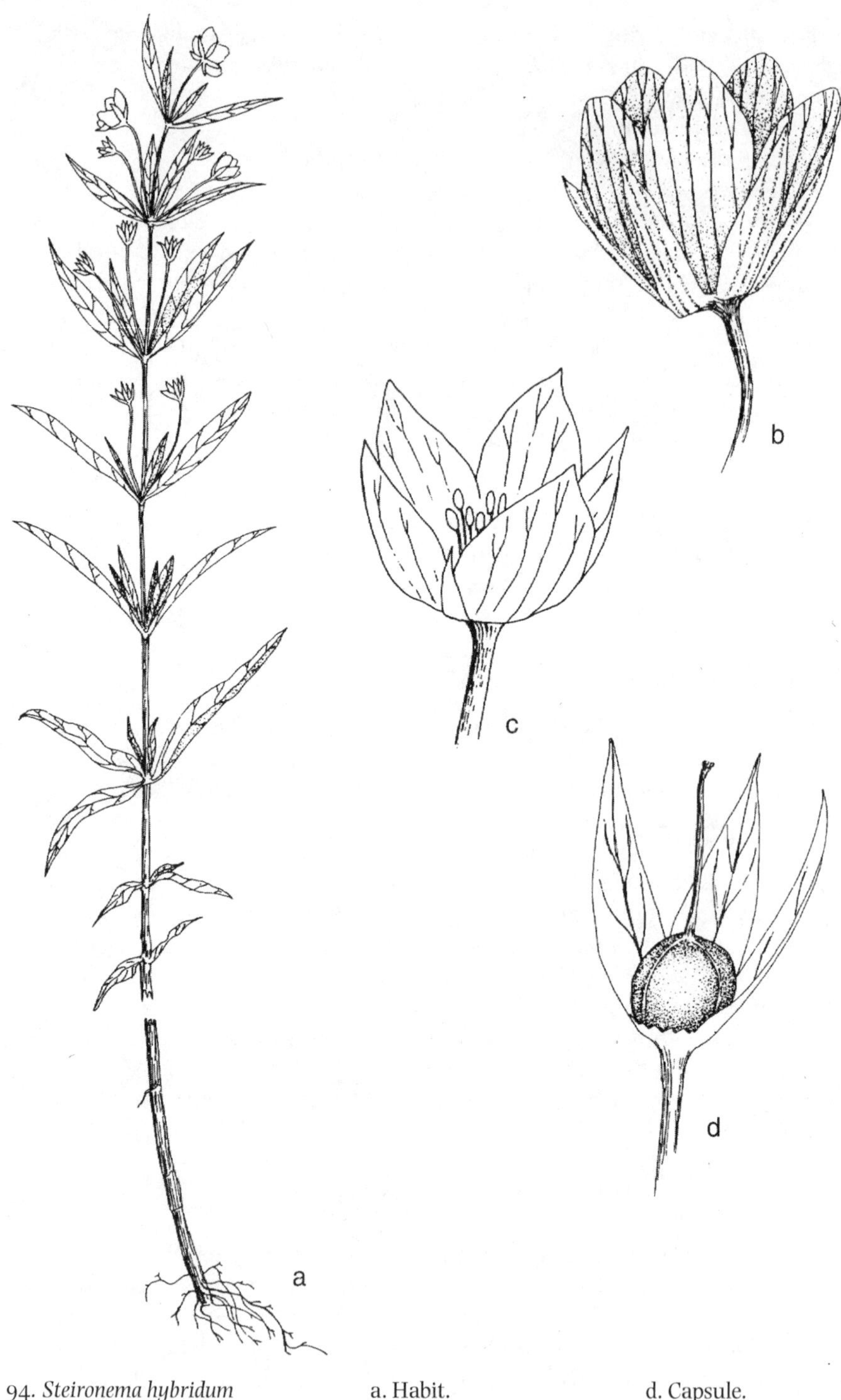

94. *Steironema hybridum* (Yellow loosestrife).

a. Habit.
b, c. Flowers.
d. Capsule.

3. **Steironema lanceolatum** (Walt.) Gray, Proc. Am. Acad. 12:63. 1877. Fig. 95.
Lysimachia lanceolata Walt. Fl. Carol. 92. 1788.
Lysimachia angustifolia Lam. Tabl. Encycl. 1:440. 1792.
Lysimachia heterophylla Michx. Fl. Bor. Am. 1:127. 1803.
Lysimachia lanceolata Walt. var. *angustifolia* (Lam.) Gray, Man. Bot., ed. 2, 273. 1856.

Herbaceous perennial from few, slender rhizomes; stems erect, sometimes branched, usually angular above, glabrous or very sparsely glandular-puberulent, to 75 cm tall; basal leaves rosulate, broadly elliptic, glabrous or nearly so, on petioles mostly shorter than the blades; cauline leaves opposite, narrowly elliptic to lanceolate, to 18 cm long, to 3.5 cm wide, acute to sub-acuminate at the apex, rarely obtuse, more or less tapering to the sessile or short-petiolate base, entire except for ciliation near the base, otherwise glabrous, green above, paler below; flowers mostly solitary from the upper axils, on essentially glabrous pedicels up to 5 cm long; calyx 5-cleft nearly to the base, the lobes lanceolate, acuminate, glabrous, 4–8 mm long, less than half as wide; corolla yellow, rotate, the lobes free nearly to the base, obovate, to 8 mm long, to 6 mm wide, with yellow glands on the inner surface; stamens 5, with the filaments usually minutely glandular, 2–3 mm long; capsules subglobose, 3–4 mm in diameter, with several triangular seeds 1.7–2.0 mm long. May–September.

Moist woods, streambanks, thickets.

IA, IL, IN, KY, MO, OH (FAC), as *Lysimachia lanceolata*.

Yellow loosestrife.

Steironema lanceolatum is segregated from *S. hybridum* on the basis of its slender rhizomes and its paler lower leaf surfaces. From *S. ciliatum* it differs by its cuneate leaf base and generally narrower leaves. The cilia on the petioles are usually confined to the base of the petioles.

This species is often place in the genus *Lysimachia*.

4. **Steironema quadriflorum** (Sims) Hitchc. Trans. Acad. Sci. St. Louis 5: 506. 1891. Fig. 96.
Lysimachia quadriflora Sims, Bot. Mag. 17:pl. 660. 1803.
Lysimachia longifolia Pursh. Fl. Am. Sept. 1:135. 1814.
Lysimachia revoluta Nutt. Gen. N. Am. Pl. 1:122. 1818.
Steironema longifolium (Pursh) Raf. ex Gray, Proc. Am. Acad. 12:63. 1876.

Herbaceous perennial from few, slender rhizomes; stems erect, sometimes branched, 4-angled, glabrous, to 90 cm tall; basal leaves rosulate, elliptic to obovate, long-petiolate, to 3 cm long, to 1 cm wide, usually withering by flowering time; cauline leaves opposite, linear, to 9 cm long, to 6 mm wide, acute to obtuse and sometimes mucronate at the apex, narrowed to the sessile base, entire but revolute along the margins, glabrous except at the ciliate base, sometimes with fascicles of smaller leaves in the axils; flowers in clusters of 1–4 in the upper axils, on glabrous pedicels to 3 cm long; calyx 5-cleft nearly to the base, the lobes lanceolate, acuminate, glabrous, 4–6 mm long, 1.5–2.0 mm wide; corolla yellow, rotate, the lobes free nearly to the base, obovate, to 12 mm long, to 9 mm wide, with gland-tipped hairs on the inner surface; stamens 5, with the filaments minutely

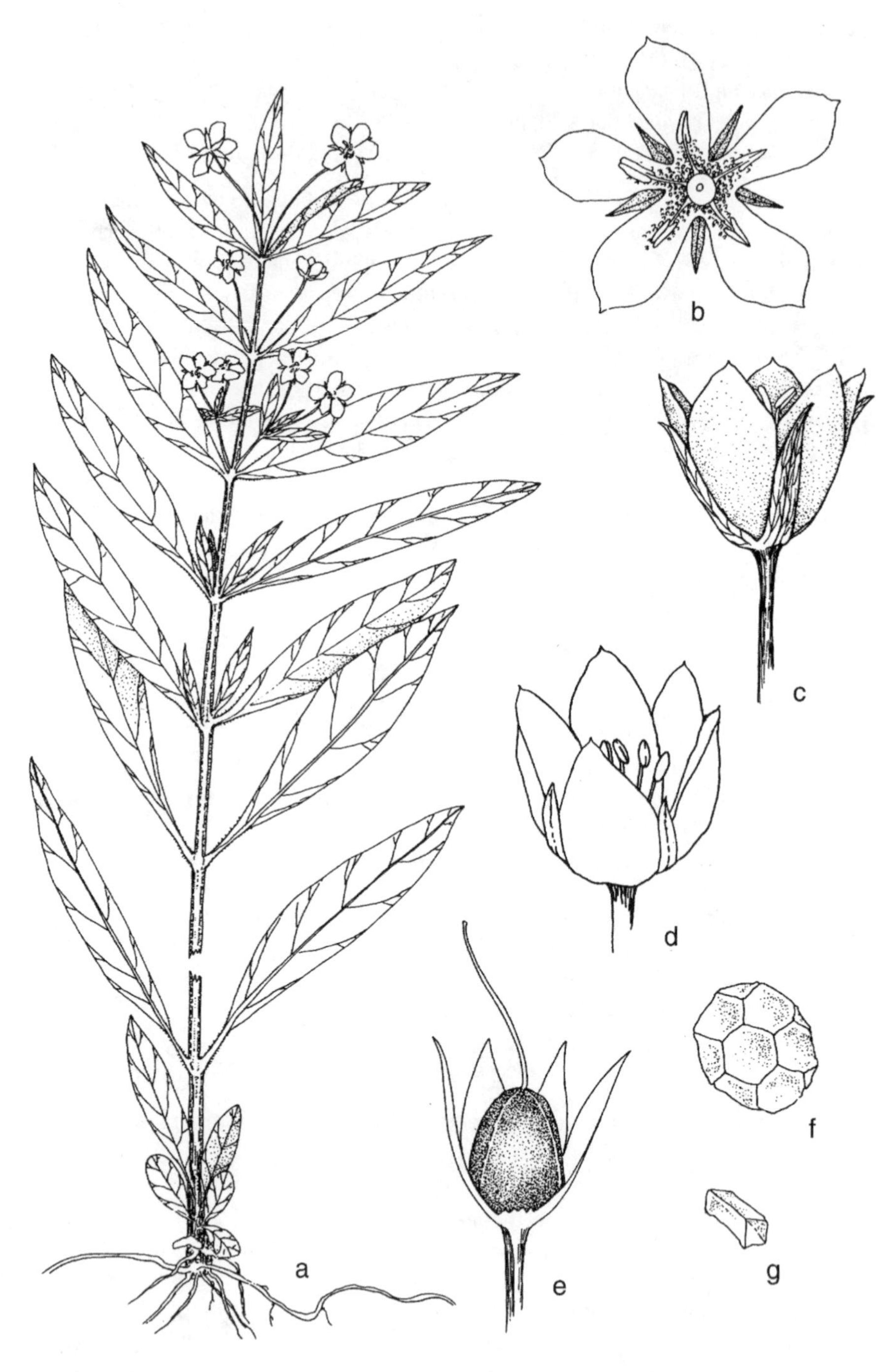

95. *Steironema lanceolatum* (Yellow loosestrife).

a. Habit.
b, c, d. Flowers.
e. Capsule.

f. Capsule.
g. Seed.

pubescent, about 2 mm long; capsule subglobose, 3–4 mm in diameter, with several flattish seeds 1.0–1.2 mm long. June–August.

Marshes, moist prairies, bogs.

IA, IL, IN, MO (OBL), KY, OH (FACW+).

Narrow-leaved loosestrife.

The narrow cauline leaves, which do not exceed a width of 6 mm, is distinctive for this species.

96. *Steironema quadriflorum* (Narrow-leaved loosestrife).
Habit. Flower (lower right).

5. **Steironema radicans** (Hook.) Gray, Proc. Am. Acad. 12:63. 1876. Fig. 97.
Lysimachia radicans Hook. Comp. Bot. Mag. 1:177. 1836.

Herbaceous perennial; stems weak, reclining, to 1 m tall, rooting at the nodes, somewhat 4-angular, glabrous or minutely pubescent; basal leaves rosulate, ovate to lanceolate, petiolate, to 3 cm long, to 1 cm wide, usually withering by flowering time; cauline leaves opposite, lanceolate to ovate, to 9 cm long, to 3 cm wide, acuminate at the apex, rounded at the base, glabrous, entire but with minute papillae along the margins, the petioles 1–4 cm long, ciliate near the base; flowers solitary in axils of the smaller, upper leaves, on glabrous pedicels to 2.5 cm long; calyx 5-cleft nearly to the base, the lobes lanceolate, acuminate, glabrous, 3–4 mm long, up to half as wide; corolla yellow, rotate, the lobes free nearly to the base, obovate, 3–5 mm long, 2–3 mm wide, with gland-tipped hairs on the inner surface; stamens 5, with the filaments minutely pubescent, up to 2 mm long; capsules subglobose, about 3 mm in diameter, with several triangular seeds 1.0–1.5 mm long. June–August.

Swampy woods.

IL, KY, MO (OBL), as *Lysimachia radicans*.

Creeping loosestrife.

The arching or reclining stems of this species are distinctive, although some specimens of *S. hybridum* may become somewhat reclining. *Steironema radicans* may be distinguished from *S. hybridum* by its broader leaves.

97. *Steironema radicans* (Creeping loosestrife). a. Habit. b. Flower. c. Capsule.

102. RANUNCULACEAE—BUTTERCUP FAMILY

Chiefly perennial herbs (vines in some *Clematis*), simple or compound; stipules absent; flowers usually perfect, actinomorphic or zygomorphic; sepals 3–15, free, often petallike; petals up to 15, or absent, usually free; stamens numerous, spirally arranged, free; pistils few to several, rarely 1, the ovary superior, 1–locular; fruit an achene, follicle, or berry.

This family contains about forty genera and approximately fifteen hundred species, mostly in north temperate regions. Certain genera, such as *Hydrastis*, show a striking similarity to the Berberidaceae.

1. Leaves evergreen, basal, ternately divided; petals with a hollow center.................... 4. *Coptis*
1. Leaves deciduous, cauline or, if basal, not ternately divided; petals, when present, without a hollow center.
  2. Vines; leaves opposite.. 3. *Clematis*
  2. Stems upright, not climbing; leaves various, but not all opposite.
    3. Leaves all basal.
      4. Leaves nearly suborbicular, crenate, cordate; flowers yellow .... 2. *Caltha*
      4. Leaves linear, entire, tapering at the base; flowers greenish...................... 5. *Myosurus*
    3. Leaves cauline as well as sometimes basal, usually not cordate.
      5. Cauline leaves opposite .... 1. *Anemone*
      5. Cauline leaves alternate.
        6. Leaves simple.
          7. Flowers white... 8. *Trauttvetteria*
          7. Flowers yellow.. 6. *Ranunculus*
        6. Leaves compound.
          8. Flowers yellow.. 6. *Ranunculus*
          8. Flowers white.
            9. Plants aquatic; flowers perfect.. 6. *Ranunculus*
            9. Plants not aquatic; flowers unisexual .... 7. *Thalictrum*

1. **Anemone** L.—Anemone

Perennial herbs with slender or tuberlike roots or rhizomes; leaves all basal, palmately lobed or divided; flowers solitary or in umbels, borne from a whorl of involucral leaves; sepals 4–numerous, free, petallike; petals absent; stamens numerous, free; pistils 2–several, free, the ovaries superior, each with a single ovule; fruit a cluster of achenes.

Anemone is a genus of nearly one hundred species in the temperate and subarctic regions of the Northern and Southern Hemispheres.

1. **Anemone canadensis** L. Syst. ed. 12, 3:App. 231. 1768. Fig. 98.
Anemone pensylvanica L. Mant. 2:247. 1771.

Herbaceous perennial from slender rhizomes; basal leaves palmately 3- to 7-divided, the divisions usually 3-cleft and coarsely toothed, conspicuously reticulate-veined, pilose on the veins beneath, to 15 cm long, about as wide, on long, sparsely hairy petioles; flowering stem stout, to 70 cm tall, sparsely pubescent, with both primary and secondary involucres, the involucral leaves opposite, similar to the basal leaves, except sessile; flowers 1–6, to 3 cm broad, on a more or less pubescent peduncle; sepals 5, white, oblong to obovate, to 2.5 cm long; petals absent; stamens numerous, all anther-bearing; pistils numerous, pubescent; fruiting heads globose, to 1.5 cm in diameter, with numerous achenes, the achenes nearly orbicular, subulate-beaked, pubescent, flat, up to 6 mm across, the beak 2–5 mm long. May–July.

Wet ground, wet prairies.

IA, IL, IN, KS, KY, MO, NE, OH (FACW).

Meadow anemone.

The meadow anemone is distinguished by its sessile involucral leaves and its subulate-beaked achenes.

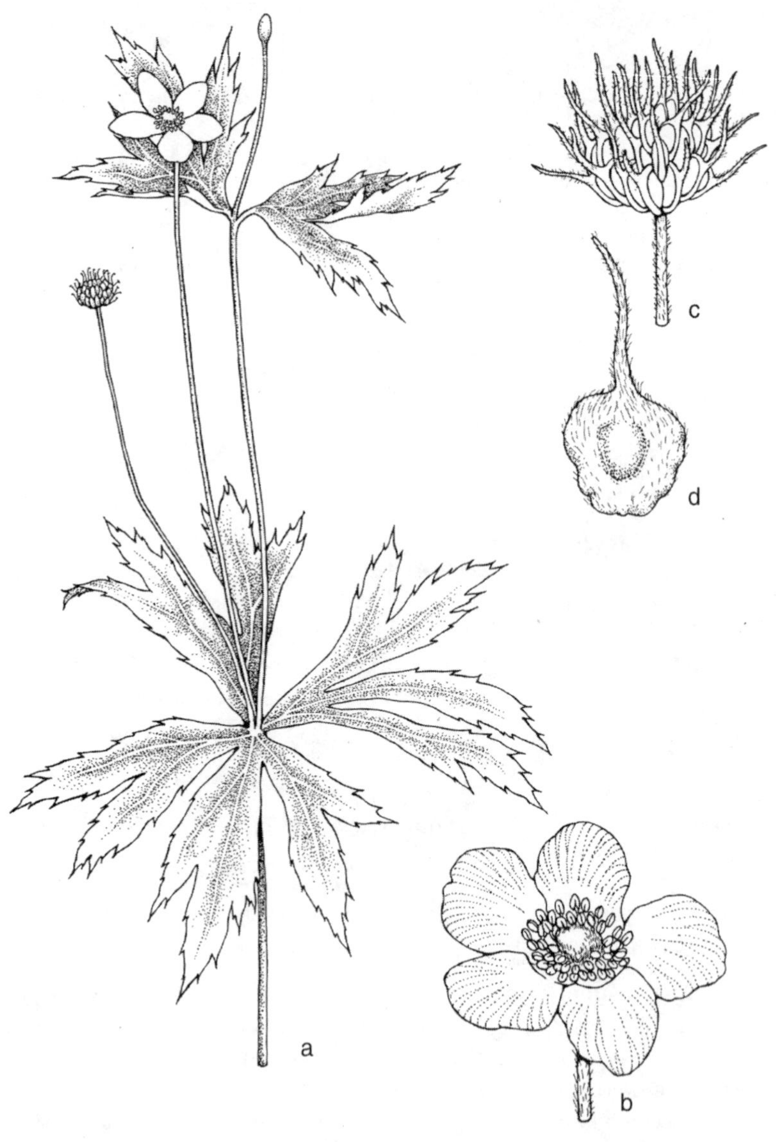

98. *Anemone canadensis* (Meadow anemone). a. Upper part of plant. b. Flower. c. Fruiting head. d. Achene.

2. **Caltha** L.—Marsh Marigold

Perennial herbs; stems floating, creeping, erect, or ascending, sometimes hollow; leaves basal and cauline, alternate, simple, entire or toothed; flowers yellow, white, or pink, usually solitary from the upper axils, on stout peduncles; sepals 5–9, free, petallike, deciduous; petals absent; stamens numerous, free; pistils 3–15, free, with several ovules in two rows; fruit a cluster of many-seeded follicles.

This is a genus of about fifteen wet ground or aquatic species, found primarily in the temperate and arctic regions of the world.

Several of the primitive characters of the family Ranunculaceae are exhibited well by this genus, such as free flower parts, numerous stamens and pistils, petallike sepals, and follicular fruits.

1. **Caltha palustris** L. Sp. Pl. 558. 1753. Fig. 99.

Rather stout perennial from fibrous roots; stems ascending to erect, to 75 cm tall, glabrous, hollow, shallowly furrowed; basal leaves suborbicular to reniform, obtuse at the apex, cordate at the base, entire to dentate, glabrous, to 15 cm long and wide, the glabrous petioles longer than the blades; upper leaves similar but subcordate to truncate at the base, on very short petioles or sessile; flowers yellow, to 3.4 cm across, solitary from the axils, on stout, glabrous peduncles to 10 cm long; sepals 5–9, yellow, oval, obtuse, to 2.5 cm long, to 2 cm broad; petals none; stamens numerous; pistils 3–12; follicles compressed, recurved-ascending, glabrous, to 1.5 cm long, prominently beaked, many-seeded. March–June.

Wet meadows, moist woods, springy calcareous fens.

IA, IL, IN, KY, MO, NE, OH (OBL).

Marsh marigold.

The marsh marigold is readily distinguished by the simple, cordate leaves and the bright yellow, solitary flowers on stout axillary peduncles.

Pioneers in bygone days used the young shoots of this plant, known to them as cowslip, for a spring vegetable.

3. **Clematis** L.—Clematis

Climbing vines or erect perennial herbs; leaves opposite, simple or pinnately compound; flowers in cymose panicles or solitary, bisexual, or unisexual; sepals 4–5, free, petal-like; petals absent; stamens numerous, free; pistils numerous, free; fruit a head of achenes with persistent, plumose styles.

Clematis is a genus of about fifty species, occurring primarily in the Northern Hemisphere of both the Old and New Worlds.

Some botanists choose to segregate those erect species with solitary flowers into the genus *Viorna*.

1. Stems terete; leaves glabrous on both sides; tails of fruits not plumose; upper one-half of the sepals recurved .... 1. *C. crispa*
1. Stems more or less 6-angular; leaves puberulent beneath; tails of the fruits densely plumose; only the tips of the sepals recurved .... 2. *C. viorna*

1. **Clematis crispa** L. Sp. Pl. 543. 1753. Fig. 100.
Clematis simsii Sweet, Hort. Brit. 1. 1826.
Viorna crispa (L.) Small, Fl. S.E. U.S. 437. 1903.

Perennial climbing herb; stems more or less terete, glabrous, to 4 m long; leaves 3- to 9-foliolate, rarely simple on the upper branches; leaflets rather thin, entire or 3-lobed, ovate to lanceolate, acute, subcordate to cuneate at the base, not promi-

99. *Caltha palustris* (Marsh marigold). a. Habit. b. Flower. c. Fruit. d. Achene.

nently reticulate, glabrous; peduncles usually bractless, with 1 nodding flower; sepals 4, forming a campanulate calyx, rather thin, bluish purple, oblong-lanceolate, undulate, 2.5–4.0 cm long, acuminate, recurved in their upper half; petals absent; stamens numerous, the anthers and filaments about equal in length; achenes ovoid, brownish, strigose, the persistent styles silky but not plumose, to 2.5 cm long. April–July.

Wet woods.

IL, MO (OBL), KY (FACW).

Blue jasmine.

This southern species is the only solitary-flowered *Clematis* with thin sepals. It is similar to *C. pitcheri,* a non-wetland species, because of its nonplumose persistent styles, but *C. pitcheri* has leathery sepals.

It differs from *C. viorna* by its densely plumose fruits.

100. *Clematis crispa* (Blue jasmine). Habit, with flower. Achene (upper right).

2. **Clematis viorna** L. Sp. Pl. 543. 1753. Fig. 101.
Viorna viorna (L.) Small, Fl. S.E. U.S. 439. 1903.
Viorna ridgwayi Standl. Smithson. Misc. Coll. 56:2, pl. 1. 1912.

Perennial climbing herb; stems 6-angled, glabrous except at the nodes, to 4 m long; leaves pinnately compound, or sometimes simple on the upper branches; leaflets rather thin, ovate to lanceolate to broadly elliptic, acute to acuminate, cuneate to rounded at the base, entire, glabrous above, puberulent beneath; peduncles usually with 2 bracts and 1 nodding flower; sepals 4, forming a campanulate calyx, very thick, purple, oblong-lanceolate, to 2.5 cm long, the tips acuminate, recurved; petals absent; stamens numerous, the anthers and filaments about equal in length; achenes ovoid, brownish, strigose, the persistent style plumose, to 2.5 cm long. May–July.

101. *Clematis viorna* (Leatherflower). a. Habit. b. Flower. c. Achene.

Along streams.

IL, KY, MO, OH. (Not listed by the U.S. Fish and Wildlife Service).

Leatherflower.

The leatherflower gets its name from the thick, leathery sepals that form a bell-shaped flower. This species differs from the other leathery-flowered species of *Clematis* by its plumose styles.

4. **Coptis** Salisb.—Goldthread

Evergreen perennial herbs; leaves basal, ternately divided; flower solitary on a leafless scape; sepals 5–7, free, petaloid; petals 5–7, free, small, with a hollow apex; stamens 15–25; pistils 3–9, free, stipitate, the ovary superior; follicles 3–9, with up to 8 seeds.

Approximately fifteen species in temperate and arctic parts of the world comprise this genus.

Only the following occurs in the central Midwest.

1. **Coptis groenlandica** (Oeder) Fern. Rhodora 31:142. 1929. Fig. 102.
Anemone groenlandica Oeder, Fl. Dan. 566. 1879.
Coptis trifolia (L.) Salisb. var. *groenlandica* (Oeder) Fassett, Trans. Wisc. Acad. Sci. 38:195. 1947.

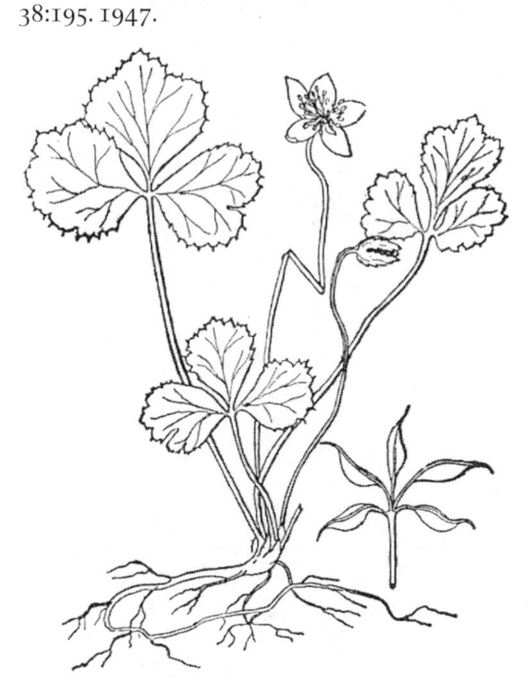

102. *Coptis groenlandica* (Goldthread). Habit (center). Fruits (lower right).

Perennial herbs from very slender, bright yellow rhizomes; leaves basal, consisting of 3 leaflets, the leaflets obovate, rounded at the apex, tapering to the base, serrate and sometimes 3-lobed, glabrous, shiny, on short petiolules; flower solitary on a leafless scape, the scape slender, to 15 cm tall, glabrous; sepals 5–7, petaloid, free, elliptic to lanceolate, to 8 mm long, obtuse or subacute at the apex, white; petals 5–7, free, fleshy, obovate, rounded and hollow at the apex, to 4 mm long; stamens 15–25; pistils 3–9, free; follicles 3–9, free, to 9 mm long, with a beak 2.5–4.0 mm long, with 4–8 seeds. May–July.

Boggy woods, swampy woods.

IA, IN, OH (FACW), as *C. trifolia*.

Goldthread.

This species is distinguished by its ternately divided evergreen leaves, its 5–7 petaloid sepals, and its small fleshy petals with a hollow apex.

5. **Myosurus** L.—Mousetail

Annual herbs from fibrous roots; leaves basal, linear, entire; flowering scape unbranched, 1-flowered; sepals 5, free, spurred at base; petals 5, free, or absent; stamens 5–20, free; pistils numerous, free, on an elongated receptacle, the ovaries superior, with 1 ovule; achenes borne on an elongated receptacle.

Myosurus is a genus of five species found in many parts of the world.

1. **Myosurus minimus** L. Sp. Pl. 284. 1753. Fig. 103.

Annual from fibrous roots; all leaves basal, linear, to 9 cm long, to 5 mm broad, obtuse to subacute, glabrous; flowering scape to 15 cm tall, unbranched, glabrous, terminated by a single flower; sepals 5 (–7), greenish-yellow, elliptic, subacute, to 5 mm long, long-spurred at the base; petals 5, greenish-yellow, linear; stamens 10–18, free; pistils numerous on an elongated receptacle, glabrous; fruiting receptacle slender, to 5 cm long, with numerous achenes, the achenes quadrate, obtuse, apiculate, glabrous. April–June.

Moist or wet ground.

IA, IL, IN, KS, MO, NE (FACW), KY, OH (FACW+).

Mousetail.

Because of its small stature, mousetail is a sometimes overlooked species occurring in moist fields or along the edges of woods. The elongated fruiting receptacle may attain a length of 5 cm but usually is much shorter.

1. **Ranunculus** L.—Buttercup

Perennial herbs with fibrous or sometimes thickened roots; stems erect or creeping and rooting at the nodes, sometimes hollow; leaves usually both basal and cauline, simple and entire or toothed or lobed, or compound; inflorescence axillary or terminal; flowers perfect, actinomorphic; sepals usually 5, spreading or reflexed; petals (1–) 5 (–numerous in "double-flowered" forms), yellow or rarely white; stamens (3–) 10–numerous; pistils (5–) 10–numerous; achenes crowded into fruiting heads, plump or flattened, smooth or papillate, beaked.

There are about 250 species of *Ranunculus* distributed widely in temperate or subarctic areas. In the tropics, this genus is generally restricted to the mountains. Plants with small flowers are usually referred to as crowfoots or spearworts, while those with larger flowers are called buttercups.

In order to make positive identification of some species of *Ranunculus*, it is necessary to have mature achenes available. Reliance on vegetative characters is not practical in distinguishing *R. abortivus* from *R. micranthus*, or in distinguishing members of the *R. hispidus-R. fascicularis-R. septentrionalis-R. carolinianus* complex.

A few taxa, such as *R. trichophyllus*, *R. longirostris*, *R. flabellaris*, and *R. gmelinii* var. *hookeri*, grow in water. Most other taxa are associated with low, wet soil. At the other extreme are *R. rhomboideus*, a prairie species, and *R. micranthus* and *R. hispidus*, species of usually dry woodlands. These latter are not included in this work.

1. Petals white; achenes covered by horizontal wrinkles.
  2. Leaves remaining firm after removal from water.
    3. Fruiting heads with 7–25 achenes, the beak of the achene about 1 mm long .... 8. *R. longirostris*
    3. Fruiting heads with 30–80 achenes, the beak of the achene about 0.5 mm long .... 17. *R. subrigidus*
  2. Leaves becoming limp after removal from water .... 18. *R. trichophyllus*
1. Petals yellow; achenes smooth or variously marked but not with horizontal wrinkles.
  4. At least some of the leaves simple and unlobed.
    5. All leaves simple and unlobed.
      6. Leaves reniform, ovate, or cordate .... 4. *R. cymbalaria*
      6. Leaves linear to lanceolate, tapering at the base.

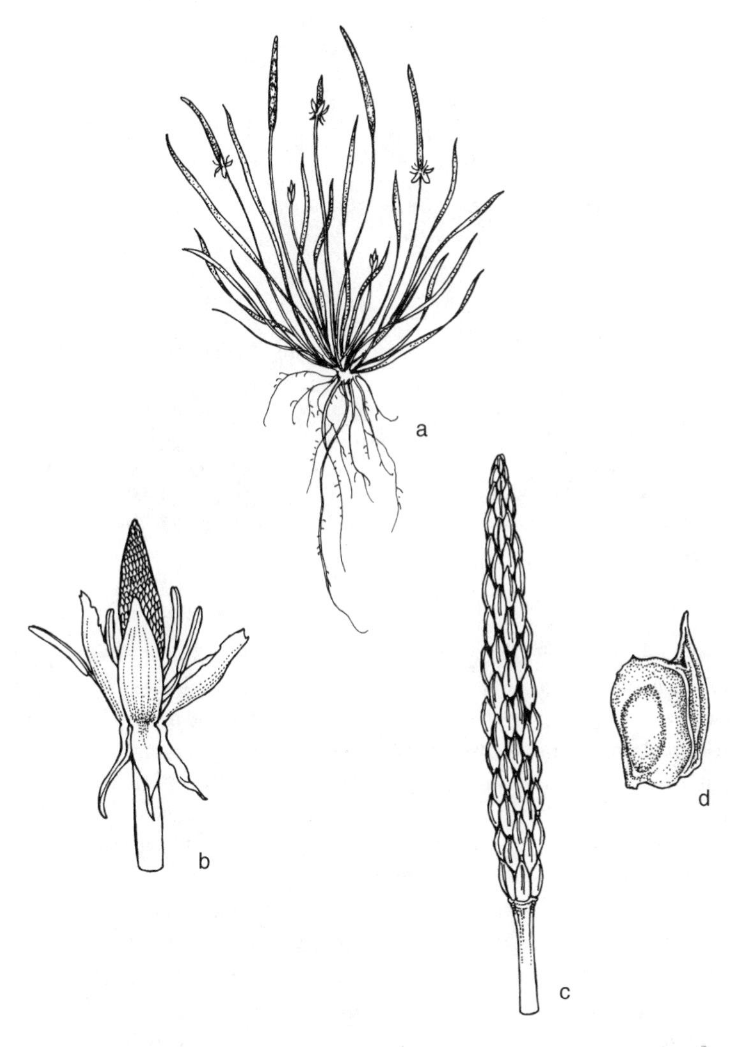

103. Myosurus *minimus* (Mousetail). a. Habit. b. Flower. c. Fruiting head. d. Achene.

7. Petals 5–7 in number, 3–9 mm long; stamens 25–50.
8. Perennial; stamens 25–50; achenes flattened, about 2 mm long....2. *R. ambigens*
8. Annual; stamens 12–25; achenes plump, about 1 mm long.......... 7. *R. laxicaulis*
7. Petals 1–3 in number, 1.0–2.5 mm long; stamens 3–10 .... 11. *R. pusillus*
5. At least some of the leaves lobed or divided.
9. Plants more or less fleshy; achenes with corky thickenings at base; stems hollow .... 5. *R. sceleratus*
9. Plants not fleshy; achenes without corky thickenings at base; stems not hollow .... 1. *R. abortivus*
4. None of the leaves simple and unlobed.
10. Plants not truly aquatics; leaves not finely dissected.
11. Petals up to 6 mm long.
12. Petals about equaling the sepals; achenes flat, with strongly recurved beaks; terminal lobe of leaves not stalked .... 12. *R. recurvatus*
12. Petals distinctly shorter than the sepals; achenes not flat, with nearly straight beaks; terminal lobe of leaves stalked.................................. 10. *R. pensylvanicus*
11. Petals 6 mm long or longer.
13. Achenes papillate on the sides.. 14. *R. sardous*
13. Achenes smooth on the sides.
14. Petals at least ½ as broad as long....................................... 3. *R. carolinianus*
14. Petals less than ½ as broad as long.
15. Achenes plump, the beak up to 1.5 mm long....................... 13. *R. repens*
15. Achenes flattened, the beak 1.5–3.0 mm long.
16. Achenes up to 3.5 (–4.5) mm long, with a low narrow keel near the margin... 16. *R. septentrionalis*
16. Achenes 3.5–5.0 mm long, with a high broad keel near the margin... .... 3. *R. carolinianus*
10. Plants aquatic or, if creeping in mud, some of the leaves finely dissected.
17. Achenes rugose on the sides, corky-thickened at base; beak of achenes about 1.5 mm long.. 5. *R. flabellaris*
17. Achenes smooth on the sides, not corky-thickened at base; beak of achenes up to 0.8 mm long...6. *R. gmelinii*

1. **Ranunculus abortivus** L. Sp. Pl. 551. 1753. Fig. 104.
Ranunculus abortivus L. f. *acrolasius* Fern. Rhodora 40:418, 1938

Perennial herb from slender, unthickened fibrous roots; stems erect, moderately branched, glabrous or pilose, to 75 cm tall; basal leaves simple or sometimes 3-cleft, the simple ones ovate to suborbicular, mostly rounded at the apex, cordate to subcordate at the base, crenate, glabrous or pilose, to 5 cm long, often nearly as wide or even slightly wider, the 3-cleft leaves usually tapering to the base, the divisions mostly oblanceolate to obovate, the petioles of both kinds of leaves glabrous or puberulent, to 10 cm long; cauline leaves simple or 3-cleft, elliptic to oblanceolate, sessile or on petioles up to 3 cm long; flowers several to many, to 7 mm across; sepals 5, elliptic, subacute to acute, spreading to reflexed, glabrous, yellow-green, 3–4 mm long; petals 5, yellow, usually lustrous, oblong to oval, 2.0–3.5 mm long, more than half as broad, shorter than the sepals; stamens 15–20; receptacle pubescent; fruiting head ellipsoid to short-ovoid, to 7 mm long, to 4 mm thick, with 10–35 achenes; achenes obovoid, plump, 1.4–1.7 mm long, glabrous, lustrous, the slender beak about 0.2 mm long. April–June.

Wet woods, fields.

IA, IL, IN, KY, MO, OH (FACW−), KS, NE (FACW).

Small-flowered crowfoot.

This plant occupies a wide variety of habitats from undisturbed to disturbed areas. It is frequent in wet woods.

Plants with pilose stems and leaves may be called f. *acrolasius.*

Ranunculus abortivus is distinguished from other wetland species of the genus by its undivided basal leaves, its usually cleft cauline leaves, its small flowers, and its short-beaked achenes.

104. *Ranunculus abortivus* (Small-flowered crowfoot). a. Habit (right). b. Flower (lower center). c. Fruiting head. (lower left)

2. **Ranunculus ambigens** Wats. Bibl. Ind. N. Am. Bot. 1:16. 1878. Fig. 105.
Ranunculus obtusiusculus Raf. Med. Repos. N.Y. II, 5:359. 1808, *nomen confusum.*
Ranunculus flammula L. var. *major* Hook. Fl. Bor. Am. 1:11. 1829.

Perennial; stems rooting at the lower nodes, ascending, sparsely branched except near the inflorescence, hollow, to 75 cm tall, glabrous; basal leaves absent; cauline leaves simple, lanceolate, acuminate, cuneate, remotely denticulate to entire, glabrous, to 12 cm long, to 2 (–3) cm wide, the petioles dilated; flowers few to several in a corymb, 1.2–2.0 cm across; sepals 5, elliptic, acute to subacute, spreading, glabrous, 4–7 mm long, shorter than the petals; petals 5–6, yellow, oblanceolate, 5–8 mm long, up to 3 mm broad; stamens 25–40; receptacle glabrous; fruiting head globose to ovoid, to 7 mm long, with 10–25 achenes; achenes obovoid, compressed, 1.5–2.5 mm long, glabrous, minutely reticulate, the beak 1.0–1.3 mm long, straight. June–September.

105. *Ranunculus ambigens* (Spearwort). a. Habit. b. Flower. c. Fruiting head. d. Achene.

Swampy woods, wet ditches.

IL, IN, KY, OH (OBL).

Spearwort.

This species, which has no divided leaves, may be confused with *R. laxicaulis*, but it differs by its perennial habit, its more numerous stamens, and its larger, compressed achenes.

3. **Ranunculus carolinianus** DC. Syst. 1:292. 1818. Fig. 106.
Ranunculus septentrionalis Poir. var. *pterocarpus* L. Benson, Bull. Torrey Club. 68:486. 1941.

Perennial herbs with prominent stolons, creeping and rooting at the nodes; stems trailing to ascending, branched, generally appressed-pubescent to nearly glabrous, to 75 cm long; all leaves similar, deeply 3-parted or trifoliolate, the terminal segment glabrous or appressed-pubescent, borne on appressed-pubescent to glabrous petioles up to 25 cm long, the uppermost leaves becoming progressively shorter, petiolate; flowers 1–20 per stem, up to 2.5 cm across, on appressed-pubescent or glabrous pedicels to 5 cm long; sepals 5, narrowly ovate, acute, spreading or reflexed, greenish yellow, glabrous or puberulent, 3.5–5.0 mm long; petals 5, bright yellow, oblong to obovate, 8–12 mm long, sometimes more than and sometimes less than half as wide, usually more than twice as long as the sepals; stamens usually 40 or more; receptacle hispidulous; fruiting head globose to ovoid, 7–14 mm long, nearly as thick, with (5–) 10–20 achenes; achenes obovoid to nearly globose, flattish, 3.5–5.0 mm long, glabrous, with a high broad keel near the margin, the beak straight, 1.5–2.0 mm long. April–June.

Low woods, marshes.

IL, IN, KY, MO, OH (FACW), NE (NI).

Carolina buttercup.

This buttercup is similar to and readily confused with *R. septentrionalis*. Both species root at the nodes and have similar-appearing leaves and flowers.

The major distinction between these two species lies in the fruits. The achenes of *R. carolinianus* are very large (over 3.5 mm long) and have a conspicuous broad, high keel near the margin.

4. **Ranunculus cymbalaria** Pursh, Fl. Am. Sept. 392. Fig. 107.
Ranunculus cymbalaria Pursh var. *typicus* L. Benson, Am. Midl. Nat. 40:215. 1948.

Tufted perennial with slender stolons; leaves mostly all basal; blades ovate to reniform, cordate, crenate, sometimes with three rounded lobes at apex, glabrous or nearly so, to 25 mm long, nearly as wide; petioles to 5 cm long, glabrous or nearly so; stipules to 1 cm long; flowering scape to 20 cm long, with 1–10 flowers, the flowers 6–10 mm across; sepals 5, elliptic, acute to subacute, spreading, glabrous, to 5 mm long, nearly as long as to slightly longer than the petals; petals usually 5, bright yellow, narrowly obovate, to 5 mm long, to 3 mm broad; stamens 15–25; receptacle pubescent; fruiting head short-cylindric, to 10 (–12) mm long, with 40–150 achenes; achenes oblongoid, 1.3–1.6 mm long, nerved on each face, glabrous, the beak up to 0.3 mm long, straight, or slightly curved. May–August,

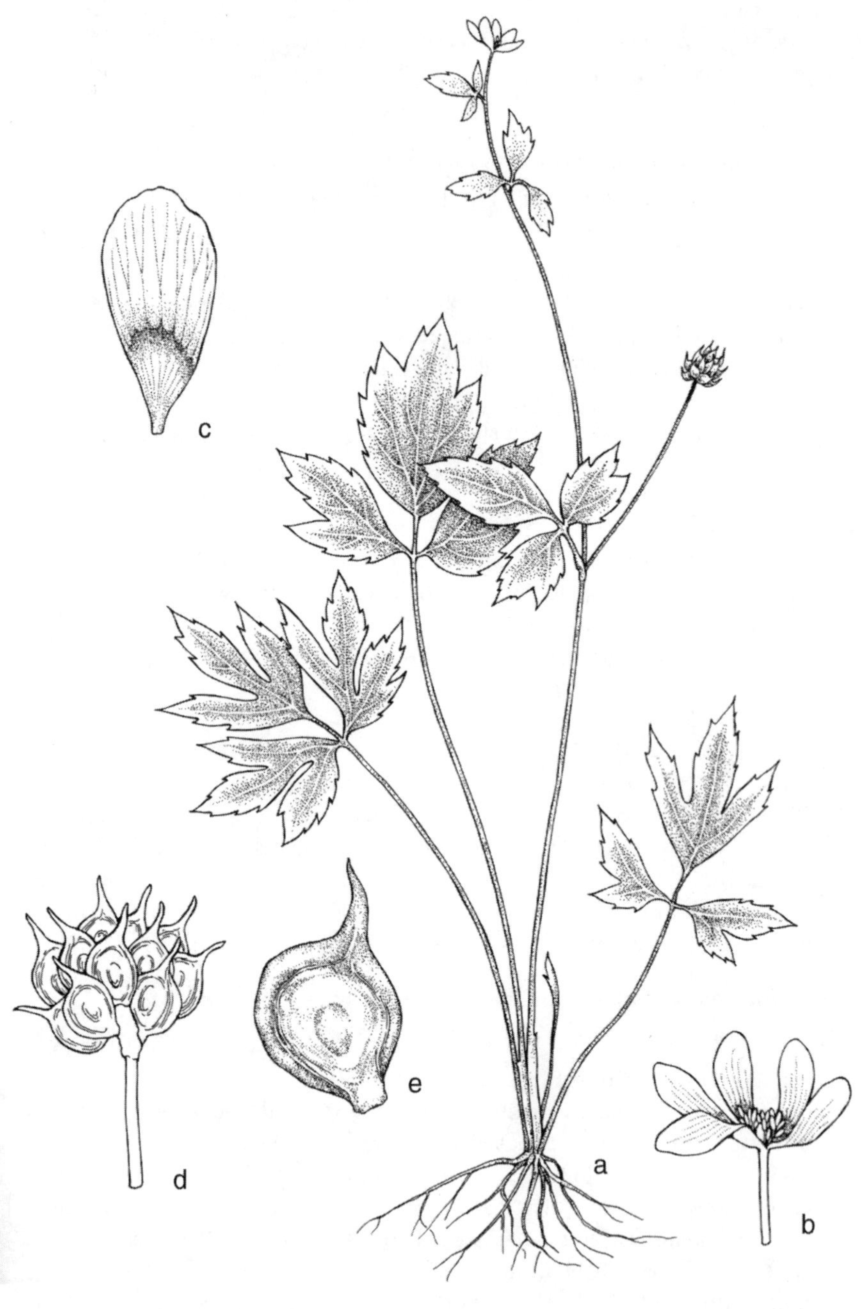

106. *Ranunculus carolinianus* (Carolina buttercup).

a. Habit.
b. Flower.
c. Petal.
d. Fruiting head.
e. Achene.

Wet soil, swamps, marshes, wet ditches, ponds, streams.

IA, IL, KS, MO, NE (OBL).

Seaside crowfoot.

This species is distinguished by its simple leaves, veiny achenes, and elongated receptacles.

5. **Ranunculus flabellaris** Raf. *apud* Bigel. Am. Monthly Mag. 2:344. March, 1818. Fig. 108.
Ranunculus multifidus Pursh, Fl. Am. Sept. 2:736. 1814, *non* Forsk. (1775).
Ranunculus delphinifolius Torr. in Eat. Man. Bot. ed. 2, 395, May 1818.
Ranunculus multifidus Pursh. var. *terrestris* Gray, Man. ed. 5, 41. 1867.
Ranunculus delphinifolius Torr. f. *rosiflorus* Clute, Am. Bot. 34:106. 1928.

Submersed or emergent aquatic perennials; stems hollow, floating or reclining, rooting at the lower nodes, much branched; submersed leaves ternately

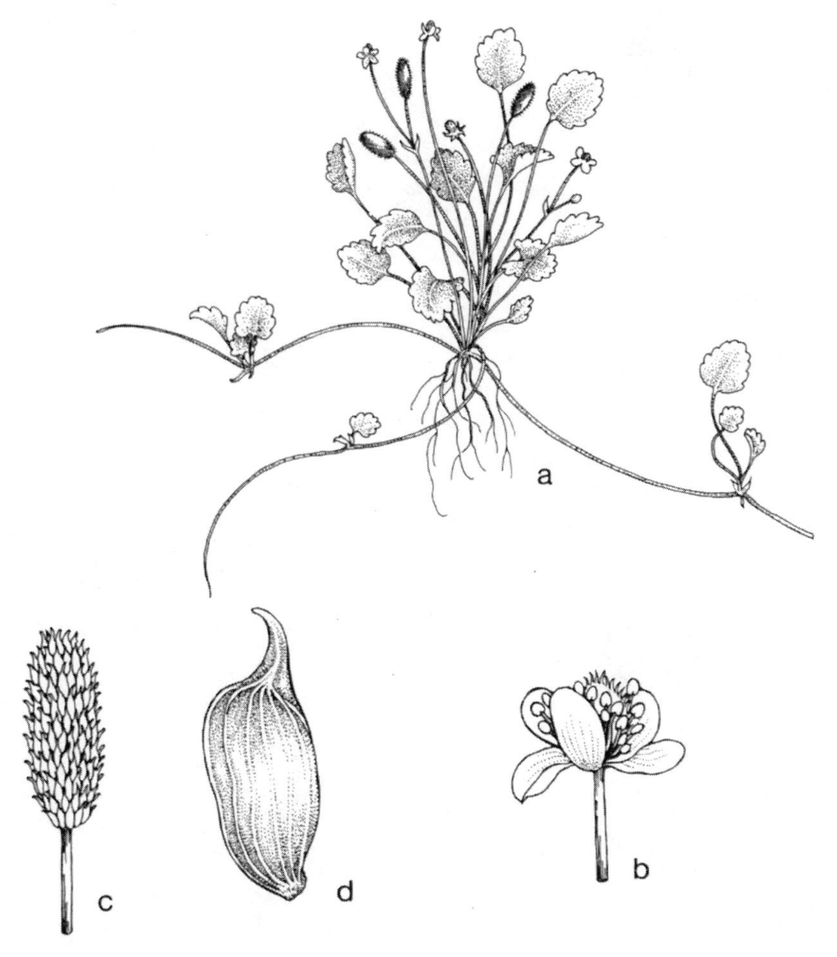

107. *Ranunculus cymbalaria* (Seaside crowfoot). a. Habit. b. Flower. c. Fruiting head. d. Achene.

compound, to 10 cm long, about as wide, the linear divisions 1.0–1.5 mm wide, glabrous; emersed leaves (when present) three-cleft, with each division 3-parted, glabrous; flowers several, to 3 cm across, on stout pedicels up to 6 cm long; sepals 5, ovate, acute, spreading, yellow-green, glabrous or nearly so, 5–8 mm long, about half as long as the petals; petals 5, yellow or rarely roseate, obovate, 7–14 mm long; stamens 50 or more; receptacle pubescent; fruiting head subglobose, to 1.5 cm in diameter, with 50 or more achenes; achenes obovoid, plump, 1.5–2.5 mm long, rugose on the sides, with the margin corky-thickened, glabrous, the broad, flat beak about 1.5 mm long. April–July.

Swamps, ponds, quiet pools, marshes.

IA, IL, IN, KS, KY, MO, NE, OH (OBL)

Yellow water crowfoot.

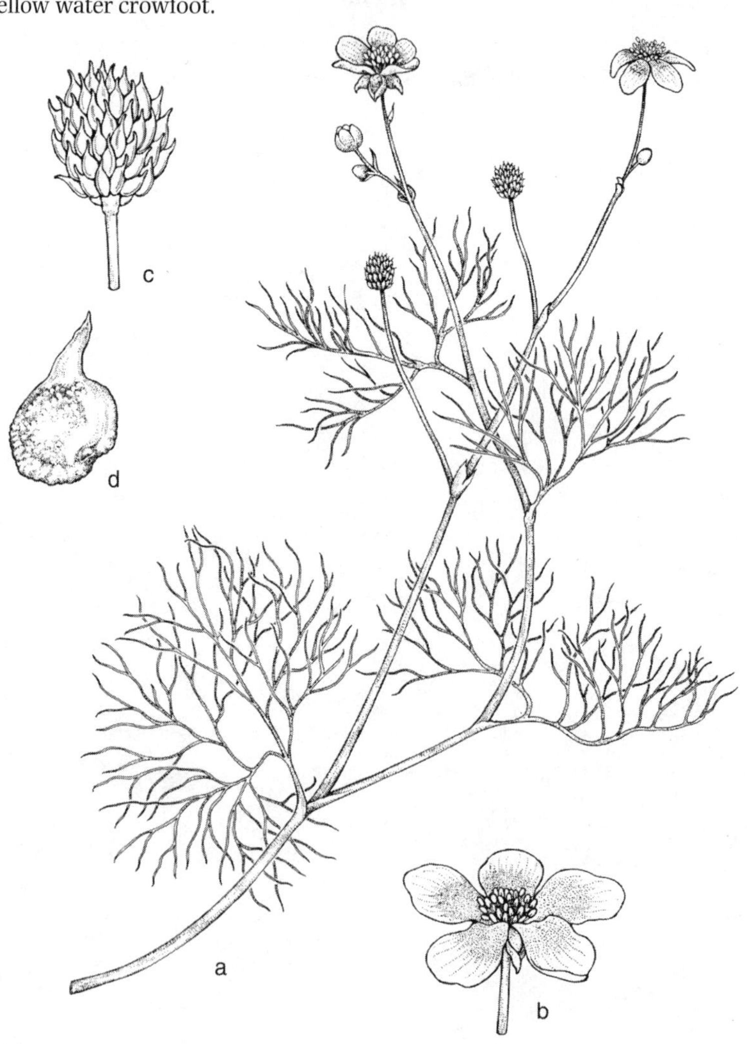

108. *Ranunculus flabellaris* (Yellow water crowfoot). a. Habit. b. Flower. c. Fruiting head. d. Achene.

This species usually grows in standing water where all its leaves are submersed and its flowers float on the surface of the water atop thickened pedicels. Occasionally plants may be found stranded on land. If these terrestrial forms persist, they usually develop less intricately divided leaves. Such obvious ecological forms are scarcely worth recognition in a different taxonomic rank.

The flowers are a waxy, golden yellow, although a roseate form was described by Clute from Illinois. I have observed that petals on dried material sometimes have a faint rosy tint.

6. **Ranunculus gmelinii** DC. var. **hookeri** (D. Don) L. Benson, Am. Midl. Nat. 40:209. 1948. Fig. 109.
Ranunculus purshii Richards. Bot. App. Frankl. 1st Journey 751. 1823.
Ranunculus purshii Richards. var. *hookeri* D. Don in G. Don, Gen. Syst. Gard. 1:33. 1831.

Submersed or emergent aquatic perennials; stems hollow, floating or reclining, rooting at the lower nodes, much branched; submersed leaves either deeply dissected with many linear divisions or 3- to 5-cleft into obovate, cuneate lobes, glabrous or pilose, to 10 cm long, about as wide; emersed leaves (when present) three-cleft, with each division 3-parted, glabrous or pilose; flowers 1–4, to 1.8 cm across, on moderately stout pedicels up to 3 cm long; sepals 5, suborbicular to ovate, rounded or subacute at the apex, spreading, yellow-green, 3–5 mm long, half to nearly as long as the petals; petals 5, yellow, obovate, 4–7 mm long; stamens 20–40; receptacle pubescent; fruiting head subglobose, to 6 mm in diameter, with 50 or more achenes; achenes obovoid, somewhat flattened, 1.0–1.5 mm long, without corky-thickened margins, glabrous, the broad, flat beak 0.4–0.8 mm long. May–July.

Ponds, marshes, wet ditches, pools, streams.

IA, IL (FACW).

Small yellow water crowfoot.

The small yellow-flowered crowfoot is smaller in all respects than *R. flabellaris*. In addition, most of the leaves of *R. gmelinii* var. *hookeri* are divided into broader segments than those of *R. flabellaris*.

The binomial *R. purshii* Richards. is the correct one for this taxon at the rank of species.

7. **Ranunculus laxicaulis** (Torr. & Gray) Darby, Bot. S. States II, 4. 1841. Fig. 110.
Ranunculus flammula L. var. *laxicaulis* Torr. & Gray, Fl. N. Am. 1:16. 1838.
Ranunculus texensis Engelm. *apud* Engelm. & Gray, Bost. Journ. Nat. Hist. 5:210. 1845.

Annual from fibrous roots; stems rooting at the lower nodes, erect to ascending, much branched, to 65 cm tall, glabrous; basal leaves simple, oblong to elliptic to ovate, obtuse, cordate to truncate to cuneate, entire or denticulate, to 4 cm long, to 3 cm wide, glabrous, the petioles to 10 cm long; cauline leaves linear to elliptic to lanceolate, acute to subacute, cuneate, entire or denticulate, to 6 cm long, to 1.2 cm wide, glabrous, sessile or short-petiolate; flowers few to several, 5–15 mm across; sepals 5, ovate, acute, spreading, 2–3 mm long, glabrous or nearly so; petals 5–7, yellow, obovate, 3–9 mm long, up to 2.5 mm broad; stamens 20–25; receptacle glabrous; fruiting head hemispherical, up to 5 mm in diameter, with 15–50 achenes;

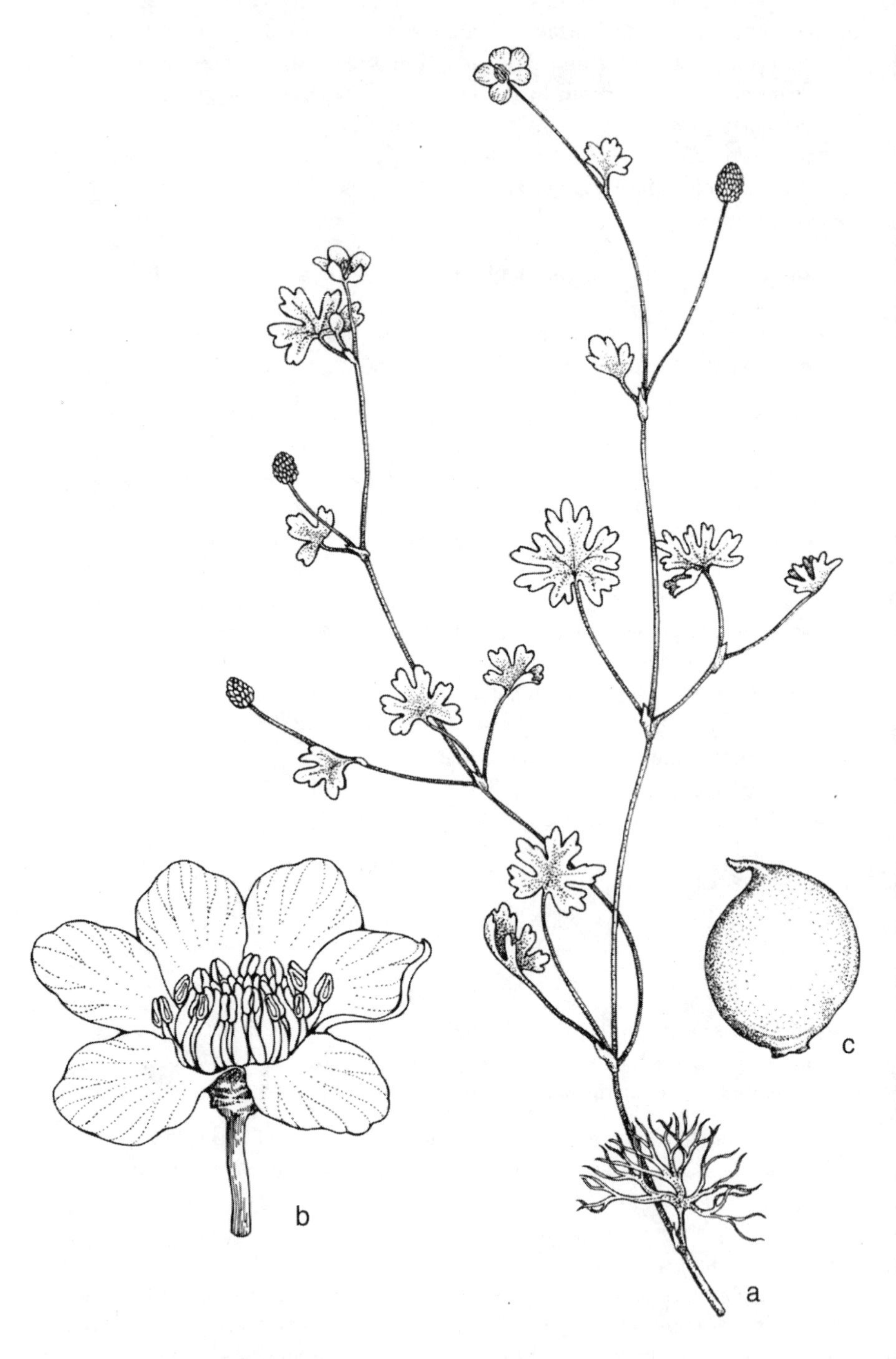

109. *Ranunculus gmelinii* (Small yellow water crowfoot). a. Habit. b. Flower. c. Achene.

achenes obovoid, plump, 1.0–1.3 mm long, glabrous, the beak 0.1–0.2 mm long. May–July.

Wet woods, wet ditches, marshes, around ponds.

IL, IN, KY, MO (OBL), KS (not listed).

Spearwort.

This species shows considerable variation in the number of petals and stamens. Its tiny, plump achenes serve best to distinguish it from *R. ambigens*.

110. *Ranunculus laxicaulis* (Spearwort).

a. Habit.
b. Lower leaf.
c. Flower.
d. Fruiting head.
e. Seed.

8. **Ranunculus longirostris** Godr. Mem. Roy. Soc. Nancy 39. 1839. Fig. 111.
Batrachium longirostre (Godr.) F. Schultz, Arch. Fl. France et All. 1:71. 1842.
Ranunculus aquatilis L. var. *longirostris* (Godr.) Laws. Trans. Roy. Soc. Can. 2:45. 1884.

Aquatic perennials; stems floating, rooting at the lower nodes, sparsely branched, up to 3 mm in diameter, generally glabrous; leaves floating or submersed, repeatedly dissected into filiform segments, remaining firm after removal from the water, to 2 cm long, glabrous, the petiole up to 1 cm long, adnate nearly the entire length to the stipule, the stipule sheathing and broad at the base; flowers borne at the surface of the water, 1–2 cm across; sepals 5, elliptic, acute to subacute, spreading, glabrous, 3–4 mm long, about half as long as the petals; petals 5, white, obovate, 4–10 mm long, up to 5 mm wide; stamens 10–20; receptacle hispid; fruiting heads globose, up to 7 mm in diameter, with 7–25 achenes; achenes obovoid,

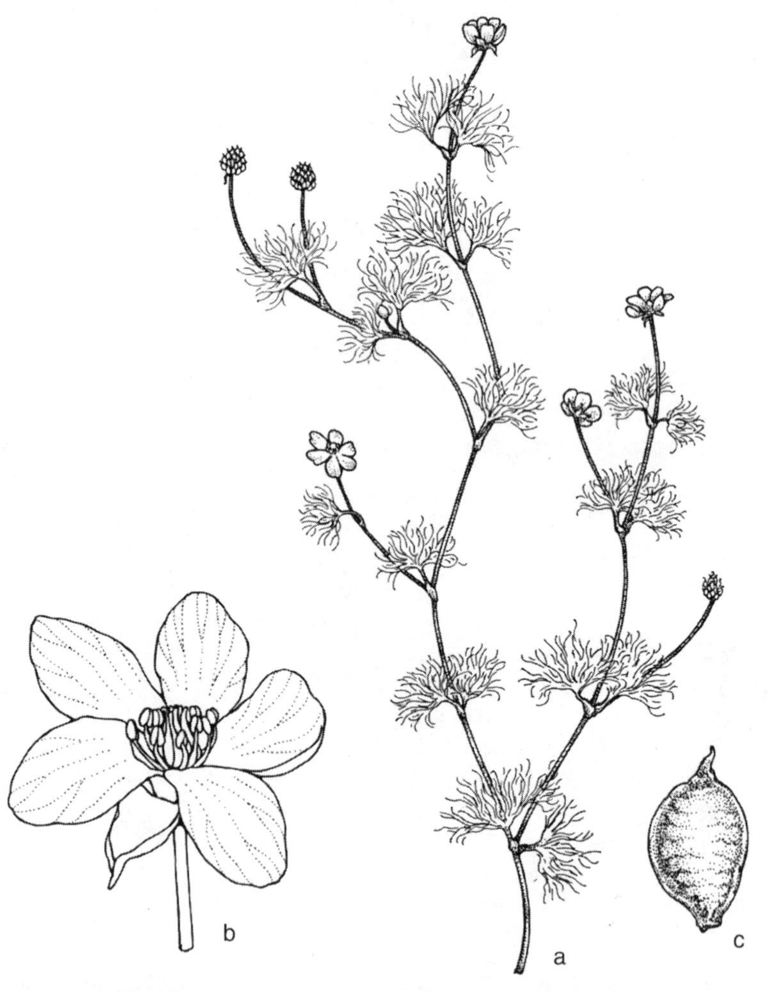

111. *Ranunculus longirostris* (White water crowfoot). a. Habit. b. Flower. c. Achene.

1.2–1.7 mm long, rugose, glabrous or rarely hispidulous, the beak 0.7–1.2 mm long. May–August.

Ponds, slow streams, wet ditches, marshes.

IA, IL, IN, KS, KY, MO, NE, OH (OBL).

White water crowfoot.

Some botanists do not distinguish this species from the Eurasian *R. circinatus.*

Ranunculus longirostris differs from *R. trichophyllus* by its leaves, which remain firm after removal from water, by its longer beak of the achene, and by its petioles adnate to the stipules. It differs from *R. flabellaris* by its white flowers.

9. **Ranunculus macounii** Britt. Trans. N.Y., Acad. Sci. 12:3, 1892. Fig. 112.

112. *Ranunculus macounii* (Marsh crowfoot). Habit (center). Flower (upper right). Achene (lower right).

Perennial herb from thickened roots and leafy stolons; stems trailing to ascending to erect, to 1 m long, hollow, hirsute; basal leaves ternately compound, the leaflets sparsely pubescent to hispid, ovate or obovate, subacute at the apex, tapering or sometimes rounded at the base, deeply toothed to 2- or 3-cleft, petiolulate; cauline leaves similar but smaller and nearly sessile; flowers 2–many, actinomorphic, perfect; sepals 5, free, green, 4–6 mm long, hispid, reflexed at anthesis; petals 5, free, golden yellow, about as long as the sepals, up to 5 mm wide; stamens numerous; pistils 10–many; fruiting heads ovoid to nearly spherical, 5–10 mm across, the achenes thin, minutely pitted, 2.5–3.5 mm long, narrowly winged, with a straight, flat beak 1.0–1.5 mm long; receptacle elongated, to 7 mm long, hirsute. June–August.

Wet woods, wet meadows, ponds, streams.

IA, NE (OBL).

Marsh crowfoot.

This species has compound leaves and small flowers with petals up to 6 mm long. It differs from *R. pensylvanicus* by its trailing habit and the presence of stolons, and from *R. recurvatus* by the straight beak of the achenes.

10. **Ranunculus pensylvanicus** L. f. Suppl. 272. 1781. Fig. 113.

Annual or sometimes perennial herbs from slender, fibrous roots; stems erect, not swollen at the base, hollow, densely villous-hirsute, somewhat branched, to nearly 1 m tall; basal leaves usually absent at flowering time, deeply 5- to 7-parted, borne on hirsute petioles to 15 cm long; cauline leaves deeply 3-cleft, with each division 3-lobed, at least the terminal division stalked, spreading hirsute; flowers few to several, to 8 mm across; sepals 5, narrowly elliptic, acute to subacute, reflexed, yellow-green, hirsutulous, 3–5 mm long, somewhat longer than the petals; petals 5, pale yellow, oblong to obovate, 2–4 mm long; stamens 15–20; receptacle hirsute; fruiting head oblongoid to cylindric, to 1.5 cm long, to 1 cm across, with 60 or more achenes; achenes obovoid, not conspicuously flattened, 2.0–2.5 mm long, glabrous, with prominent narrow margins, the beak nearly straight, up to 1 mm long. July–September.

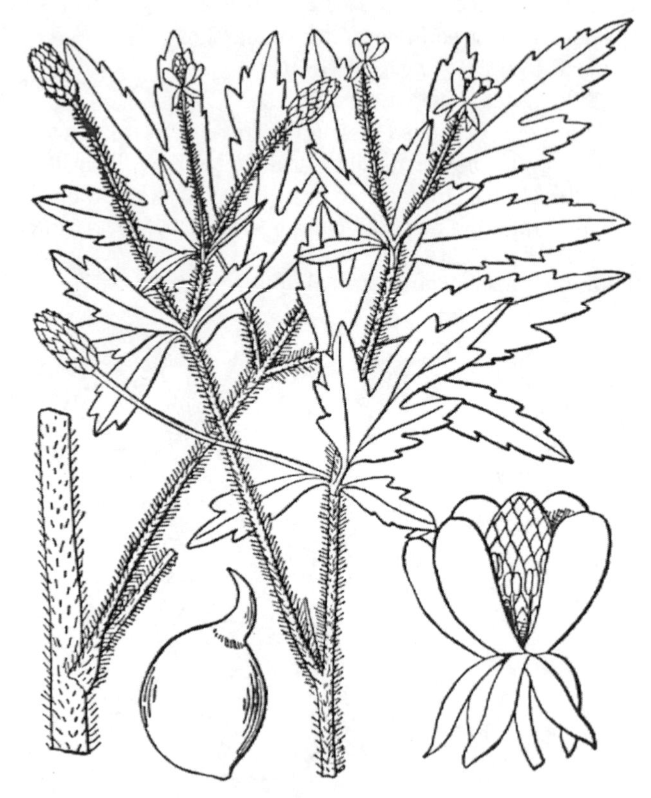

113. *Ranunculus pensylvanicus* (Bristly crowfoot). Habit (center). Stem (left). Achene (bottom center). Flower (lower right).

Wet ditches, wet meadows, marshes.

IA, IL, IN, NE, OH (OBL).

Bristly crowfoot.

The species most nearly related to *R. pensylvanicus* apprarently is *R. recurvatus.* This latter species is distinguished, however, by its very recurved beaks of the achenes, its less deeply divided leaves, and its less pubescent stems and leaves.

The basal leaves of *R. pensylvanicus*, which are divided into more lobes than are the cauline leaves, usually are withered by the time of flowering.

11. **Ranunculus pusillus** Poir. ex Lam. Encycl. 6:99. 1804. Fig. 114.
Ranunculus oblongifolius Ell. Sketch. 2:58. 1816.
Ranunculus pusillus Poir. var. *typicus* L. Benson, Am. Midl. Nat. 40:197. 1948.

Annual from fibrous roots; stems rooting at the lower nodes, erect or ascending, sparsely branched, to 50 cm tall, glabrous; basal leaves simple, oblong to ovate, acute to subacute, usually rounded at the base, entire or nearly so, to 3 cm long, to 1.5 cm broad, glabrous, the petioles to 6 cm long; cauline leaves linear to lanceolate, acute to subacute, cuneate, entire or sparsely denticulate, to 5 cm long, to 5 mm

114. *Ranunculus pusillus* (*Dwarf* spearwort).

a. Habit.
b. Flower.
c. Fruiting head.
d. Achene.

wide, glabrous, sessile or nearly so; flowers few, to 5 mm across; sepals 5, ovate, acute, spreading, glabrous, 1.0–1.5 mm long; petals 1–3 (–5), yellow, obovate, 1.0–1.5 mm long, about 1 mm broad; stamens 3–10; receptacle glabrous; fruiting head hemispherical, up to 4 mm in diameter, with 15–100 achenes; achenes oblongoid, plump, 0.9–1.1 mm long, glabrous, the beak 0.1–0.2 mm long. April–June.

Wet ditches, wet woods, swamps.

IL, IN, KY, MO, OH (OBL).

Dwarf spearwort.

This usually delicate annual occupies the same habitats as *R. laxicaulis* and sometimes is confused with it. The petals in *R. pusillus* number from one to three, occasionally 5, while the stamens never exceed ten in number. It is the smallest flowered species of *Ranunculus*.

12. **Ranunculus recurvatus** Poir. in Lam. Encyc. Meth. 6:125. 1804. Fig. 115.

Perennial from slender, fibrous roots; stems erect, often somewhat swollen at the base, branched, sparsely villous-hirsute, to 75 cm tall; basal leaves simple, cordate, shallowly to deeply 3-cleft, with each division shallowly lobed or crenate, more or less hirsutulous or pilose above and below, to 7 cm long, nearly as wide or slightly wider, the petioles to 20 cm long, villous-hirsute; cauline leaves similar to the basal leaves but smaller; flowers few to several, to 1 cm across; sepals 5, ovate, acute, reflexed, green, pilose, 3–6 mm long; petals 5, pale yellow, narrowly elliptic, 2–5 mm long, usually slightly shorter than the sepals; stamens 10–25; receptacle hispid; fruiting head globose to ovoid, 5–7 mm in diameter, with 10–25 achenes; achenes obovoid, flattened, 1.5–2.0 mm long (excluding the beak), glabrous, minutely pitted, the margin sharply defined, the beak strongly hooked, 1.0–1.5 mm long. April–June.

Wet woods, fens.

IA, IL, IN, MO (FACW), KS, NE (FACW−), KY, OH (FAC+).

Rough crowfoot; recurved buttercup.

This species is recognized by its dense short pubescence, its small, pale yellow petals, its broadly lobed leaves, and its achenes with hooked beaks. The somewhat similar *R. pensylvanicus* has more deeply cleft leaves and shorter, straight beaks on the achenes, and longer pubescence.

13. **Ranunculus repens** L. Sp. Pl. 554. 1753. Fig. 116.

Perennial stoloniferous herbs with fibrous roots; stems trailing or ascending, rooting at the nodes, branched, to 85 cm long, glabrous to hirsute; all leaves similar, deeply 3-parted or trifoliolate, the terminal segment stalked, glabrous or appressed-pubescent petioles up to 25 cm long, often mottled with white; flowers several, up to 2 cm across, on appressed-pubescent or nearly glabrous pedicels up to 10 cm long; sepals 5, ovate, acute, spreading, greenish, pilose, 5–7 mm long; petals 5–numerous, bright yellow, obovate, 6–15 mm long, over half as wide, usually about twice as long as the sepals; stamens 5–many; receptacle pubescent; fruiting head subglobose, 6–10 mm in diameter, with 20–25 achenes; achenes obovoid, plump, 2.0–3.5 mm long, glabrous, narrowly margined, the beak recurved, 0.8–1.5 mm long. May–August.

115. *Ranunculus recurvatus* (Rough crowfoot).

a. Habit.
b. Flower.
c. Fruiting head.
d. Achene.

Wet soil.

Native of Europe; IA, IL, IN, MO (FAC+), KY, OH (FAC), NE (not listed).

Creeping buttercup.

The creeping buttercup is similar to both *R. septentrionalis* and *R. carolinianus* in that the stems are often creeping and rooting at the nodes. *Ranunculus repens* differs from the other two by its plump achenes, which have a beak never more than 1.5 mm long.

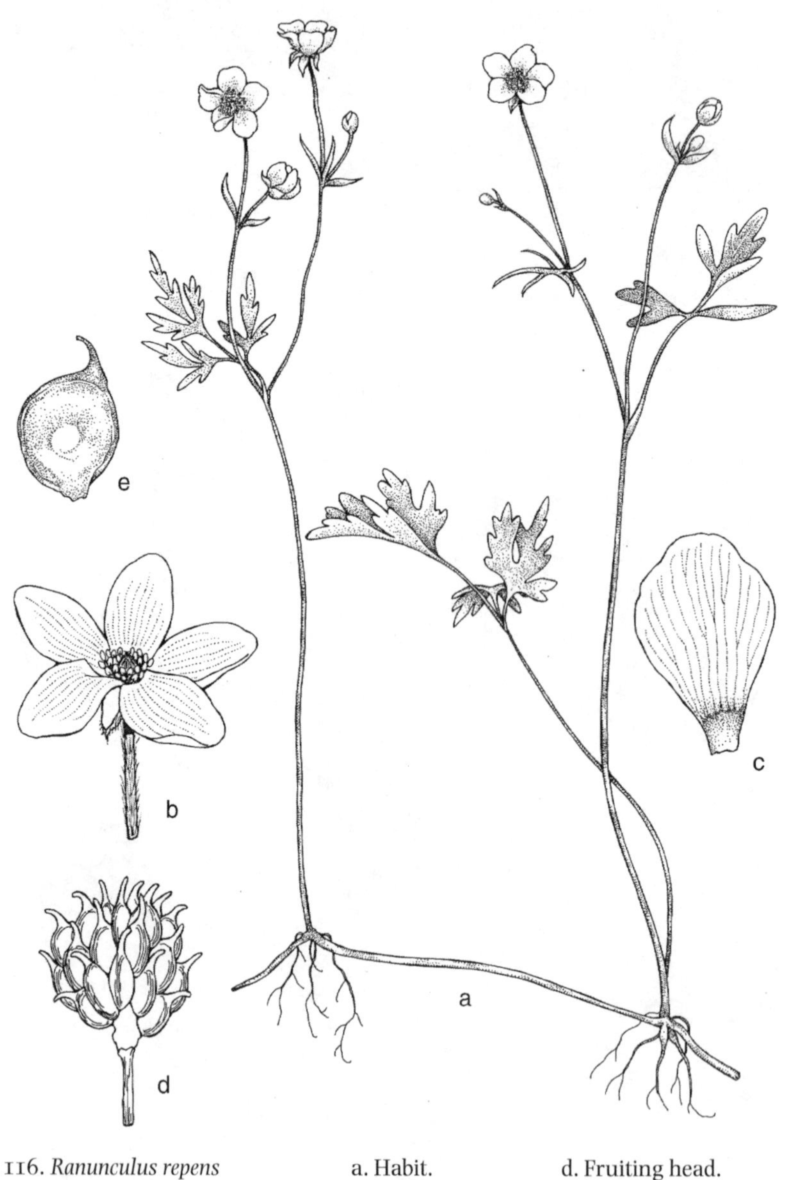

116. *Ranunculus repens* (Creeping buttercup).

a. Habit.
b. Flower.
c. Petal.
d. Fruiting head.
e. Achene.

14. **Ranunculus sardous** Crantz. Stirp. Austr. ed. 1, fasc. 2:84. 1763. Fig. 117.
Ranunculus parvulus L. Mant. Pl. 79. 1767.

Annual with a sometimes swollen base and numerous fibrous roots; stems erect, much branched, hirsute to appressed-pilose, to 50 cm tall; basal leaves trifoliolate, the terminal segment stalked, pilose, on hirsute petioles up to 15 cm long; cauline leaves similar but usually with narrow divisions and with the uppermost becoming sessile; flowers numerous, to nearly 2 cm across, on slender strigose pedicels to 5 cm long; sepals 5, ovate, acute, reflexed, greenish yellow, pilose, 2–5 mm long; petals 5,

117. *Ranunculus sardous* (Buttercup).

a. Habit.
b. Flower.
c. Petal.
d. Fruiting head.
e. Achene.

bright yellow, obovate, 8–12 mm long, nearly as broad, usually less than twice as long as the sepals; stamens 25 or more; receptacle pilose; fruiting head globose, to 8 mm in diameter, with 10 or more achenes; achenes nearly orbicular, flat, 2–3 mm long, the sides papillate or rarely smooth, distinctly narrow-margined, the beak curved, up to 0.5 mm long. April–May.

Wet fields.

IL, KS, MO (FAC), KY (UPL).

Buttercup.

In its growth form, *R. sardous* most nearly resembles *R. bulbosus*, even to the extent that some specimens of *R. sardous* are conspicuously swollen at the base, but the achenes are papillate.

A few specimens of *R. sardous*, however, are only sparsely papillate on the sides of the achenes.

Despite its non-wetland designations, this plant is an inhabitant of wet fields.

15. **Ranunculus sceleratus** L. Sp. Pl. 551. 1753. Fig. 118.

Rather fleshy annual with fleshy, fibrous roots; stems erect, seldom rooting at the lower nodes, much branched, hollow, glabrous, to 75 cm tall; basal leaves simple, somewhat succulent, reniform in outline, 3-cleft, the divisions often further divided, the ultimate lobes broadly rounded, the blades cordate at the base, glabrous, to 4 cm long, nearly as wide or wider, the petioles to 15 cm long, glabrous; cauline leaves smaller, usually cleft into three linear-oblong divisions; flowers numerous, to 1 cm across; sepals 5, ovate, acute, spreading, pilose or glabrate, yellow-green, 2–4 mm long, over half as broad; petals 5, pale yellow, obovate, 1.5–3.5 mm long, slightly shorter than the sepals; stamens 10–25; receptacle glabrous or puberulent; fruiting head cylindrical, to 1 cm long, to 6 mm thick, with 30 or more achenes; achenes obovoid, plump, 0.8–1.2 mm long, the surface sometimes with microscopic depressions, glabrous, the minute beak about 0.1 mm long. May–August.

Ponds, wet depressions in woods, marshes.

IA, IL, IN, KS, KY, MO, NE, OH (OBL).

Cursed crowfoot.

This buttercup is the most succulent species of *Ranunculus*. The lower leaves are fleshy, and the hollow stem sometimes measures up to 2 cm in diameter at the base.

The pattern of leaf-cutting is very distinctive for this species. The common name is derived from the fact that this species may completely occupy the surface of small ponds.

16. **Ranunculus septentrionalis** Poir. in Lam. Encyc. Meth. 6:125. 1803. Figs. 119, 120.

Ranunculus caricetorum Greene, Pittonia 5:194. 1903.

Ranunculus septentrionalis Poir. var. *caricetorum* (Greene) Fern. Rhodora 38:177. 1936.

Perennial herbs with prominent stolons creeping and rooting at the nodes; stems trailing to ascending, branched, sometimes hollow, glabrous or nearly so to appressed-pubescent to densely retrorse-hirsute, to 85 cm long; all leaves similar, deeply 3-parted or trifoliolate, the terminal segment stalked, appressed-pubescent to hirsute, borne on appressed-pubescent to retrorse-hirsute petioles up to 30 cm long,

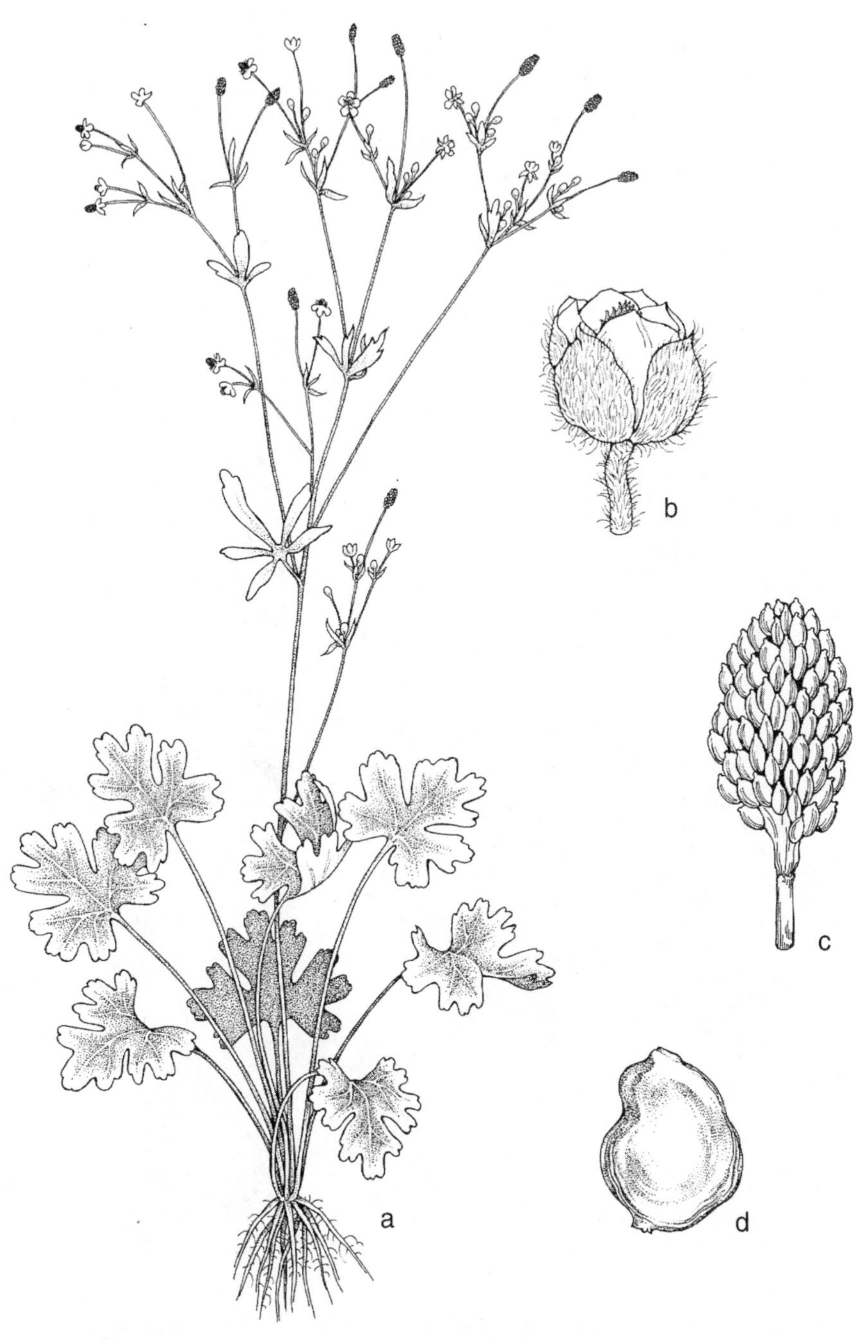

118. *Ranunculus sceleratus* (Cursed crowfoot).

a. Habit.
b. Flower.

c. Fruiting head.
d. Achene.

the uppermost leaves often shorter petiolate; flowers 1–10 per stem, up to 2.5 cm across, on appressed- or spreading-hispid pedicels to 6 cm long; sepals 5, narrowly ovate, acute, spreading, greenish yellow, nearly glabrous to pilose to hispidulous, 6–10 mm long; petals 5, bright yellow, mostly obovate, 8–15 mm long, usually over half as wide, usually less than twice as long as the sepals; stamens usually 40 or more; receptacle hispidulous; fruiting head globose to ovoid, 8–14 mm long, nearly as thick, with 15–30 achenes; achenes obovoid, flattish, to 3.5 mm long, glabrous, with a low narrow keel near the margin, the beak straight or curved, 1.8–3.0 mm long. March–June; September–October.

119. *Ranunculus septentrionalis* (Marsh buttercup). a. Habit. b. Flower. c. Petal. d. Fruiting head. e. Achene.

Wet woods, wet meadows, swamps, springs, marshes.

IA, IL, IN, MO (FACW+), KS, KY, NE, OH (OBL).

Marsh buttercup.

Plants that are densely retrorse-pubescent are known as var. *caricetorum*.

The typical variety of *R. septentrionalis* is not conspicuously pubescent and therefore is readily distinguished from hairy forms of *R. hispidus* and *R. fascicularis*. There is close similarity, however, between var. *septentrionalis* and *R. hispidus* var. *marilandicus*. The creeping stems of *R. septentrionalis* are distinctive, although this character often is not apparent in most herbarium specimens. Sparsely pubescent forms of *R. fascicularis* are distinguished by their tuberous-thickened roots.

Ranunculus carolinianus is a similar species, differing by its achene, which is larger and which has a high broad keel near the margin.

120. *Ranunculus septentrionalis* var. *caricetorum* (Hairy marsh buttercup). Habit.

17. **Ranunculus subrigidus** Drew, Rhodora 38:39–42. 1936. Fig. 121.

Aquatic perennial herb; stems floating, branched or unbranched, glabrous, up to 1 m long; leaves remaining fairly stiff upon removal from water, deeply divided, up to 3 cm long, nearly as wide, the divisions toothed, glabrous; flowers axillary, on pedicels up to 8 cm long; sepals 5, free, purple-green, 2–4 mm long; petals 5, free, white but often with a yellowish base, 4–10 mm long, without a nectary at the base; stamens 10–12; achenes 30–80 in a globose head up to 5 mm in diameter, each achene obovoid, transversely ridged, 1.0–1.5 mm long, the beak up to 0.5 mm long, the pedicels recurved in fruit. May–August.

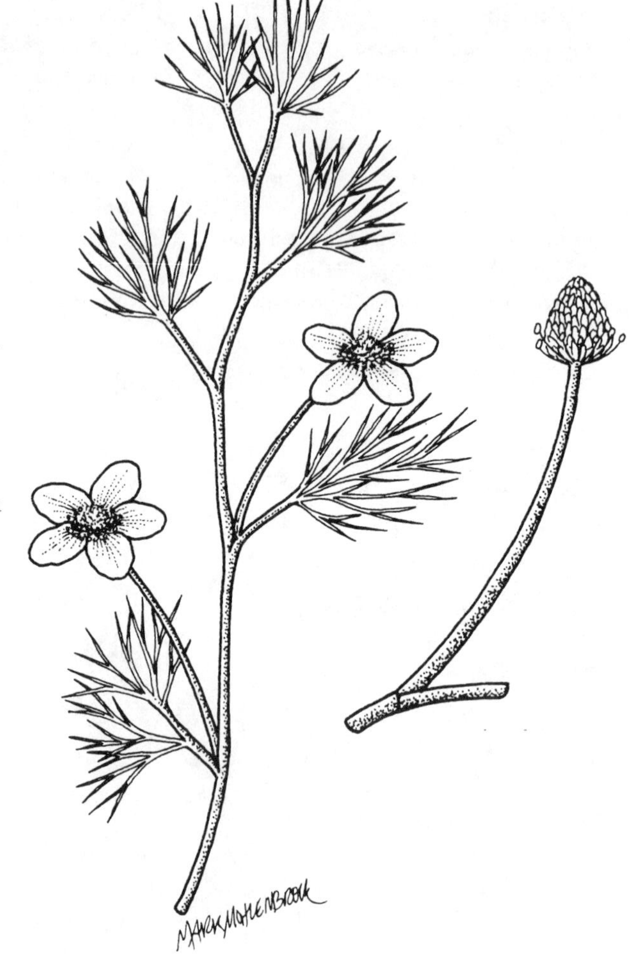

121. *Ranunculus subrigidus* (White water crowfoot). Habit (left). Fruiting head (right).

Ponds, marshes, wet ditches, usually in shallow water.

IA (OBL).

White water crowfoot.

This species has fairly firm leaves when removed from water. It differs from *R. longirostris* by more achenes (30–80) in a head and a shorter beak (up to 0.5 mm long) on each achene.

18. **Ranunculus trichophyllus** Chaix in Vill. Hist. Pl. Dauph. 1:335. 1786. Fig. 122.
Ranunculus capillaceus Thuill. Fl. Par. ed 1, 1:278. 1799.
Ranunculus aquatilis L. var. *capillaceus* (Thuill.) DC. Prodr. 1:26. 1824.
Batrachium trichophyllum (Chaix) F. Schultz, Arch. Fl. France et All. 1:107. 1848.
Ranunculus aquatilis L. var. *trichophyllus* (Chaix) Gray, Man. Bot. ed. 5, 40. 1867.

Aquatic perennials; stems submersed, rooting at the lower nodes, much branched, up to 2.5 mm in diameter, generally glabrous; floating leaves absent; submersed leaves repeatedly dissected into filiform segments, becoming limp after removal from the water, to 5 cm long, glabrous, the petiole up to 2 cm long, free from the stipule, the stipule sheathing and broad at the base; flowers borne at the surface of the water, 0.8–1.5 cm across; sepals 5, ovate, acute, spreading, glabrous, 2–5 mm long, about half as long as the petals; petals 5, white, obovate, 4–8 mm long, up to 2.5 mm wide; stamens 10–25; receptacle hirsutulous with tufts of hairs; fruiting head globose, to 1 cm in diameter, with (10–) 15–25 achenes; achenes obovoid, 1.0–1.5 mm long, rugose, glabrous or nearly so, the beak up to 0.3 mm long. May–August.

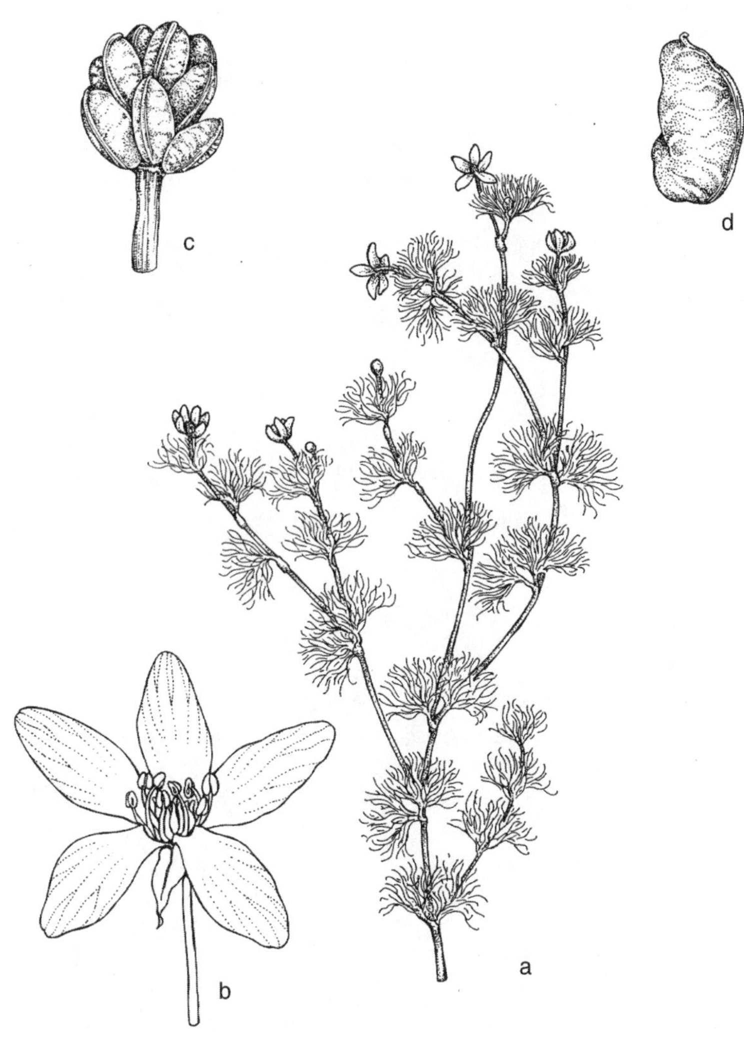

122. *Ranunculus trichophyllus* (White water crowfoot). a. Habit. b. Flower. c. Fruiting head. d. Achene.

Ponds, slow streams, swamps.

IA, IL, IN, KS (OBL).

White water crowfoot.

This species is best distinguished from *R. longirostris* by its leaves, which become limp after their removal from water. The beak of the achene in *R. trichophyllus* is much shorter than the beak in *R. longirostris.*

Benson (1948) and others believe that material assignable to *R. trichophyllus* should not be considered specifically distinct from the Old World *R. aquatilis* L. They would call our plants *R. aquatilis* L. var. *capillaceus* (Thuill.) DC. Since the Old World *R. aquatilis* possesses floating leaves, petals 1 cm long or longer, about 30 stamens, and achenes 2 mm long, I prefer to give *R. trichophyllus* species status.

7. **Thalictrum** L.—Meadow Rue

Perennial herbs, usually from creeping, scaly rootstocks; leaves basal and cauline, 2–3 ternately compound, on petioles dilated at the base; flowers in corymbs, racemes, or panicles, perfect, polygamous, or dioecious; sepals 4–5, free, petallike, caducous; petals absent; stamens numerous, free; pistils up to 15, free; ovaries superior, each with 1 ovule; achenes up to 15, beaked.

Thalictrum is a genus of nearly one hundred species, most of which are found in the north temperate regions of the world. About twenty-five species occur in the United States. Boivin (1957) has merged the genus *Anemonella* with *Thalictrum.*

1. Leaves with glandular hairs on the lower surface .... 3. *T. revolutum*
1. Leaves glabrous or pubescent on the lower surface but without glandular hairs.
    2. Achenes not stipitate; anthers not constricted at the tip; stigmas more or less straight...... .... 1. *T. dasycarpum*
    2. Achenes stipitate; anthers constricted at the tip; stigmas curved.................2. *T. pubescens*

1. **Thalictrum dasycarpum** Fisch. & Lall. Ind. Sem. Hort. Petrop. 8:72. 1842. Figs. 123, 124.

Thalictrum hypoglaucum Rydb. Brittonia 1:88. 1931.

Thalictrum dasycarpum Fisch. & Lall. var. *hypoglaucum* (Rydb.) Boivin, Rhodora 46:482. 1944.

Perennial herb from short, thick rootstocks; stems erect, to nearly 1 m tall, usually purplish, glabrous or pubescent; leaves ternately decompound, the lower ones on glabrous or puberulent petioles much longer than the upper petioles; leaflets firm or thin, oblong to obovate, mostly 3-lobed near the apex, weakly reticulate-nerved, to 4 cm long, sometimes nearly as broad, dark green and glabrous or sparsely puberulent above, somewhat paler and puberulent or glabrous beneath, the margins flat on more or less revolute; flowers whitish, mostly unisexual; sepals 4–5, lanceolate to narrowly obvate, glabrous, to 5 mm long; petals absent; stamens numerous, exserted, drooping shortly after the flowers open; pistils glabrous or pubescent; achenes up to 15 in a cluster, lanceoloid to ovoid, short-stipitate at the base, glabrous or pubescent, ridged, to 5 mm long, including the slender beak. May–July.

Wet meadows, moist wooded ravines.

IA, IL, IN, MO (FACW−), KS, KY, NE, OH (FACW).

Purple meadow rue.

Typical var. *dasycarpum* has firm leaflets that are pubescent on the lower surface. Plants with thin leaflets that are glabrous on the lower surface are known as var. *hypoglaucum*.

a

b

c

d

123. *Thalictrum dasycarpum* (Purple meadow rue).

a. Upper part of staminate plant.
b. Staminate flower.
c. Pistillate inflorescence.
d. Fruit.

124. *Thalictrum dasycarpum* var. *hypoglaucum.* (Purple meadow rue).

a. Upper part of staminate flower.
b. Staminate flower.
c. Pistillate inflorescence.
d. Fruit.

2. **Thalictrum pubescens** Pursh, Fl. Am. Sept. 2:388. 1814. Fig. 125.

125. *Thalictrum pubescens* (Tall meadow rue). Leaf and flowering branch (center). Cluster of fruits (bottom center). Stamen (lower right).

Perennial herb from short, thick rootstocks; stems erect, to 2.5 m tall, usually pubescent but occasionally glabrous, eglandular; leaves ternately decompound, the lower ones on pubescent petioles, the upper ones sessile or nearly so; leaflets firm, obovate to nearly orbicular, often shallowly lobed, the lobes mucronate, to 7 cm long, nearly as wide, pubescent or less commonly glabrous on the lower surface, flat or slightly revolute; flowers usually whitish, unisexual, numerous in panicles; sepals 4–5 (–6), free, elliptic, obtuse at the apex, white or occasionally purplish, 2.0–3.5 mm long; petals absent; stamens numerous, usually exserted, white to purplish, ascending; pistils numerous, pubescent or glabrous; achenes many in a cluster, stipitate, conspicuously veiny, ellipsoid, 3–5 mm long, the beak about ½ as long as the body, glabrous or pubescent. June–August.

Wet meadows, swamps.

IL, IN (FAC), KY, OH (FACW+).

Tall meadow rue.

This very tall species closely resembles *T. dasycarpum*, differing by its ascending rather than drooping stamens and the beak of the achene only about half as long as the body.

In the past, this species has usually been known as *T. polygamum*, but this latter binomial is invalid.

3. **Thalictrum revolutum** DC. Syst. 1:173. 1818. Fig. 126.
Thalictrum revolutum DC. f. *glabrum* Pennell, Bartonia 12:12. 1931.
Thalictrum purpurascens DC. var. *ceriferum* Austin ex Gray, Man. Bot., ed. 5, 39. 1867.

Perennial herb from cordlike rootstocks; stems erect, to nearly 1 m tall, often purplish, glabrous or nearly so; leaves ternately decompound, the lower ones on glabrous petioles much longer than the upper petioles, strongly scented; leaflets firm, ovate to obovate, mostly 3-lobed near the apex, strongly reticulate-nerved, to 3 cm long, sometimes nearly as wide, dark green and more or less glabrous beneath, the margins revolute; flowers white, polygamous or dioecious, the pistillate flowers usually with some stamens; sepals 4–5, white, ovate to oblong, glabrous, to 4 mm long; petals absent; stamens numerous, exserted, drooping shortly after the flowers open, the anthers linear; pistils glandular-pubescent; achenes up to 15 in a cluster, narrowly ovoid, short-stipitate at the base, puberulent, ridged, to 6 mm long, including the slender, sometimes bent, beak. May–June.

Wet prairies, wet meadows.

IA, IL, IN, MO (FAC), KY, OH (UPL).

Waxy meadow rue.

This species owes its waxy appearance to glandular pubescence on the lower leaflet surface. These glands make the plant strongly aromatic. Specimens with glands lacking may be called *f. glabrum*.

The pistillate flowers usually have a few stamens present and also seem to possess slightly shorter sepals than the other flowers.

126. *Thalictrum revolutum* (Waxy meadow rue).

a. Upper part of staminate plant.
b. Leaflet.
c. Staminate flower.
d. Pistillate inflorescence.
e. Fruit.

8. **Trautvetteria** Fisch. & Mey.—False Bugbane

Perennial herb; leaves basal and cauline, palmately lobed; inflorescence corymbose, with several small flowers; sepals 3–5, free, petallike, caducous; petals absent; stamens numerous, free; pistils numerous, free; ovaries superior, each with 1 ovule; achenes numerous, beaked.

Trautvetteria is composed of one species in the eastern United States, one in the western United States, and one in Asia. It has no close relatives among other members of the Ranunculaceae.

1. **Trautvetteria caroliniensis** (Walt.) Vail, Mem. Torrey Club 2:42. 1890. Fig. 127.
Hydrastis caroliniensis Walt. Fl. Car. 156. 1788.
Cimicifuga palmata Michx. Fl. Bor. Am. 1:316. 1803.
Trautvetteria palmata (Michx.) Fisch. & Mey. Ind. Sem. Petr. 1:22. 1834.
Trautvetteria palmata (Michx.) Fisch, & Mey. ß *coriacea* Huth in Engl. Bot. Jahrb. 16:288. 1892.

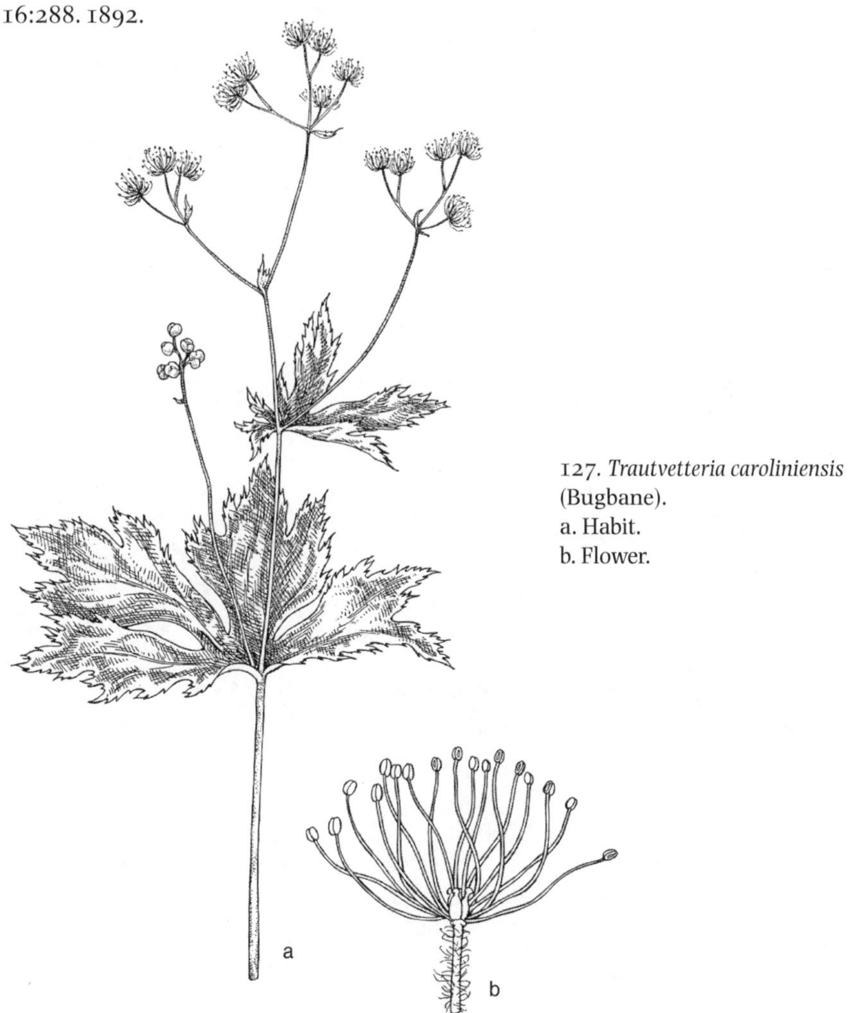

127. *Trautvetteria caroliniensis* (Bugbane).
a. Habit.
b. Flower.

Perennial herb from fibrous roots; stems stout, erect, glabrous, branched, to nearly 1 m tall; basal leaves palmately 5- to 11-lobed, to 35 cm long, usually a little wider, glabrous above, more or less puberulent beneath, the lobes sharply toothed, acute; petioles glabrous, to 50 cm long; cauline leaves alternate, similar to the basal leaves except for the progressively shorter petioles; flowers white, to 1.3 cm across; sepals (3–) 4 (–5), white, broadly ovate, concave, acute at the tip, glabrous, to 1 cm long, falling away as the flower opens; petals absent; stamens numerous, with slender filaments and oblong anthers; achenes numerous in small heads, beaked, more or less 4-angular, inflated, to 8 mm long. June–July.

Along rivers and streams.

IL, IN, MO (FAC−), KY (FACW−).

False bugbane.

This species has flowers that lack petals, as in *Thalictrum*, but the leaves are simple and palmately lobed.

103. RHAMNACEAE—BUCKTHORN FAMILY

Trees or shrubs, sometimes climbing, sometimes with thorns; leaves alternate, simple, stipulate; flowers usually perfect, rarely unisexual and dioecious, variously arranged; sepals 4- to 5-lobed; petals (4–) 5, free, sometimes absent; stamens (4–) 5, free, borne from a disk, the disk fleshy; pistil 1, the ovary superior, 2- to 5-locular, with 1 (–2) ovules per locule; fruit a berry, drupe, capsule, or samara.

This family is composed of about forty-five to fifty genera and nearly nine hundred species found throughout most of the World.

1. Woody vines .... 1. *Berchemia*
1. Trees or shrubs.
   2. Winter buds without scales; flowers perfect; nutlets without a groove .... 2. *Frangula*
   2. Winter buds scaly; flowers unisexual; nutlets with a groove .... 3. *Rhamnus*

1. **Berchemia** Neck.—Supple-jack

Climbing woody shrubs without thorns; leaves alternate, simple, pinnately nerved; flowers perfect, in panicles; calyx 5-lobed, petals 5, free; stamens 5, free; disk filling the calyx tube but not attached to the ovary; pistil 1, the ovary superior but covered by the disk; fruit a 1-seeded drupe.

Ten species make up the genus, with all but ours native to Asia and Africa.

1. **Berchemia scandens** (Hill) K. Koch, Dendr. 1:602. 1869. Fig. 128.
Rhamnus scandens Hill, Hort. Kew. 453. 1768.

High-climbing woody vine; stems terete, slender, tough, glabrous; buds lanceoloid, glabrous, closely appressed to the stem; leaf scars round, elevated, with one bundle scar; leaves ovate-oblong, obtuse to acute at the apex, rounded at the base, serrulate or undulate, glabrous on both surfaces, dark green above, paler below, to 5 cm long, to 2.5 cm wide, the petioles to 1 cm long; flowers to 3 mm across, greenish white, perfect, in axillary or terminal panicles; calyx 5-lobed, the lobes acute; petals 5, acute; stamens 5, shorter than the petals; drupe ellipsoid, blue, to 8 mm long, 1-seeded. April–June.

Swamps, wet woods.

IL, MO (FAC+), KY (FACW).

Supple-jack; rattan vine.

This high-climbing vine may have stem diameters in excess of 15 cm. It occurs only in the southern part of our range.

2. **Frangula** Mill.—Buckthorn

Trees or shrubs, sometimes with thorns; winter buds without scales, hairy; leaves alternate, simple, entire, pinnately nerved; flowers perfect, actinomorphic, appearing after the leaves, in cymes, panicles, or racemes; calyx campanulate, 4- or

128. *Berchemia scandens* (Supple-jack). a. Leafy branch with inflorescences. b. Flower, partially removed. c. Fruit.

5-lobed; petals 4–5, free, clawed, or petals absent; stamens 4–5; disk fleshy, not attached to the ovary; style undivided; drupe berrylike with 2–4 nutlets.

Although species in this genus have often been included in *Rhamnus*, it seems best to recognize *Frangula* as a separate genus because of the perfect flowers that appear after the leaves expand, the undivided style, and the hairy, scaleless winter buds.

Frangula consists of about fifty species, most of them native to Europe and Asia. Only the following occurs in central Midwest wetlands.

1. **Frangula alnus** Mill. Gard. Dict., ed. 8, 1768, Fig. 129.
Rhamnus frangula L. Sp. Pl. 1753.
Rhamnus frangula L. var. *angustifolia* Loud. Arb. & Fruct. 2:537. 1838.

Shrub or small tree to 5 m tall, without thorns; twigs slender, gray to brown, puberulent; buds ovoid, acute, tomentose, without scales, 4–5 mm long; leaf scars

129. *Frangula alnus* (Glossy buckthorn). a. Leafy branch, with flowers. b. Flower. c. Fruit. d. Nutlet. e. var. *angustifolis.*

oval, slightly elevated, with 3 bundle scars; leaves alternate, simple, obovate to narrowly lanceolate, obtuse to acute at the apex, rounded to cuneate at the base, entire or sparsely crenulate, glabrous, to 7 cm long, to 4 cm broad, the glabrous petioles to 1 cm long; flowers perfect, solitary or few-flowered in axillary, sessile umbels, greenish, to 2 mm broad; calyx campanulate, 5-lobed, the lobes acute, glabrous; petals 5, free, bifid; stamens 5, free; drupe globose, red at first, becoming black, to 8 mm in diameter, with 3 ungrooved nutlets. May–July.

Bogs, disturbed areas.

IA, IL, IN (FAC+), KY, NE (FAC), OH (not listed), as *Rhamnus frangula.*

Glossy buckthorn; shiny buckthorn.

Rare specimens with narrowly lanceolate leaves may be called var. *angustifolia*.

This species differs from *Rhamnus cathartica* by all its leaves alternate and with teeth.

3. **Rhamnus** L.—Buckthorn

Trees or shrubs, sometimes with thorns; winter buds with scales; leaves alternate, simple, entire, pinnately nerved; flowers unisexual, appearing with the leaves, the plants monoecious or dioecious, actinomorphic, in cymes, panicles, or racemes; calyx campanulate, 4- or 5-lobed; petals 4–5, free, clawed, or petals absent; stamens 4–5; disk fleshy, not attached to the ovary; styles 3- or 4-cleft; drupe berrylike, with 2–4 nutlets.

In this work, *Rhamnus* includes only those species with unisexual flowers, winter buds with scales, and divided styles. As treated, *Rhamnus* consists of about fifty species in the Northern Hemisphere.

1. Leaves with 3–4 pairs of distinct lateral veins, obtuse to abruptly acuminate at the apex, not as much as six times as long as the petioles ....2. *R. cathartica*
1. Leaves with (4–) 5 or more pairs of distinct lateral veins, acuminate at the apex, more than six times longer than the petioles.
   2. Leaves less than 8 cm long and less than 3.3 cm wide.................................3. *R. lanceolata*
   2. Leaves more than 8 cm long or more than 3.3 cm wide or both...................... 1. *R. alnifolia*

1. **Rhamnus alnifolia** L'Hér. Sert. Angl. 3. 1788. Fig. 130.

Shrub to 75 cm tall, without thorns; twigs slender, red or brown, puberulent; buds ovoid, subacute, to 5 mm long; leaf scars half-round, slightly elevated, with 3 bundle scars; leaves alternate, simple, elliptic to oval, obtuse to acute at the apex, rounded to cuneate at the base, serrulate, glabrous or nearly so, to 10 cm long, up to half as wide, on glabrous petioles to 1.5 cm long; flowers mostly unisexual, dioecious, 1–3 per leaf axil; calyx campanulate, with 5 acute, deltoid, glabrous lobes; petals absent; stamens 5, free; drupes globose, black, to 6 mm in diameter, with 3 shallowly grooved, flat nutlets. May–July.

Fens, bogs, swamps, woods.

IA, IL, IN, KY, OH (OBL).

Alder-leaved buckthorn.

This is the lowest growing species of *Rhamnus*, never reaching a height of one meter. It is also the only *Rhamnus* that lacks petals.

2. **Rhamnus cathartica** L. Sp. Pl. 193. 1753. Fig. 131.

Shrub or small tree to 6 m tall, with spine-tipped branches; twigs slender, gray or brown, glabrous, with a short spine at the tip; buds lance-ovoid, acute, to 5 mm long; leaf scars half-elliptic, slightly elevated, with 3 bundle scars; leaves alternate or opposite, simple, elliptic to ovate, obtuse to acute at the apex, rounded or subcordate at the base, dull on the upper surface, crenulate, glabrous, to 5.5 cm long, to 2.5 cm wide, on glabrous petioles to 1 cm long; flowers mostly unisexual, dioecious,

130. *Rhamnus alnifolia* (Alder-leaved buckthorn). a. Branch, with leaves and fruits. b, c. Flowers. d. Fruit. e. Seed.

greenish, to 2 mm across, 1–4 per leaf axil; calyx campanulate, with 4 acute lobes; petals 4, free, linear-lanceolate; stamens 4, free; drupe globose, black, to 8 mm in diameter, with 3–4 deeply grooved nutlets. May–June.

Several habitats including wetlands.

Native of Europe; IA, IL, IN, MO (UPL), KS, KY, NE, OH (FACU).

European buckthorn

In the central Midwest, this is an obnoxious weed in disturbed woods and along fences. Its spread is partly attributed to seed dispersal by birds.

The fruits have a strong laxative effect.

131. *Rhamnus cathartica* (European buckthorn). a. Leafy branch, with flowers. b. Staminate flower. c. Fruit. d. Nutlet.

3. **Rhamnus lanceolata** Pursh, Fl. Am. Sept. 166. 1814. Fig. 132.
Rhamnus lanceolata Pursh var. *glabrata* Gl. Phytologia 2:288. 1947.

Shrub to 5 m tall, without thorns; twigs slender, gray, puberulent; buds ovoid, acute, to 5 mm long; leaf scars half-round, slightly elevated, with 3 bundle scars; leaves alternate, simple, lanceolate to lance-ovate, obtuse to acute to acuminate at the apex, rounded to cuneate at the base, serrulate, glabrous or nearly so on the upper surface, glabrous or puberulent on the lower surface, to 8 cm long, to 2.5 cm wide, on puberulent to nearly glabrous petioles to 1 cm long; flowers bisexual, of two kinds on separate plants, one with an included style, one with an exserted style, in axillary clusters of 1–3 per leaf axil, greenish-yellow, to 3 mm across; calyx campanulate, with 4 acute lobes; petals 4, free, bifid; stamens 4, free; drupe globose, black, to 8 mm in diameter, with 2 deeply grooved nutlets. April–June.

Riverbanks, cliffs, calcareous fens.

IA, IL, IN, KS, KY, MO, NE, OH (NI).

Lance-leaved buckthorn.

The flowers of this species appear as the leaves begin to unfold.

132. *Rhamnus lanceolata* (Lance-leaved buckthorn). a. Leafy branch with flowers. b. Staminate flower. c. Pistillate flower. d. Fruit. e. Nutlet.

104. ROSACEAE—ROSE FAMILY

Herbs, trees, shrubs, or rarely woody vines, sometimes bearing spines or thorns; leaves alternate, simple or compound, usually with stipules; inflorescence various; flowers usually perfect, actinomorphic; sepals (4–) 5, often attached to the rim of a floral cup; petals (4–) 5, free; stamens (10–) 15 or more; pistils 1–many, the ovary superior to inferior; fruit various.

This large and diverse family consists of approximately one hundred genera and three thousand species. Several genera and species occur in wetlands in the central Midwest.

1. Plants woody, either trees, shrubs, or woody arching brambles.
  2. Leaves simple.
    3. Ovary or ovaries superior.
      4. Ovary 1; fruit fleshy .... 15. *Prunus*
      4. Ovaries 3–8; fruit dry.
        5. Leaves palmately lobed; ovaries 3–5 .... 13. *Physocarpus*
        5. Leaves unlobed; ovaries 5–8........ 19. *Spiraea*
    3. Ovary inferior, enclosed, at least in part, by the hypanthium.
      6. Plants without thorns.
        7. Petals at least twice as long as broad; ovary 6- to 10-locular; leaves often subcordate at base .... 2. *Amelanchier*
        7. Petals less than twice as long as broad; ovary 2- to 5-locular; leaves seldom if ever subcordate at base.
          8. Petals 1 cm long or longer; small trees; midrib of leaves eglandular..... 10. *Malus*
          8. Petals less than 1 cm long; shrubs; midrib of leaves usually glandular .... 12. *Photinia*
      6. Plants with thorns.
        9. Styles united at base; fruit a small apple........ 10. *Malus*
        9. Styles free to base; fruit a haw .... 5. *Crataegus*
  2. Leaves compound.
    10. Plants spiny or bristly, usually arching or trailing.
      11. Leaves palmately compound; flowers usually white; fruit a blackberry or raspberry ....17. *Rubus*
      11. Leaves pinnately compound; flowers usually pink or rose; fruit enclosed in a hypanthium........16. *Rosa*
    10. Plants without spines or bristles, erect........11. *Pentaphylloides*
1. Plants herbaceous.
  12. Leaves all basal.
    13. Flowers of two types, some with petals and no pistils, others without petals but with pistils; leaves simple........ 6. *Dalibarda*
    13. Flowers all alike; leaves trifoliolate........ 8. *Fragaria*
  12. Leaves cauline.
    14. Flowers pink, rose, or purple.
      15. Flowers pink, many in a panicle, each flower less than 1 cm across; stems glabrous; terminal leaflet several-lobed; plants 1–2 m tall .... 7. *Filipendula*
      15. Flowers purple or pinkish purple, solitary or few in a cyme or corymb, each flower more than 1 cm across; stems pubescent; terminal leaflet usually not lobed; plants up to 1 m tall.
        16. Flowers erect; leaflets 5–7; style not persistent on the achene, not plumose..... ....4. *Comarum*

16. Flowers often nodding; leaflets 7–17; styles persistent on the achene, plumose. .... 9. *Geum*

14. Flowers white, cream, or yellow.

17. Petals absent; calyx 4-parted .... 18. *Sanguisorba*

17. Petals present; calyx 5- (6-) parted.

18. Petals white.

19. Plants trailing or ascending, nearly glabrous; leaves pedately 3- or 5-foliolate .... 17. *Rubus*

19. Plants erect, pubescent; leaves pinnately 3- to 11-foliolate.

20. Style persistent on the achene; petals 3–5 mm long; stems not glandular-villous; uppermost leaf often unlobed .... 9. *Geum*

20. Style not persistent on the achene; petals 5–8 mm long; stems glandular-villous; uppermost leaf usually lobed .... 14. *Potentilla*

18. Petals yellow.

21. All leaves trifoliolate .... 14. *Potentilla*

21. At least some of the leaves 5-foliolate or more.

22. Leaves pinnately compound.

23. Bractlets present between calyx lobes; ovaries at least 30 per flower.

24. Style persistent on the achene .... 9. *Geum*

24. Style deciduous from the achene.

25. Leaflets silvery-silky beneath; flower solitary ... 3. *Argentina*

25. Leaflets not silvery-silky beneath; flowers in cymes .... 14. *Potentilla*

23. Bractlets absent between calyx lobes; ovaries 1–4 per flower .... 1. *Agrimonia*

22. Leaves palmately compound.

26. Style deciduous from the achene .... 14. *Potentilla*

26. Style persistent on the achene .... 9. *Geum*

1. **Agrimonia** L.—Groovebur; Agrimony

Perennial herbs; leaves alternate, pinnately compound with small intercalary leaflets mixed with regular leaflets, stipulate; flowers in spikelike racemes, actinomorphic, perfect, with a hypanthium, subtended by a 3-lobed bract; calyx tubular, turbinate, bristly at the throat, 5-lobed; petals 5, free, yellow; stamens 5–15; pistils 2, the ovaries hidden by the hypanthium; fruit a pair of achenes enclosed by the calyx, the calyx with hooked bristles.

Agrimonia consists of fifteen species in north temperate regions of the world.

1. Leaflets (excluding smaller interposed ones) 11–23 .... 1. *A. parviflora*

1. Leaflets (excluding smaller interposed ones) 5–9 .... 2. *A. pubescens*

1. **Agrimonia parviflora** Ait. Hort. Kew. 2:130. 1789. Fig. 133.

Perennial herbs from fibrous roots; stems erect, sometimes robust, to 2 m tall, densely villous to densely hirsute; leaves alternate, pinnately compound, with 11–23 leaflets and several smaller leaflets intermixed, the larger leaflets lanceolate to lance-ovate, acuminate at the apex, tapering to the sessile base, sharply serrate, glabrous above, pubescent on the veins beneath, glandular on the lower surface, to 2.5 cm long, to 2.0 cm wide, sessile; flowers in an interrupted spikelike raceme, the axis of the inflorescence glandular and finely pubescent, each flower perfect, actinomorphic, on short pedicels, bracteolate; hypanthium 2–3 mm long, with stiff bristles; petals 5, free, yellow, 1–5 (–6) mm long; stamens 5–10; fruit turbinate, 4–5 mm long, with numerous hooked bristles, the achenes 2.0–2.5 mm long. July–September.

Swampy woods, marshes.

IA, IL, IN, MO (FAC+), KS, KY, NE, OH (FAC).

Swamp groovebur; swamp agrimony.

With 11–23 major leaflets per leaf, this species has more leaflets than any other *Agrimonia* in the central Midwest. It is usually the most wetland of any species of *Agrimonia* in the central Midwest.

133. *Agrimonia parviflora* (Swamp groovebur). Leafy branch with flowers (center). Fruit (above).

2. **Agrimonia pubescens** Wallr. Beitr. Bot. 1:45–46. 1846. Fig. 134.

Perennial herbs from tuberous thickened roots; stems erect, stout, to 1.2 m tall, densely hirsute; leaves alternate, pinnately compound, with 5–9 (–13) major leaflets and numerous small intercalary leaflets, the major leaflets lanceolate to elliptic, acute at the apex, tapering to the base, coarsely serrate, glabrous or pubescent on the upper surface, densely velvety-pubescent on the lower surface, to 3.5 cm long, to 2.2 cm wide; stipules foliaceous, coarsely serrate to laciniate, to 2 cm long; flowers in an interrupted, spikelike raceme to 20 cm long, the axis densely pubescent with eglandular hairs; flowers perfect, actinomorphic, 4–7 mm across; sepals 5, forming a hypanthium 2.5–3.0 mm long, strigose and with numerous short, stiff hairs near the base; petals 5, free from each other, yellow, 2.0–3.5 mm long; stamens 5–15, free; pistils 2, the ovaries within the hypanthium; fruit an achene. June–August.

134. *Agrimonia pubescens* (Hairy groovebur). Flower (upper left). Inflorescence (center). Leaf (next to right). Base of stein (lower right).

Moist or dry woods.

IA, IL, IN, KS, KY, MO, NE, OH (UPL).

Hairy groovebur; hairy agrimony.

This species is distinguished from other species of *Agrimonia* by its 5–9 velvety-pubescent major leaflets and its densely hirsute stems. It rarely occurs in swampy woods. In DuPage County, Illinois, it grows at the edge of a fen.

2. **Amelanchier** Medic.—Serviceberry; Shadbush

Shrubs or small trees; leaves alternate, simple, toothed, usually slightly subcordate at the base, without stipules; inflorescence usually racemose, with white or pink flowers; flowers perfect, actinomorphic, sometimes opening before the leaves expand; calyx 5-parted, united below; petals 5, free; stamens usually about 20; ovary superior, 5-locular; fruit a berrylike pome.

This genus of usually attractive shrubs or small trees consists of about twenty species in the Northern Hemisphere. Several species occur in dry habitats in the central Midwest, with the following sometimes found in wetlands.

1. Leaves broadly rounded at the apex.
   2. Veins of leaves curving upward, forked; inflorescence up to 3 cm long; petals 6.0–7.5 mm long .... 1. *A. alnifolia*
   2. Veins of leaves straight, not forked; inflorescence more than 3 cm long; petals 10–15 mm long .... 4. *A. sanguinea*

1. Leaves acute to acuminate at the apex.
   3. Leaves with up to 10 pairs of veins per blade; petals obovate .... 2. *A. interior*
   3. Leaves with 12 or more pairs of veins per blade; petals narrowly oblong........... 3. *A. laevis*

1. **Amelanchier alnifolia** (Nutt.) Nutt. in Roemer, Syn. Man. 3:147. 1847. Fig. 135.
Aronia alnifolia Nutt. Gen. 1:306. 1818.

135. *Amelanchier alnifolia* (Alder-leaved shadbush). Fruit (left). Leaves and flowers (right).

Shrub to 7 m tall, with extensive stolons, usually forming colonies; leaves alternate, simple, plicate before expanding fully, suborbicular, broadly rounded at the apex, rounded and usually slightly subcordate at the base, to 4 (–5) cm long, nearly as wide, coarsely serrate, the primary veins forked, with the end of the forks extending into the teeth, yellow-tomentose, becoming glabrous or nearly so at maturity, on slender petioles ¼–⅓ as long as the blades; inflorescence a few-flowered raceme to 3 cm long, the axis of the inflorescence sericeous; flowers perfect, actinomorphic, blooming as the leaves begin to unfold, pedicellate, the lowest on pedicels up to 1 cm long; sepals 5, deltate, united below, green, 2.5–3.0 mm long, with recurved tips; petals 5, free, white, 6–8 mm long; stamens about 20; ovary tomentose at apex; fruit globose to obovoid, blue or purplish, about 1 cm in diameter, the flesh juicy and sweet. April–May.

Along streams.

IA (FACU), NE (FAC−).

Alder-leaved shadbush.

This northern species barely enters our range in northwestern Iowa and Nebraska where it forms colonies along streams. It is distinguished by its broadly rounded leaves with primary veins forking. The blue or purplish fruits are sweet and edible.

2. **Amelanchier interior** Nielsen, Am. Midl. Nat. 22:185. 1939. Fig. 136.

Shrub or small tree to 10 m tall, not forming stolons nor colonies; leaves alternate, simple, elliptic to broadly ovate, acute at the apex, rounded or slightly subcordate at the base, to 7 cm long, to 5 cm wide, flowering after the leaves expand, finely serrate, glabrous at maturity, with 8–10 pairs of veins per leaf, the petioles ¼–⅓ as long as the blades; inflorescence racemose, the racemes pendulous, 3–7 cm long, the axis glabrous; flowers perfect, actinomorphic, the lowest on pedicels to 45 mm long; sepals 5, united at the base, glabrous except at the tip, reflexed, 2.5–3.5 mm long; petals 5, free, white, obovate, 1.3–1.5 cm long; stamens about 20; fruit globose, purple-black, 6–8 mm in diameter. June.

Banks of streams.

IA, IL, IN (UPL).

Northwestern shadbush.

This species is distinguished by its acute leaves with 8–10 pairs of veins and its obovate petals. Although primarily an upland species, *A. interior* sometimes grows along streams.

136. *Amelanchier interior* (Northwestern shadbush). Fruiting branch (left). Flowering branch (right).

3. **Amelanchier laevis** Wieg. Rhodora 14:154–158. 1912. Fig. 137.

Shrub or usually small tree to 15 m tall; leaves alternate, simple, elliptic to ovate, acute to acuminate at the apex, rounded or slightly subcordate at the base, expanding before the flowers open, to 6 cm long, to 4 cm wide, glabrous, with 12–17 pairs of veins, more or less glaucous below, the petioles 1/3–1/2 as long as the blades; inflorescence racemose, usually pendulous, 3–7 cm long, the axis glabrous; flowers perfect, actinomorphic, the lowest on pedicels up to 5 cm long; sepals 5, united at the base, glabrous, reflexed, 2.5–4.0 mm long; petals 5, free, white, narrowly oblong, 10–18 mm long; stamens usually 20; summit of ovary glabrous; fruit globose, glaucous, purple-black, 8–12 mm in diameter, sweet, juicy. March–June.

Swampy woods, mesic woods, dry woods.

IA, IL, IN, KY, OH (UPL).

Smooth shadbush.

This small tree is distinguished by its acute to acuminate, glabrous leaves with more than 12 pairs of veins and its narrowly oblong petals.

This species occasionally grows in swampy woods.

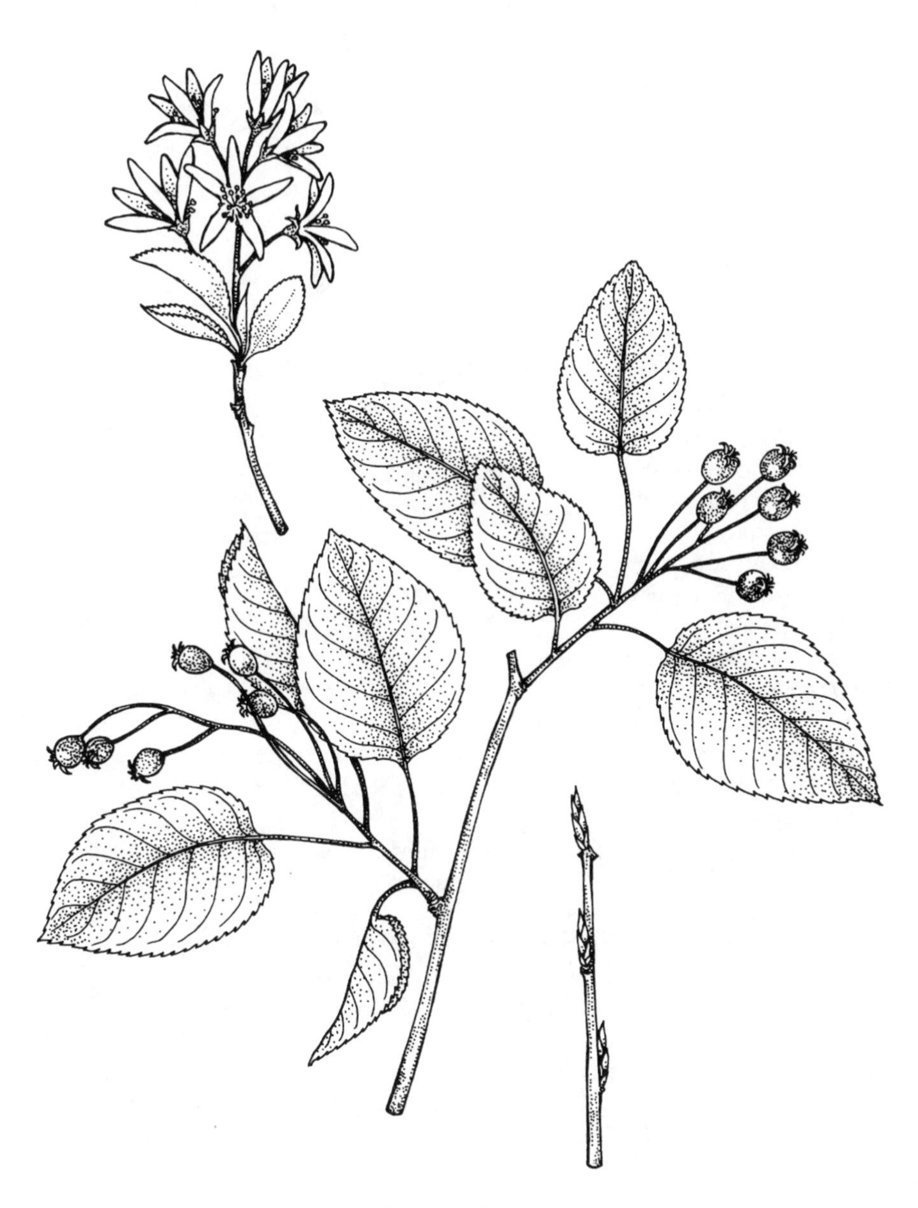

137. *Amelanchier laevis* (Smooth shadbush). Habit. Flowers (upper left). Twig (lower right).

4. **Amelanchier sanguinea** (Pursh) DC. Prodr. 2:633. 1825. Fig. 138.
Pyrus sanguinea Pursh, Fl. Am. Sept. 1:340. 1813.

Shrub or small tree to 3.5 m tall, usually with several stems together; leaves alternate, simple, broadly oblong to nearly orbicular, rounded but usually mucronulate at the apex, rounded or less commonly slightly subcordate at the base, about half grown when the flowers open, to 6 cm long, to 5 cm wide, usually glabrous above, tomentose beneath, serrate except near the base, usually with 6–8 pairs of veins, the veins extending to the teeth; inflorescence racemose, ascending; flowers perfect, actinomorphic, the lowest on pedicels up to 3.5 cm long; sepals 5, united at the base, tomentose, 4–5 mm long; petals 5, free, white, oblong to narrowly ovate, 10–15 mm long; stamens usually 20; fruit globose, purple-black, 7–10 mm in diameter. May–June.

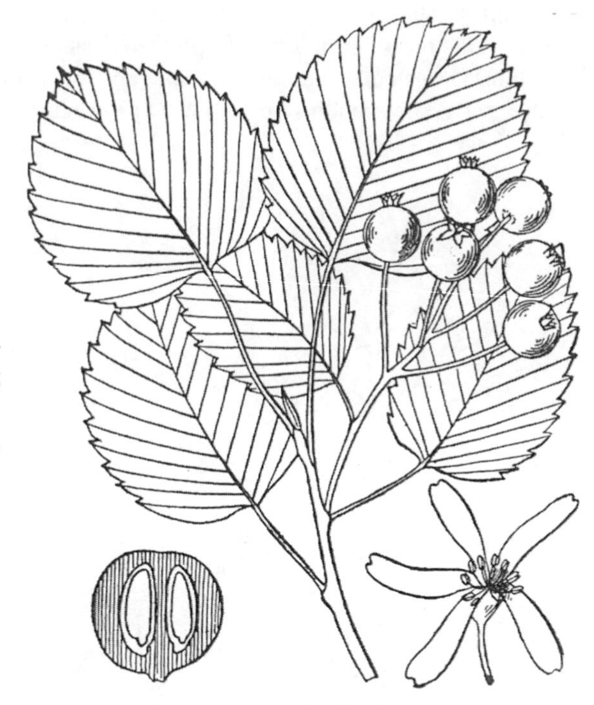
138. *Amelanchier sanguinea* (Shadbush). Leaves and fruits (center). Section of fruit (lower left). Flower (lower right).

Often in wet ground.

IA, IL, IN, OH (UP).

Shadbush.

This species is recognized by its many clumped stems up to 3.5 m tall and its leaves that are tomentose on the lower surface.

3. **Argentina** Lam.—Silver Cinquefoil

Herbaceous perennials with stolons; leaves basal, pinnately compound, with 9 or more leaflets; flower solitary on leafless peduncles, perfect, actinomorphic, yellow, bracteolate; sepals 5, united below; petals 5, free, longer than the calyx lobes; stamens about 20; pistils numerous, the styles lateral; achenes corky, deeply grooved.

This genus consists of eight species. Traditionally the members of this genus are included within *Potentilla*.

1. **Argentina anserina** (L.) Rydb. Mem. Dept. Bot. Col. Univ. 2:159. 1898. Fig. 139.
Potentilla anserina L. Sp. Pl. 1:495. 1753.
Argentina vulgaris Lam. Fl. Franc. 3:119. 1778.

Tufted perennial herb with slender stolons up to 1 m long; leaves basal, erect, pinnately compound, with up to 25 leaflets, stipulate; leaflets variable in size, oblong to oblanceolate to obovate, acute or obtuse at the apex, tapering to the sessile base, sharply toothed, glabrous or nearly so, at least at maturity, silvery tomentose and shiny below, up to 4.5 cm long, up to 1.2 cm wide; flower solitary on a leafless peduncle, the peduncle up to 30 cm long, the flower up to 2.5 cm across, subtended by cleft bracteoles; sepals 5, green, united at the base, 8–12 mm long; petals 5, free, yellow, obovate, 10–12 mm long; stamens about 20; style lateral, filiform; receptacle villous; achenes ovoid to subglobose, corky, deeply grooved, 2.2–2.6 mm long. May–September.

Sandy shores, sandy beaches.

IA, IL, IN (FACW+), OH (OBL).

Silverweed.

This is a species of sandy shores, readily distinguished by its silvery tomentose leaves and large solitary flower on a leafless peduncle.

139. *Argentina anserina* (Silverweed). Habit (center). Flower (lower left).

4. **Comarum** L.—Marsh Cinquefoil

Suffruticose plant with rhizomes; leaves alternate, pinnately compound, with 5 or 7 leaflets, stipulate; flowers solitary or in cymes, perfect, actinomorphic, erect, pink, bracteolate; sepals 5, united below; petals 5, free, shorter than the calyx lobes; stamens numerous, attached to a disk; pistils numerous, the style lateral; receptacle spongy during fruiting; achenes glabrous.

Only one species comprises this genus. It is often included within *Potentilla* from which it differs by its lateral styles and pink petals.

1. **Comarum palustre** L. Sp. Pl. 1:502. 1753. Fig. 140.
Potentilla palustris (L.) Scop. Fl. Carn., ed. 2, 1:359. 1772.

Perennial herb, often woody at the base, with long rhizomes; stems decumbent to ascending, usually reddish brown, to 60 cm tall; leaves alternate, pinnately compound, on petioles to 15 cm long, with 5 or 7 leaflets, the leaflets oblanceolate to elliptic to oblong, to 8 (–10) cm long, to 3.0 (–3.5) cm wide, usually glabrous or nearly so, rarely sericeous, glaucous, sharply serrate; inflorescence leafy, bearing 1–10 erect flowers; flowers perfect, actinomorphic, up to 2 cm across, bracteolate; sepals 5, united below, green, to 10 mm long; petals 5, free, pink to red-purple, to 5 mm long, about half as long as the sepals; stamens numerous, inserted on a pubescent disk; styles lateral; fruit a cluster of achenes attached to a spongy receptacle, each achene smooth, brown, with a beak attached to one side. June–August.

140. *Comarum palustre* (Marsh cinquefoil). Habit (center). Flower (upper left).

Wet meadows, bogs, fens, around lakes, often in standing water.

IA, IL, IN, OH (OBL).

Marsh cinquefoil.

This species is readily distinguished by its pinnately compound leaves, pink to red-purple flowers, and spongy receptacle in fruit.

5. **Crataegus** L.—Hawthorn

Usually spiny shrubs or trees; leaves alternate, simple, toothed and sometimes lobed; inflorescence corymbose, with 1–several flowers; flowers perfect, actinomorphic; sepals 5, free or united below, green; petals 5, free, usually white; stamens 5, 10, 15, or 20, with oblong anthers pink, yellow, red, or white; ovary inferior, with 1–5 carpels and 1–5 free styles persistent on the fruit; fruit a pome, usually globose, red or yellow or greenish, with 1–5 1-seeded nutlets.

The number of species in this genus is undetermined due to the variation and hybridization of most of the species.

For most of my botanical career, I have treated the genus *Crataegus* the way many botanists do by simply calling most specimens *C. mollis, C. viridis, C. crus-galli,* and a few others, or by just ignoring them completely.

Since I started working intensively in wetlands since 1989, I have noticed that many specimens actually do not fit my general concept of *C. mollis, C. viridis,* and *C. crus-galli*. As a result, I began observing in detail the leaves, flowers, and fruits of the hawthorns I have come in contact with in the central Midwest.

During the last twenty years, my concept of Midwestern wetland hawthorns has changed considerably, and I believe I have a better understanding of this group. There is much variation within the species, due for a large part to crossing and back-crossing, but several species may be separated rather satisfactorily. One problem is that flowers and fruits are often necessary to be absolutely certain for positive identification.

I have tried in the key below to use fairly reliable vegetative characteristics where possible. This key should enable the user to identify most of the wetland *Crataegus* that occur in the central Midwest.

Key to Wetland Species of *Crataegus* in the Central Midwest

Note: In order to identify species of *Crataegus* as accurately as possible, it is necessary to have both flowering and fruiting material available. However, since having both of these is difficult and often inconvenient, I have based the following key primarily on mature leaves first, followed by flowering and fruiting characteristics.

1. Veins of the larger leaves running to the sinuses as well as the point of each tooth.
    2. Leaves obovate to spatulate, longer than broad, unlobed or 3-lobed near the apex; fruit subglobose .... 31. *C. spathulata*
    2. Leaves broadly obovate to deltoid, broader than long, deeply several-lobed; fruit oblongoid .... 16. *C. marshallii*
1. Veins of the larger leaves running only to the point of each tooth.
    3. Leaves tapering to the base.
        4. Leaves unlobed or occasionally less commonly shallowly lobed above the middle.
            5. Leaves glossy on the upper surface.
                6. Leaves yellow-green at maturity.
                    7. Leaves with strongly impressed veins on the upper surface.
                        8. Mature leaves glabrous on both surfaces .... 10. *C. hannibalensis*
                        8. Mature leaves villous on the veins on the lower surface .... 33. *C. vallicola*
                    7. Leaves not strongly-impressed veined on the upper surface.
                        9. Mature leaves pubescent on the lower surface .... 8. *C. dispessa*
                        9. Mature leaves glabrous on the lower surface.

10. Trees to 10 m tall; flowers 10–15 mm across; branches of inflorescence glabrous (villous in var. *insignis*); stamens 10 or 15, the anthers pale yellow ... . 1. *C. acutifolia*

10. Shrubs to 5 m tall; flowers 16–18 mm across; branches of inflorescence villous; stamens 20, the anthers dark red .... 30. *C. simulata*

6. Leaves dark green to blue-green to bright green at maturity.

11. Mature leaves pubescent on the veins on the lower surface; flowers blooming from late May to mid-June .... 32. *C. vailiae*

11. Mature leaves glabrous on the veins on the lower surface; flowers blooming from late April to mid-May.

12. Young branches and branches of inflorescence villous .... 11. *C. incaedua*

12. Young branches and branches of inflorescence glabrous or sparsely villous.

13. Leaves thin, nearly membranaceous .... 28. *C. regalis*

13. Leaves thick, subcoriaceous to coriaceous.

14. Leaves broadest above the middle.

15. Leaves bright green; petioles minutely glandular .... 9. *C. fecunda*

15. Leaves dark green to blue-green; petioles eglandular.

16. Flowers 5–12 in a corymb .... 7. *C. disperma*

16. Flowers 10–20 in a corymb .... 5. *C. crus-galli*

14. Leaves broadest at or below the middle.

17. Leaves strongly impressed-veined on the upper surface, paler on the lower surface .... 24. *C. peoriensis*

17. Leaves not impressed-veined on the upper surface, scarcely paler on the lower surface.

18. Leaves obovate to elliptic; stamens 10; fruit 6–10 mm in diameter, red.

19. Anthers pale yellow; fruit 6–8 mm in diameter .... 22. *C. palmeri*

19. Anthers pink; fruit 8–10 mm in diameter .... 7. *C. disperma*

18. Some or all the leaves ovate; stamens 18–20; fruit 8–13 mm in diameter, orange-red .... 29. *C. reverchonii*

5. Leaves dull on the upper surface.

20. Leaves thin, the veins not impressed on the upper surface.

21. Leaves oblong to obovate; flowers 15–18 mm across; petioles 2 cm long or longer; thorns on branches gray .... 35. *C. viridis*

21. Leaves ovate; flowers 10–13 mm across; petioles up to 2 cm long; thorns on branches dark purple .... 21. *C. ovata*

20. Leaves thick, the veins impressed on the upper surface.

22. Leaves obtuse to acute at the apex .... 13. *C. lettermanii*

22. Leaves acute to acuminate at the apex.

23. Leaves villous on the veins on the lower surface; stamens 20; fruit 10–13 mm in diameter, the flesh firm .... 34. *C. verruculosa*

23. Leaves glabrous or slightly pubescent on the veins on the lower surface; stamens 10 or 15, rarely 20; fruit 8–10 mm in diameter, the flesh mealy ... . 4. *C. collina*

4. Most of the mature leaves distinctly lobed above the middle.

24. Petioles glandular.

25. Leaves thick, shiny on the upper surface; fruit dark red .... 19. *C. nitida*

25. Leaves thin, dull on the upper surface; fruit bright red or orange-red.

26. Leaves acute at the apex.

27. Leaves puberulent on the veins on the lower surface; flowers 12–15 mm across; stamens 20, the anthers pale yellow; fruit 5–7 mm in diameter, bright red .... 35. *C. viridis*
27. Leaves glabrous on the veins on the lower surface; flowers 16–18 mm across; stamens 10, the anthers pink or white; fruit 9–12 mm in diameter, orange-red.. 18. *C. neobushii*
26. Leaves, or most of them, obtuse at the apex .... 21. *C. ovata*
24. Petioles eglandular .... 12. *C. laxiflora*
3. Leaves rounded or truncate or somewhat cordate at the base.
28. Some or all of the mature leaves cordate at the base .... 3. *C. coccinioides*
28. All mature leaves rounded or truncate at the base.
29. Leaves thin, not firm, sometimes nearly membranaceous.
30. Some of the leaves deeply lobed; branches of the inflorescence glabrous; fruit dry .... 18. *C. neobushii*
30. Leaves merely shallowly lobed, or occasionally unlobed; branches of the inflorescence villous; fruit juicy (except in *C. pringlei*).
31. Shrubs; leaves blue-green; flowers 12–15 mm across .... 2. *C. apiomorpha*
31. Small trees; leaves green or yellow-green; flowers 15–20 mm across.
32. Branches after the second year dark gray-brown; flowers 15–18 mm across; anthers pink; fruit mealy or juicy .... 15. *C. lucorum*
32. Branches after the second year bright orange-brown; flowers 18–20 mm across; anthers purple; fruit dry .... 25. *C. pringlei*
29. Leaves thick or, if thin, then firm.
33. Mature leaves 8 cm long or longer, nearly as wide, densely tomentose throughout .... 17. *C. mollis*
33. Mature leaves less than 8 cm long, less than 8 cm wide, glabrous or variously pubescent, but not densely tomentose.
34. Some of the mature leaves deeply lobed.
35. Mature leaves with some pubescence; flowers 20–24 mm across; stamens 20, the anthers pink or pale yellow .... 14. *C. locuples*
35. Mature leaves glabrous or nearly so; flowers 16–20 mm across; stamens 10 or, if 20, the anthers red.
36. Leaves dark green; fruit 7–10 mm in diameter .... 23. *C. pedicellata*
36. Leaves yellow-green; fruit 12–15 mm in diameter .... 27. *C. putnamiana*
34. Mature leaves unlobed or sometimes shallowly lobed.
37. Mature leaves pubescent.
38. Leaves dark green, up to 4 cm long, up to 3 cm wide; flowers 10–15 mm across; stamens 20; nutlets pitted.......................... 32. *C. vailiae*
38. Leaves yellow-green, 4–7 cm long, 3–6 cm wide; flowers 20–23 mm across; stamens 10; nutlets not pitted....................... 20. *C. noelensis*
37. Mature leaves glabrous or nearly so.
39. Leaves usually unlobed, yellow-green; branches of the inflorescence villous; stamens 5 or 10 .... 27. *C. putnamiana*
39. Leaves, or some of them, shallowly lobed, dark green; branches of the inflorescence glabrous; stamens 10 or 20.
40. Flowers 15–17 mm across; stamens 10; fruit red .... 26. *C. populnea*
40. Flowers 18–20 mm across; stamens 10 or 20; fruit greenish red .... 6. *C. disjuncta*

1. **Crataegus acutifolia** Sarg. Bot. Gaz. 31:217–218. 1901. Fig. 141.
Crataegus insignis Sarg. Trees & Shrubs 1:107. 1903.
Crataegus acutifolia Sarg. var. *insignis* (Sarg.) Palmer, Brittonia 5:482. 1946.

Small tree to 10 m tall; branches stout, forming a rounded crown, dark gray or brown at maturity, with a few straight gray thorns, the trunk up to 50 cm in diameter; mature leaves alternate, simple, oval to oblong, acute to acuminate at the apex, cuneate at the base, broadest at or slightly above the middle, serrate except near the base, unlobed or occasionally shallowly lobed above the middle, thin but firm, shiny on the upper surface, yellow-green, glabrous on both surfaces, to 6 cm long, to 5 cm wide, the petioles up to 1 cm long; inflorescence a corymb of several flowers, the branches glabrous or villous, the flowers 10–15 mm across; stamens 10 (15), the anthers pale yellow; fruit oblongoid, dull red, 8–12 mm in diameter, dry, with 2–4 nutlets. April–May.

Low woods.

IL, IN, MO (not listed).

Hawthorn.

The distinguishing features of this hawthorn are the combination of its slightly lobed oval to oblong leaves, its completely glabrous leaves, and its 10 stamens with pale yellow anthers. Plants with branches of the inflorescence villous may be called var. *insignis*.

When I was conducting a survey for the St. Louis District, U.S. Army Corps of Engineers, I came across a small colony of this plant in a low woods along the Cahokia Canal east of the village of Brooklyn.

2. **Crataegus apiomorpha** Sarg. Bot. Gaz. 35:386–387. 1903. Fig. 142.

Shrub to 5 m tall, with slender branches forming a narrow crown, the branchlets dark gray to brown and with a few slender light gray thorns up to 3 cm long, the thorns sometimes absent, the trunk at maturity broken up into plates, gray, up to 15 cm in diameter; mature leaves alternate, simple, oblong to ovate, acute to

141. *Crataegus acutifolia* (Hawthorn). Flowering branch (left). Fruiting branch (right).

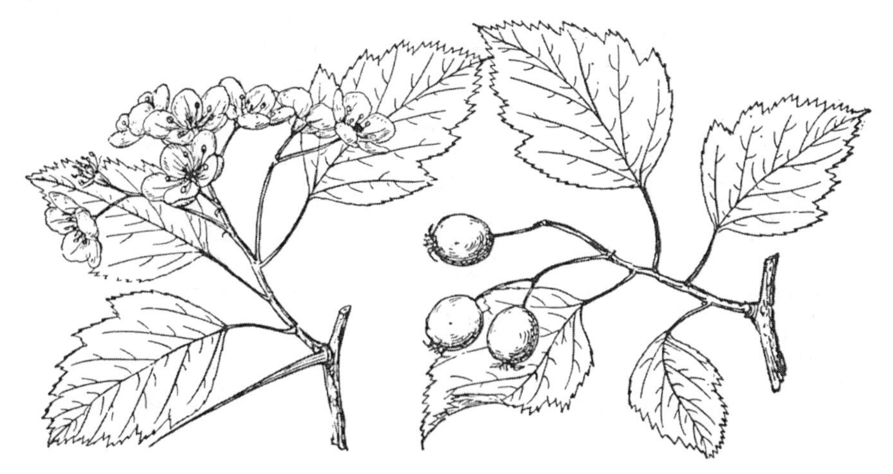

142. *Crataegus apiomorpha* (Hawthorn). Flowering branch (left). Fruiting branch (right).

acuminate at the apex, more or less rounded at the base, broadest at or below the middle, sharply serrate except near the base, some of them shallowly lobed, thin, shiny, blue-green above, paler beneath, glabrous above, glabrous or with some pubescence on the veins beneath, to 6 cm long, to 3 cm wide, the winged petioles up to 2 cm long; inflorescence a corymb of several flowers, the branches villous, the flowers 12–15 mm across; stamens 5 (10), the anthers pink; fruit obovoid, bright red, 10–14 mm in diameter, juicy, with 3–5 nutlets. April–May.

Along streams.

IL, KS, OH (not listed).

Hawthorn

This shrubby species was originally found in northeastern Illinois. Specimens that appear to be this species still occur in that area. I have also seen specimens from Kansas and Ohio that I would assign to this species.

3. **Crataegus coccinioides** Ashe, Journ. Elisha Mitchell Sci. Soc. 16:74–75. 1902. Fig. 143.

Small tree to 7 m tall; branches stout, spreading to form a broad crown, brown at maturity, with thick reddish purple thorns up to 4 cm long, the trunk up to 25 cm in diameter; mature leaves alternate, simple, broadly ovate, acute at the apex, rounded or truncate or subcordate at the base, broadest below the middle, serrate, occasionally with short lobes above the middle, thin but firm, dull, dark green on the upper surface, paler on the lower surface, glabrous above without strongly appressed veins, sparsely pubescent beneath, to 7 cm long, to 5 cm wide, the petioles glandular, up to 2 cm long; inflorescence a corymb of up to 7 flowers, the branches glabrous to sparsely pubescent, the flowers 14–16 mm across; stamens 20, the anthers dark pink to rose; fruit subglobose, dark red, shiny, dotted on the surface, 16–18 mm in diameter, with 5 nutlets. April–May.

Usually dry woods, but occasionally found along streams.

IA, IL, KS, KY, MO (not listed).

143. *Crataegus coccinioides* (Scarlet hawthorn). Flowering branch (left). Fruiting branch (right).

Scarlet hawthorn.

This rather widespread hawthorn, usually found in dry woods, sometimes occurs in bottomland forests along streams in the central Midwest. It is distinguished by its broadly ovate leaves, some of them subcordate at the base.

4. **Crataegus collina** Chapm. Fl. S. U. S. Suppl. 2:684. 1892. Fig. 144.

Small tree to 8 m tall, the branches stout, spreading, gray at maturity, with short, shiny thorns up to 5 cm long, the trunk up to 30 cm in diameter; mature leaves alternate, simple, obovate, acute at the apex, cuneate at the base, broadest above the middle, serrate except at the base, unlobed, thick, with impressed veins on the upper surface, dull, yellow-green, glabrous above, glabrous or sparsely

144. *Crataegus collina* (Hawthorn). Flowering branch (left). Fruiting branch (right).

pubescent beneath, to 5 cm long, to 3.5 cm wide, on slender petioles up to 15 mm long; inflorescence a corymb of many flowers, the branches villous, the flowers 15–20 mm across, on short pedicels; stamens 10 or 15, rarely 20, the anthers pale yellow; fruit subglobose, red, 8–10 mm in diameter, the flesh mealy, with 3–5 nutlets. April–May.

Along streams in bottomland forests.

IL, IN, MO (not listed).

Hawthorn.

This species is common in the southeastern United States, barely reaching our area in southern Missouri, southern Illinois, and southern Indiana. In Jackson County, Illinois, it occurs in a bottomland forest where other species more common in the southeastern United States have been found.

Crataegus collina is similar to *C. punctata*, differing by its more subglobose fruits, its stouter flowering and fruiting stalks, and its usually unlobed leaves.

5. **Crataegus crus-galli** L. Sp. Pl. 1:476. 1753. Fig. 145.
Crataegus crus-galli L. var. *pyracanthifolia* Ait. Hort. Kew. 2:170. 1789.

Small tree to 8 m tall, with stout spreading branches forming a broad crown, the branchlets brown or gray, with straight or curved brown or gray thorns up to 4 cm long, the trunk scaly at maturity, up to 30 cm in diameter; mature leaves alternate, simple, obovate, obtuse or rarely acute at the apex, cuneate at the base, unlobed, serrate, shiny, thick, dark green to blue-green, glabrous on both surfaces, to 5 cm long, to 2.5 cm wide, the petioles stout, up to 15 mm long; inflorescence a corymb of 10–20 flowers, the branches glabrous, the flowers 10–15 mm across, on slender pedicels; stamens 10, the anthers rose, pink, or white; fruit subglobose, red, 8–10 mm in diameter, with 1–3 nutlets.

May–June.

Woods, pastures, along streams, bottomland forests.

IA, IL, IN, MO (FAC), KS, KY, OH (FACU).

Cock-spur thorn.

This is one of the more common and widespread species in the central Midwest. However, most of the other obovate-leaved species of *Crataegus* in the area are often placed in *C. crus-galli*.

Plants whose mature leaves rarely are wider than 15 mm may be called var. *pyracanthifolia*.

6. **Crataegus disjuncta** Sarg. Trees and Shrubs 1:109. 1903. Fig. 146.

Shrub or small tree to 6 m tall, with stout spreading branches forming an irregular crown, the branchlets gray to pale brown, with straight purple thorns up to 6 cm long, the trunk dark gray, scaly at maturity, up to 15 cm in diameter; mature leaves alternate, simple, broadly ovate, acute to acuminate at the apex, truncate at the base, broadest below the middle, shallowly lobed, serrate, thin but firm, dull, dark blue-green on the upper surface, paler beneath, glabrous on both surfaces, to 6 cm long, to 5 cm wide, on slender petioles up to 2.5 cm long; inflorescence a corymb of 3–6 flowers, the branches glabrous, the flowers 18–20 mm across, on stout

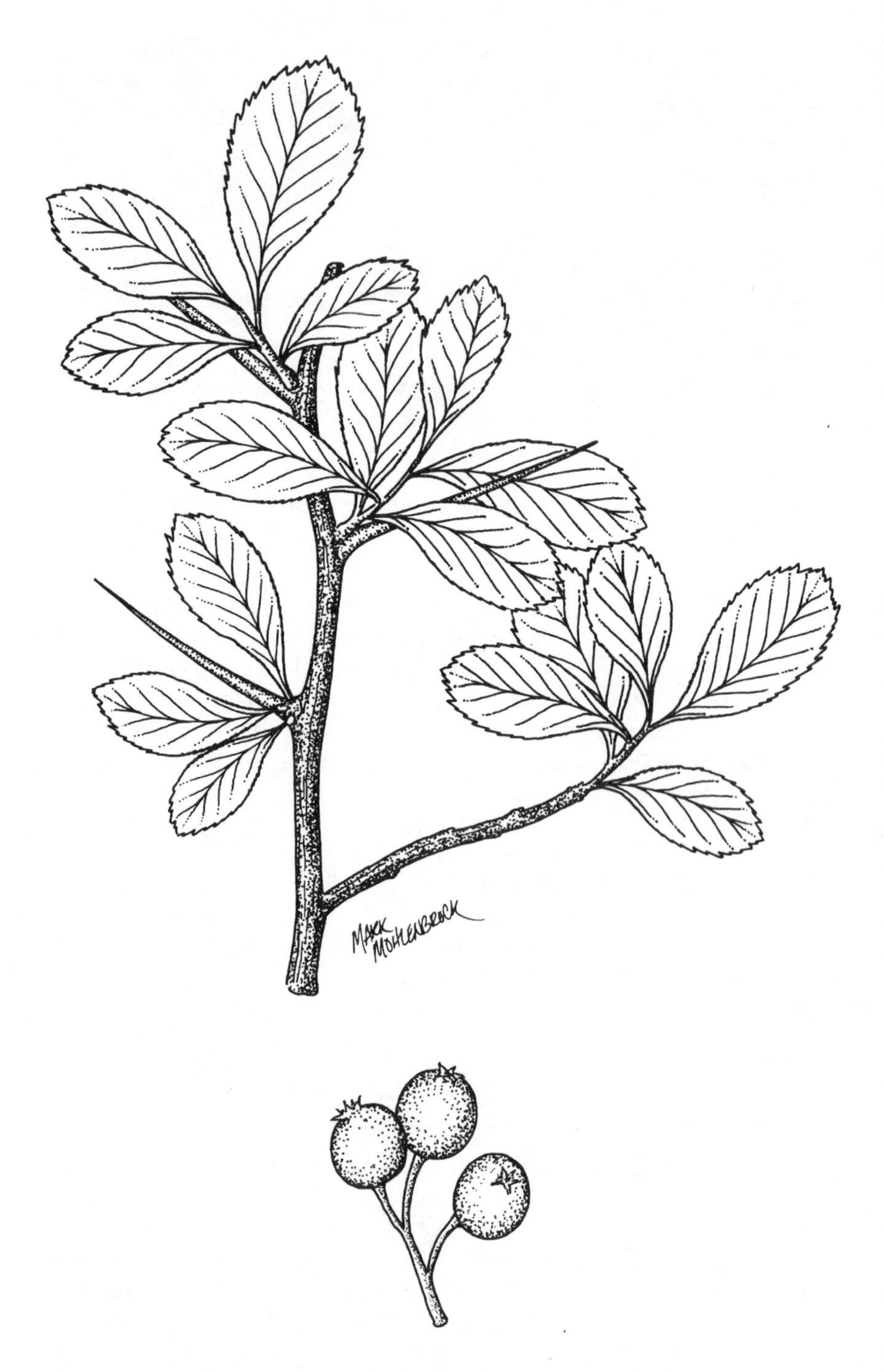

145. *Crataegus crus-galli* (Cock-spur thorn). Habit (center). Fruits (below).

146. *Crataegus disujuncta* (Hawthorn). Flowering branch (left). Fruiting branch (right).

pedicels; stamens 10–20, the anthers white or dark pink; fruit subglobose, green to dull red, 10–15 mm in diameter, the flesh dry, with 2–5 nutlets. April–May.

Along streams.

KY, MO, OH (not listed).

Hawthorn.

This species is very similar to *C. populnea*, differing by its larger flowers, often 20 stamens, and greenish red fruits.

Nearly all of the plants of this species occur along the banks of rocky, clear streams.

7. **Crataegus disperma** Ashe, Journ. Elisha Mitchell Sci. Soc. 17:14–15. 1900. Fig. 147.

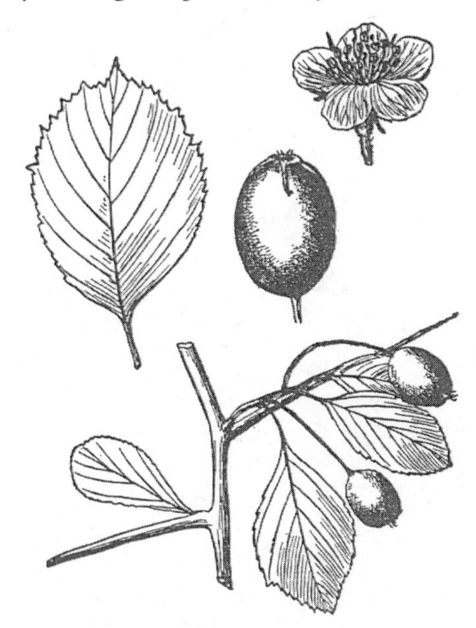

147. *Crataegus disperma* (Spreading hawthorn). Branch with leaves, spines, and fruits (below). Leaf (upper left). Fruit (upper center). Flower (upper right).

Small tree to 8 m tall, with spreading branches forming an irregular crown, the branchlets gray, with numerous straight or slightly curved purple thorns up to 4 cm long, the trunk up to 20 cm in diameter, gray and slightly scaly at maturity; mature leaves alternate, simple, obovate to elliptic, acute to obtuse at the apex, cuneate at the base, broadest above or at the middle, sometimes shallowly lobed above, serrate, firm, glossy, dark green above, paler beneath, glabrous on both surfaces, to 5 cm long, to 3.5 cm wide, on slender winged petioles up to 1.5 cm long; inflorescence a corymb of 5–12 flowers, the branches glabrous, the flowers 13–18 mm across, with 10 stamens, the anthers pink; fruit subglobose to oblongoid, 8–10 mm in diameter, red, with dry flesh and 2–3 nutlets. April–May.

Along streams.

IA, IL, IN, KS, MO, OH (not listed).

Spreading hawthorn.

This species resembles *C. crus-galli* since most of the leaves are broadest above the middle, or *C. palmeri* if the leaves are broadest at the middle. It differs from *C. crus-galli* buy its fewer-flowered corymbs and from *C. palmeri* by its larger fruits and its pink anthers.

The fact that Ashe described two species with similar spellings—*C. disperma* and *C. dispessa*—has been a source of confusion. In his Manual of the Trees of North America, Sargent inadvertently used the binomial *C. dispersa* for *C. dispessa*.

8. **Crataegus dispessa** Ashe, Journ. Elisha Mitchell Sci. Soc. 19:17. 1903. Fig. 148.

Tree to 10 m tall, with spreading branches forming a broad crown, the branchlets light brown, slender, pubescent, with straight light brown spines up to 3 cm long, the trunk up 30 cm in diameter, gray; mature leaves alternate, simple, oval to elliptic, obtuse at the apex, cuneate at the base, broadest at the middle, sometimes shallowly lobed above, serrate, thin but firm, glossy, yellow-green above, paler and pubescent beneath, not impressed-veiny, to 10 cm long, to 7.5 cm wide, on slender, pubescent petioles up to 2.5 cm long; inflorescence a corymb of several flowers, the branches villous, the flowers 20–23 mm across, with 10 (12) stamens, the anthers rose or yellow; fruit obovoid, 10–12 mm in diameter, bright red, juicy, with 4–5 deeply grooved nutlets. April–May.

Bottomland forests, often along streams.

IL, IN, MO (not listed).

Mink hawthorn.

This species has attractive large flowers and bright red juicy fruits. It differs from other hawthorns by its glossy yellow-green leaves that do not have impressed veins but are pubescent on the lower surface.

148. *Crataegus dispessa* (Mink hawthorn). Flowering branch (left). Fruiting branch (right).

9. **Crataegus fecunda** Sarg. Bot. Gaz. 37:111–113. 1902. Fig. 149.

Small tree to 8 m tall, with spreading branches forming a broad crown, the branchlets gray, stout, glabrous, with straight, chestnut-brown thorns to 4 cm long, the trunk up to 20 cm in diameter, dark brown, scaly; mature leaves alternate, simple, obovate, obtuse at the apex, cuneate at the base, broadest above the middle, unlobed, serrate, thick, firm, glossy, bright green above, paler below, glabrous, not impressed-veiny, to 8 cm long, to 6.5 cm wide, on slender puberulent glandular petioles up to 1.5 cm long; inflorescence a corymb of many flowers, the branches glabrous or sparsely villous, the flowers 13–16 mm across, the stamens 10 or 20, with pink or rose anthers; fruit subglobose, 10–14 mm in diameter, orange-red, with black dots, the flesh thick, with 2–3 nutlets. May.

Bottomland forests, usually along streams.

IL, MO (not listed).

Hawthorn.

This hawthorn is fairly common in bottomland forests in southern Illinois and southern Missouri. It is distinguished by its unlobed, obovate leaves that are thick and bright green and with glandular petioles.

149. *Crataegus fecunda* (Hawthorn). Flowering branch (left). Fruiting branch (right).

10. **Crataegus hannibalensis** Palmer, Journ. Arn. Arb. 16:353–355. 1925. Fig. 150.

Small tree to 8 m tall, with spreading branches forming a broad crown, the branchlets slender, gray, glabrous, with numerous short, stout thorns to 3.5 cm long, the trunk up to 30 cm in diameter, gray; mature leaves alternate, simple, obovate, obtuse at the apex, cuneate at the base, broadest above the middle, unlobed, thick, firm, glossy, yellow-green above, paler below, with strongly impressed veins, glabrous, to 5 cm long, to 2.5 cm wide, on slender, glabrous petioles up to 1.5 cm long; inflorescence a corymb of several flowers, the branches glabrous, the flowers 14–16 mm across, the stamens 10, with pale yellow anthers; fruit oblongoid, 7–8 mm in diameter, greenish red, with dry flesh and 1–3 nutlets. May.

Moist woods, sometimes along streams.

IA, IL, MO (not listed).

Hannibal hawthorn.

The yellow-green, obovate leaves that are strongly impressed veiny are distinctive for this species. The greenish red fruits are among the smallest in the genus. This species appears to be confined to the upper part of the central Midwest.

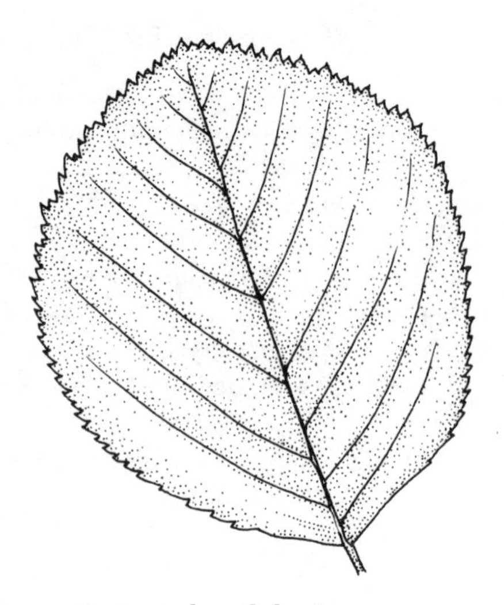

150. *Crataegus hannibalensis* (Hannibal hawthorn). Leaf.

11. **Crataegus incaedua** Sarg. Trees and Shrubs 2:3. 1907. Fig. 151.

Small tree to 7 m tall, with spreading branches forming a rounded crown, the branchlets slender, often zigzag, gray, villous, with purplish or grayish thorns up to 5 cm long, the trunk up to 20 cm in diameter, gray; mature leaves alternate, simple, obovate to broadly oval, acute to obtuse at the apex, cuneate at the base, broadest at or above the middle, unlobed, serrate, rather thick, firm, glabrous, dark green above, paler below, not glossy, without impressed veins, to 5 cm long, to 4 cm wide, on stout, winged petioles up to 1.2 cm long; inflorescence a corymb of many flowers, the branches villous, the flowers 13–15 mm across, the stamens 10, with pale yellow anthers; fruit subglobose, 10–12 mm in diameter, yellowish to red, with firm flesh and 2–3 nutlets. April–May.

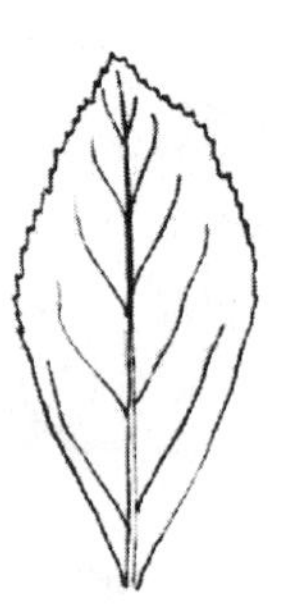

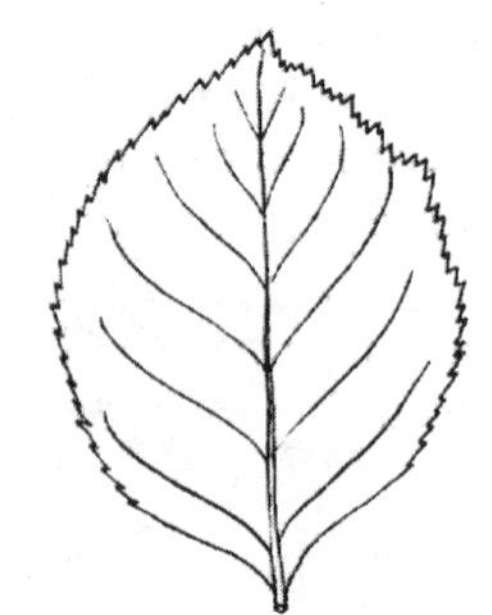

151. *Crataegus incaedua* (Hawthorn). Leaves.

Moist woods, usually along streams.

IN, MO (not listed).

Hawthorn.

This species is possibly a hybrid between *C. collina* and *C. calpodendron*. It is distinguished by its obovate to broadly oval dark green leaves and its villous branchlets and branches of the inflorescence.

12. **Crataegus laxiflora** Sarg. Bot. Gaz. 35:400–401. 1902. Fig. 152.

Small tree to 8 m tall, with spreading branches forming an irregular crown, the branchlets slender, more or less zigzag, glabrous, gray, with numerous stout, straight, purple thorns to 4 cm long, the trunk up to 20 cm in diameter, gray;

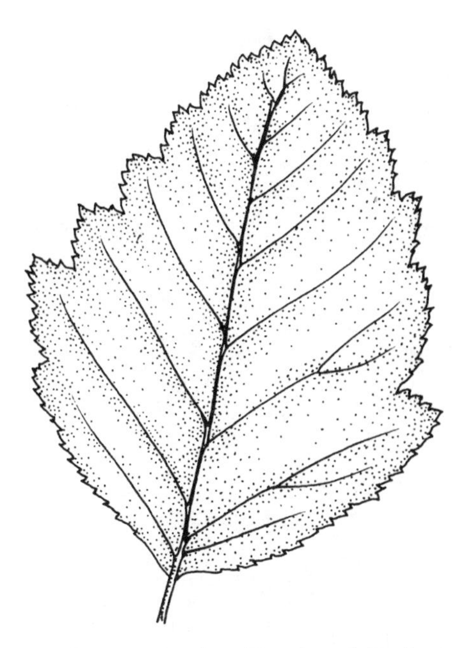

152. *Crataegus laxiflora* (Hawthorn). Leaf.

mature leaves alternate, simple, broadly oval, obtuse to acute at the apex, cuneate at the base, broadest at or above the middle, serrate, shallowly lobed above the middle, glabrous, glossy, thick, firm, dark green above, paler below, to 6 cm long, to 5 cm wide, the petioles slender, glabrous, eglandular, to 1.5 cm long; inflorescence a corymb of many flowers, the branches villous, the flowers 10–13 mm across, the stamens 20, with dark red anthers; fruit subglobose, 7–8 mm in diameter, bright red, with dry flesh, with 2–3 nutlets. May.

Along streams in woods.

IL (not listed).

Hawthorn.

Plants conforming to the description given above occur along several woodland streams in the northern one-fourth of Illinois.

This species is similar in appearance to *C. neobushii, C. nitida, C. ovata,* and *C. viridis* but differs from these by its glandless petioles.

13. **Crataegus lettermanii** Sarg. Bot. Gaz. 31:220–221. Fig. 153.

Small tree to 7 m tall, with spreading branches forming a broad crown, the branchlets slender, reddish brown, tomentose when young, becoming glabrous, with branched shiny reddish brown thorns up to 4 cm long, the trunk up to 25 cm in diameter, dark brown, scaly; mature leaves alternate, simple, obovate, obtuse to acute at the apex, cuneate at the base, broadest above the middle, serrate, rarely with a few short lobes above the middle, thick, more or less dull, yellow-green and scabrous on the upper surface, pubescent at least on the veins on the lower surface, the veins on the upper surface impressed, to 5 cm long, to 3 cm wide, the petioles stout, slightly winged, to 1.5 cm long; inflorescence a corymb of many flowers, the branches villous, the flowers 15–18 mm across, the stamens 10, 15, or 20, with pale yellow or pink anthers; fruit subglobose, 8–10 mm in diameter, dark red with numerous black dots, the flesh firm, with 3–5 prominently keeled nutlets. May.

Bottomland forests, frequently inundated for short periods.

MO (not listed).

Letterman's hawthorn.

This bottomland forest species survives periodic inundation where it is known only from several areas west of St. Louis. It is distinguished by the combination of obovate leaves that are dull and impressed-veiny on the upper surface, pubescent veins on the lower surface, and mostly obtuse leaves. It is most similar to *C. verruculosa*, but this latter species has acute to acuminate leaves.

153. *Crataegus lettermanii* (Letterman's hawthorn). Flowering branch (left). Fruiting branch (right).

14. **Crataegus locuples** Sarg. Crataegus in Missouri 97–98. 1908. Fig. 154.

Small tree to 8 m tall, with spreading branches forming an irregular crown, the branches rather stout, dark gray, sometimes zigzag, very thorny, the thorns straight, up to 3.5 cm long, the trunk up to 20 cm in diameter, dark gray, more or less scaly; mature leaves alternate, simple, ovate, acute at the apex, rounded or truncate or occasionally subcordate at the base, broadest below the middle, serrate, lobed, thin but firm, yellow-green and glabrous above, pubescent below, at least on the veins, to 5 cm long, to 4 cm wide, the petioles slender, puberulent, to 3 cm long; inflorescence a corymb of up to 10 flowers, the branches pubescent, the flowers 20–24 mm across, the stamens 20, with pale yellow or pink anthers; fruit subglobose, 10–12 mm in diameter, red, the flesh dry, with 4–5 nutlets. April–May.

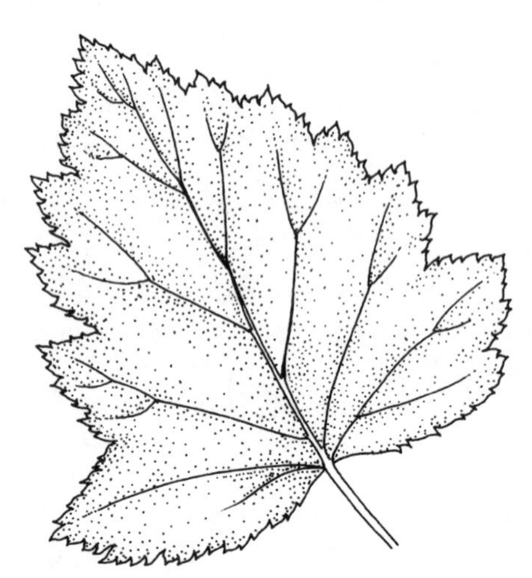

154. *Crataegus locuples* (Hawthorn). Leaf.

Along streams in woods.

KY, MO, OH (not listed).

Hawthorn.

This species has some of the largest flowers in the genus. Vegetatively it is similar to *C. putnamiana*, but the latter species has glabrous leaves, smaller flowers, and larger fruits.

15. **Cratraegus lucorum** Sarg. Bot. Gaz. 31:227–229. 1904. Fig. 155.

Small tree to 8 m tall, the branches ascending to form a narrow crown, the branchlets slender, dark gray-brown, glabrous at maturity, the trunk up to 20 cm in diameter, dark brown; mature leaves alternate, simple, ovate, acute at the apex, rounded at the base, broadest below the middle, serrate, shallowly lobed, dull, thin, green or yellow-green and glabrous on the upper surface, paler and glabrous on the lower surface, to 6 cm long, to 4.5 cm wide, the petioles slender, up to 2.5 cm long; inflorescence a corymb of several flowers, the branches slightly villous, the flowers 15–18 mm across, the stamens 5, 10, or 20, with pink anthers; fruit obovoid or oblongoid, 8–15 mm in diameter, red, with black dots, the flesh mealy or juicy, with 4–5 nutlets. May.

Rich woods, particularly along streams.

IL, OH (not listed).

Hawthorn.

The distinguishing features of this species are its thin leaves with rounded bases and its dark gray-brown branchlets.

155. *Crataegus lucorum* (Hawthorn). Flowering branch (left). Fruiting branch (right).

16. **Crataegus marshallii** Eggl. Rhodora 10:79. 1908. Fig. 156.

Small tree to 7 m tall, the branches spreading to form an irregular crown, the branchlets slender, somewhat zigzag, glabrous at maturity, light brown to gray, with stout, straight, brown thorns up to 4 cm long; mature leaves alternate, simple, broadly ovate, obtuse to acute at the apex, truncate at the base, broadest below the middle, serrate, deeply lobed, somewhat glossy, bright green on the upper surface, paler below, glabrous, the veins of the larger leaves running to the sinuses as well as the point of each tooth, to 5 cm long, to 5 cm wide, the petioles slender, up to 5 cm long, sparsely pubescent; inflorescence a corymb of many flowers, the branches villous, the flowers 10–12 mm across, the stamens 20, with red anthers; fruit obovoid to oblongoid, 4–8 mm in diameter, bright red, the flesh dry, with 1–3 nutlets. April–May.

Mesic woods, bottomland forests, swampy woods, along streams.

IL, MO (FACW), KY (FACU+).

Parsley hawthorn.

This species is readily distinguished by its deeply lobed leaves. It and *C. spathulata* are the only two species of *Crataegus* in the central Midwest in which the veins of the larger leaves run to the sinuses as well as to the point of each tooth.

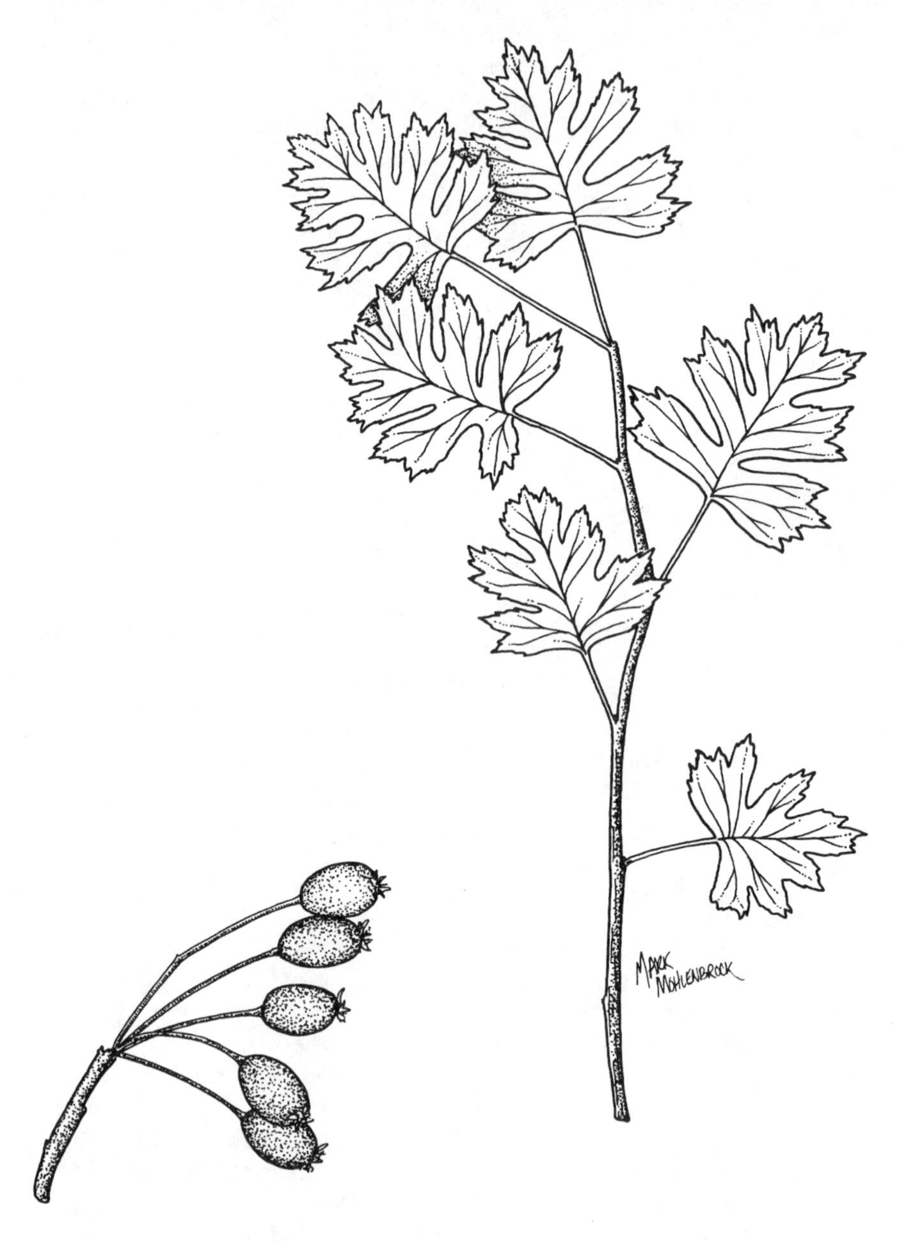

156. *Crataegus marshallii* (Parsley hawthorn). Leafy branch (right). Fruits (lower left).

17. **Crataegus mollis** (Torr. & Gray) Scheele, Linnaea 21:569. 1848. Fig. 157.
Crataegus coccinea L. var. *mollis* Torr. & Gray, Fl. N. Am. 1:465. 1848.

Small to medium tree to 12 m tall, with spreading branches forming a broad crown, the branches stout, gray, pubescent when young but becoming glabrous, with shiny chestnut-brown thorns up to 4 cm long, the trunk up to 25 cm in diameter, becoming scaly with age; leaves alternate, simple, broadly ovate, more or less obtuse at the apex, rounded or truncate at the base, broadest below the middle, sharply serrate and with shallow lobes at or above the middle, thin, yellow-green above, paler beneath, pubescent on both surfaces, dull, to 10 cm long, to 6 cm wide, the pubescent petioles up to 2.5 cm long; inflorescence a corymb of many flowers, the branches tomentose, the flowers 17–22 mm across, the stamens 20, with pale yellow anthers; fruit subglobose, 10–15 mm in diameter, red and often dark-spotted, dry and mealy, with 4–5 nutlets. April–May.

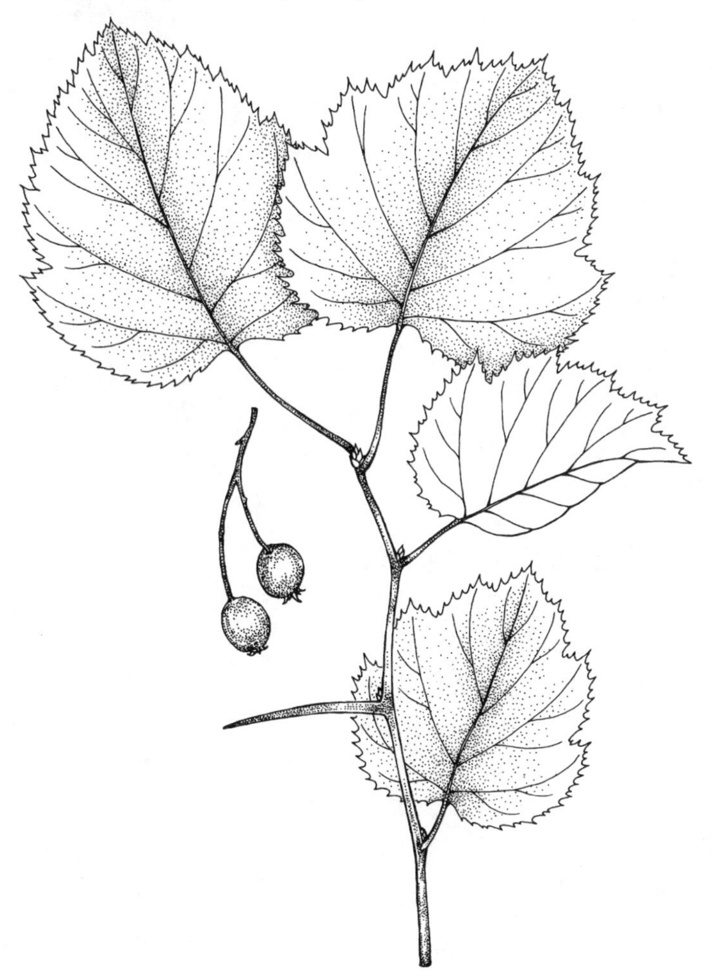

157. *Crataegus mollis* (Red haw). Habit (right). Fruit (left).

Bottomland forests.

IA, IL, IN, MO (FACW), KS, NE (FAC), KY, OH (FACU).

Red haw.

This species is readily recognized by its densely pubescent leaves that are broadly ovate and shallowly lobed. Fruits of this species, particularly in pioneer days, was used in the making of jelly.

18. **Crataegus neobushii** Sarg. Trees & Shrubs 2:9. 1907. Fig. 158.

Shrub to 10 m tall, with spreading branches forming an irregular crown, the branches rather thin, gray, with or without short thorns; leaves alternate, simple, ovate, acute at the apex, rounded at the base, broadest at or near the middle, serrate, some of them rather deeply lobed, thin, not firm, glabrous at maturity, to 6 cm long, to 3.5 cm wide, the petioles slender, glabrous, to 3 cm long; inflorescence a corymb of up to 8 flowers, the branches glabrous, the flowers 16–18 mm across, the stamens 10, with yellow or pink anthers; fruit subglobose, 8–13 mm in diameter, red or greenish, the flesh dry, with 2–3 nutlets. April–May.

Along streams; dry thickets.

IL, MO (not listed).

Bush's hawthorn.

This shrubby hawthorn is distinguished by its ovate, often rather deeply lobed leaves and its rather small flowers borne on glabrous branches of the inflorescence. The woody branches are often without thorns.

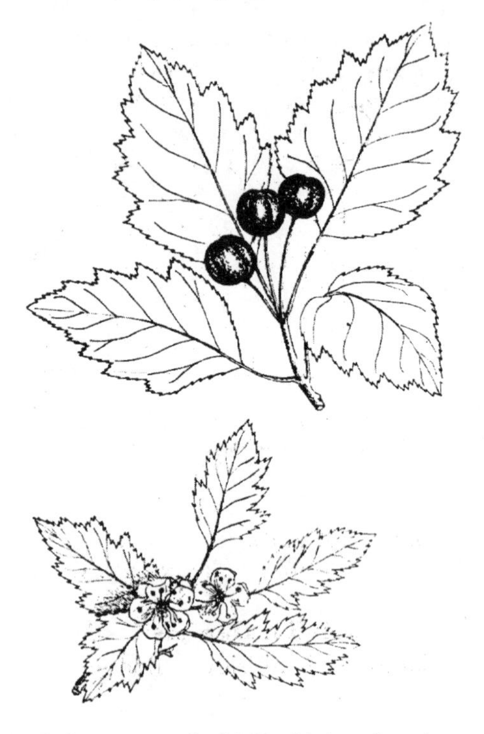

158. *Crataegus neobushii* (Bush's hawthorn). Flowering branch (below). Fruiting branch (above).

19. **Crataegus nitida** (Engelm. ex Britt. & A. Br.) Sarg. Bot. Gaz. 31:231. 1901. Fig. 159.
Crataegus viridis L. var. *nitida* Engelm. ex Britt. & A Br. Ill. Fl. N. U. S. 2:242. 1897.
Crataegus atrorubens Ashe, Journ. Elisha Mitchell Sci. Soc. 16:78–79. 1899.

Small tree to 10 m tall, with spreading branches forming an irregular crown, the branches slender, glabrous, gray, with bright chestnut-brown, shiny thorns up to 3 cm long, or the thorns sometimes absent, the trunk up to 20 cm in diameter, dark gray, scaly at maturity; leaves alternate, simple, mostly obovate, acute at the apex, tapering to the base, broadest usually above the middle, serrate, sometimes shallowly lobed above the middle, thick, coriaceous, shiny, dark green above, paler beneath, up to 6 cm long, up to 3 cm wide, on glabrous petioles up to 2 cm long; inflorescence a corymb of many flowers, the branches glabrous, the flowers 15–17 mm across, the stamens (15) 20, with pale yellow anthers; fruit oblongoid, dark red, glaucous, 8–10 mm in diameter, with 3–5 nutlets. April–May.

Bottomland forests.

IL, MO (OBL), OH (FACU+).

Shiny green haw.

This species resembles *C. viridis* but differs by its thicker, shiny leaves and its dark red fruits. It is a species of bottomland hardwood forests.

Crataegus atrorubens, known from bottomland forests on either side of the Mississippi River near St. Louis, seems to be a broader-leaved variation of *C. nitida*.

159. *Crataegus nitida* (Shiny green haw). Flowering branch (left). Fruiting branch (right).

20. **Crataegus noelensis** Sarg. Journ. Arn. Arb. 1:353. 1920. Fig. 160.

Small tree to 25 m tall, with spreading branches forming a broad crown, the branches reddish brown becoming gray, often zigzag, thorny, the thorns straight, purple, shiny, to 5 cm long, the trunk gray and beset with branched thorns; leaves alternate, simple, ovate, acute at the apex, rounded at the base, serrate, scarcely lobed, broadest near the base, thick, yellow-green, glabrous on the upper surface, densely pubescent on the veins beneath, up to 7 cm long, up to 6 cm wide, on slender, pubescent to glabrous petioles up to 3 cm long; inflorescence a corymb of 5–10 flowers, the branches pubescent, the flowers 20–23 mm across, the stamens 10, with rose anthers; fruit subglobose, orange-red, 10–12 mm in diameter, with soft flesh and 3–5 unpitted nutlets. April–May.

Along streams.

MO (not listed).

Ozark hawthorn.

This species is distinguished by its ovate, unlobed leaves with dense pubescence on the veins beneath, its large flowers, and its unpitted nutlets.

160. *Crataegus noelensis* (Ozark hawthorn). Flowering branch (left). Fruiting branch (right).

21. **Crataegus ovata** Sarg. Man. Trees N. Am. 402–403. 1905. Fig. 161.
Crataegus viridis L. var. *ovata* (Sarg.) Palmer, Ann. Mo.Bot. Gard. 22:561. 1935.

Small tree to 10 m tall, with spreading branches forming an irregular crown, the branches gray, slender, glabrous, with a few dark purple, slender, curved thorns, the trunk up to 30 cm in diameter, gray, smooth; leaves alternate, simple, ovate, acute at the apex, rounded at the base, broadest at or below the middle, serrate, rarely with short lobes, thin, dark green and dull above, paler beneath, glabrous above, usually with a tuft of hairs in the axils of the veins beneath, to 6 cm long, to 4.5 cm wide, the petioles slender, glabrous, up to 2 cm long; inflorescence a corymb of many flowers, the branches glabrous, the flowers 10–13 mm across, the stamens 20, with yellow anthers; fruit subglobose, 5–8 mm in diameter, orange-red, with 5 nutlets. April–May.

161. *Crataegus ovata* (Ovate-leaved hawthorn). Flowering branch (left). Fruiting branch (right).

Along streams.
IL, MO (not listed).
Ovate-leaved hawthorn.
Some botanists believe this merely a variety of *C. viridis*, but it differs by its more ovate leaves and subglobose fruits.

22. **Crataegus palmeri** Sarg. Trees & Shrubs 1:57. 1903. Fig. 162.

Small tree up to 8 m tall, with spreading branches forming a broad crown, the branches light gray to reddish brown, glabrous, with thorns up to 7 cm long, the trunk up to 20 cm in diameter, gray; leaves alternate, simple, obovate to elliptic, rounded at the apex, tapering to the base, broadest at the middle, serrate, without lobes, dark green and shiny above, slightly paler beneath, firm, glabrous, to 5 cm long, to 3 cm wide, the petioles slender, glabrous, to 3 cm long; inflorescence a corymb of many flowers, the branches glabrous, the flowers 12–15 mm across, the stamens 10, with yellow anthers; fruit subglobose, 6–8 mm in diameter, red or greenish red, dotted, with 3 grooved nutlets. April–May.
Along streams.
KS, MO (not listed).
Palmer's hawthorn.
This species is similar to *C. disperma*, differing by its smaller fruits and yellow anthers.

23. **Crataegus pedicellata** Sarg. Bot. Gaz. 31:226–227. 1901. Fig. 163.
Crataegus albicans Ashe, Journ. Elisha Mitchell Sci. Soc. 17:20. 1901.
Crataegus pedicellata Sarg. var. *albicans* (Ashe) Palmer, Fl. Vermont, ed. 3, 154. 1937.

Small tree to 8 m tall, the branches forming a spreading crown, the branchlets slender, reddish brown becoming gray, glabrous, with straight shiny brown thorns to 4 cm long, the trunk up to 25 cm in diameter, scaly, red-brown; leaves alternate,

162. *Crataegus palmeri* (Palmer's hawthorn). Flowering branch (left). Fruiting branch (right).

163. *Crataegus pedicellata* (Pedicellate hawthorn). Flowering branch (left). Fruiting branch (right).

simple, ovate, acute to acuminate at the apex, rounded or somewhat tapering at the base, serrate, usually with several short lobes, dark green, thin but firm, paler beneath, broadest at or below the middle, glabrous at maturity, to 5.5 cm long, to 4.5 cm wide, the petioles slender, glabrous, to 6 cm long; inflorescence a corymb of many flowers, the branches pubescent, the flowers 17–20 mm across, the stamens 10, with pink anthers; fruit obovoid, 7–10 mm in diameter, bright red, with 3–5 deeply grooved nutlets. May.

Along streams.

IL, IN, KY, OH (not listed).

Pedicellate hawthorn.

This species is characterized by its ovate, dark green, glabrous, several-lobed leaves and its obovoid fruits 7–10 mm in diameter. Some plants from northern Illinois with flowers 17–18 mm across have been called var. *albicans*.

24. **Crataegus peoriensis** Sarg. Bot. Gaz. 31:5–6. 1901. Fig. 164.

Small tree to 8 m tall, with rather stout, spreading, orange-brown branches forming a flat crown, with more or less straight brown thorns up to 5 cm long, the trunk up to 20 cm in diameter, dark gray, scaly; leaves alternate, simple, obovate to elliptic, obtuse to acute at the apex, tapering to the base, serrate, usually unlobed, firm, dark green, shiny, glabrous, with deeply impressed veins, to 5 cm long, to 2 cm wide, the petioles sparsely pubescent, to 6 mm long; inflorescence a corymb of several flowers, the branches glabrous, the flowers 10–15 mm across, the stamens 10, with red anthers; fruit obovoid to subglobose, 9–12 mm in diameter, bright red, dotted, the flesh thick, with 2–3 nutlets. May.

Along streams.

IL (not listed).

164. *Crataegus peoriensis* (Peoria hawthorn). Flowering branch (left). Fruiting branch (right).

Peoria hawthorn.

The distinguishing features of this species are its obovate to elliptic leaves that are firm, dark green, shiny, with deeply impressed veins. This small tree occurs along several streams in the Peoria, Illinois, area.

25. **Crataegus populnea** Ashe, Ann. Carnegie Mus. 1:395. 1902. Fig. 165.

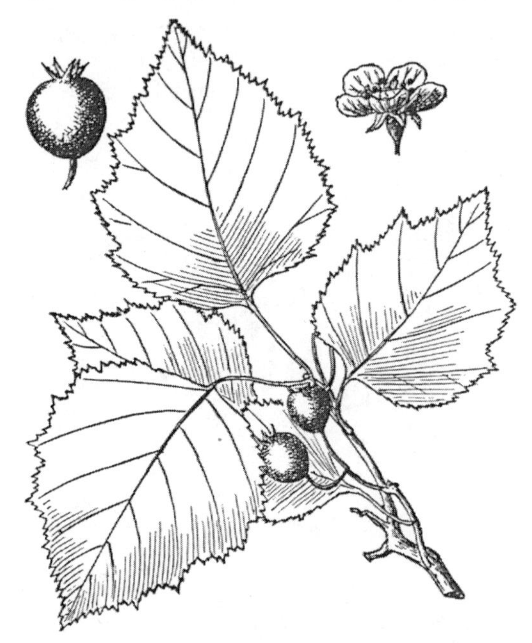

165. *Crataegus populnea* (Hawthorn). Fruiting branch (center). Fruit (upper left). Flower (upper right).

Shrub or small tree to 6 m tall, the branches slender, gray, very thorny, with gray thorns up to 3 cm long, the trunk up to 15 cm in diameter, gray; leaves alternate, simple, ovate or oval, acute at the apex, rounded at the base, broadest below the middle, serrate, often shallowly lobed, thick, dark green, glabrous, to 6 cm long, to 5 cm wide, the petioles slender, glabrous, to 3.5 cm long; inflorescence a corymb of few to several flowers, the branches glabrous, the flowers 15–17 mm across, with 10 stamens, the anthers pink; fruit subglobose, 9–12 mm in diameter, red, with 3–4 nutlets. May.

Along streams.
OH (not listed).
Hawthorn.
The thick, ovate, glabrous leaves and the very thorny short branches distinguish this shrubby species.

26. **Crataegus pringlei** Sarg. Rhodora 3:21–22. 1901. Fig. 166.

Small tree to 8 m tall, the branches spreading forming an irregular crown, the branches brown, rather stout, often zigzag, with usually straight brown thorns up to 3 cm long, the trunk up to 20 cm in diameter, dark brown, scaly; leaves alternate, simple, ovate to oval, obtuse at the apex, rounded at the base, serrate, sometimes with short lobes, thin, yellow-green, glabrous at maturity, to 5 cm long, nearly as wide, the petioles slender, glabrous, to 2 cm long; inflorescence a corymb of many flowers, the branches villous, the flowers 18–20 mm across, the stamens 5 or 10, with purple anthers; fruit obovoid, 8–12 mm in diameter, dark red, with thick, yellow flesh, with 3–5 nutlets. May.
Along streams.
IL, OH (not listed).
Pringle's hawthorn.
This species is recognized by its ovate, thin leaves that are rounded at the base and the rather large flowers.

27. **Crataegus putnamiana** Sarg. Journ. Arn. Arb. 4:102–103. 1923. Fig. 167.

Small tree to 10 m tall, with spreading branches forming a broad crown, the branches slender, gray, with usually straight thorns up to 3 cm long, the trunk up to 30 cm in diameter, gray; leaves alternate, simple, ovate, acute at the apex, rounded at the base, broadest below the middle, serrate, usually with shallow lobes, glabrous at maturity, thin but firm, yellow-green, to 6 cm long, to 5 cm wide, the petioles rather stout, glabrous at maturity, to 4 cm long; inflorescence a corymb of several

166. *Crataegus pringlei* (Pringle's hawthorn). Flowering branch (left). Fruiting branch (right).

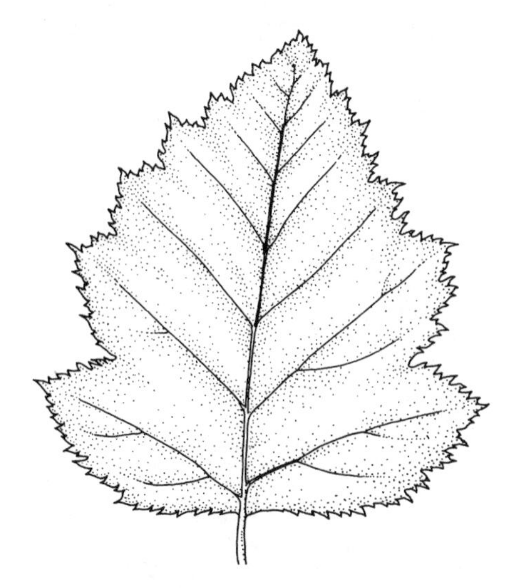

167. *Crataegus putnamiana* (Putnam's hawthorn). Leaf.

flowers, the branches glabrous, the flowers 16–18 mm across, the stamens 20, with red anthers; fruit subglobose, 12–15 mm in diameter, bright red, with 4–5 nutlets. May.

Along streams.

IL, IN, KY, OH (not listed).

Putnam's hawthorn.

This species is similar to *C. pedicellata,* differing by its yellow-green leaves and larger fruits.

28. **Crataegus regalis** Beadle, Biltmore Bot. Stud. 1:134–135. 1906. Fig. 168.

Small tree to 8 m tall, with ascending to spreading branches forming a broad crown, the branches stout, glabrous, orange-brown, with stout, straight thorns up to 5 cm long, the trunk up to 20 cm in diameter; leaves alternate, simple, broadly elliptic, acute at the apex, tapering to the base, serrate, without lobes, thin but firm, glabrous, shiny, bright green, to 5 cm long, to 3.5 cm wide, the petioles stout, glabrous, to 2 cm long; inflorescence a corymb of many flowers, the branches glabrous, the flowers 12–15 mm across, the stamens 10, with pale yellow anthers; fruit oblongoid, 6–8 mm in diameter, green with a reddish tint, with 2–3 nutlets. May.

Along streams.

IN, MO (not listed).

Royal hawthorn.

The bright green, shiny, broadly elliptic, unlobed leaves are distinctive for this species.

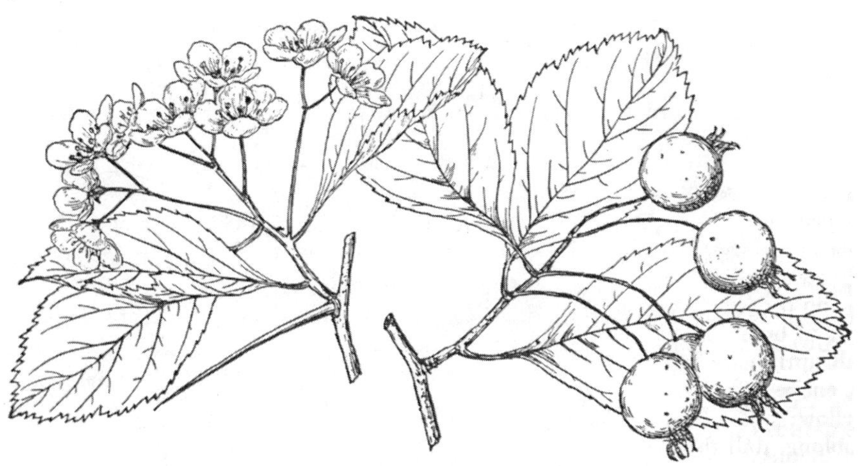

168. *Crataegus regalis* (Royal hawthorn). Flowering branch (left). Fruiting branch (right).

29. **Crataegus reverchonii** Sarg. var. **discolor** (Sarg.) Palmer, Brittonia 5:482. 1946. Fig. 169.
Crataegus discolor Sarg. Crataegus in Missouri 44–45. 1908.

Small tree to 8 m tall, with spreading branches forming a broad crown, the branches slender, gray, glabrous, with many straight thorns up to 3 cm long, the trunk up to 20 cm in diameter, scaly, gray; leaves alternate, simple, obovate to ovate, acute at the apex, tapering to the base, firm, thick, shiny, dark green above, paler beneath, glabrous, to 5 cm long, to 3.5 cm wide, the petioles slender, glabrous, to 3.5 cm long; inflorescence a corymb of many flowers, the branches glabrous, the flowers 12–17 mm across, the stamens 18 or 20, with pink or pale yellow anthers; fruit subglobose, 8–13 mm in diameter, orange-red, with 3–5 nutlets. May.

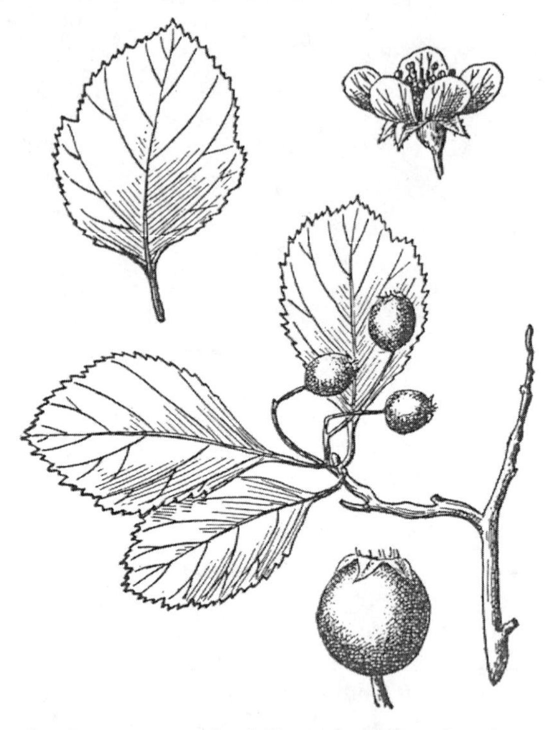

169. *Crataegus reverchonii* (Reverchon's hawthorn). Leaf (upper left). Flower (upper right). Branch with leaves and fruits (center). Fruit (below).

Along streams.

KS, MO (not listed).

Reverchon's hawthorn.

This plant is sometimes considered to be a distinct species called *C. discolor*.

The larger leaves and numerous stamens distinguish this plant from the similar appearing *C. disperma* and *C. palmeri*.

30. **Crataegus simulata** Sarg. Ann. Rep. Mo. Bot. Gard. 22:82–83. 1912. Fig. 170.

Shrub to 5 m tall, with spreading branches forming an irregular crown, the branches gray, glabrous, with numerous stout thorns to 3 cm long; leaves alternate, simple, ovate, acute at the apex, tapering to the base, to 6 cm long, to 5 cm wide, broadest below the middle, serrate, mostly unlobed, yellow-green, shiny, glabrous at maturity, the petioles rather stout, glabrous, to 3 cm long; inflorescence a corymb of few to several flowers, the branches villous, the flowers 16–18 mm across, with 20 stamens, the anthers dark red; fruit subglobose, 7–8 mm in diameter, red, with 3 nutlets. May.

Along streams.

KS, MO (not listed).

Hawthorn.

This shrubby species occurs in the Joplin, Missouri, area in both Kansas and Missouri. It is distinguished by its yellow-green, shiny, ovate leaves that taper to the base and by its villous branches of the inflorescence.

31. **Crataegus spathulata** Michx. Fl. Bor. Am. 1:288. 1803. Fig. 171.

Small tree to 10 m tall, with spreading branches forming a broad crown, the branches slender, red-brown, glabrous, without thorns, or with a few stout straight thorns up to 3 cm long; leaves alternate, simple, obovate to spatulate, obtuse to acute at the apex, tapering to the base, to 2 cm long, to 1 cm wide, usually with 3 short lobes at the tip, serrate except near the base, firm, green, glabrous, with the veins of the larger leaves running to the sinuses as well as to the point of each tooth, the petioles slender, glabrous, to 5 mm long; inflorescence a corymb of many flowers, the branches glabrous, the flowers 6–8 mm across, with 20 stamens, the anthers pale yellow; fruit subglobose, 4–7 mm in diameter, bright red, shiny, with 3–5 nutlets. April–May.

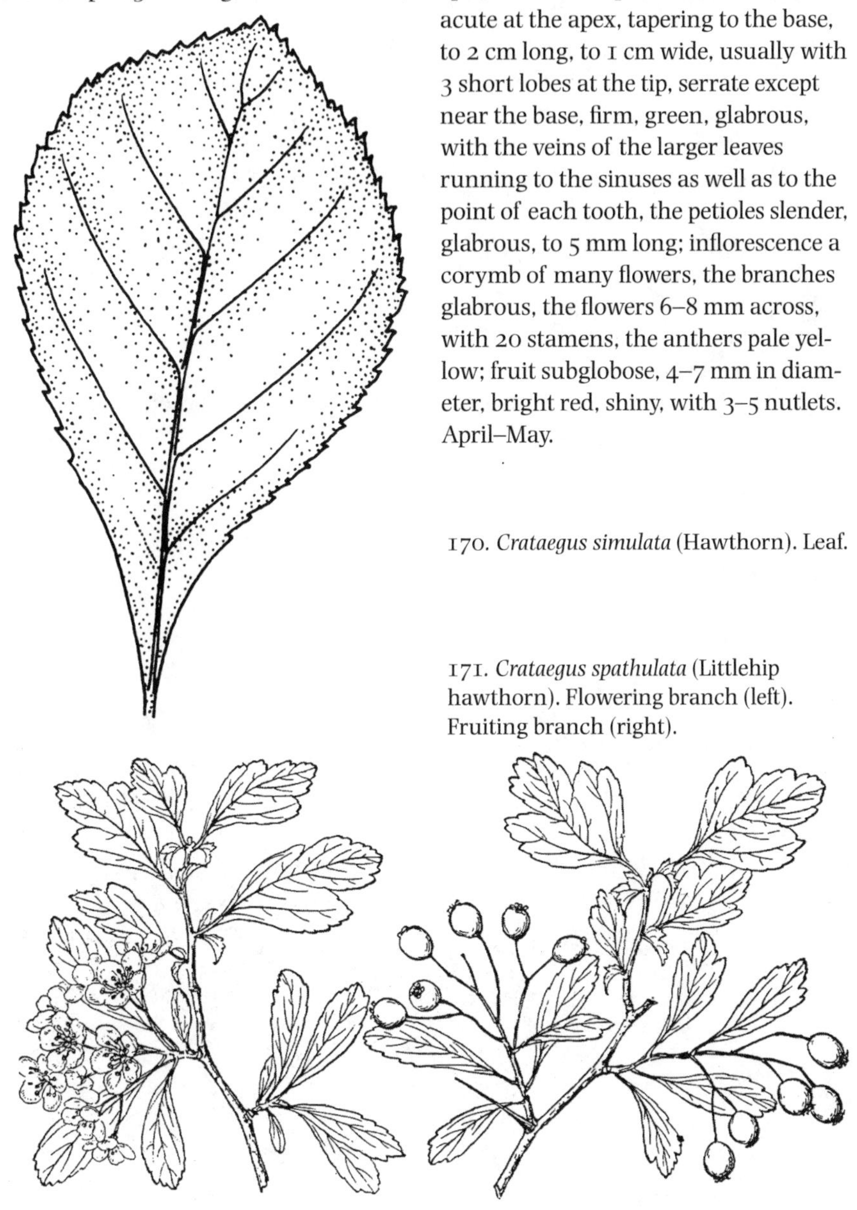

170. *Crataegus simulata* (Hawthorn). Leaf.

171. *Crataegus spathulata* (Littlehip hawthorn). Flowering branch (left). Fruiting branch (right).

Swampy woods.

IL, MO (FACW), KY (FAC).

Littlehip hawthorn.

Crataegus spathulata and *C. marshallii* are the only species of *Crataegus* in the central Midwest whose larger leaves have their veins running to the sinuses as well as to the point of each tooth. The leaves of *C. spathulata* are very distinctive because of the three small shallow lobes near the apex of the blade.

32. **Crataegus vailiae** Britt. Bull. Torrey Club 24:53. 1897. Fig. 172.

Shrub or small tree to 5 m tall, with spreading branches forming an irregular crown, the branches slender, gray, pubescent, with several thorns to 2.5 cm long; leaves alternate, simple, obovate to elliptic, obtuse to acute at the apex, tapering to the base, to 5 cm long, to 3.5 cm wide, serrate, without lobes, dark green, thick, rather shiny, pubescent on the veins beneath, the petioles stout, winged, to 1 cm long; inflorescence a corymb of a few flowers, the branches tomentose, the flowers 10–15 mm across, the stamens 20, with red or whitish anthers; fruit subglobose, 8–10 mm in diameter, red, pubescent when young, with 3–5 pitted nutlets. May–June.

Along streams.

MO (not listed).

Miss Vail's hawthorn.

The unusual feature of this species is its late flowering period, beginning toward the end of May and continuing into early June. This species is also fairly distinct because of its thick, dark green leaves with the veins on the lower surface pubescent and by its tomentose branches of the inflorescence.

172. *Crataegus vailiae* (Miss Vail's hawthorn). Flowering branch (left). Leaves (right).

33. **Crataegus vallicola** Sarg. Crataegus in Missouri 74–75. 1908. Fig. 173.

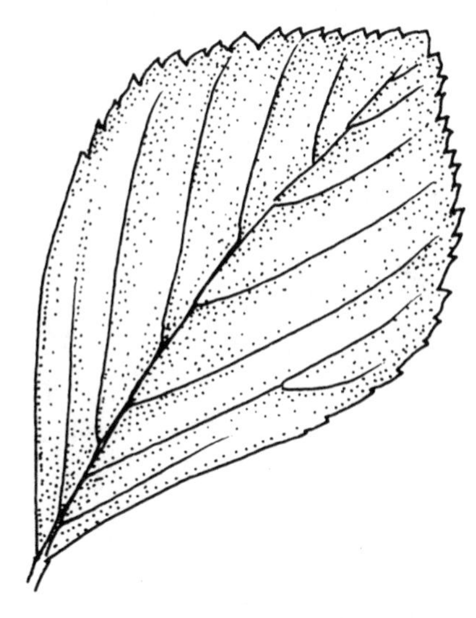

173. *Crataegus vallicola* (Hawthorn). Leaf.

Small tree to 8 m tall, with spreading branches forming a broad crown, the branches slender, gray, pubescent at first, becoming glabrous, with several short, stout thorns to 3 cm long, the trunk up to 20 cm in diameter, gray; leaves alternate, simple, obovate, obtuse at the apex, tapering to the base, broadest above the middle, unlobed, thick, firm, glossy, yellow-green above, paler beneath, with strongly impressed veins above, with puberulent veins beneath, to 5 cm long, to 3 cm wide, on slender, glabrous petioles up to 2 cm long; inflorescence a corymb of several flowers, the branches villous, the flowers 14–16 mm across, the stamens 10, with pale yellow anthers; fruit subglobose, 12–15 mm in diameter, greenish red, the flesh dry, with 1–3 nutlets. May.

Along streams.

MO, OH (not listed).

Hawthorn.

The leaves of this plant are similar to those of *C. hannibalensis*, except the veins on the lower surface of the leaves are pubescent in *C. vallicola*. The rounder, much larger fruits in *C. vallicola* are very different from the fruits in *C. hannibalensis*.

34. **Crataegus verruculosa** Sarg. Man. Trees N. Am. 394. 1905. Fig. 174.

Small tree to 8 m tall, with spreading branches forming a broad crown, the branches stout, red-brown, glabrous but warty, with straight, shiny, brown thorns up to 2 cm long; leaves alternate, simple, obovate, acute to acuminate at the apex, tapering to the base, serrate, unlobed or with very few shallow lobes above the middle, dull, thick, yellow-green, strongly impressed-veiny on the upper surface, villous on the veins on the lower surface, to 4.5 cm long, to 3 cm wide, the petioles stout, winged, pubescent, to 1.2 cm long; inflorescence a corymb of several flowers, the branches villous, the flowers 18–23 mm across, the stamens 20, with pink anthers; fruit subglobose, 10–13 mm in diameter, dark red, with 3–5 nutlets and firm flesh. April–May.

Along streams.

KY, MO (not listed).

Warty hawthorn.

This species receives its specific epithet and common name from the small warty outgrowths on the branches. The leaves are distinctive in being obovate, acute to acuminate at the apex, dull, thick, with impressed veins above and villous veins beneath. The flowers are among the largest in the genus.

174. *Crataegus verruculosa* (Warty hawthorn). Flowering branch (left). Fruiting branch (right).

35. **Crataegus viridis** L. Sp. Pl. 1:476. 1753. Fig. 175.

Small to medium tree to 12 m tall, with spreading branches forming a rounded crown, the branches slender, gray to reddish brown, glabrous, with slender straight thorns up to 3.5 cm long, the trunk up to 40 cm in diameter, gray; leaves alternate, simple, obovate to oblong, acute at the apex, tapering to the base, serrate, rarely with a few shallow lobes at and above the middle, to 4 cm long, to 3 cm wide, dark green above, paler beneath, shiny, thin, glabrous at maturity, the petioles slender, glabrous, to 2 cm long; inflorescence a corymb of many flowers, the branches glabrous, the flowers 15–18 mm across, with 20 stamens, the anthers pale yellow; fruit subglobose, 9–12 mm in diameter, red or orange, with 5 nutlets. April–May.

Swampy woods, along streams.

IL, IN, MO (FACW), KS (FAC).

Green haw.

This is one of the more common species of *Crataegus* in some parts of the central Midwest wetlands. Similar species that may be confused with *C. viridis* are *C. nitida*, which has thicker leaves usually with a few shallow lobes and glandular petioles, and *C. ovata*, which has leaves obtuse at the apex.

6. **Dalibarda** L.—Dewdrop; Dalibarda

Small tufted perennial herb; leaves basal, simple, on long petioles, stipulate; flower solitary on leafless peduncles, of 2 kinds, some sterile with petals and some fertile without petals; sepals 5–6, united below; petals 5, free, caducous; stamens numerous; pistils 5–10, free, the style terminal; fruit a cluster of nearly dry drupelets enclosed by the persistent sepals.

Only the following species comprises the genus.

175. *Crataegus viridis* (Green haw). Habit (center). Flower (lower right).

1. **Dalibarda repens** L. Sp. Pl. 1:491. 1753. Fig. 176.

Small tufted perennial herb; leaves simple, basal, broadly ovate to orbicular, obtuse to subacute at the apex, cordate at the base, to 5 cm long, nearly as wide, crenate, pubescent above and below, on long, slender, downy petioles usually as long as the blades, with setaceous stipules; flower solitary, of 2 kinds, those elevated on long peduncles usually sterile, with 5–6 green sepals united near the base and unequal in size, 5 white petals, free, 4–8 mm long, and numerous stamens, those on shorter curved peduncles fertile, with 5–6 green sepals united near the base and unequal in size, no petals, numerous stamens, and 5–10 pistils, each with a terminal style and a villous ovary; fruit a cluster of nearly dry drupelets, each drupelet oblongoid, roughened, 3–4 mm long. June–September.

Swampy areas, moist woods.

OH (not listed).

Robin-run-away; dewdrop; false violet.

This species has the growth form of stemless violets, and even has both fertile and sterile flowers of stemless violets, but the presence of numerous stamens and several pistils clearly places it in the Rosaceae.

176. *Dalibarda repens* (Robin-run-away). Habit (center). Longitudinal section of flower (upper left).

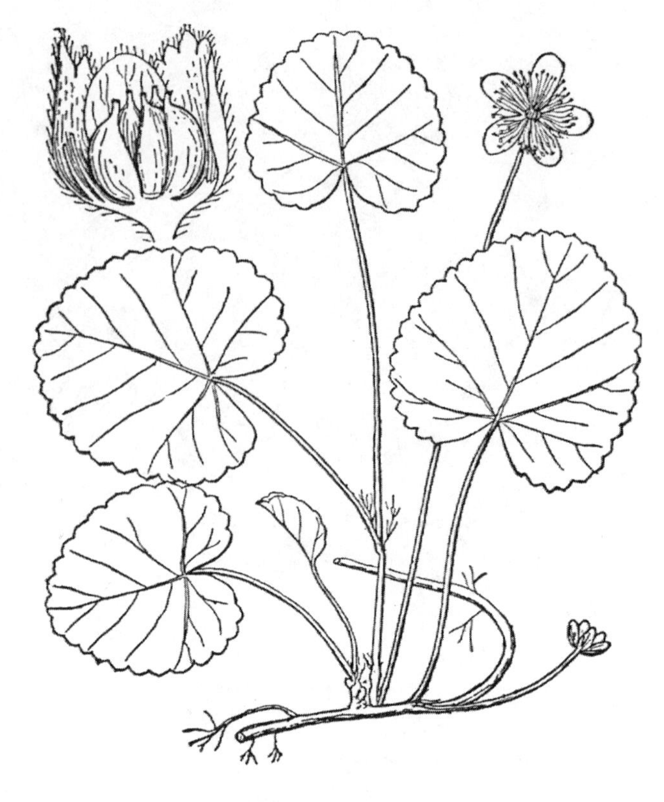

7. **Filipendula** Mill.—Queen-of-the-prairie

Perennial herb with rhizomes; leaves alternate, pinnately divided, stipulate; inflorescence a large panicle of many small flowers; flowers actinomorphic, perfect, pedicellate, with a hypanthium; calyx (4-) 5-parted, green; petals (4-) 5, free; stamens 20 or more, arranged in 10 rows; pistils 5–15, free, arranged in a circle, each style curved; fruit 1-seeded, indehiscent, twisted in some species.

About ten species are in this genus, either in North America or Eurasia.

Filipendula is distinguished by its once-pinnate cleft or coarsely toothed leaves, its 5–15 free pistils, and its non-prickly fruits.

Only the following species occurs in wetlands in the central Midwest.

1. **Filipendula rubra** (Hill) B. L. Robins. Rhodora 8:204. 1906. Fig. 177.
Ulmaria rubra Hill, Hort. Kew. 214. 1769.
Spiraea rubra (Hill) Britt. Bull. Torrey Club 18:270. 1891.

Robust perennial herb; stems erect, branched, grooved, glabrous, up to 2.5 m tall; leaves alternate, pinnately compound, to 1 m long, with 3–7 major leaflets and often several small intercalary leaflets intermixed, with large, toothed or cleft stipules 8–16 mm long, the leaflets more or less obovate, serrate and usually 3- to 9-lobed, the terminal leaflet the largest, green on both sides, glabrous or puberulent, sessile; inflorescence a large panicle of cymes up to 25 cm across; flowers perfect, actinomorphic, fragrant, 7–10 mm across; calyx 5-parted, the sepals ovate, acute at the apex, green, 2–4 mm long; petals 5, free, pink or rose, 6–9 mm long; stamens at least 20; pistils 5–15, free; fruit a cluster of achenes, compressed, lanceolate to oblong, often twisted, glabrous, 6–8 mm long. June–July.

Wet prairies, wet meadows, calcareous fens.

IA, IL, IN, MO (FACW+), KY, OH (FACW).

Queen-of-the-prairie.

When in flower, this is a very handsome species. It is recognized by its large inflorescence of pink to rose flowers with about 20 stamens and 5–15 pistils. The leaves have conspicuous stipules at the base.

177. *Filipendula rubra* (Queen-of-the-prairie). Leaves (center). Flower (upper left). Fruits (lower left). Inflorescence (right).

8. **Fragaria** L.—Strawberry

Perennial herbs with stolons; leaves basal, trifoliolate; flowers few to several on a leafless peduncle, actinomorphic, perfect, bracteate; sepals 5, free or united at base, green; petals 5, free, white; hypanthium present; stamens numerous; pistils numerous, attached to the enlarged receptacle; fruit of numerous minute achenes attached to the enlarged receptacle, the calyx and bracts persistent.

Fragaria consists of about thirty-five species in North and South America.

1. **Fragaria virginiana** Mill, Gard. Dict., ed. 8, Fragaria no. 2. 1768. Fig. 178.
Fragaria virginiana Mill. var. *illinoensis* Prince ex A. Gray, Man. Bot., ed. 5, 155. 1867.

Herbaceous perennial with slender stolons; stems pubescent, sometimes with spreading hairs; leaves basal, trifoliolate, the leaflets ovate, acute at the apex, tapering to the petiolulate base, glabrous or sericeous, sharply serrate, to 4 cm long, to 3 cm wide; flowers few to several on a leafless peduncle, the peduncle with appressed or spreading hairs; sepals 5, free or attached at base, green, 3–5 mm long; petals 5, free, white, 7–10 mm long; stamens numerous; pistils numerous, attached to an enlarged receptacle; fruit of numerous small achenes attached to the large, fleshy, red receptacle, the receptacle sweet, juicy. April–June.

Various habitats, from dry woods and prairies to wet woods and meadows.

IA, IL, IN, MO (FAC−), KS, KY, NE, OH (FACU).

Wild strawberry.

178. *Fragaria virginiana* (Wild strawberry). Habit (center). Fruit (lower right). Flower (upper right).

This variable species is easily recognized by its basal trifoliolate leaves and its juicy red fruits. Plants with numerous spreading hairs on the stems and peduncles have been called var. *illinoensis*.

9. **Geum** L.—Avens

Perennial herbs from a thickened caudex or rhizomes; leaves basal and cauline, serrate, the basal simple, ternately compound, or pinnately compound, on long petioles, the cauline leaves alternate, variously compound below, sometimes simple above, with stipules; flowers solitary or in cymes or corymbs, bracteate; sepals 5, free or united at base, usually reflexed during fruiting; petals 5, free, shorter than to a little longer than the sepals; stamens 20 or more; pistils several, free from each other, attached to a glabrous or pubescent receptacle; fruit a cluster of achenes, each with a persistent, often geniculate style.

Geum consists of about fifty species, nearly all of them in the Northern Hemisphere. Only the following may occur in wetlands in the central Midwest.

1. Flowers purple .... 4. *G. rivale*
1. Flowers yellow or white.
   2. Petals white.
      3. Most of the petals 5 mm long or longer; peduncles puberulent; fruiting receptacle densely white-villous .... 2. *G. canadense*
      3. Petals up to 5 mm long; peduncles hirsute; fruiting receptacle glabrous or nearly so .... 3. *G. laciniatum*
   2. Petals yellow .... 1. *G. aleppicum*

1. **Geum aleppicum** Jacq. Icon. Pl. Rar. 1:10. 1786. Fig. 179.
Geum strictum Ait. Hort. Kew. 2:217. 1789.
Geum aleppicum Jacq. var. *strictum* (Ait.) Fern. Rhodora 37:294. 1935.

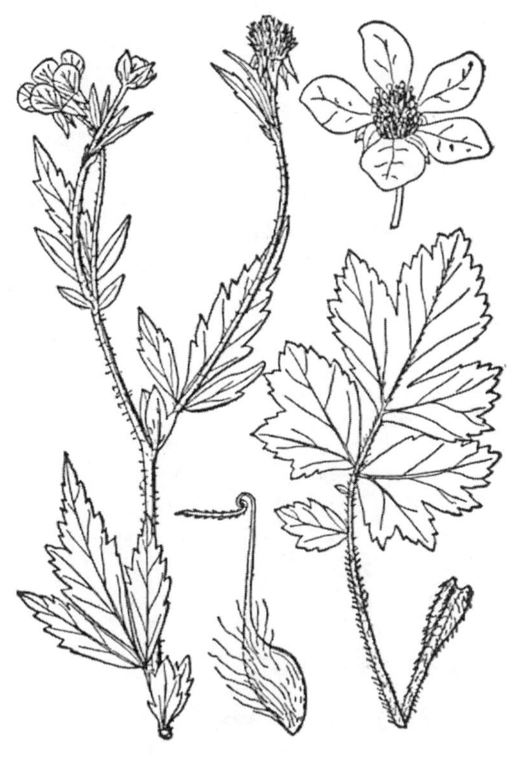

179. *Geum aleppicum* (Yellow avens). Habit (left). Fruit (bottom center). Flower (upper right). Leaf (lower right).

Perennial herb with rhizomes; stems ascending to erect, branched, hirsute, to 1.7 m tall; basal leaves pinnate, with 5–7 leaflets, on hirsute petioles; cauline leaves alternate, variable, the lower usually with 3 or 5 leaflets and nearly sessile, the upper usually simple and sessile, hirsute, serrate, often the pinnate leaves with small intercalary leaflets intermixed; stipules foliaceous; flowering branches on hirsute peduncles; flowers actinomorphic, perfect, bracteate; sepals 5, united below, lanceolate, green, 5–8 mm long; petals 5, free, orange or yellow, obovate to suborbicular, 5–10 mm long, as long as or barely longer than the sepals; stamens 20; fruiting heads globose, 15–22 mm in diameter, consisting of several achenes, the achenes glabrous or short-pilose, 3–5 mm long, attached to a pubescent receptacle, each achene with a curved beak. June–August.

Swampy woods, bogs, moist thickets, wet meadows.

IA, IL, IN, MO (FAC+), NE (FACU), OH (FAC).

Yellow avens.

European plants of this species have densely pubescent achenes. American plants, which have glabrous or short-pilose achenes, are sometimes segregated as var. *strictum*, or as a distinct species known as *G. strictum*.

Geum aleppicum differs from other wetland species of *Geum* in the central Midwest by its yellow or orange flowers.

2. **Geum canadense** Jacq. Hort. Bot. Vindob. 2:82. 1782. Fig. 180.
Geum carolinianum Walt. Fl. Carol. 150. 1788.
Geum album Gmel. Syst. 2:861. 1791.
Geum camporum Rydb. N. Am. Fl. 22:403. 1913.
Geum canadense Jacq. var. *camporum* (Rydb.) Fern. & Weatherby, Rhodora 24:49. 1922.
Geum canadense Jacq. var. *grimesii* Fern. & Weatherby, Rhodora 24:49–50. 1922.

Perennial herb with rhizomes; stems erect, branched, pubescent to glabrous, sometimes glandular, to 1 m tall; basal leaves simple or 3-parted, serrate, usually pubescent, on long petioles; upper leaves three-parted or simple, sessile or subsessile, serrate, pubescent or glabrous; stipules lance-ovate, 1–2 cm long; flowers usually solitary at the ends of branchlets, actinomorphic, perfect, 10–15 mm across, on velvety pedicels; calyx 5-parted, the lobes lanceolate, acute to acuminate at the apex, green, 4–8 mm long; petals 5, free, white, oblong, 5–10 mm long, about as long as or slightly longer than the calyx lobes; stamens 20; fruiting heads globose to obovoid, sessile, 1.2–1.8 cm in diameter, consisting of numerous achenes, the achenes usually glabrous below, pubescent above, 2.5–3.5 mm long, attached to a densely white-villous receptacle, each achene with a hook-tipped persistent style. May–August.

Mesic woods, wet woods.

IA, IL, IN, MO (FAC), KS, KY, NE, OH (FACU).

White avens.

180. *Geum canadense* (White avens). Habit (center). Flower (right).

This common species, found in various kinds of woodland habitats, is distinguished by its white flowers, velvety-pubescent pedicels, and densely white-villous receptacle.

Plants with achenes hispid above and glabrous below have been called var. *grimesii*. Plants with more than 60 achenes per fruiting head may be called var. *camporum*.

3. **Geum laciniatum** Murr. Novi Comm. Gott. 5:30. 1775. Fig. 181.
Geum laciniatum Murr. var. *trichocarpum* Fern. Rhodora 37:283. 1935.

Perennial herb with rhizomes; stems erect, branched, villous to hirsute, to 1 m tall; basal leaves simple or 3-parted, serrate, pubescent, on long petioles; upper leaves 3-parted or 3-lobed, serrate, pubescent, sessile or nearly so; flower usually solitary at the ends of the branchlets, actinomorphic, perfect, 7–12 mm across, on hirsute pedicels; calyx 5-parted, the lobes lanceolate, acute to acuminate at the apex, green, 4–9 mm long; petals 5, free, white, 3–5 mm long, much shorter than the calyx lobes; stamens 20; fruiting heads globose, sessile, 1.8–2.5 cm in diameter, consisting of numerous achenes, the achenes glabrous or puberulent below, glabrous or sparsely bristly above, 3–5 mm long, attached to a glabrous receptacle, each achene with a hook-tipped persistent style. May–July.

Wet woods, wet meadows.

IA, IL, IN, MO (FACW), KS, NE (NI), KY, OH (FAC+).

Cleft-leaved white avens; rough avens.

This species differs from the white-flowered *G. canadense* by its smaller flowers, hirsute pedicels, and glabrous receptacles.

181. *Geum laciniatum* (Cleft-leaved white avens). Habit (left). Flower (right).

4. **Geum rivale** L. Sp. Pl. 501. 1753. Fig. 182.

Perennial herb with a short, stout rhizome; stems erect, usually unbranched, hirsutulous, to (0.8–) 1 m tall; basal leaves deeply lobed to pinnately divided into 3 or 5 leaflets, sometimes with tiny leaflets intermixed, hirsutulous, the terminal leaflet broadly obovate, serrate and sometimes 3-lobed as well, the lateral leaflets narrowly obovate; cauline leaves smaller, 3-foliolate or simple and 3-lobed, or simple and merely serrate; flowers several in a nodding cluster, on hirsute pedicels up to 1 cm long, subtended by linear bractlets 3–4 mm long, actinomorphic, perfect; calyx 5-parted, campanulate, the lobes lanceolate, acute to acuminate at the apex, purple, 7–10 mm long; petals 5, free, yellow with purple veins, obovate, 5–8 mm long, a little shorter than the calyx; stamens 20; fruiting head globose, pedunculate, 2.0–2.5 cm in diameter, often purplish, consisting of numerous spreading achenes, the achenes hirsute, 2.5–4.0 mm long, attached to the receptacle, each achene with a hook-tipped persistent style. May–August.

Swamps, swampy woods, wet meadows, bogs, occasionally in shallow water.

IL, IN, OH (OBL).

Swamp avens; water avens; purple avens.

The pendulous flowers have purple sepals and yellow petals with purple veins.

182. *Geum rivale* (Swamp avens). Habit (center). Leaf (right). Fruit (lower left).

10. **Malus** Mill.—Apple

Shrubs or trees, sometimes thorny; leaves alternate, simple, toothed or lobed; flowers in clusters, pedicellate; sepals 5, free or united at base; petals 5, free, with a basal claw; hypanthium present; stamens numerous; ovary inferior, 3- to 5-locular; fruit a pome.

This genus, sometimes placed in *Pyrus*, consists of about fifty species.

1. Petioles, pedicels, calyx lobes, and hypanthia densely and permanently tomentose; leaves thick, impressed-veined .... 3. *M. ioensis*
1. Petioles, pedicels, calyx lobes, and hypanthia glabrous or sparsely pilose; leaves thin, not impressed-veined.
    2. Leaves oblong to narrowly elliptic .... 1. *M. angustifolia*
    2. Leaves broadly lanceolate to ovate .... 2. *M. coronaria*

1. **Malus angustifolia** (Ait.) Michx. Fl. Bor. Am. 1:292. 1803. Fig. 183.
Pyrus angustifolia Ait. Hort. Kew. 2:176. 1789.

Small tree to 7 m tall, with a trunk diameter up to 20 cm, the crown spreading; bark reddish brown, deeply furrowed, scaly; branches slender, reddish brown or pale brown, smooth, sometimes spur-like, the leaf scars alternate, narrow, curved, with 3 bundle traces; leaves alternate, simple, oblong to narrowly elliptic, obtuse to acute at the apex, tapering to the petiolate base, serrate and rarely shallowly lobed, usually pubescent at first but becoming glabrous, to 5.5 cm long, to 3.5 cm wide, the petiole up to 2 cm long, glabrous or pubescent; flowers few in a cluster, showy, up to 2.2 cm across, each on a pedicel up to 2 cm long; sepals 5, free or united at base, green, 2–4 mm long; petals 5, free, rose, 8–12 mm long; stamens numerous; hypanthium glabrous or pubescent; ovary inferior; fruit a globose pome up to 2.5 cm in diameter, yellow-green, edible. May–June.

183. *Malus angustifolia* (Narrow-leaved crab apple). Flowering branch (right). Fruiting branch (center).

Woods, often along streams in low woods.
IL, KS, KY, MO, OH (not listed).
Narrow-leaved crab apple.
This species differs from the other species of *Malus* in the central Midwest by its narrower leaves and its usually glabrous or sparsely pubescent ovaries. It is found in low woods where it grows along streams.

2. **Malus coronaria** (L.) Mill. Gard. Dict. ed. 8, no. 2. 1768. Fig. 184.
Pyrus coronaria L. Sp. Pl. 1:480. 1753.

Small tree to 8 m tall, the trunk up to 30 cm in diameter, the crown spreading; bark gray-brown to red-brown, with rather deep furrows between the scales; branches moderately stout, reddish brown, often spurlike, sometimes spiny, usually glabrous at maturity, the leaf scars alternate, narrow, curved, with 3 bundle traces; leaves alternate, simple, lanceolate to ovate, obtuse to acute at the apex, tapering or rounded at the petiolate base, coarsely serrate and sometimes slightly lobed, yellow-green and glabrous on the upper surface, paler on the lower surface, to 3.2 cm long, to 2.0 cm wide, the petioles stout, up to 4 cm long, glabrous or pubescent; flowers few in a cluster, showy, up to 3.1 cm across, on pedicels up to 4 cm long; sepals 5, free or united at base, green, 2–4 mm long; petals 5, free, white or pinkish, 9–13 mm long; stamens numerous; hypanthium glabrous or pubescent; ovary inferior; fruit a pome up to 2.4 cm in diameter, yellow-green, edible. May–June.
Woods, often along streams, edge of fields, edge of prairies.
IL, IN, KS, KY, MO, OH (UPL).
Prairie crab apple.
This species differs from *M. angustifolia* by its broader leaves and from *M. ioensis* by its usually fewer lobed leaves.

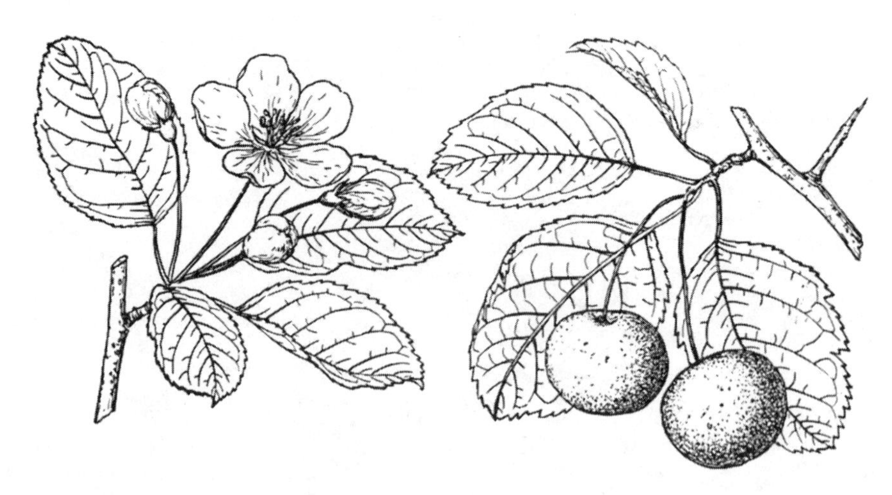

184. *Malus coronaria* (Prairie crab apple). Flowering branch (left). Fruiting branch (right).

3. **Malus ioensis** (Wood) Britt. Ill. Fl. N. US. 2:235. 1897. Fig. 185.
Pyrus coronaria L. *var. ioensis* Wood, Class-book 333. 1860.
Pyrus ioensis (Wood) Carruth, Trans. Kans. Acad. Sci. 5:48. 1877.

Small tree to 8 m tall, the trunk diameter up to 30 cm, the crown spreading; bark reddish brown, scaly; branches moderately stout, reddish brown, sometimes spiny, usually somewhat pubescent at maturity, the leaf scars alternate, narrow, curved, with 3 bundle traces; leaves alternate, simple, elliptic to ovate, thick, impressed-veined, acute to obtuse at the apex, tapering or rounded at the petiolate base, coarsely serrate and often shallowly lobed, up to 10 cm long, up to 4.2 cm wide, dark green and glabrous on the upper surface, yellow-green and usually somewhat pubescent on the lower surface, the petioles stout, up to 2.5 cm long, pubescent; flowers few in a cluster, showy, up to 4 cm across, on tomentose pedicels up to 4.5 cm long; sepals 5, free or united at the base, green, tomentose; petals 5, free, white or rose; hypanthium tomentose; stamens numerous; ovary inferior, tomentose; fruit a pome up to 3.5 cm in diameter, yellow-green, edible. May–June.

Along streams, edge of fields, edge of prairies.

IA, IL, IN, KS, KY, MO, NE, OH (UPL).

Iowa crab apple.

The fruits make delicious jelly. This species is distinguished from the other crab apples by the preponderance of lobed leaves and the tomentose pedicels, sepals, hypanthia, and ovaries.

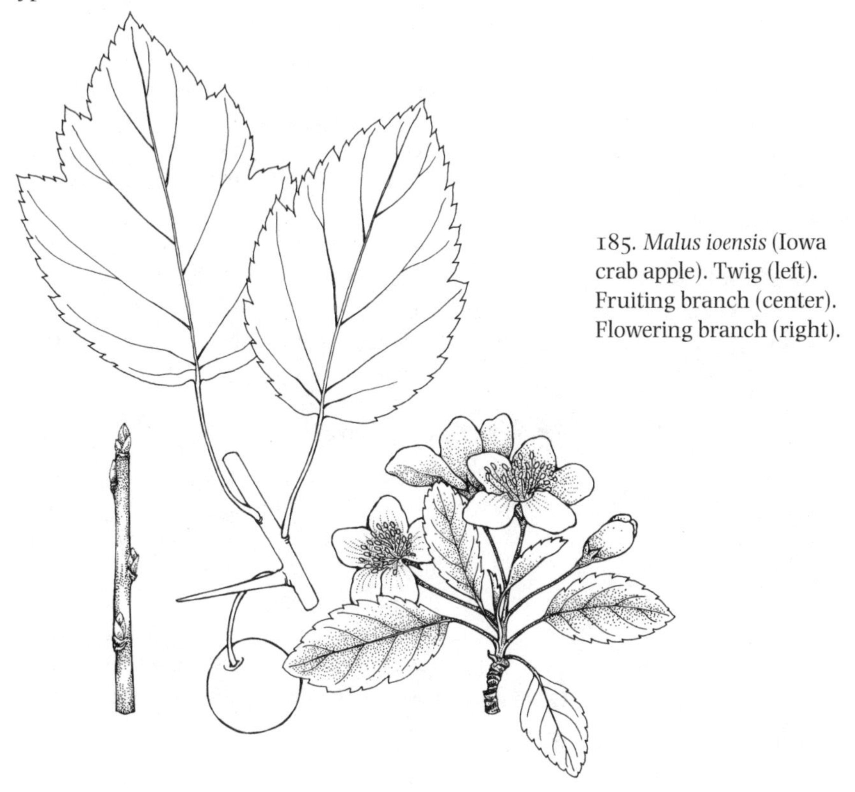

185. *Malus ioensis* (Iowa crab apple). Twig (left). Fruiting branch (center). Flowering branch (right).

11. **Pentaphylloides** Duhamel—Shrubby Cinquefoil

Much branched shrubs; leaves pinnately compound, with entire leaflets; flowers 1–few, perfect, actinomorphic, bracteate; sepals 5, united below, green; petals 5, free, yellow; stamens numerous; carpels several, pubescent, the style attached below the middle; achenes pubescent.

Often included in the genus *Potentilla*, the species of *Pentaphylloides* seem to be placed better in their own genus. The genus consists of seven species. Only the following occurs in the central Midwest.

1. **Pentaphylloides floribunda** (Pursh) A. Love, Svensk. Bot. Tidsk. 48:223. 1954. Fig. 186.
Potentilla floribunda Pursh, Fl. Am. Sept. 1:255. 1813.
Pentaphylloides fruticosa (L.) O. Schwarz, Mitt. Thuring. Bot. Gaz. 1:105. 1949.
Dasiphora fruticosa (L.) Kartesz ssp. *floribunda* (Pursh) Kartesz, Syn. N. Am. Fl. version 10, item no. 4. 1999.

Much branched shrub to 1.2 m tall, with a shreddy outer bark when old; leaves alternate, pinnately divided, with 5–7 leaflets; leaflets narrowly oblong to lanceolate, acute to subacute at the apex, tapering or more or less rounded at the base, glabrous to densely white-villous, entire, often revolute, to 3 cm long, 7–10 mm wide; flowers 1–few, bright yellow, up to 3 cm across, with narrow bracteoles as long as or longer than the sepals; sepals 5, united below, green, villous, to 15 mm long; petals 5, yellow, 15–20 mm long, broadly elliptic, more or less rounded at the apex; stamens numerous; carpels several, densely villous, the styles attached below the middle of the carpels, often immersed in the pubescence; achenes villous. June–August.

Interdunal ponds, boggy fens, hill prairies.

IA, IL, IN, OH (FACW).

Shrubby cinquefoil.

The distinguishing features of this species are its shrubby habit, its 5–7 entire leaflets per leaf, and its bright yellow flowers up to 3 cm across. This species is so handsome that it is a popular ornamental.

This species was originally placed in the genus *Potentilla*, but its shrubby habit and floral structures seem to justify its placement in a separate genus.

12. **Photinia** Lindl.—Chokeberry

Shrubs without spines or thorns (in our species); leaves simple, toothed, glandular along the midvein on the upper surface (in our species); flowers in small, rounded clusters, perfect, actinomorphic, more than 1 cm across; sepals 5, glandular on the margins; petals 5, free, spreading; hypanthium obconic; stamens usually 20; pistil 1, the ovary woolly at the tip, with 5 long-persistent styles; fruit a small pome.

The species of *Photinia* in the central Midwest have usually been placed in the genus *Aronia*, but Robertson and Phipps have presented evidence that they should be included within *Photinia*. According to Robertson and Phipps, *Photinia* consists of sixty-five species.

186. *Pentaphylloides floribunda* (Shrubby cinquefoil). Habit (center). Calyx (lower left).

1. Lower surface of leaves, branchlets, and pedicels glabrous; fruit black.........2. *P. melanocarpa*
1. Lower surface of leaves, branchlets, and pedicels softly pubescent; fruit red or dark purple.
  2. Sepals with conspicuous glands on the margin; fruit red................................ 3. *P. pyrifolia*
  2. Sepals with obscure glands on the margin; fruit dark purple........................1. *P. floribunda*

1. **Photinia floribunda** (Lindl.) Robertson & Phipps, Syst. Bot. 16:391. 1991. Fig. 187.
Pyrus floribunda Lindl. Bot. Reg. 12. 1826.
Aronia floribunda (Lindl.) Spach, Hist. Veg. 2:89. 1839.
Aronia prunifolia (Marsh.) Rehder, Journ. Arn. Arb. 19:74. 1938, *non Photinia prunifolia* (Hook.& Arn.) Lindl. (1822).

Colonial shrubs from extensive rhizomes, up to 3 m tall; branchlets softly pubescent; leaves alternate, simple, narrowly oblanceolate to elliptic, acute or abruptly acuminate at the apex, tapering to the base, 3–8 cm long, up to 3 cm wide, glandular along the midvein of the upper surface, softly pubescent on the lower surface, serrulate to crenulate, not turning red in the autumn, on petioles up to 1 cm long; flowers several in a cluster, 1.0–1.5 cm across, on softly pubescent pedicels; sepals 5, green, with inconspicuous glands along the margin or sometimes eglandular; petals 5, white or pinkish tinged, 4–7 mm long; fruit dark purple, glabrous, 8–10 mm in diameter. May–June.

187. *Photinia floribunda* (Purple chokeberry). Leafy branch, with fruits (left). Flower (lower right).

Bogs, swamps, occasionally on sandstone ledges.

IL, IN, KY, OH (FACW).

Purple chokeberry.

This species is sometimes combined with *P. melanocarpa*, but *P. floribunda* has softly pubescent leaves, branchlets, and pedicels and a dark purple fruit. In the past is has usually been called *Aronia prunifolia*. The usual habitat for this species is in bogs and swamps, but it also occurs on sandstone bluffs in southern Illinois.

2. **Photinia melanocarpa** (Michx.) Robertson & Phipps, Syst. Bot. 16:391. 1991. Fig. 188.
Mespilus arbutifolia (L.) Ell. var. *melanocarpa* Michx. Fl. Bor. Am. 1:292. 1803.
Pyrus melanocarpa (Michx.) Willd. Enum. Pl. Hort. Berol. 525. 1809.
Aronia melanocarpa (Michx.) Ell. Bot. S. Car. & Ga. 1:557. 1821.

Colonial shrubs from extensive rhizomes, up to 3 m tall; branchlets glabrous or nearly so; leaves alternate, simple, narrowly oblanceolate to elliptic, acute to abruptly acuminate at the apex, tapering to the base, 3–8 cm long, up to 3 cm wide, glandular along the midvein of the upper surface, otherwise glabrous or nearly so on both surfaces, serrulate to crenulate, not turning red in autumn, on petioles up to 1 cm long; flowers several in a cluster, 1.0–1.5 cm across, on glabrous pedicels; sepals 5, green, with inconspicuous glands on the margin or occasionally eglandular; petals 5, free, white or sometimes pinkish tinged, 4–7 mm long; fruit black, globose, 7–10 mm in diameter, usually withering by winter. April–July.

Bogs, swamps, wet woods.

IA, IL, IN, MO (FACW−), KY, OH (FAC).

Black chokeberry.

This is the only species of *Photinia* with glabrous lower leaf surfaces, branchlets, and pedicels. It has usually been called *Aronia melanocarpa*.

188. *Photinia melanocarpa* (Black chokeberry). Habit (left). Leaf and fruits (lower right).

3. **Photinia pyrifolia** (Lam.) Robertson & Phipps, Syst. Bot. 16:391. 1991. Fig. 189.
Crataegus pyrifolia Lam. Encycl. Meth. Bot. 1:83. 1783.
Aronia arbutifolia (L.) Persoon, Syn. Pl. 2:39. 1806, *non Photinia arbutifolia* Lindl. (1822).

Colonial shrubs from extensive rhizomes, up to 4 m tall; branchlets softly pubescent, sometimes becoming glabrate; leaves alternate, simple, narrowly oblanceolate to elliptic, acute to abruptly acuminate at the apex, tapering to the base, 3–8 cm long, up to 3 cm wide, glandular on the midvein of the upper surface, the upper surface otherwise dark green and glabrous, the lower surface paler and densely tomentose, serrulate to crenulate, turning red in autumn, on petioles up to 1 cm long; flowers several in a cluster, 1.0–1.2 cm across, on softly pubescent pedicels; sepals 5, green, with conspicuous glands on the margin; petals 5, free, white or sometimes pinkish tinged, 4–6 mm long; fruit red, obovoid to nearly globose, 4–10 mm in diameter, persisting into the winter. April–July.

189. *Photinia pyrifolia* (Red chokeberry). Leafy branch (left). Fruits (right).

Wet woods, swamps, bogs.
KY (FACW), MO (NI).
Red chokeberry.
This species, which barely enters the central Midwest, is distinguished by its conspicuous glandular margin of the sepals, its red fruits, and its softly pubescent lower leaf surfaces, branchlets, and pedicels.
It has usually been called *Aronia arbutifolia*.

13. **Physocarpus** Maxim.—Ninebark

Shrubs; leaves alternate, simple, palmately lobed, with caducous stipules; inflorescence racemose or corymbose, with numerous flowers; flowers perfect, actinomorphic, pedicellate, with a hypanthium; sepals 5, united at the base, persistent on the fruit; petals 5, free, attached to the calyx; stamens 20–40; pistils (1–) 3–5, free or nearly so, the ovaries superior; fruit a cluster of follicles, splitting down two sides, with 2–4 seeds.
About twleve species are in the genus, all but one in the United States.

1. **Physocarpus opulifolius** (L.) Maxim. Trudy Imp. St. Petersb. Bot. Sada 6:220. 1879. Fig. 190.
Spiraea opulifolia L. Sp. Pl. 489. 1753.
Opulaster opulifolius (L.) Kuntze, Rev. Gen. Pl. 949. 1891.
Opulaster intermedius Rydb. in Britt. Man. 492. 1901.
Physocarpus intermedius (Rydb.) Schneid. Handb. Laubh. 1:807. 1906.
Physocarpus opulifolius (L.) Maxim. var. *intermedius* (Rydb.) B. L. Robins. Rhodora 10: 32. 1908.

Shrub to 3 m tall, branched, the branchlets glabrous but the bark shredding into long strips at maturity; leaves alternate, simple, ovate to obovate to suborbicular, obtuse to acute at the apex, rounded or cordate at the base, often shallowly 3-lobed, serrate, glabrous, to 6 cm long, often nearly as broad, the petioles glabrous, up to 2 cm long; inflorescence a crowded, round-topped corymb, with numerous flowers; flowers actinomorphic, perfect, to 10 mm across, pedicellate, the pedicels slender, glabrous or puberulent, 1–2 cm long; calyx 5-parted, green, ovate, glabrous or pubescent; petals 5, free, 3–5 mm long, white, obovate, broadly rounded at the apex; stamens 20–40; pistils 3–5, free; follicles 3–5, free, inflated, broadly oblongoid with a slender terminal beak, shiny, pale brown or occasionally with a purple tinge, 1.5–2.0 mm long, glabrous or pubescent, the hairs, when present, sometimes stellate. May–July.
Along streams, rocky banks, sometimes on exposed blufftops.
IA, IL, IN, KY, MO, OH (FACW−), KS, NE (FACU).
Ninebark.
Although this shrub commonly inhabits rocky streambanks, it also may be found on exposed sandstone blufftops. It is attractive when in flower and resembles some species of *Spiraea*.
The often 3-lobed leaves are distinctive for this shrub.

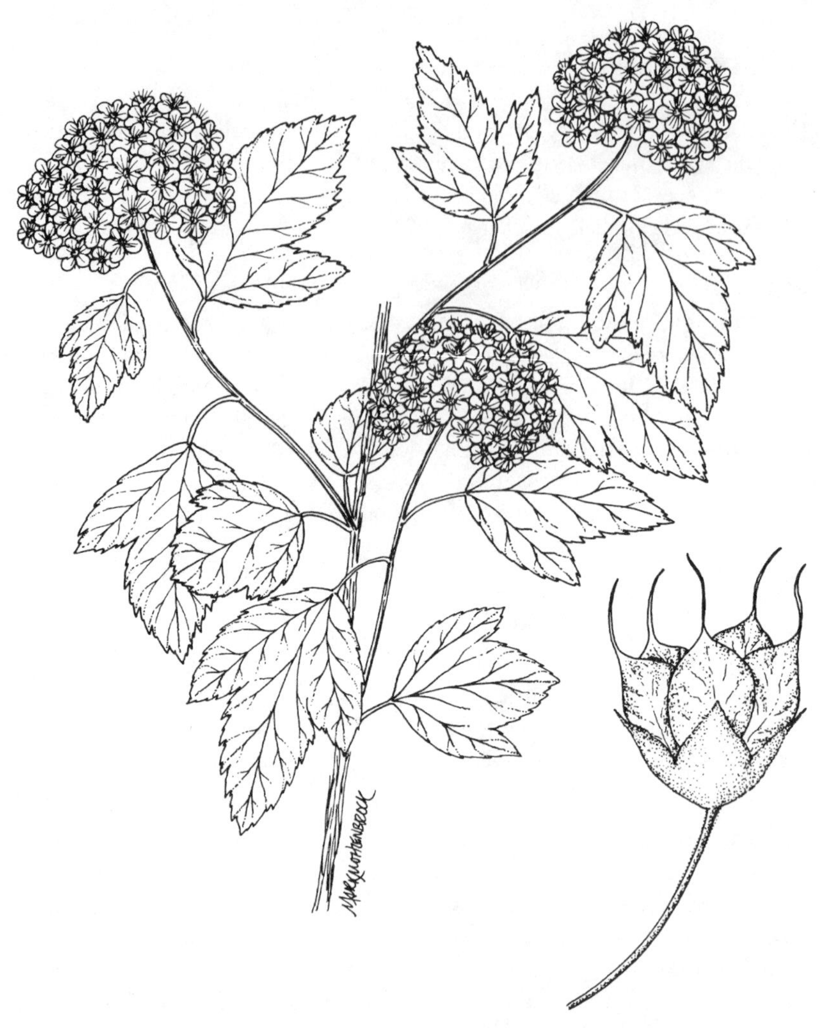

190. *Physocarpus opulifolius* (Ninebark). Habit with flowers (center). Fruits (lower right).

14. **Potentilla** L.—Cinquefoil

Herbaceous annuals or perennials; leaves pinnately or palmately compound, basal and alternate; flowers few, actinomorphic, perfect, usually in cymes, yellow; sepals united below, 5-lobed, subtended by 5 bracts; petals 5, free; stamens 5–numerous, free; pistils few to numerous, free; fruit a cluster of achenes, each achene usually beaked by the persistent style.

Nearly three hundred species comprise this genus of mostly north temperate plants. Several species of *Potentilla* that occur in the central Midwest are found in dry habitats. One shrubby species is now in the genus *Pentaphylloides*, another species with pink flowers is in the genus *Comarum*, and one species with lateral rather than terminal styles is in the genus *Argentina*.

1. Lower leaves ternate or digitate.
  2. All leaves with 3 leaflets, softly villous; stamens about 10 .... 1. *P. millegrana*
  2. Leaves with 3 and 5 leaflets, hirsute; stamens 5 .... 3. *P. pentandra*
1. Lower leaves pinnate.
  3. Cauline leaves pinnate, with (5–) 7–11 leaflets; achenes ribbed .... 2. *P. paradoxa*
  3. Cauline leaves palmate, with 3 or 5 leaflets; achenes smooth .... 4. *P. rivalis*

1. **Potentilla millegrana** Engelm. in Lehm. Ind. Sem. Hamb. 1849:add. 12. 1849. Fig. 191.
Potentilla rivalis Nutt. var. *millegrana* (Engelm.) Wats. Proc. Am. Acad. 8:553. 1873.

191. *Potentilla millegrana* (Many-seeded cinquefoil). Habit (center). Flower (upper center). Fruit (lower left).

Annual or biennial herb with fibrous roots; stems spreading to ascending, much branched, weak, softly villous to nearly glabrous, to 1.5 m tall; leaves alternate and basal, digitately compound, all 3-foliolate, the leaflets oblong to oblanceolate, acute to acuminate at the apex, tapering to the base, serrate, usually softly villous, to 5 cm long, to 3 cm wide; flowers in cymes, bracteate, actinomorphic, perfect; sepals 5, united below, 4–7 mm long, about twice as long as the petals; petals 5, free, 2–4 mm long, yellow; stamens 10 or 15; achenes 0.5–1.0 mm long, smooth, pale brown. June–August.

Low woods, wet prairies, along rivers and streams.

IA, IL, MO (OBL), KS, NE (FACW).

Many-seeded cinquefoil.

This species is similar to *P. pentandra* and *P. rivalis*, but all of the leaves are trifoliolate.

2. **Potentilla paradoxa** Nutt. ex Torr. & Gray, Fl. N. Am. 1:437. 1840. Fig. 192.
Potentilla supina L. var. *nicolletii* S. Wats. Proc. Am. Acad. Arts 8:553. 1873.
Potentilla nicolletii (S. Wats.) E. Sheld. Minn. Bot. Stud. 1:16. 1894.

Annual or perennial herbs; stems spreading to ascending, branched, villous to nearly glabrous, to 45 cm tall; leaves basal and cauline, the lower pinnately divided with (5–) 7–11 leaflets, the cauline ones often 5-parted and digitate, the leaflets oblong to obovate, serrate or crenate, villous to nearly glabrous, to 3 cm long, to 1.5 cm wide; flowers several in cymes, bracteate, actinomorphic, perfect; sepals 5, united at base, green, usually glabrous, 4–5 mm long, about as long as the petals;

petals 5, free, yellow, 4–5 mm long; stamens 10–20; achenes 0.8–1.5 mm long, ribbed and with a corky ridge on one side, beaked. May–August.

Along rivers.

IA, IL, IN, MO (FACW+), KS, NE (FAC), KY, OH (OBL).

Riverbank cinquefoil.

This is the only species of *Potentilla* in wetlands in which most of the leaves are usually with 7–11 leaflets. Plants in the western part of our range with the uppermost leaves ternate or digitate and with 10 or 15 stamens have been called *P. nicolletii*.

192. *Potentilla paradoxa* (Riverbank cinquefoil). Habit (center). Leaf (lower left). Back of flower (lower right). Fruit (center right).

3. **Potentilla pentandra** Engelm. ex Torr. & Gray, Fl. N. Am. 1:447. 1840. Fig. 193.

Annual or biennial herb; stems spreading to ascending, much branched, hirsute, to 1.2 m tall; leaves basal and alternate, the basal digitately 5-parted, the cauline trifoliolate, the leaflets oblong to oblanceolate, acute at the apex, tapering to the base, serrate, hirsute, to 5 cm long, to 3 cm wide; flowers in cymes, bracteate, actinomorphic, perfect; sepals 5, united at base, green, 4–6 mm long, about twice as long as the petals; petals 5, free, yellow; stamens 5; achenes 0.5–1.0 mm long, smooth, pale brown. June–September.

Low prairies, along streams.

KS, MO, NE (FACW+).

Five-stamened cinquefoil.

This is the only wetland *Potentilla* that has five stamens per flower. Otherwise it is much like *P. rivalis*, which has 10 or 20 stamens, but the leaflets of *P. pentandra* are hirsute, while those of *P. rivalis* are softly villous.

4. **Potentilla rivalis** Nutt. Fl. N. Am. 1:437. 1840. Fig. 194.

Annual or biennial herb; stems spreading to ascending to erect, much branched, to 1.5 m tall, softly villous; leaves basal and alternate, the basal pinnately com-

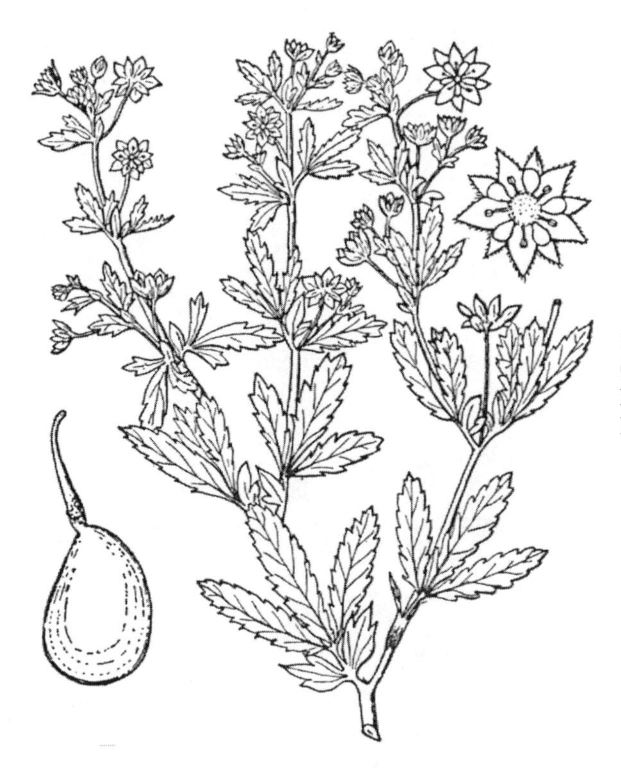

193. *Potentilla pentandra* (Five-stamened cinquefoil). Habit (center). Fruit (lower left). Flower (upper right).

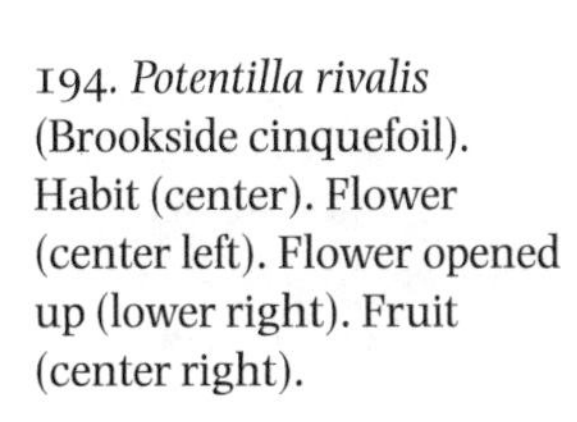

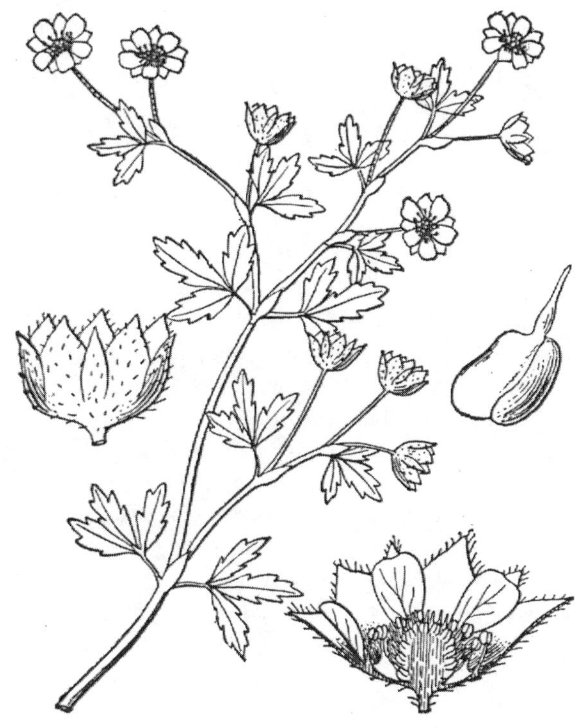

194. *Potentilla rivalis* (Brookside cinquefoil). Habit (center). Flower (center left). Flower opened up (lower right). Fruit (center right).

pound with 3 or 5 leaflets, the cauline palmately compound with 3 or 5 leaflets, the leaflets oblong to oblanceolate, acute at the apex, tapering to the base, usually softly villous, to 5 cm long, to 2.7 cm wide; flowers several in a much branched cyme, bracteate, actinomorphic, perfect; sepals 5, united at base, green, pilose, 5–8 mm long, much longer than the petals; petals 5, free, yellow, 2–3 mm long; stamens 10 or 20; achenes 0.8–1.5 mm long, smooth, brown. June–September.

Along rivers, low prairies, low woods.

IA, IL (adventive), KS, MO, NE (FACW+).

Brookside cinquefoil.

The distinguishing characteristics of this species are the pinnately compound basal leaves, the palmately compound cauline leaves, and the softly villous leaflets.

15. **Prunus** L.—Plum; Cherry

Shrubs or trees; leaves alternate, simple, serrate, usually with glands on the petiole; flowers solitary in the axils of the leaves or several in umbels or racemes, actinomorphic, perfect; sepals 5, free or united at base; petals 5, free; stamens 20; hypanthium cupular to urceolate; ovary superior, attached to the bottom of the hypanthium, bearing 2 ovules; fruit a drupe.

More than two hundred species comprise this genus, almost all of them in north temperate regions of the world. Most species in the central Midwest do not occur in wetland habitats.

1. Flowers in racemes; medium to tall trees; fruit a cherry ....4. *P. serotina*
1. Flowers in small clusters, sometimes solitary; shrubs to short trees; fruit a plum.
    2. Lobes of calyx glandular along the margins; flowers 10–15 mm across; leaves not folded lengthwise, most of them 5 cm long or longer.
        3. Flowers appearing before the expansion of the leaves; teeth of leaves triangular, each with a terminal gland ....2. *P. hortulana*
        3. Flowers appearing after the expansion of the leaves; teeth of leaves low and rounded, the gland in the sinus between the teeth... 3. *P. munsoniana*
    2. Lobes of calyx eglandular along the margins; flowers 8–10 mm across; leaves usually folded lengthwise, all or most of them 3–5 cm long.. 1. *P. angustifolia*

1. **Prunus angustifolia** Marsh. Arb. Am. 111. 1785. Fig. 195.

Shrub or small tree to 5 m tall, often forming colonies; branchlets much branched, glabrous, brown, very thorny; leaves alternate, simple, unfolding as the flowers begin to open, folded lengthwise, lanceolate to narrowly elliptic, acute to acuminate at the apex, tapering or somewhat rounded at the base, serrulate with a gland in the sinus between the teeth, glabrous, shiny, 3–5 (–6) cm long, 2–4 cm wide, the petioles up to 2.5 cm long; flowers in small clusters, each flower 8–10 mm across, on pedicels up to 8 mm long; sepals 5, glabrous on the outer face, pubescent near the base on the inner face, 2–3 mm long, eglandular along the margin; petals 5, free, white, 4–6 mm long; stamens 20; drupe subglobose, red, tart, 1–2 cm in diameter, the stone not compressed. April–May.

Damp thickets, often along streams, sandy soil.

IL, IN, KS, KY, MO, OH (UPL).

Chickasaw plum.

This usually very thorny shrub often grows in dense thickets along streams in the lower part of the central Midwest. It has smaller fruits and smaller flowers than the other wild plums in the region. Its sepals also lack glands along the margin. The flowers and leaves expand at about the same time. The leaves are usually folded lengthwise.

195. *Prunus angustifolia* (Chickasaw plum). Leafy branch (center). Fruit (left). Flowering branch (right).

2. **Prunus hortulana** Bailey, Gard. & Forest 5:90. 1892. Fig. 196.

Small tree to 6 m tall, with a trunk diameter up to 20 cm, the crown broad and rounded; bark gray or brown, becoming scaly at maturity; branchlets slender, red-brown, smooth, the leaf scars alternate, half-round, elevated, with 3 bundle traces; leaves alternate, simple, oblong to oval, acute at the apex, rounded or tapering at the base, up to 15 cm long, up to 3 cm wide, serrate with large, triangular teeth, the teeth gland-tipped, green and usually glabrous on the upper surface, paler and sometimes pubescent on the lower surface, the petioles slender, up to 2.2 cm long, with 1–several glands, glabrous or sparsely pubescent; flowers in small clusters, opening before the leaves appear, each flower 10–15 mm across, on pedicels up to 15 mm long; sepals 5, green, usually somewhat pubescent, glandular along the margin, 2.5–3.0 mm long; petals 5, free, white, 5–7 mm long; stamens 20; fruit globose, red or yellowish, 2–3 cm in diameter, the stone not compressed. March–April.

196. *Prunus hortulana* (Wild goose plum). Fruiting branch.

Thickets, often along streams.

IL, IN, KS, KY, MO, OH (UPL).

Wild goose plum.

This wild plum flowers before the leaves appear, thereby distinguishing it from other wild plums in the central Midwest. It also differs form the similar appearing *P. munsoniana* by its larger triangular teeth of the leaves.

3. **Prunus munsoniana** Wight & Hedrick, N. Y. Ag. Exp. Sta. Rep. 1910, part 2:88. 1911. Fig. 197.

Shrub or small tree to 5 m tall, with a trunk diameter up to 18 cm, the crown broad and rounded; bark gray or brown, scaly; branchlets brownish, glabrous; leaves alternate, simple, not folded lengthwise, lanceolate to narrowly oblong, acute to acuminate at the apex, usually rounded at the base, serrulate with low teeth with a gland in the sinus between the teeth, to 15 cm long, to 5 cm wide, glabrous on the upper surface, usually with some pubescence below, on petioles to 1.5 cm long; flowers in small clusters, opening after the leaves have developed, each flower 10–15 mm across, on pedicels up to 12 mm long; sepals 5, green, usually pubescent, glandular along the margin, 2.5–3.0 mm long; petals 5, free, white, 4–7 mm long; stamens 20; fruit globose to slightly oblongoid, red, 1.8–2.2 cm in diameter, sweet, juicy, the stone not compressed. March–April.

Thickets, often along streams.

IL, IN, KS, KY, MO, OH (UPL).

Munson's plum.

This species is distinguished from *P. hortulana* by its flowers that open after the leaves unfold and the low teeth of the leaves.

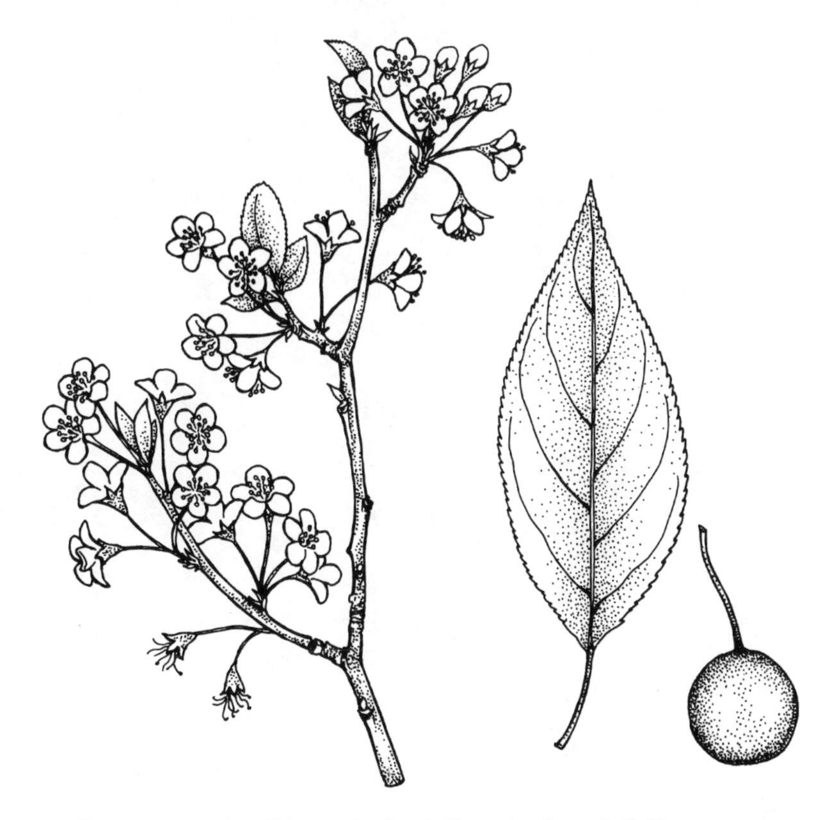

197. *Prunus munsoniana* (Munson's plum). Flowering branch (left). Leaf (center). Fruit (right).

4. **Prunus serotina** Ehrh. Beitr. Naturk. 3:20. 1788. Fig. 198.

Medium to tall tree up to 25 m tall, with a trunk diameter up to 1 m, the crown rounded; bark thin, smooth, reddish brown at first, becoming deeply furrowed and black; branchlets rigid, slender, glabrous, dark brown, the leaf scars half-round, each with 3 bundle traces; leaves alternate, simple, oblong to oval, acute at the apex, tapering to the base, finely serrulate along the margin, green, glabrous and shiny on the upper surface, paler on the lower surface, sometimes with rusty or white hairs along the midvein below, the petioles slender, up to 2 cm long, glabrous, with 1 or more reddish glands near the top; flowers in pendulous, elongated racemes up to 15 cm long, appearing when the leaves are partially grown, each flower 3–5 mm across; sepals 5, green; petals 5, free, white; stamens 20; fruit globose, dark purple, up to 1 cm in diameter, juicy. April–May.

198. *Prunus serotina* (Wild black cherry). Leafy branch with flowers (left). Cluster of fruits (center). Twig (right).

Woods, along streams, sometimes in bottomland forests.

IA, IL, IN, KS, KY, MO, NE, OH (FACU).

Wild black cherry.

The usual presence of rusty or white matted hairs along the lower part of the midvein on the lower surface of the leaves is distinguishing.

16. **Rosa** L.—Rose

Usually prickly shrubs; leaves alternate, pinnately compound, with 3–11 leaflets, usually with stipules adnate to the petiole; flowers large, variously colored, the pedicels usually with aromatic glands; hypanthium urceolate to globose, narrowed at the apex; sepals 5, often narrowed to a prolonged tip, usually persistent on the fruit; petals 5, free from each other but attached to the hypanthium; stamens numerous, attached to the hypanthium; ovaries several, pubescent; fruit fleshy, containing achenes.

This familiar genus contains more than one hundred wild species and numerous ornamental hybrids and varieties. Some of the wild species may occur in wetlands.

1. At least some of the leaves with 3 leaflets .... 3. *R. setigera*
1. All the leaves with 5–11 leaflets.
    2. Stipules laciniate; flowers white or pale pink, 2–4 cm across; prickles narrow throughout .... 1. *R. multiflora*
    2. Stipules with smooth margins; flowers rose-pink, 4–7 cm across; prickles broad-based .... 2. *R. palustris*

1. **Rosa multiflora** Thunb. Fl. Jap. 214. 1784. Fig. 199.

199. *Rosa multiflora* (Multiflora rose). Flowering branch.

Trailing or ascending prickly shrubs to 3 m tall, the prickles straight and narrow throughout; leaves alternate, pinnately compound, with (5–) 7–9 leaflets, with laciniate stipules; leaflets elliptic to obovate, acute to obtuse at the apex, sharply serrate, glabrous or nearly so on both surfaces, to 4 cm long, to 2.5 cm wide; flowers numerous, 2–4 cm across, white or pale pink, on glandular-hispid pedicels; sepals 5, green, tapering to a caudate tip, rarely reflexed at anthesis, 5–8 mm long; petals 5, white or pink, free, 1–2 cm long; stamens numerous; styles connate into a column exserted beyond the hypanthium; fruits red, ellipsoid to globose, 4–8 (–10) mm in diameter. May–June.

Thickets, rarely in standing water, often planted and becoming aggressive and widespread.

IA, IL, IN, KY, MO, OH (FACU), KS, NE (UPL).

Multiflora rose.

This species, native to Asia, was introduced into the United States as a wildlife plant because of its edible fruits. However, its aggressive nature has allowed this species to invade natural habitats. In recent years, multiflora rose has moved into wetlands. At one pond at the Miscatatuck National Wildlife Refuge in Indiana, is actually grows in shallow standing water.

2. **Rosa palustris** Marsh. Arb. Am. 135. 1785. Fig. 200.

Arching prickly shrub to 3 m tall, the prickles recurved and broad-based; leaves alternate, pinnately compound, with 5–9 leaflets, with smooth-margined stipules; leaflets oblong to elliptic, obtuse to acute at the apex, glabrous on the upper surface, puberulent on the lower surface, up to 6 cm long, up to 2.5 cm wide; flowers few in corymbs, 4–7 cm across, rose-pink, on glandular-hispid pedicels; sepals 5, green, eventually reflexed, 10–15 mm long; petals 5, free, 2–3 cm long; stamens numerous; styles not connate into a column; fruits red, subglobose, 8–12 mm in diameter. June–August.

200. *Rosa palustris* (Swamp rose). Habit, with flower (above). Fruit (below).

Swamps, bogs, moist thickets.

IA, IL, IN, KS, KY, MO, NE, OH (OBL).

Swamp rose.

This species is by far the most common rose found in wetlands and standing water. It differs from *R. multiflora* by its large, rose-pink flowers, smooth-margined stipules, and its styles not connate into a column. From *R. setigera*, it differs by its leaflets 7–9 in number and its styles not connate into a column.

3. **Rosa setigera** Michx. Fl. Bor. Am. 1:295. 1803. Fig. 201.

Climbing prickly shrubs to 8 m tall, the prickles broad-based; leaves alternate, pinnately compound, with 3 or 5 leaflets, with narrow, glandular-ciliate stipules; leaflets ovate to oblong, acuminate at the apex, sharply serrate, glabrous and lustrous on the upper surface, paler and sometimes tomentose on the veins beneath, to 10 cm long, to 4 cm wide, the terminal leaflet long-petiolulate, the lateral leaflets subsessile; flowers several in a corymb, each flower 4–8 cm across, rose-pink, on glandular-hispid pedicels; sepals 5, green, reflexed, attenuate at the apex, 12–15 mm long; petals 5, free, 2–4 cm long; stamens numerous; styles connate into a column about as long as the stamens; fruits red, subglobose, 8–12 mm in diameter. June–July.

201. *Rosa setigera* (Prairie rose). Habit (center). Flower (lower left).

Woods, thickets, less commonly in wetland situations.

IA, IL, IN, MO (FACU+), KS, KY, OH (FACU).

Prairie rose; climbing rose.

This species is the only wild rose in the central Midwest in which some of the leaves have only three leaflets. In addition, it differs from *R. palustris* and *R. multiflora* by its styles connate into a column.

While this is usually a species of relatively dry habitats, it occurs in shallow standing water at Eagle Lake in Indianapolis and a pond at the Miscatatuck National Wildlife Refuge in Indiana. At both locations, it climbs high into trees adjacent to the water.

17. **Rubus** L.—Blackberry; Raspberry

Shrubs or rarely herbs, usually beset with prickles; leaves alternate, compound or less commonly simple; flowers variously arranged, actinomorphic, usually perfect; sepals 5, green, usually free; petals 5, free, white or pink or purple; stamens numerous; hypanthium usually flat; pistils numerous, attached to a convex receptacle; fruit a dense cluster of drupelets.

Rubus consists of many species, the number variable depending upon the view of each individual botanist. Most species are upland, forming thickets, and some of these upland plants may be found in wetlands. Those that are almost always confined to uplands are not considered in this work.

1. Leaves simple; petals purple .... 3. *R. odoratus*
1. Leaves compound; petals white or pinkish.
  2. Stems without prickles or bristles, herbaceous .... 4. *R. pubescens*
  2. Stems prickly, bristly, or both, shrubby.
    3. Leaves whitened on the lower surface; plants arching to erect; fruit a raspberry.
      4. Stems strongly glaucous; fruit purple-black at maturity .... 2. *R. occidentalis*
      4. Stems not glaucous, or only sparsely so, usually purplish ; fruit red at maturity .... 5. *R. strigosus*
    3. Leaves green on the lower surface; plants trailing; fruit a blackberry or dewberry.
      5. Stems beset with numerous glandular bristles .... 6. *R. trivialis*
      5. Stems beset with non-glandular bristles .... 1. *R. hispidus*

1. **Rubus hispidus** L. Sp. Pl. 1:493. 1753. Fig. 202.
Rubus obovalis Michx. Fl. Bor. Am. 1:298. 1803.
Rubus hispidus L. var. *obovalis* (Michx.) Fern. Rhodora 42:281. 1940.

Stems woody, trailing on the ground or sometimes low-arching, rooting at the tip with weak, eglandular, straight bristles that are sometimes reflexed and up to 4 mm long, with short glandular hairs also usually present; leaves alternate, palmately compound with 3 or 5 leaflets, the leaflets shiny, leathery, narrowly ovate to ovate, acute to obtuse at the apex, usually tapering to the base, serrate, glabrous, the terminal leaflet up to 6 cm long, up to 5 cm wide; flowers actinomorphic, perfect, in a loose corymb, the pedicels pilose; sepals 5, pubescent, eglandular, green; petals 5, free, white, 5–9 (–11) mm long; stamens numerous; fruit blackish, up to 1.5 cm in diameter, sour, with many seeds.

Bogs, marshes, damp thickets, black oak woods.

IA, IL, IN, KY, MO, OH (FACW).

Swamp dewberry.

This trailing species is distinguished from *R. trivialis*, the other trailing dewberry that is sometimes found in wetlands, by its nearly straight, eglandular bristles on the stems. The fruits are sour.

Plants with obtuse leaflets and petals 9–11 mm long have been called var. *obovalis*.

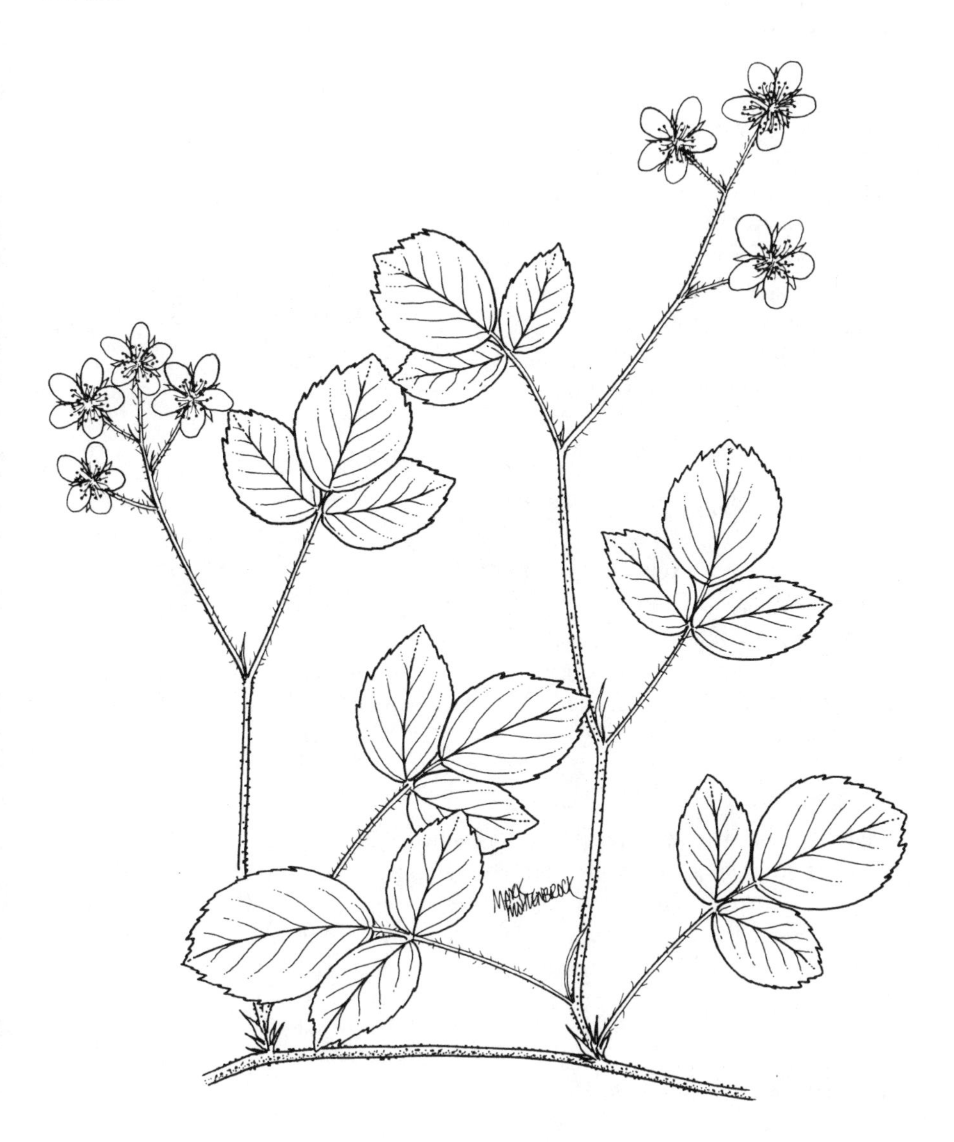

202. *Rubus hispidus* (Swamp dewberry). Flowering branch.

2. **Rubus occidentalis** L. Sp. Pl. 1:493. 1753. Fig. 203.

Stems woody, arching to erect, sometimes rooting at the tip, strongly glaucous, with stout, straight or hooked prickles; leaves alternate, usually trifoliolate but sometimes 5-foliolate, the terminal leaflet broadly ovate, acute at the apex, rounded or subcordate at the base, sharply serrate, green above, gray to white beneath, to 9 cm long, to 6 cm wide; inflorescence cymose, with few to several flowers, the flowers on pedicels with hooked prickles; sepals 5, free, green, 5–8 mm long; petals 5, free, white, 2–4 mm long, much shorter than the sepals; stamens numerous; sepals purple-black, oblongoid to nearly globose, 0.8–1.4 cm in diameter, sweet. May–June.

203. *Rubus occidentalis* (Black raspberry). Habit (center). Flower (lower right).

Thickets, edge of woods and fields, boggy sites.

IA, IL, IN, KS, KY, MO, NE, OH (UPL).

Black raspberry.

The strongly white-backed leaflets and the strongly glaucous stems are distinctive for this species. Although *R. occidentalis* is common in upland habitats, it sometimes is found in wetter sites, particularly where *R. strigosus* is present. The fruits are edible.

3. **Rubus odoratus** L. Sp. Pl. 1:494. 1753. Fig. 204.

Shrubs to 2 m tall, the stems much branched, without prickles or bristles, but bearing numerous glands; leaves alternate, simple, suborbicular to nearly deltoid, 3- or 5-lobed, 10–20 cm long, nearly as wide, the lobes acute at the apex, rounded or nearly subcordate at the base, the lobes coarsely but irregularly serrate, glabrous or pilose above and below, sometimes velvety below; flowers few to as many a 75 in an open cyme, actinomorphic, perfect, 3.5–5.5 cm across; sepals 5, united at base, the lobes acuminate, green with purple glandular hairs; petals 5, free, purple or rose, 1.5–2.5 cm long; stamens 20; fruit rather dry, 1.0–1.2 cm in diameter, the drupelets falling separately. June–August.

Moist or wet woods.

OH (UPL).

Flowering raspberry; thimbleberry.

204. *Rubus odoratus* (Flowering raspberry). Habit (center). Fruit (lower right).

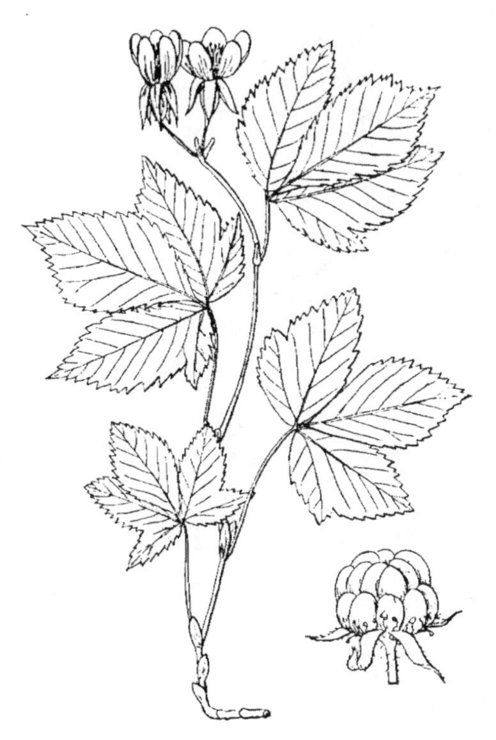

This raspberry is unique because of its simple leaves, large flowers with purple or rose petals, and the absence of prickles and bristles. It sometimes occurs in shaded areas in moist or wet woods.

4. **Rubus pubescens** Raf. Med. Rep. ser. 3, 2:333. 1811. Fig. 205.

Stems herbaceous, trailing along the ground or occasionally just below the surface of the soil, usually rooting at the tips, without prickles but sometimes with a few bristles, up to 50 cm long; leaves alternate, compound, with 3 leaflets, subtended by oblanceolate stipules; leaflets ovate to obovate, acute to acuminate at the apex, tapering to the base, serrate except near the base, glabrous or nearly so, to 8 cm long, to 6 cm wide; flowers usually 1–3 in a cluster, actinomorphic, perfect, up to 1.5 cm across, the pedicels with glandular hairs; sepals 5, free or nearly so, the lobes linear, green, 3–7 (–10) mm long; petals 5, free, white to less commonly pink, 6–10 mm long; stamens 20; fruit up to 1 cm in diameter, dark red, juicy. May–July.

Bogs, wet woods, wet flatwoods, swampy forests.

IA, IL, IN (FACW+), OH (FACW).

Dwarf raspberry.

This is the most delicate species of *Rubus* in the central Midwest, having herbaceous stems that trail on or just below the surface of the ground. The stems are usually unarmed, but some of them may have a few bristles.

205. *Rubus pubescens* (Dwarf raspberry). Flowering branch (left). Fruit (right).

5. **Rubus strigosus** Michx. Fl. Bor. Am. 1:297. 1803. Fig. 206.
Rubus idaeus L. var. *strigosus* (Michx.) Maxim. Bull Acad. Imp. Sci. St. Petersb. 17:161. 1872.

Stems more or less woody, spreading, arching to erect, to 2 m tall, often pinkish purple, with slender prickly bristles and glandular bristles; leaves alternate, compound, with 3 or 5 leaflets, the petioles sometimes prickly, the leaflets ovate to lanceolate, acute to acuminate at the apex, rounded or subcordate at the base, sharply serrate, to 6 cm long, to 4.5 cm wide, gray-pubescent on the lower surface; flowers 2–5 in a cyme, up to 1.2 cm across, the pedicels with bristles and glandular hairs; sepals 5, free or nearly so, green, reflexed; petals 5, free, white, 4–5 mm long; stamens 20; fruit 1.0–1.5 cm in diameter, red, falling as a unit. May–July.

Moist or wet areas, boggy sites, swampy woods.

206. *Rubus strigosus* (Red raspberry). Habit, in fruit (center). Stem showing glandular hairs (upper left).

IA, IL, IN (FACW−), OH (NI).

Red raspberry.

Some botanists consider this species to be the same as the European and domestic *Rubus idaeus*, some consider it an American variety of *R. idaeus*, and others believe it to be a distinct species, as treated in this work.

Rubus strigosus differs from *R. occidentalis* by its pink-purple, scarcely glaucous stems with slender prickles and bristles, its smaller flowers, and its red fruits. It is a common species in wetland habitats in the northern parts of the central Midwest.

6. **Rubus trivialis** Michx Fl. Bor. Am. 1:296. 1803. Fig. 207.

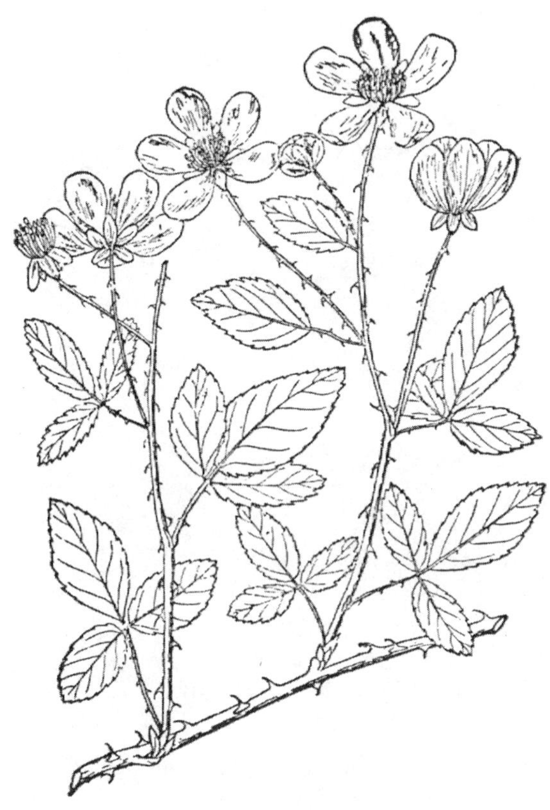

207. *Rubus trivialis* (Southern dewberry). Habit.

Stems woody, trailing along the ground, rooting at the tip, with short, stout, recurved prickles and numerous red-glandular stiff bristles; leaves alternate, compound, with 3 or 5 leaflets, the leaflets oblong to elliptic, acute at the apex, more or less rounded at the base, serrate, green on both sides, the terminal leaflet up to 10 cm long, up to 3 cm wide, on prickly petiolules, the lateral leaflets sessile or nearly so; flowers 1 (–3) in a cluster, on bristly or glandular or glabrous pedicels; sepals 5, free or united below, green, glabrous or glandular; petals 5, free, white, up to 10 mm long; stamens 10; fruit subglobose or oblongoid, black. April–May.

Fields, thickets, sometimes along rivers and streams.

IL, MO (FACU+), KY (FACU).

Southern dewberry; scratch-ankle.

This species differs from *R. flagellaris*, the northern dewberry, by the presence of numerous red stiff bristles on the stem.

When this species flowers and fruits in the spring near Milton, Florida, the city holds a festival known as the Scratch-ankle Festival because of the prickly stems growing at ground level.

18. **Sanguisorba** L.—Burnet

Annual or perennial herbs; leaves alternate, pinnately compound, the leaflets serrate to pinnatifid; flowers perfect (in our species), actinomorphic, borne in heads or spikes, with an urceolate hypanthium; sepals 4, free, petaloid; petals absent; stamens 2 or 4; pistils 1 or 2, the ovary superior; fruit an achene surrounded by the persistent hypanthium.

Approximately twenty-five species are in this genus, either in North America, Europe, or Asia. Only the following species is native in the upper part of the central Midwest.

1. **Sanguisorba canadensis** L. Sp. Pl. 1:116. 1753. Fig. 208.

Perennial herb from a thickened rhizome; stems erect, sometimes branched above, glabrous, to 1.5 m tall; leaves alternate, pinnately compound with 7–15 leaflets, the leaflets elliptic to ovate, obtuse to acute at the apex, somewhat rounded or tapering to the base, sharply serrate, glabrous, to 8 cm long, to 6 cm wide; flowers in 1 or more spikes, the spikes to 12 cm long, on long peduncles, each flower 3.5–5.5 mm across; sepals 4, free, white, elliptic, spreading, 2–3 mm long; petals absent; stamens 4; pistils 1 or 2; fruit an achene enclosed by the hypanthium. July–September.

Wet meadows, wet prairies, marshes.

IN, KY, OH (FACW+).

American burnet.

This species is unique in the rose family because of its 4 sepals, 4 stamens, and no petals. Its usually obtuse leaflets are also distinctive.

208. *Sanguisorba canadensis* (American burnet). Inflorescence (left). Leaves (right). Achene (lower left). Achene with 4-winged calyx (lower center). Flower (upper right).

19. **Spiraea** L.—Spiraea

Shrubs; leaves simple, alternate, serrate, without stipules; flowers in terminal or lateral corymbs or panicles, white or pink or rose; sepals 5, united below; petals 5, free from each other; stamens 10–many; hypanthium present; pistils 5 (–8), free from each other; fruit a cluster of follicles with linear seeds.

This genus consists of about seventy species in the Northern Hemisphere. Several species are grown as ornamentals.

1. Petals white; inflorescence broad; leaves glabrous or nearly so beneath .... 1. *S. alba*
1. Petals rose; inflorescence narrow; leaves densely white-tomentose beneath ... 2. *S. tomentosa*

1. **Spiraea alba** DuRoi, Harbk. Beumz. 2:430. 1772. Fig. 209.

209. *Spiraea alba* (Meadowsweet). Habit (center). Flower (above right). Fruit (below right).

Shrubs to 2 m tall, the branchlets yellow-brown, glabrous; leaves alternate, simple, lanceolate to narrowly oblong, acute at the apex, finely serrate, glabrous or nearly so on both surfaces, 3–7 cm long, 1–2 cm wide; inflorescence a broad, terminal panicle with numerous flowers; flowers white, 4–8 mm across; sepals 5-parted, green, persistent on the fruit; petals 5, free, 2–4 mm long, suborbicular; stamens 10–50; follicles glabrous, 4–6 mm long. July–August.

Wet soil, marshes, bogs, fens.

IA, IL, IN, MO (FACW+).

Meadowsweet.

This species is readily distinguished by its glabrous, finely serrate leaves and its panicles of small white flowers.

2. **Spiraea tomentosa** L. Sp. Pl. 489. 1753. Fig. 210.

Spiraea rosea Raf. New Fl. & Bot. N.A. 3:62. 1936.

Spiraea tomentosa L. var. *rosea* (Raf.) Fern. Rhodora 14:190. 1912.

Shrubs to 2 m tall, the branchlets brown, tomentose; leaves alternate, simple, ovate to oblong to lanceolate, acute or subacute at the apex, serrate, white- or rufous-tomentose on the lower surface, 3–5 cm long, 1–2 cm wide; inflorescence a spirelike terminal panicle to 15 cm long, with numerous flowers; flowers pink or rose-pink, 3–4 mm across; sepals 5-parted, green, persistent and reflexed on the fruit; petals 5, free, 1.5–2.0 mm long, suborbicular; stamens 10–50; follicles tomentose to glabrate, 3–5 mm long. July–August.

Bogs, swamps, wet meadows, moist thickets.

IL, IN, MO, OH (FACW).

Hardhack; steeplebush.

Our plants with rose flowers have been called var. *rosea*. Typical var. *tomentosa*, with paler flowers, occurs to the north of our area.

This shrub is readily distinguished by its densely white- or rufous-tomentose lower leaf surfaces.

When early settlers came to the central Midwest, they had trouble clearing the land of this species because of its very dense wood. To cut down these shrubs, the settlers would say "the task was a hard hack," hence the common name. The other common name, steeplebush, comes from the spirelike inflorescence.

210. *Spiraea tomentosa* (Hardhack). Habit (center). Flower (right).

105. RUBIACEAE—MADDER FAMILY

Herbs, shrubs, or trees (in the tropics and subtropics); leaves simple, entire, opposite or whorled, with stipules when opposite; flowers actinomorphic, perfect, axillary or in terminal inflorescences; calyx 4-parted, small, sometimes nearly lacking, attached to the ovary; petals 4 (–5), attached to each other; stamens twice as many as the lobes of the corolla and attached to the corolla tube; ovary inferior, with 2–4 (–5) locules; fruit dry or fleshy.

Although this family of about sixty-five hundred species occurs worldwide, it is most abundant in the tropics and subtropics where its species are frequently woody. Among the plants in this family of commercial importance are *Coffea* and several ornamental genera.

1. Shrubs; some of the leaves whorled; flowers in dense, globose, pedunculate heads .... 1. *Cephalanthus*
1. Herbs; leaves either all opposite or all whorled; flowers solitary or 1–3 together in cymes or glomerules.
    2. All leaves whorled .... 3. *Galium*
    2. All leaves opposite.
        3. Flowers in pairs, united by their hypanthia; stems creeping; fruit a berry..... 5. *Mitchella*
        3. Flowers not in pairs; stems decumbent, ascending, or erect; fruit a capsule or a pair of nutlets.
            4. Flowers sessile.
                5. Flowers 1–3 per axil .... 2. *Diodia*
                5. Flowers several in an axillary glomerule .... 7. *Spermacoce*
            4. Flowers pedicellate.
                6. Corolla salverform or funnelform; flowers solitary or in cymes, panicles, or corymbs; seeds up to 20 per locule........ 4. *Houstonia*
                6. Corolla rotate; flowers in glomerules; seeds numerous in each locule .... 6. *Oldenlandia*

1. **Cephalanthus** L.—Buttonbush

Shrubs or small trees; leaves simple, opposite and whorled on the same plant, entire, stipulate; flowers perfect, in dense, spherical, pedunculate heads; calyx united below, with 4 lobes; corolla funnelform, united below, with 4 lobes, white; hypanthium present; stamens 8; ovary inferior, with an exserted, filiform style; fruits dry, 1-seeded, crowded into a dense, pedunculate head.

There are six species of *Cephalanthus* in the world.

1. **Cephalanthus occidentalis** L. Sp. Pl. 95. 1753. Fig. 211.
Cephalanthus occidentalis L. var. *pubescens* Raf. Med. Fl. 1:101. 1828.

Shrub to 3 m tall or small tree to 10 m tall; twigs glabrous or pubescent; leaves simple, opposite and whorled on the same plant, oblong-ovate, rarely lanceolate, acuminate at the apex, more or less rounded at the base, entire, glabrous or sometimes pubescent, to 15 cm long, to 8 cm wide, with deltate stipules glandular-toothed, 2–3 mm long; inflorescence a globose head up to 3 cm in diameter, long-pedunculate; flowers white, subtended by clavate bractlets; sepals 4-parted, green; corolla 4-parted, funnelform, white, 5–8 mm long; stamens 8; ovary inferior; fruits in a globose head, each fruit 1-seeded. June–August.

Swamps, often in standing water.

IA, IL, IN, KS, KY, MO, NE, OH (OBL).

Buttonbush.

This is the most common shrub in swamps and standing water in the central Midwest. Most plants will have some branches bearing whorled leaves.

Some variation occurs within the species. Plants with densely pubescent twigs and leaves occur in the more southern part of the central Midwest; these may be referred to as var. *pubescens*.

211. *Cephalanthis occidentalis* (Buttonbush). Branch with fruit (center). Flower (lower right).

2. **Diodia** L.—Buttonweed

Annual or perennial herbs; leaves opposite, simple, entire, sessile, stipulate; flowers in the axils of the leaves, actinomorphic, perfect; calyx 2- or 4-parted, with minute teeth between the lobes; corolla 4-parted; stamens 4, attached to the corolla; ovary inferior, 2-locular, with one ovule per locule, the stigmas 1–2; fruit separating into two nutlets.

Approximately fifty species comprise this genus.

The sessile flowers and the setaceous stipules are distinctive.

1. Corolla 4–6 mm long, pink; fruits 2.5–4.0 mm long; leaves linear to linear-lanceolate .... 1. *D. teres*
1. Corolla 7–10 mm long, white; fruits 7–10 mm long; leaves lanceolate to narrowly oblong ... 2. *D. virginiana*

1. **Diodia teres** Walt. Fl. Carol. 87. 1788. Fig. 212.
Diodia teres Walt. var. *setifera* Fern. & Grisc. Rhodora 39:307. 1937.

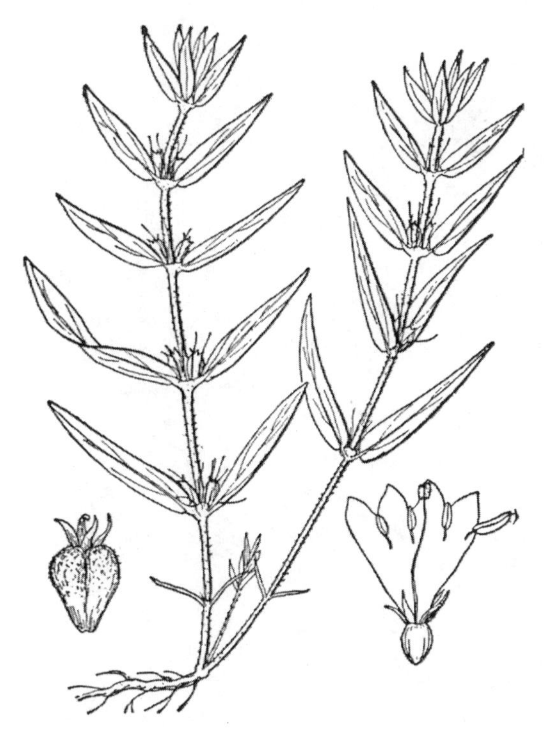

212. *Diodia teres* (Rough buttonweed). Habit (center). Fruit (lower left). Flower (lower right),

Annual herb from fibrous roots; stems erect to spreading, branched, stiff, rough-pubescent, rarely glabrous, to 75 cm tall, often 4-sided; leaves opposite, simple, linear to linear-lanceolate, acute at the apex, tapering to the sessile base, stiff, scabrous, sometimes with bristles at the tip, to 3.5 cm long, to 8 mm wide, revolute when dry, the 5–8 stipules setaceous, 4–10 mm long; flower solitary in the leaf axils; calyx 4-parted, lanceolate, to 4 mm long, green; corolla funnelform, 4-parted, usually pink, 4–6 mm long; stamens 4; stigma 1; fruit obovoid, hispid, 3–6 mm long, the calyx persistent at the tip. July–August.

Usually dry soil, sandy clearings, moist woods.

IA, IL, IN, KS, MO, NE (FACU), KY, OH (UPL).

Rough buttonweed.

This species differs from *D. virginiana* by its smaller pink flowers, narrower leaves, and larger fruits. Plants with hirsute stems and leaves with bristles near the tip may be known as var. *setifera*.

Although this plant is common in several upland habitats, it sometimes may occur at the edges of wetlands.

2. **Diodia virginiana** L. Sp. Pl. 104. 1753. Fig. 213.

Annual herb with fibrous roots; stems decumbent to ascending, branched, hispid or less commonly nearly glabrous, to 80 cm long, usually 4-angled; leaves opposite, simple, lanceolate to narrowly oblong, acute at the apex, tapering to the sessile base, pubescent to nearly glabrous, to 6 (–8) cm long, to 1.5 (–2.0) cm wide, the 5–8 stipules linear, 3–5 mm long; flower solitary in the leaf axils; calyx 2-parted, linear to linear-lanceolate, to 6 mm long, green; corolla salverform, 4-parted, white, short-hairy on the upper surface, 7–10 mm long; stamens 4; stigmas 2; fruit ellipsoid or ovoid, glabrous or hirsute, 6–8 mm long, the calyx persistent at the tip. June–August.

Swamps, wet woods, wet ground.

IL, IN, KY, MO, OH (FACW).

Virginia buttonweed.

This species is distinguished by its short-hairy white petals, its 2 sepals, and its ovoid to ellipsoid axillary fruits.

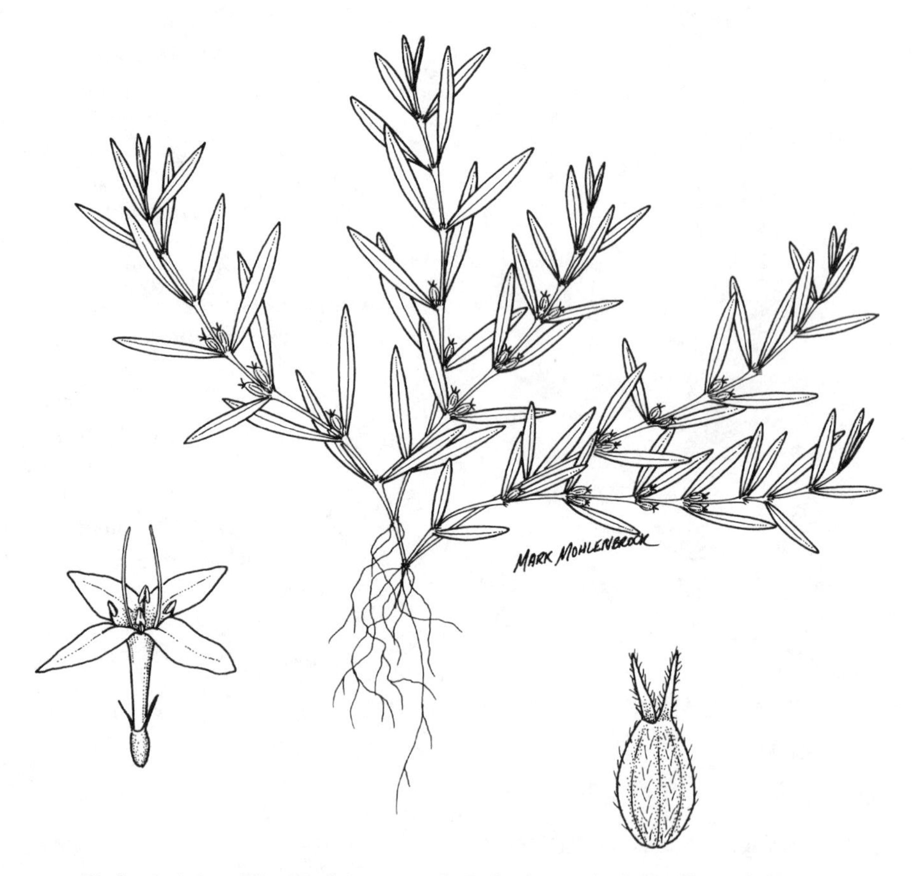

213. *Diodia virginiana* (Virginia buttonweed). Habit (center). Flower (lower left). Fruit (lower right).

3. **Galium** L.—Bedstraw

Often weak annual or perennial herbs; leaves simple, whorled, entire, without stipules; flowers in small cymes or leafy panicles, without bracts; sepals usually absent; corolla rotate, with (3–) 4 (–5) lobes, the lobes equaling or longer than the tube; stamens (3–) 4 or 8; ovary inferior, with 2 styles; fruit dry or somewhat fleshy, globose, twins, separating into two 1-seeded nutlets.

This genus consists of nearly three hundred species worldwide. In the central Midwest, about half of the species are upland and half are wetland.

Some botanists consider the species in this genus have opposite leaves with two or more leaflike stipules, a view not followed here.

1. Some of the leaves in whorls of 7 or 8; plants annual .... 1. *G. aparine*
1. All leaves in whorls of 3, 4, 5, or 6; plants perennial.
  2. Leaves all in whorls of 4.
    3. Plants erect; flowers in dense panicles; leaves more or less acute at the apex .... 3. *G. boreale*
    3. Plants spreading to ascending; flowers in cymes; leaves usually obtuse at the apex.
      4. Leaves recurved to reflexed, up to 3 mm wide .... 4. *G. labradoricum*
      4. Leaves spreading to ascending, most or all of them more than 3 mm wide.
        5. All corollas 4-parted; stems glabrous .... 5. *G. obtusum*
        5. Some of the corollas 3-parted; stems scabrous .... 7. *G. trifidum*
  2. Some of the leaves in whorls of 6.
    6. Leaves all in whorls of 6, conspicuously mucronate; fruit bristly or pubescent.. 8. *G. triflorum*
    6. Leaves in whorls of 3, 4, 5, and 6, not conspicuously mucronate; fruits glabrous.
      7. Some of the corollas 3-lobed; upper part of stems scabrous but not retrorsely bristly.
        8. Pedicels smooth; stems and leaves glabrous or barely scabrous....... 6. *G. tinctorium*
        8. Pedicels scabrous; stems and leaves scabrous .... 7. *G. trifidum*
      7. Corollas all 4-lobed; stems retrorsely bristly .... 2. *G. asprellum*

1. **Galium aparine** L. Sp. Pl. 108. 1753. Fig. 214.
Galium agreste var. *echinospermon* Wallr. Sched. Crit. 59. 1822.
Galium aparine L. var. *echinospermum* (Wallr.) Farw. Rep. Mich. Acad. Sci. 19:260. 1817.

Annual herb with weak, scrambling stems to 1.5 m long, the stem much branched and with retrorse prickles; leaves simple, in whorls of (6–) 8, linear-lanceolate to oblanceolate, acute and cuspidate at the apex, tapering to the sessile base, with prickly margins and veins, 1–6 cm long, 4–12 mm wide; inflorescence small cymes with 1–3 flowers from the upper axils on ascending pedicels; calyx absent; corolla 4-parted, white, 3–4 mm long; stamens 8; fruit dry, covered with hooked bristles, 3.0–4.5 mm in diameter. May–June.

Woods and thickets, moist or dry.

IA, IL, IN, KS, KY, MO, NE, OH (FACU).

Cleavers; sticky bedstraw.

This weak-stemmed species is the only species of *Galium* with more than 6 leaves in a whorl at some time. It occurs in both wet and dry situations.

Plants with very short leaves up to 2 cm long may be called var. *echinospermum*.

214. *Callum aparine* (Cleavers). Habit (center). Fruit (right).

2. **Galium asprellum** Michx. Fl. Bor. Am. 1:78. 1803. Fig. 215.

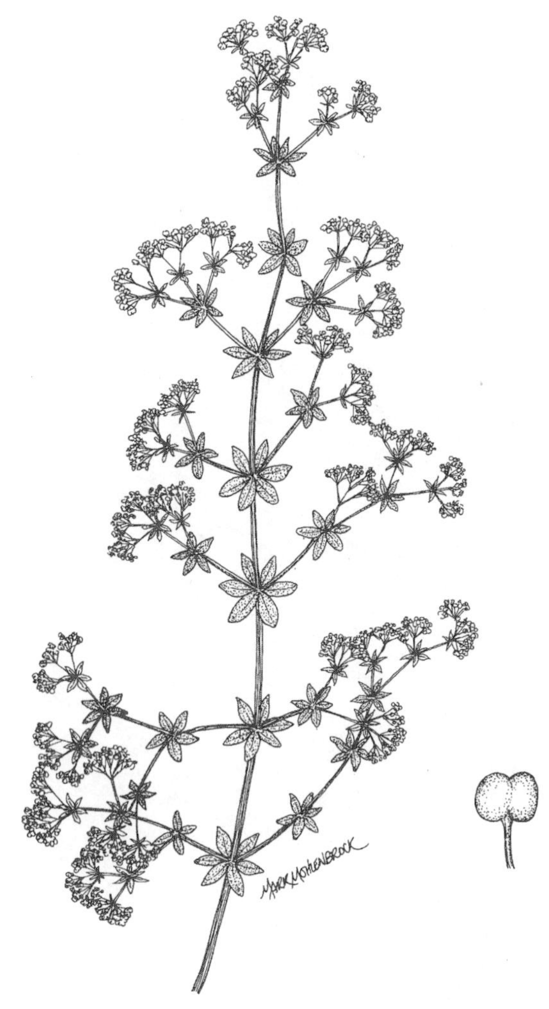

215. *Galium asprellum* (Rough bedstraw). Habit (center). Fruits (lower right).

Perennial herbs with spreading or scrambling stems up to 2 m long, the stems much branched and with retrorse prickles; leaves mostly in whorls of 6, occasionally with some 4's and 5's, ovate to lanceolate, acute and cuspidate at the apex, with prickly margins and veins, to 20 mm long, to 6 mm wide; inflorescence a leafy panicle with several small flowers on spreading or ascending pedicels; calyx absent; corolla 4-parted, white, about 3 mm across; stamens 8; fruit dry, globose, about 2 mm in diameter, smooth. August.

Wet woods, wet thickets, marshes, calcareous fens, rarely in standing water.

IA, IL, IN, MO, OH (OBL).

Rough bedstraw.

This bedstraw has extremely scabrous leaves that are mostly all in whorls of six. It characteristically leans on other vegetation.

3. **Galium boreale** L. Sp. Pl. 108. 1753. Fig. 216.

Perennial herbs with erect, branched or unbranched, glabrous stems; leaves in whorls of 4, spreading or ascending, linear to lanceolate, acute at the apex, tapering to the sessile base, 3-nerved, glabrous, 2–5 cm long, 3–6 mm wide; inflorescence a dense, many-flowered panicle; calyx absent; corolla 4-parted, white, 2–3 mm across; stamens 8; fruit dry, hispid but sometimes becoming nearly glabrous, about 2 mm in diameter. May–July.

Moist meadows, prairies, fields, pastures.

IA, IL, IN, MO (FAC), OH (FACU).

Northern bedstraw.

Northern bedstraw is recognized by its dense, many-flowered panicles and its acute leaves in whorls of four.

216. *Galium boreale* (Northern bedstraw). Habit (center). Flower (lower right).

4. **Galium labradoricum** (Wieg.) Wieg. Rhodora 6:2. 1904. Fig. 217.
Galium tinctorium L. var. *labradoricum* Wieg. Bull. Torrey Club 24:398. 1897.

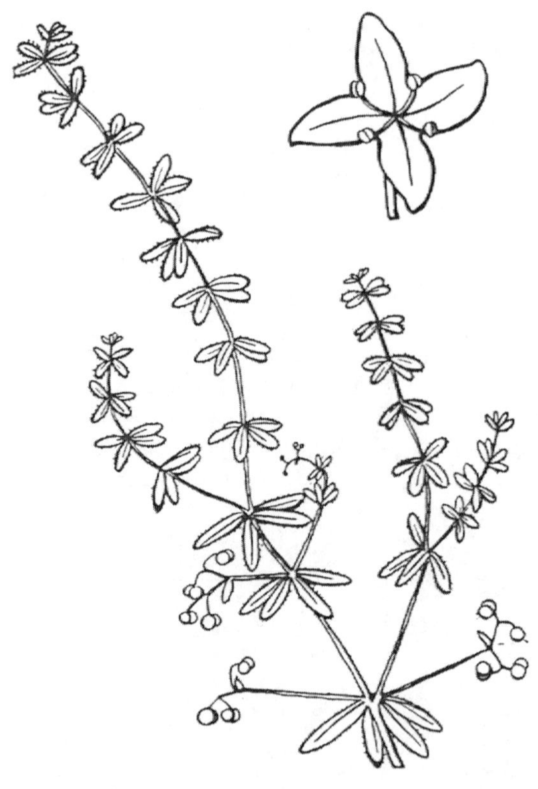

217. *Galium labradoricum* (Labrador bedstraw). Habit (center). Flower (upper right).

Perennial herbs with slender, erect to ascending, smooth stems; leaves in whorls of 4, recurved, narrowly oblanceolate, more or less obtuse at the apex, scabrous on the veins below and sometimes on the margins, up to 15 mm long, up to 3 mm wide; inflorescence a terminal cyme, usually 3-flowered; calyx absent; corolla 4-parted, white, up to 3 mm across; stamens 8; fruit dry, globose, 1–3 mm in diameter, smooth. June–August.

Bogs, marshes, swamps, wet thickets, calcareous fens.

IA, IL, IN, OH (OBL).

Labrador bedstraw.

This northern species barely reaches the northern edge of our range. It differs from the similar appearing *G. obtusum* by its scabrous margins and veins of the leaves.

5. **Galium obtusum** Bigel. Fl. Bost., ed. 2, 55. 1824. Fig. 218.

Perennial herbs with erect, much branched stems, the stems glabrous except at the nodes, to 75 cm long; leaves in whorls of 4, spreading to ascending, elliptic to oblanceolate, obtuse at the apex, slightly scabrous or smooth on the margins and veins, to 15 mm long, to 6 mm wide; inflorescence of terminal cymes with several flowers; calyx absent; corolla 4-parted, white, 2–3 mm across; stamens 8; fruit dry, globose, smooth, 3–5 mm in diameter. May–July.

Moist ground, low woods, swamps, calcareous fens, prairies, sometimes in standing water.

IA, IL, IN, KY, MO, OH (FACW+), KS, NE (FACW).

Blunt-leaved bedstraw.

This species is recognized by its smooth, obtuse leaves in whorls of four and its smooth stems.

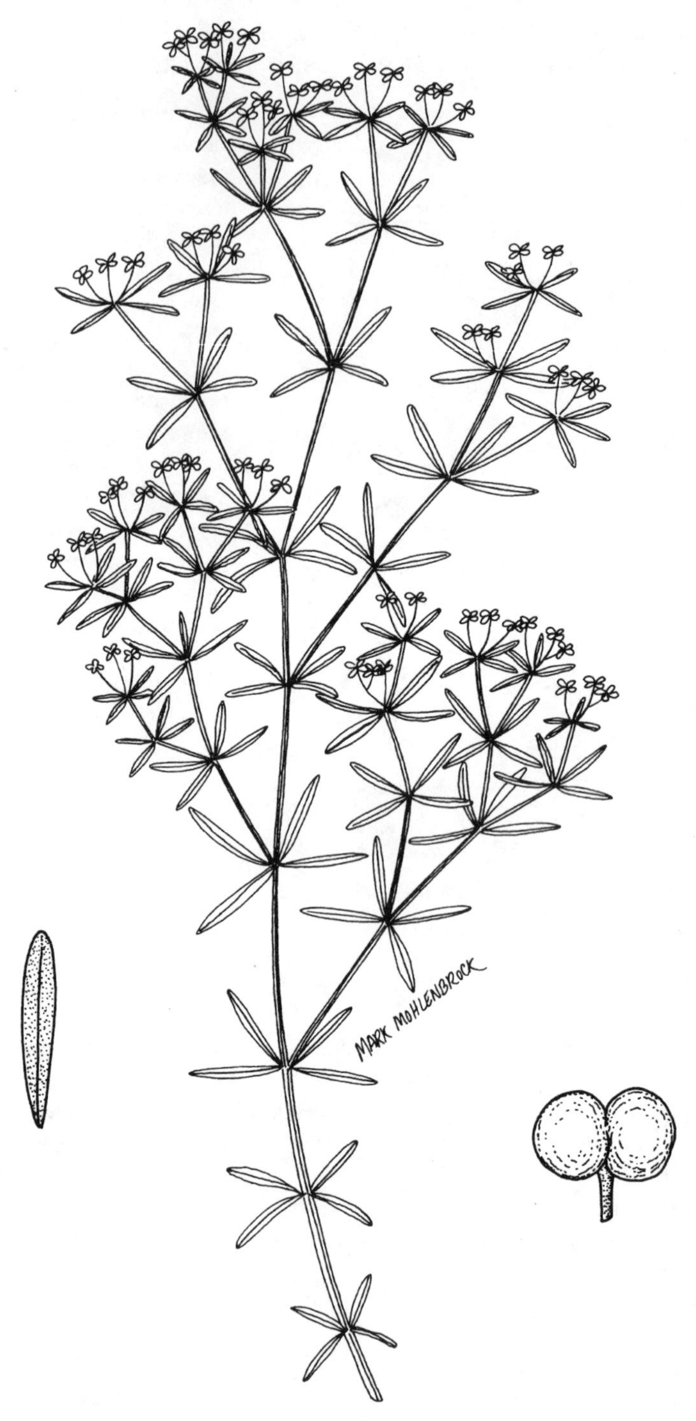

218. *Galium obtusum* (Blunt-leaved bedstraw). Habit (center). Leaf (lower left). Fruits (lower right).

6. **Galium tinctorium** L. Sp. Pl. 106. 1753. Fig. 219.

Weak, sprawling perennial; stems retrorse-scabrous, at least when young, to 75 cm long; leaves in whorls of 4, 5, and 6, linear to oblanceolate, more or less obtuse at the apex, slightly scabrous, to 2 cm long, to 4 mm wide; inflorescence a cyme with stiff branches and several flowers on spreading pedicels; calyx absent; petals (3–) 4 (–5) parted, white, 2–3 mm across; stamens 8; fruit dry, globose, smooth, 2–3 mm in diameter. May–September.

Swamps, marshes.

IA, IL, IN, KY, MO, OH (FACW+), KS, NE (FACW).

Swamp bedstraw.

This species is interesting in that on the same plant are leaves in whorls of 4, 5, and 6, and corolla lobes that are 3, 4, and 5.

The roots contain a colored sap that was used as a dye by pioneers.

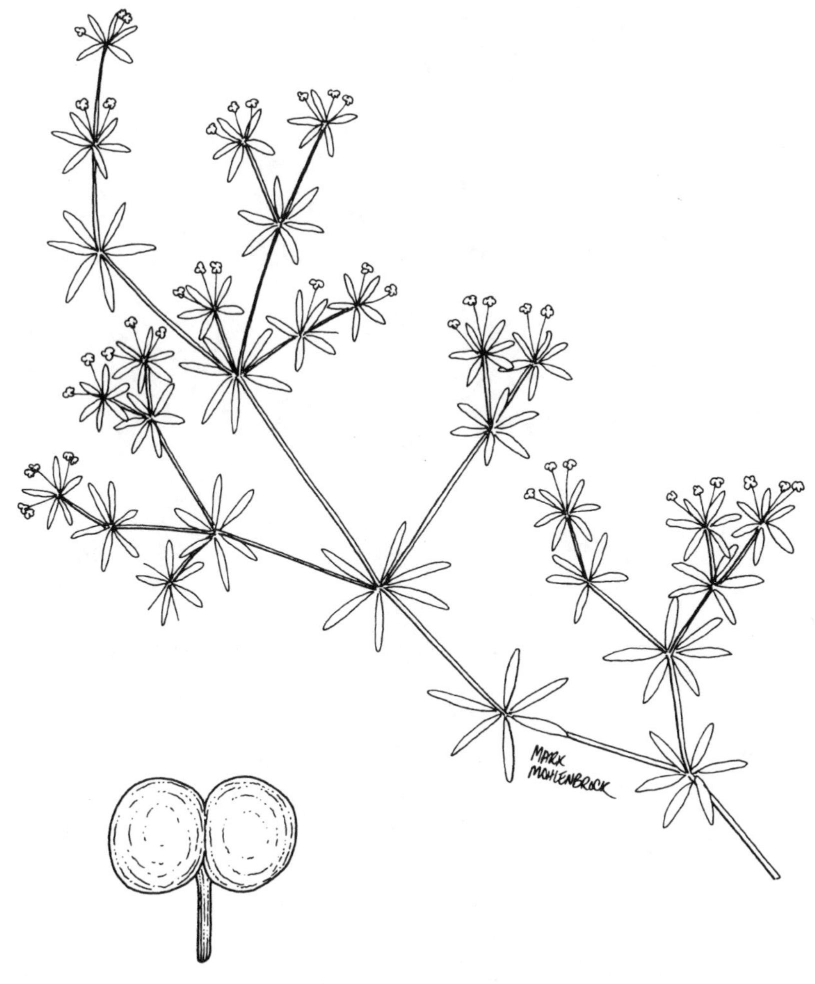

219. *Galium tinctorium* (Swamp bedstraw). Habit (above). Fruits (below).

7. **Galium trifidum** L. Sp. Pl. 105. 1753. Fig. 220.

Weak, sprawling perennial herbs; stems much branched, scabrous, up to 75 cm long; leaves mostly in whorls of 4, occasionally in 5 or 6, linear to narrowly oblanceolate, more or less obtuse at the apex, retrorsely scabrous on the margins, up to 2 cm long, up to 4 mm wide; inflorescence a terminal cyme, usually 3-flowered, on scabrous pedicels; calyx absent; corolla (3-) 4-parted, white, 1.5–3.0 mm across; stamens 8; fruit dry, globose, 1–2 mm in diameter. July–August.

Moist soil, swamps, calcareous fens, marshes, bogs.

IA, IL, IN, OH (FACW+), NE (OBL).

Rough northern bedstraw.

This species is distinguished by its leaves mostly in whorls of 4, its scabrous stems, and its very small flowers.

220. *Galium trifidum* (Rough northern bedstraw). Habit (left). Close-up of stem, leaves, and fruits (lower right).

8. **Galium triflorum** Michx. Fl. Bor. Am. 1:80. 1803. Fig. 221.

221. *Galium triflorum* (Fragrant bedstraw). Habit (center). Fruit (upper left). Flower (upper right).

Perennial herbs with spreading to ascending, much branched, glabrous stems up to 15 cm long, fragrant when dry; leaves simple, in whorls of 6, oval to oblanceolate, acute and mucronate at the apex, tapering to the sessile base, glabrous or nearly so, to 7.5 cm long, to 1.2 cm wide; flowers 1–3 in terminal and axillary cymes; calyx absent; corolla 4-lobed, greenish white to white, 2.5–4.0 mm across; stamens 8; fruit dry, covered with hooked bristles, 2.5– 3.5 mm in diameter. May–September.

Moist woods, bogs, calcareous fens.

IA, IL, IN, MO (FACU+), KS, KY, NE, OH (FACU).

Fragrant bedstraw.

The most distinguishing feature of this species is the mucronate tip of each of the six leaves in a whorl. The common name is derived from the odor emitted by the plant when dry.

4. **Houstonia** L.—Bluets

Annual or perennial herbs; leaves opposite, simple, entire, usually stipulate; flowers 1–4 in small cymes; calyx 4-lobed, green; corolla rotate, funnelform, or salverform, 4-lobed, blue, lavender, or white; stamens 4, attached to the corolla; stigmas 2; ovary inferior, 2-locular, each locule with numerous ovules; fruit a partly inferior capsule.

Several species in the central Midwest occur in uplands. In recent years, the central Midwest species of *Houstonia* were transferred to *Hedyotis*. However, in the strict sense, species of *Hedyotis* have calyx lobes that are shorter than the lower part of the ovary, fruits longer than broad, and ciliate or bristlelike stipules. The only central Midwest species I still recognize in *Hedyotis* is the upland *Hedyotis nigricans*.

1. Plants annual; corolla rotate or salvervorm.
    2. Corolla lobes 2.5–5.0 mm broad, very pale blue .... 1. *H. caerulea*
    2. Corolla lobes up to 2.5 mm broad, blue to dark blue.......... 3. *H. pusilla*
1. Plants perennial; corolla funnelform.......... 2. *H. purpurea*

1. **Houstonia caerulea** L. Sp. Pl. 105. 1753. Fig. 222.
Hedyotis caerulea (L.) Hook. Bor. Am. 1:286. 1833.

222. *Houstonia caerulea* (Bluets). Habit (center). Fruit (lower left). Flowers (right).

Annual herbs from fibrous roots; stems erect, slender, glabrous or nearly so, up to 15 cm tall; leaves opposite, simple, the lowermost in a basal rosette, spatulate to oblanceolate, obtuse to acute at the apex, tapering to the sessile or short-petiolate base, usually ciliate, sometimes hispid, to 12 mm long, to 8 mm wide; flower solitary on terminal and axillary pedicels. pale blue with a yellow center; calyx 4-lobed, green, the lobes linear; corolla salverform, 4-lobed, the lobes 2.5–5.0 mm wide; stamens 4; capsules bilobed, compressed, shorter than the calyx lobes. March–April.

Fields, prairies, woods, often in moist habitats; sand prairies.

IL, IN, MO (FAC), KY, OH (FACU).

Bluets.

This dainty wildflower occurs in a variety of habitats including moist fields, prairies, and woods.

It is the only *Houstonia* with a flower that has a yellow center.

2. **Houstonia purpurea** L. Sp. Pl. 105. 1753. Fig. 223.
Hedyotis purpurea (L.) Torr. & Gray, Fl. N. Am. 2:40. 1841.

Tufted perennial herbs from thickened roots; stems erect, branched or unbranched, glabrous or pubescent, up to 50 cm tall; leaves opposite, simple, narrowly to broadly ovate, acute at the apex, rounded at the sessile or short-petiolate base, 3- or 5-nerved, glabrous or somewhat pubescent, 2.0–4.5 cm long, (0.8–) 1.0–1.5 cm wide, ciliate along the margins; flowers 1–4 in terminal or axillary cymes, on pedicels 2–6 mm long; calyx 4-lobed, the lobes 4–6 mm long; petals 4, blue or lavender or less commonly white; stamens 4; capsules nearly globose but more or less compressed at the summit, 2–3 mm in diameter, about half as long as the calyx lobes. May–June.

Moist woods and fields; limestone glades.

IA, IL, IN, KY, MO, OH (UPL).

Broad-leaved bluets.

This species usually has leaves at least 1 cm wide. It is similar to *H. lanceolata* because of the 3- or 5-nerved leaves, but the leaves of this latter species rarely reach a width of 1 cm.

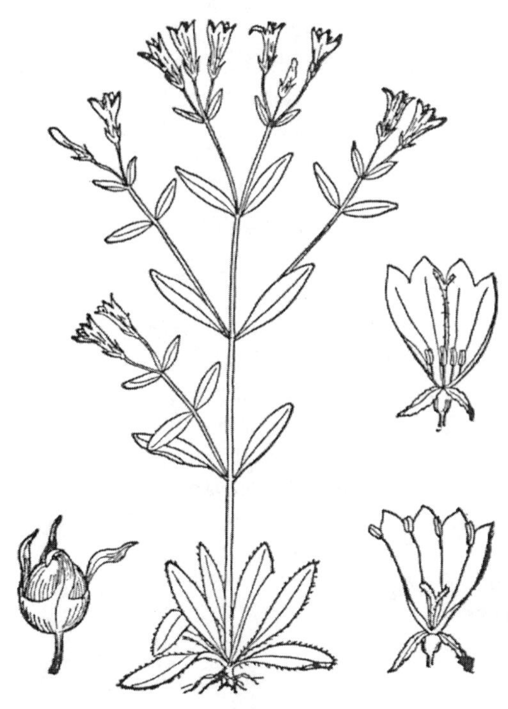

223. *Houstonia purpurea* (Broad-leaved bluets). Habit (center). Fruit (lower left). Flowers (right).

3. **Houstonia pusilla** Schoepf, Reise Nordamer. Staat. 2:306. 1788. Fig. 224.

Annual herbs from fibrous roots; stems spreading to erect, glabrous or somewhat scabrous, to 6 cm tall; leaves opposite and basal, simple, oval to ovate, obtuse to subacute at the apex, tapering to the sessile or short-petiolate base, entire, glabrous or nearly so, to 12 mm long, to 8 mm wide; flowers 1–few in terminal and axillary cymes, on pedicels to 2 cm long; calyx 4-lobed, green, the lobes linear; corolla salverform, blue, 4-lobed, the lobes 1–2 mm long; stamens 4; capsules 2-lobed, compressed, up to 4 mm in diameter, about equaling the calyx lobes. March–April.

Dry woods, moist depressions on cliffs, moist or dry fields.

IA, IL, IN, KY, MO, OH (UPL).

Small bluets.

This tiny species sometimes occurs in soil-filled depressions on cliffs where water stands for short periods of time.

224. *Houstonia pusilla* (Small bluets). Habit.

5. **Mitchella** L.—Partridge-berry

Creeping herbs; leaves opposite, simple, evergreen, usually entire, stipulate; flowers in pairs in the axils of the leaves, perfect, actinomorphic; calyx usually 4-lobed, green; corolla funnelform, 4-lobed, pubescent on the inner faces of the tube, the lobes recurved; stamens 4, attached to the corolla; stigmas 4; ovary inferior, 4-locular, each locule with one ovule; fruit of 2 united drupes.

This species below and one in southeast Asia comprise the genus.

1. **Mitchella repens** L. Sp. Pl. 111. 1753. Fig. 225.

Perennial herbs, rooting at the nodes; stems creeping, branched, glabrous to slightly pubescent, to 30 cm long; leaves opposite, simple, broadly ovate to orbicular, obtuse at the apex, rounded or subcordate at the base, evergreen, entire, shiny, to 2 cm long, often as wide; flowers sessile, in pairs on a common peduncle, the peduncle to 1.2 cm long; calyx 4-lobed, green, the lobes about as long as the tube; corolla funnelform, white, 4-lobed, to 1 cm long, the lobes spreading to recurved; stamens 4; drupes a little wider than tall, red, to 4 mm in diameter, persistent during the winter, edible. May–June.

Mesic woods, swamp forests, dune slopes.

IA, IL, IN, MO (FACU+), KY, OH (FACU).

Partridge-berry.

This species, inhabiting both moist and dry sites, is an evergreen creeper with pairs of white flowers in the axils of the leaves. The red drupes persist throughout most of the winter.

6. **Oldenlandia** L.—Oldenlandia

Annual or perennial herbs; leaves opposite, simple, entire, stipulate; flowers solitary or in small clusters in the axils of the leaves; calyx tubular below, 4-lobed above, green; corolla rotate, 4-lobed, usually white; stamens 4, attached to the corolla; ovary inferior, 2-locular, with numerous ovules in each locule.

Nearly two hundred species comprise this genus, most of them in tropical Asia. Several botanists include *Oldenlandia* within *Houstonia*, but *Oldenlandia* has rotate rather than salverform corollas, ovules and seeds numerous per locule rather than up to 20, and flowers often in glomerules rather than in cymes.

1. Perennials; leaves linear to linear-lanceolate; fruit smooth or slightly warty ....1. *O. boscii*
1. Annuals; leaves lanceolate; fruit pubescent .... 2. *O. uniflora*

1. **Oldenlandia boscii** (DC.) Chapm. Fl. South. U.S. 181. 1860. Fig. 226.
Hedyotis boscii DC. Prodr. 4:480. 1830.

Perennial herbs with slender rhizomes; stems prostrate to ascending, branched, glabrous, to 25 cm long; leaves opposite, simple, linear to linear-lanceolate, acute at the apex, usually tapering at the sessile or short-petiolate base, entire, usually glabrous, to 3 cm long, to 3 mm wide; flowers in small glomerules in the axils of the leaves, each flower to 2 mm across, sessile; calyx 4-lobed, the lobes lanceolate, long-acuminate, to 1 mm long, green; corolla rotate, 4-lobed, the lobes shorter than

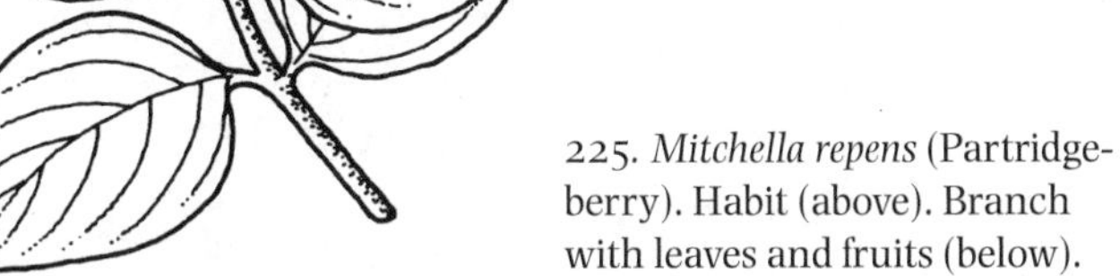

225. *Mitchella repens* (Partridge-berry). Habit (above). Branch with leaves and fruits (below).

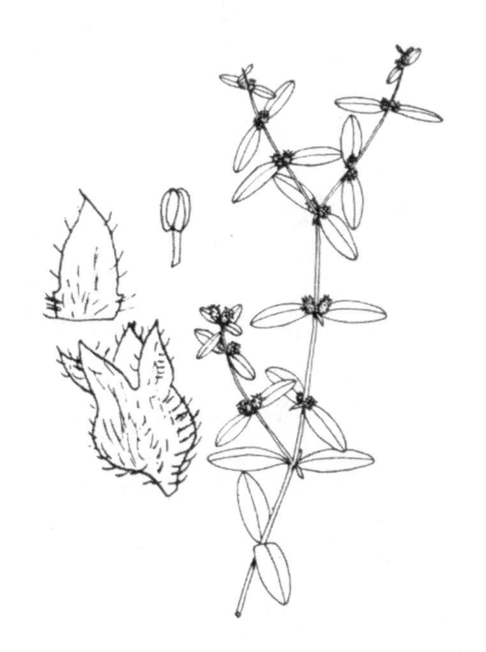

226. *Oldenlandia boscii* (Bose's oldenlandia). Habit (right). Sepal (upper left). Calyx (lower left). Stamen (upper center).

the sepals, white; stamens 4; ovary inferior; fruit ovoid, glabrous or slightly warty, 1.5–2.0 mm across, with numerous seeds. April–May.

Wet soil.

KY (FACW), MO (OBL).

Bosc's oldenlandia.

This tiny species is distinguished by its usually linear leaves and smooth or slightly warty fruits.

2. **Oldenlandia uniflora** L. Sp. Pl. 119. 1753. Fig. 227.
Hedyotis uniflora (L.) Lam. Tabl. Encycl. 1:271. 1792.

Annual herbs with fibrous roots; stems spreading to ascending, usually branched, pubescent, to 25 cm long; leaves opposite, simple, entire, lanceolate, acute at the apex, tapering at the sessile or short-petiolate base, usually glabrous, to 2.5 cm long, to 2.5 mm wide; flowers in small axillary glomerules or solitary, each flower to 2 mm across, sessile; calyx 4-lobed, hirsute, the lobes lanceolate to ovate, acute at the apex, about 1 mm long, green; corolla rotate, 4-lobed, the lobes shorter than the sepals; stamens 4; ovary inferior; fruit ovoid, densely white-hairy, 1.5–2.0 mm across, with numerous seeds. April–May.

227. *Oldenlandia uniflora* (One-flowered oldenlandia). Habit (center). Fruiting calyx (lower left). Flower (lower right).

Wet and muddy soil.

IL, MO (FACW−), KY (FACW).

One-flowered oldenlandia.

This small annual differs from the perennial *O. boscii* by its broader leaves and its densely white-hairy fruits.

7. **Spermacoce** L.—Smooth Buttonweed

Perennial herbs; leaves simple, opposite, entire, with bristle-tipped, membranous stipules; flowers crowded into sessile, axillary glomerules; sepals 4, united at the base, persistent on the fruit; corolla 4-parted, funnelform, white; stamens 8; ovary inferior, the style 2-cleft; fruit a bilocular capsule.

This genus of about one hundred species occurs mostly in subtropical regions of the New World.

Only the following species occurs in the central Midwest.

1. **Spermacoce glabra** Michx. Fl. Bor. Am. 1:82. 1803. Fig. 228.

Perennial herbs with spreading to erect stems; stems glabrous, to 50 cm long; leaves opposite, simple, oblong to lanceolate, acute to subacute at the apex, tapering to the short-petiolate base, entire, glabrous, to 8 cm long, to 3 cm wide; inflorescence a many-flowered, axillary glomerule up to 1.5 cm across; calyx 4-parted, green; corolla 4-parted, white, 2–3 mm long, slightly longer than the sepals, bearded within; stamens 8; capsules obovoid, 3–5 mm long. June–October.

Wet ground, occasionally in standing water.

IL, IN, KS, MO (FACW+), KY, OH (FACW).

Smooth buttonweed.

This species is distinguished by its bristlelike stipules, its glabrous herbage, and its axillary glomerules of white flowers.

228. *Spermacoce glabra* (Smooth buttonweed). Habit (center). Fruit (lower left).

106. SALICACEAE—WILLOW FAMILY

Trees or shrubs, mostly dioecious; leaves simple, alternate (rarely nearly opposite), with stipules; flowers unixsexual, arranged in catkins, without a perianth; staminate flowers each subtended by a bract and either a cupular disc or 1–2 glands, with 2 to several stamens, the filaments free or connate at the base, the anthers 2-celled, with vertical dehiscence; pistillate flowers each subtended by a bract and either a cupular disc or 1–2 glands, with one pistil, the ovary superior, unilocular, with 2–4 parietal placentae and numerous ovules, the style with 2–4 stigmas; fruit a capsule with 2–4 valves; seeds comose.

The Salicaceae is composed of two genera and about 350 species, distributed throughout most of the world. It differs from other catkin-bearing families by its dioecious nature, its bracteate flowers, and its comose seeds.

The absence of a perianth is interpreted by most botanists as a reduction from some ancestor having a perianth.

Many species of both *Salix* and *Populus* are grown as ornamentals.

1. Leaves never twice as long as broad; bud scales several; catkins pendulous .... 1. *Populus*
1. Leaves twice as long as broad or longer; bud scale 1; catkins not pendulous .... 2. *Salix*

1. **Populus** L.—Poplar

Dioecious trees; buds with several scales; leaves alternate, petiolate, with minute, caducous stipules; flowers unisexual, without a perianth; staminate flowers in dense, pendulous aments, from a cyathiform disc, each flower subtended by a dissected bract, with 4–60 free stamens; pistillate flowers in erect, spreading, or pendulous aments, from a cyathiform disc, each flower subtended by a dissected bract, with one pistil, the ovary superior, the stigmas 2–4, entire or 4-lobed; capsule 2- to 4-valved, with several comose seeds.

The leaves nearly as broad as long and the several-scaled buds distinguish *Populus* from *Salix*.

There are about thirty species of *Populus*, all native to the Northern Hemisphere. Several species are native to the western United States.

1. Part or all of the petiole flattened.
    2. Leaves triangular-ovate to rhombic, with incurved teeth.............................2. *P. deltoides*
    2. Leaves broadly ovate, with straight teeth .... 4. *P. tremuloides*
1. Petiole round in cross-section throughout.
    3. Leaves rounded or truncate at base or, if subcordate, the buds heavily resinous ....1. *P. balsamifera*
    3. Leaves cordate at base; buds not heavily resinous .... 3. *P. heterophylla*

1. **Populus balsamifera** L. Sp. Pl. 2:1034. 1753. Fig. 229.

Trees to 25 (–30) m tall; trunk up to 2 m in diameter, the upper part smooth; branches usually yellow-brown, spreading, glabrous, often stout; buds up to 1.2 cm long, shiny, fragrant, resinous; leaves alternate, simple, ovate to broadly lanceolate, acuminate at the apex, more or less truncate to subcordate at the base, glandular-toothed, green and glabrous on the upper surface, paler with a metallic sheen on the lower surface and sometimes pubescent on the veins, to 12 cm long to 9 cm wide, the petioles terete, up to 6 cm long, glabrous; staminate aments short-stalked, pendulous, densely flowered, to 4.5 cm long, the disk of each flower broad and oblique, with about 60 stamens; pistillate aments short-stalked, pendulous, sparsely flowered, to 7 cm long, the disk broad and more or less crenate; fruiting aments to 30 cm long, the capsules ovoid to ellipsoid, acute, 3- or 4-valved, to 8 mm long, glabrous, on glabrous pedicels up to 5.5 mm long; seeds oblongoid-obovoid, light brown, to 2 mm long, with a cottony coma. March–April.

River and stream banks.

Native in NE (FAC), adventive elsewhere.

Balsam poplar.

This species, whose natural range barely enters the northwestern part of the central Midwest, is readily distinguished by its balsamic-scented, resinous buds.

229. *Populus balsamifera* (Balsam poplar).

a. Branch with leaves and fruiting ament.
b. Branch with pistillate ament.
c. Pistillate flower with bract.

2. **Populus deltoides** Marsh. Arb. Am. 106. 1785. Fig. 230.
Populus monilifera Ait. Hort. Kew. 3:406. 1789.
Populus nigra L. β *virginiana* Castiglioni, Viagg. Stati Uniti 2:334. 1790.
Populus angulata Ait. var. *missouriensis* Henry in Elwes & Henry, Trees Gt. Brit. & Irel. 7:1811. 1913.
Populus deltoides Marsh. var. *missouriense* (Henry) Henry, Gard. Chron. ser. 3, 56:46. 1914.
Populus balsamifera L. var. *virginiana* (Castiglioni) Sarg. Journ. Arn. Arb. 1:63. 1919.
Populus balsamifera var. *missouriensis* (Henry) Rehder, Man. Cult. Trees & Shrubs 92. 1927.

Trees to 35 m tall, with a trunk diameter up to 2.5 m, the bark smooth and gray when young, becoming furrowed at maturity; leaves triangular-ovate, short-acuminate at the apex, truncate or slightly cordate at the base, ciliolate, to 12 cm long, nearly as broad or slightly broader, with 2–3 basal glands, more or less glabrous on both surfaces with numerous curved teeth, the petiole flat, up to 10 cm long, glabrous; stipules small, caducous; staminate aments short-stalked, pendulous, stout, densely flowered, to 5 cm long, borne as the leaves begin to expand, the disk of each flower broad, asymmetrical, with about 60 stamens inserted on the disk, the filaments short, distinct, the bracts light brown, scarious, glabrous, fimbriate; pistillate aments short-stalked, pendulous or spreading, sparsely flowered, to 8 cm long, borne as the leaves begin to expand, the disk of each flower broad, crenulate, with one glabrous pistil, with 3–4 laciniate-lobed stigmas, the bracts light brown, scarious, glabrous, fimbriate; fruiting aments to 12 cm long, the capsules ovoid to ellipsoid, acute, 3- to 4-valved, to 10 mm long, glabrous, on glabrous pedicels to 6 mm long, the light brown seeds oblongoid-obovoid, to 2 mm long, with a cottony coma. February–March.

Along rivers and streams.

IA, IL, IN, MO (FAC+), KS, KY, NE, OH (FAC).

Eastern cottonwood.

The cottonwood is one of the most common trees in moist situations in the central Midwest. It is extremely fast-growing, sometimes attaining heights in excess of 30 meters in about one hundred years.

This species differs from other species of *Populus* by its triangular leaves and its curved teeth of its leaves.

3. **Populus heterophylla** L. Sp. Pl. 2:1034. 1753. Fig. 231.
Populus argentea Michx. f. Hist. Arb. Am. 3:290, pl. 9. 1813.

Trees to 25 m tall, with a trunk diameter up to 0.8 m, the crown more or less open, the trunk usually straight, columnar, the bark reddish-brown, broken into elongated plates, the twigs stout, gray or brown, somewhat shiny, tomentose when young, becoming sparsely pubescent or nearly glabrous with age, the buds ovoid, acute, to 1 cm long, reddish-brown, slightly resinous; leaves broadly ovate, obtuse to subacute at the apex, truncate, rounded, or subcordate at the base, glandular-serrate, to 20 cm long, to 15 cm broad, white-tomentose when young, becoming glabrous or nearly so at maturity, dark green above, paler below, the petiole terete,

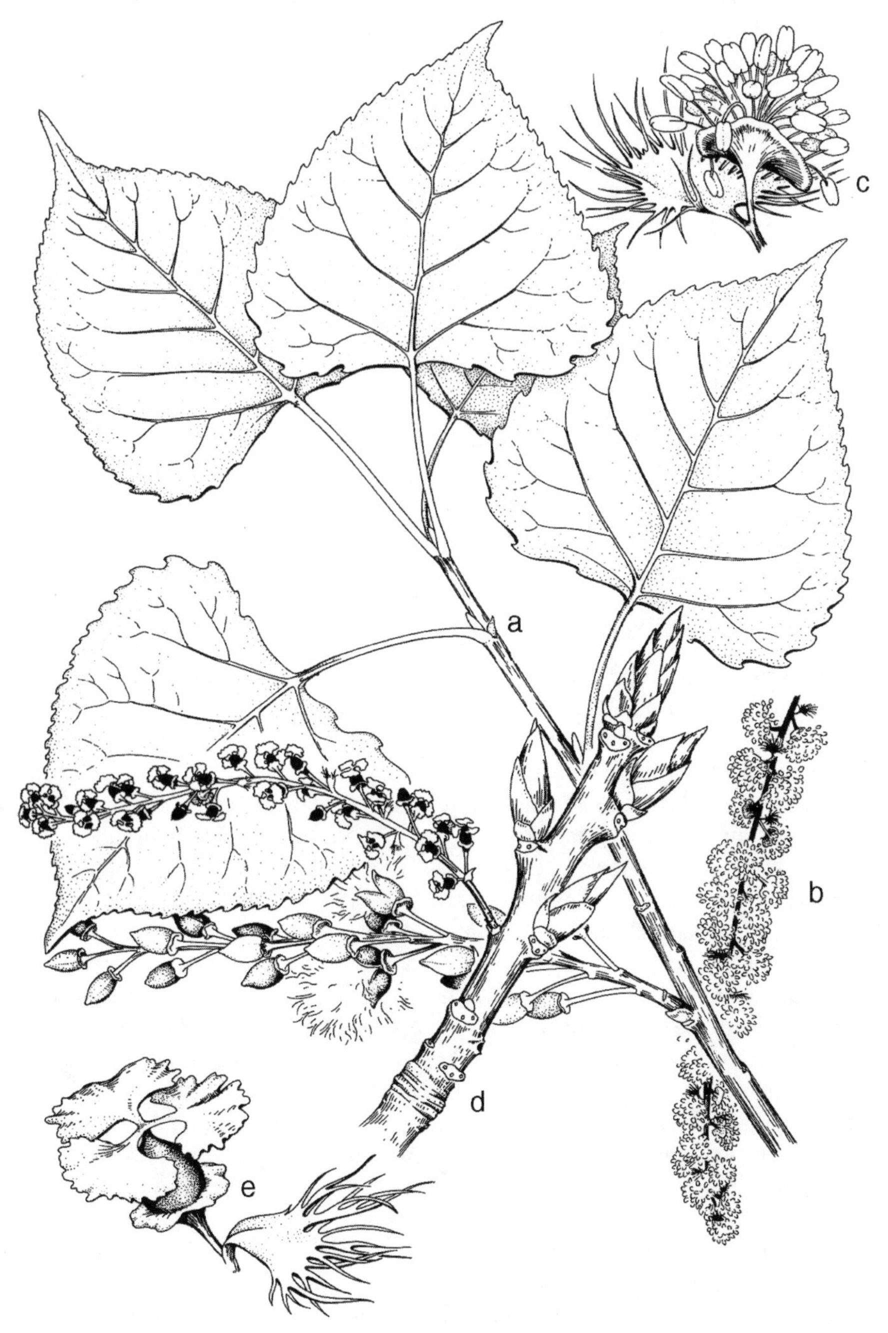

230. *Populus deltoides* (Eastern cottonwood).

a. Branch with leaves and fruits.
b. Staminate ament.
c. Staminate flower.
d. Winter twig with pistillate ament.
e. Pistillate flower with bract.

slender, yellow, more or less pubescent, to 8 cm long; stipules small, caducous; staminate aments erect at first, becoming pendulous, to 6 cm long, densely flowered, borne as the leaves begin to expand, the disk of each flower asymmetrical, entire, with 12–20 stamens, the bracts brown, glabrous or pubescent, fimbriate at the apex; pistillate aments pendulous, becoming erect by fruiting time, loosely flowered, to 5 cm long, borne as the leaves begin to expand, the disk of each flower symmetrical, long-toothed, with one glabrous pistil, the bracts brown, glabrous or pubescent, fimbriate at the upper end; fruiting aments to 6 cm long, more or less erect, the capsule ovoid, acute, 2- to 3-valved, to 12 mm long, on glabrous pedicels usually longer than the capsules, with numerous minute seeds. March–April.

Swampy woods.

IL, IN, MO (OBL), KY (FACW+).

Swamp cottonwood.

Swamp cottonwood is a common member of the bald cypress swamp community. The lobed disk of the pistillate flowers is one of its most unique characters, as are its more or less erect fruiting aments.

The wood of the swamp cottonwood is used for pulp in paper making and for the manufacture of excelsior.

4. **Populus tremuloides** Michx. Fl. Bor. Am. 2:243. 1803. Fig. 232.

Trees to 15 m tall, with a trunk diameter up to 0.4 m, the crown with a loose, rounded top, the bark pale yellow-green to white when young, becoming divided into black scaly ridges at maturity, the twigs slender, reddish-brown at first, becoming dark gray, nearly glabrous, with orange lenticels, the buds conic, acute, reddish brown, nearly glabrous, to 10 mm long; leaves broadly ovate, short-acuminate at the apex, more or less rounded at the base, crenate-serrate with 20 or more glandular teeth (averaging 31), to 8 cm long, to 7 cm broad, glabrous on both surfaces, shiny green above, dull green beneath, the petiole flat, up to 6 cm long, glabrous; stipules small, caducous; staminate aments pendulous, densely flowered, to 6 cm long, borne as the leaves begin to expand, the disk of each flower asymmetrical, entire, with 6–12 stamens, the filaments short, distinct, the bracts brown, long-pubescent, deeply lobed; pistillate aments pendulous or spreading, sparsely flowered, to 8 cm long, borne as the flowers begin to expand, the disk of each flower asymmetrical, crenulate, with one puberulent pistil, the stigmas divided, the bracts brown, long-pubescent, deeply lobed; fruiting aments to 10 cm long, the capsule conic-elongated, acuminate, 2-valved, to 6 mm long, puberulent, on puberulent pedicels to 2 mm long, the light brown seeds minute. March–April.

Low ground of woods, marshes, bogs, and fens.

IA, IL, KS, MO, NE (FAC), OH (FACU), as *P. tremula*.

Quaking aspen.

Quaking aspen has the ability to germinate readily in forests recently denuded by fires. It often forms large colonies because of its ability to produce suckers from the roots.

The wood is very light and soft, making it useful for pulp used in papermaking.

231. *Populus heterophylla* (Swamp cottonwood).

a. Branch with leaves and fruiting ament.
b. Twig with staminate aments.
c. Staminate flower.
d. Pistillate flower.

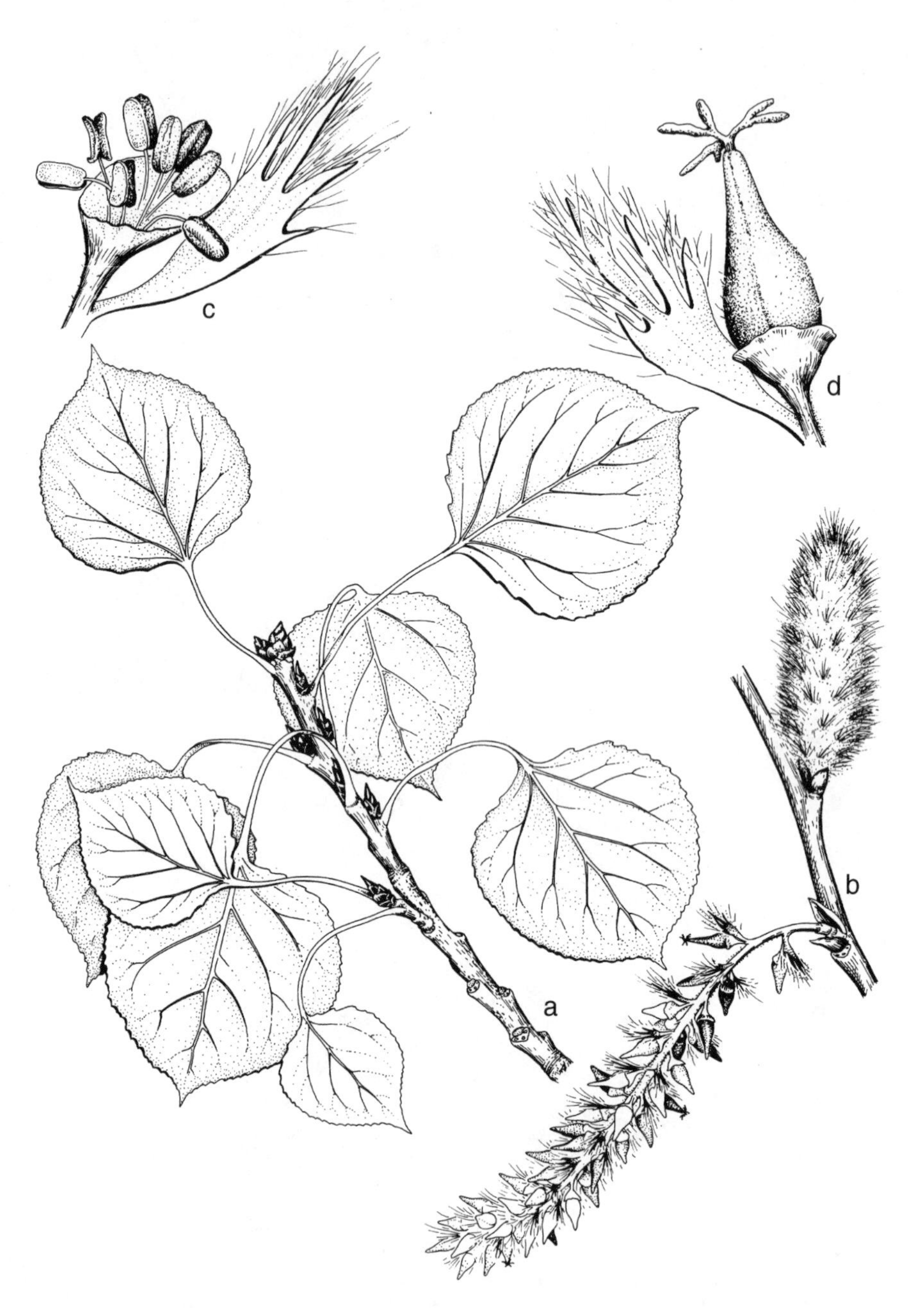

232. *Populus tremuloides* (Quaking aspen). a. Vegetative branch. b. Branch with staminate ament (above) and pistillate ament (below). c. Staminate flower. d. Pistillate flower.

2. **Salix** L.—Willow

Dioecious trees or shrubs; buds with only one scale; leaves alternate, rarely subopposite; flowers unisexual, without a perianth; staminate flowers in erect or spreading aments, each flower subtended by one bract and 1–2 glands, with (1–) 2 (–10) stamens with free or basally connate filaments; pistillate flowers in erect or spreading aments, each flower subtended by a bract and 1–2 glands, with one pistil, the ovary superior, the 2 stigmas entire or 2-cleft; capsule 2-valved, with several comose seeds.

There are about three hundred different species of *Salix*, primarily in the north temperate and arctic regions of the world. In addition, a substantial number of hybrids has been reported.

1. One or more glands present at upper end of petioles.
  2. Leaves whitish beneath.
    3. Leaves with some pubescence, usually on the nerves below; petioles and branchlets often pubescent.
      4. Teeth of leaves pointed .... 7. *S. caroliniana*
      4. Teeth of leaves blunt .... 21. *S. X rubens*
    3. Leaves glabrous; petioles and branchlets glabrous .... 23. *S. serissima*
  2. Leaves green beneath.
    5. Leaves linear-lanceolate to ovate-lanceolate, often falcate, at least five times longer than wide; buds pointed.
      6. All or most of the leaves at least 7 times longer than wide .... 15. *S. nigra*
      6. All or most of the leaves less than 7 times longer than wide .... 12. *S. X glatfelteri*
    5. Leaves lanceolate to ovate, up to 5 times longer than wide; buds rounded.
      7. Leaves with a caudate tip; stipules large and persistent .... 14. *S. lucida*
      7. Leaves acute or acuminate at the tip; stipules minute or absent.
        8. Leaves coriaceous, some of them usually at least 3 times longer than wide .... 23. *S. serissima*
        8. Leaves membranous, less than 3 times longer than wide .... 17. *S. pentandra*
1. Petioles not glandular.
  9. Leaves purplish, at least some of them opposite .... 19. *S. purpurea*
  9. Leaves not purplish, all of them alternate.
    10. Leaves glabrous or sparsely pubescent only on the veins below.
      11. Leaves remotely denticulate .... 13. *S. interior*
      11. Leaves closely serrate, crenulate, undulate, or entire.
        12. Leaves entire, revolute; some of the stems creeping .... 16. *S. pedicellaris*
        12. Leaves serrate, crenulate, or undulate, flat; all stems upright.
          13. Stipules absent or minute and falling away early on vegetative sprouts and young branchlets.
            14. Leaves ovate-lanceolate, long-attenuate to caudate at the tip; buds pointed .... 2. *S. amygdaloides*
            14. Leaves linear to lanceolate, acute to acuminate at the tip but not caudate; buds rounded.
              15. Leaves glabrous; petioles 7 mm long or longer.
                16. Leaves linear, serrulate; petioles pubescent; branches "weeping". .... 3. *S. babylonica*
                16. Leaves lanceolate, serrate; petioles glabrous; branches usually not "weeping" .... 10. *S. fragilis*
              15. Leaves usually with a few hairs on the veins below; petioles less than 7 mm long.

17. Teeth of the leaves pointed; hairs reddish........... 18. *S. petiolaris*
17. Teeth of the leaves blunt; hairs white ....21. *S. X rubens*
13. Stipules present on vegetative sprouts and young branchlets.
18. Leaves green on both sides.
19. Leaves lanceolate, rarely as much as 2 cm wide (except sometimes those on sprouts), tapering at base..................................15. *S. nigra*
19. Leaves oblong-lanceolate, some of them at least 2 cm wide, rounded or subcordate at base.. 20. *S. rigida*
18. Leaves pale or white on the lower surface.
20. Leaves irregularly crenate; capsules pubescent ....8. *S. discolor*
20. Leaves finely serrulate; capsules glabrous or granular.
21. Leaves lanceolate, often falcate ....7. *S. caroliniana*
21. Leaves oblong to narrowly ovate............... 11. *S. glaucophylloides*
10. Leaves pubescent on the lower surface (if only on veins, go back to couplet 17).
22. Young branchlets and new leaves covered with white wool..................6. *S. candida*
22. Young branchlets and new leaves not covered with white wool.
23. Leaves entire or undulate.
24. Capsules up to 6 mm long; bracts brown, with a black tip........................... .... 24. *S. X subsericea*
24. Capsules 6–10 mm long; bracts yellow.................................... 4. *S. bebbiana*
23. Leaves serrulate or crenate or denticulate.
25. Leaves silvery-silky on lower surface.
26. Petioles up to 3 mm long; margin of leaves remotely denticulate........... ....13. *S. interior*
26. Petioles 3 mm long or longer; margin of leaves finely serrulate.
27. Teeth along margin of leaves not extending all the way to the base. .... 18. *S. petiolaris*
27. Teeth along margin of leaves extending all the way to the base.
28. Young branchlets silky; flowers appearing with the leaves......... .... 1. *S. alba*
28. Young branchlets glabrous or glabrate; flowers appearing before the leaves.. 22. *S. sericea*
25. Leaves pubescent beneath, but not with silvery-silky hairs.
29. Leaves narrowed or rounded at base, not subcordate; stipules on sprouts and young branchlets inconspicuous and falling away early.
30. Upper surface of leaves shiny; flowers appearing before the leaves... ....8. *S. discolor*
30. Upper surface of leaves dull; flowers appearing with the leaves........ .... 4. *S. bebbiana*
29. Leaves subcordate at base (tapering in *S. eriocephala*); stipules on sprouts and young branchlets large and persistent.
31. Branchlets permanently tomentulose...................... 9. *S. eriocephala*
31. Branchlets glabrous or glabrate.
32. Leaves glaucous beneath, caudate at tip ....5. *S. X bebbii*
32. Leaves green beneath, acute to short-acuminate......21. *S. rigida*

1. **Salix alba** L. Sp. Pl. 2:1021. 1753. Fig. 233.
Salix vitellina L. Sp. Pl. ed. 2, 1442. 1763.
Salix alba L. var. *vitellina* (L.) Stokes, Bot. Mat. Med. 4:506. 1812.

Trees to 15 m tall, with a trunk diameter up to 0.5 m, the bark thick, rough, gray; twigs olive-green or brown or yellow, slender, pubescent or glabrous, not brittle; leaves lanceolate, acuminate at the apex, cuneate at the base, straight, serrulate, to 10 cm long, to 2.5 cm wide, sericeous or glabrous, more or less whitened on the lower surface, the petiole to 1 cm long, pubescent or glabrous, glandular at the upper end; stipules minute, caducous; aments borne as the leaves begin to expand; staminate aments erect to ascending, slender, to 3.5 cm long, the flowers spirally arranged, each flower with 2 (–3) stamens, the filaments free, pilose at the base, the bract oblong, yellow, sparsely pubescent, caducous, the glands 2; pistillate aments erect to ascending, slender, to 6 cm long, the flowers spirally arranged, each

233. *Salix alba* (White willow). a. Fruiting branch. b. Staminate flower. c. Pistillate flower.

with a glabrous ovary, the bract oblong, yellow, sparsely pubescent, caducous, the gland 1, the style less than 1 mm long; capsules ovoid-conical, 3.0–4.5 mm long, glabrous, sessile or nearly so. April–May.

Along streams; disturbed moist soil.

IA, IL, IN, KS, KY, MO, NE, OH (FACW).

White willow.

This native of Europe has been widely planted in the United States. It hybridizes readily with *S. fragilis* to form *S. X rubens*.

Plants with leaves glabrous or nearly so on the lower surface and with yellow branchlets may be called var. *vitellina*, the golden willow.

Although the yellow branchlets of the golden willow give this a strikingly different appearance from the white willow, floral and fruiting similarities indicate it to be treated best as a variety of *Salix alba*.

2. **Salix amygdaloides** Anders. in Ofv. Svensk. Vetensk. Acad. Forh. 15:114. 1858. Fig. 234.
Salix nigra Marsh. var. *amygdaloides* (Anders.) Anders. in DC. Prodr. 16:201. 1868.

Trees to 20 m tall, with a trunk diameter up to 0.6 m, the bark rough and scaly, brown or reddish-brown; twigs yellowish or brown, slender, glabrous, tough and flexible; leaves lanceolate to ovate-lanceolate, attenuate at the apex, rounded or subcuneate at the base, straight, serrulate, to 15 cm long, to 2.5 cm wide, occasionally pubescent when young, glabrous on both surfaces at maturity, green above, glaucous below, the petiole to 2 cm long, glabrous, sometimes with glands at the upper end; stipules absent or minute, rarely persistent on sprouts; aments borne before the leaves begin to expand; staminate aments erect to ascending, slender, to 7 cm long, the flowers whorled, each with 3–5 stamens, the filaments free, pilose near the base, the bracts yellow, more or less pubescent, sparingly persistent, the glands 2; pistillate aments erect to ascending, slender, loosely flowered, to 8 cm long, the flowers whorled, each with a glabrous ovary, the bract oblong, yellow, more or less pubescent, the gland one, the style less than 1 mm long; capsules narrowly ovoid to ovoid-lanceoloid, obscurely flattened, 3–4 mm long, glabrous, on pedicels 1.5–2.5 mm long, more than twice as long as the gland. April–June.

Along streams, low woods, borders of ponds.

IA, IL, IN, KS. KY, MO, NE, OH (FACW).

Peach-leaved willow.

The peach-leaved willow differs from the somewhat similar *S. caroliniana* by its nonpersistent stipules and by its long-attenuate, broader leaves.

The lateral branches of *S. amygdaloides* tend to be more or less pendulous, and are exceedingly tough and flexible.

Hybrids between *Salix amygdaloides* and *S. nigra*, called *S.* X *glatfelteri* Schneid., occur in the central Midwest.

3. **Salix babylonica** L. Sp. Pl. 2:1017. 1753. Fig. 235.
Salix annularis Forbes, Salict. Woburn. 41. 1829.

Trees to 15 m tall, with a trunk diameter up to 0.5 m, the bark thick, rough, gray; twigs yellow or brown, glabrous, pendulous; leaves linear-lanceolate, long-acuminate at the apex, cuneate at the base, straight, serrulate, to 12 cm long, to 2 cm wide, sericeous when young, becoming glabrous or nearly so at maturity, yellow-green above, whitened below, the petiole to 1.2 cm long, glabrous, with glands at the upper end; stipules absent or up to 7 mm long; aments borne as the leaves begin to expand; staminate aments erect to ascending, slender, to 4 cm long, the flowers spirally arranged, each with 2 (–5) stamens, the filaments free, pilose near the base, the bract oblong, yellow, pubescent, caducous, the glands 1–2; pistillate

234. *Salix amygdaloides* (Peach-leaved willow).

a. Fruiting branch.
b. Flowering branchlet.
c. Staminate flower.
d. Pistillate flower.
e. Seed.

aments erect to ascending, slender, to 3.5 cm long, the flowers spirally arranged, each with a glabrous ovary, the bract oblong, yellow, pubescent, caducous, the glands 1 (–2), cupular, the style less than 1 mm long; capsules ovoid-conical, 1–2 cm long, glabrous, sessile. April–May.

Along roads, around homesteads, wet ground.

IA, IL, IN, KY, MO, OH (FACW), KS, NE (FACW−).

Weeping willow.

235. *Salix babylonica* (Weeping willow).
a. Fruiting branch.
b. Leaf.
c. Staminate flower.
d. Pistillate ament.
e. Pistillate flower.
f. Capsule.

Native of Europe and Asia; escaped from cultivation in much of the eastern half of the United States. Weeping willow is a commonly planted ornamental, popular primarily because of its pendulous ("weeping") branches and rapid manner of growth.

This species may be confused with weeping forms of the crack willow but differs by its serrulate leaves, its smaller, sessile capsules, and its nonbrittle branches.

4. **Salix bebbiana** Sarg. Gard. & For. 8:463. 1895. Fig. 236.
Salix rostrata Richardson, Frankl. Journ. App. 753. 1823, *non* Thuill. (1799).
Salix starkeana Willd. ssp. *bebbiana* (Sarg.) Youngberg, Rhodora 72:549. 1970.

Shrubs or small trees to 6 m tall; twigs reddish brown to dark brown, tomentose at least when young; leaves oblanceolate to ovate-oblong, acute at the apex, cuneate to rounded at the base, entire or irregularly glandular-crenate, to 7.5 cm long, to 3.5 cm wide, pilose to tomentose when young, becoming sparsely pubescent at maturity, dull green above, glaucous and rugose beneath, the petiole to 1 cm long, pubescent, without glands at the tip; stipules small, caducous; aments borne as the leaves begin to expand; staminate aments erect to ascending, to 2.5 cm long, the flowers spirally arranged, each with 2 stamens, the filaments free, pilose at the base, the bract lanceolate, yellowish, pubescent, subpersistent, the gland 1; pistillate aments erect to spreading, to 5 cm long, loosely flowered, the flowers spirally arranged, each flower with a sericeous ovary, the bract lanceolate, yellow, pubescent, subpersistent, the gland 1, the style minute or absent; capsules conical-subulate, 5–9 mm long, pubescent, on pedicels many times longer than the gland. May.

Boggy soils.

IA, IL, IN (FACW+), OH (FACW).

Bebb's willow.

Salix bebbiana is named for Michael Shuck Bebb, an authority on *Salix*, who lived at Fountaindale, Illinois, southwest of Rockford during the last half of the nineteenth century. Considerable variation exists in the degree of pubescence on the leaves and the amount of serration on the margins. It often happens that the lowermost leaves on the twigs are entire, while the uppermost leaves are crenate.

5. **Salix X bebbii** Gand. Pl. Cret. 182. 1916. (Not illustrated).
Salix X myricoides Muhl. Neue Schrift. Ges. Nat. Fr. Berlin 4:236. 1803.
Salix cordata Muhl. var. *myricoides* (Muhl.) Carey ex Gray, Man Bot. 427. 1848.

Shrub to 3 m tall; twigs brownish or yellowish, slender, usually canescent-pubescent, brittle at the base; leaves lanceolate, long-acuminate at the apex, cuneate or subcuneate at the base, serrulate, to 10 cm long, to 3 cm wide, more or less sericeous beneath, green above, whitened below, the petiole to 1 cm long, more or less pubescent, without glands at the upper end; stipules very small, caducous; aments borne before the leaves begin to expand; staminate aments erect to ascending, slender, to 2.5 cm long, the flowers spirally arranged, each with 2 stamens, the filaments free, sparsely pubescent at the base, the bract narrowly oblong, dark brown, pilose, not persistent, the gland 1; pistillate aments erect to ascending, slender, to 4 cm long, the flowers spirally arranged, each with a thin sericeous ovary, the bract narrowly oblong, dark brown, pubescent, subpersistent, the gland 1, the style minute; capsules narrowly ovoid, 3–5 mm long, thinly sericeous, on pedicels usually less than twice as long as the gland. April–May.

Moist ground.

IL, MO (not listed).

Bebb's hybrid willow.

Salix X bebbii is a reputed hybrid between *S. rigida* and *S. sericea*, although it appears to have more of the characteristics of *S. sericea*. However, it differs from *S. sericea* by its canescent twigs and thinly sericeous capsules, while *S. sericea* has glabrous or glabrate twigs and densely sericeous capsules.

236. *Salix bebbiana* (Bebb's willow). a. Vegetative branch. b. Leaf. c. Staminate flower. d. Pistillate flower. e. Fruiting branch.

6. **Salix candida** Fluegge ex Willd. Sp. Pl. 4:708. 1806. Fig. 237.

Shrubs to 3 m tall; twigs yellow to brown, stout, whitish-tomentose when young, becoming glabrous or nearly so; leaves linear to oblong, to 10 cm long, acute at the apex, cuneate at the base, entire to undulate, revolute, densely whitish tomentose when young, remaining white-tomentose or becoming sparsely tomentose at maturity, the petiole to 1 cm long, pubescent, without glands at the

237. *Salix candida* (Hoary willow). a. Vegetative branch. b. Staminate inflorescence. c. Staminate flower. d, e. Pistillate flower.

upper end; stipules lanceolate, tomentose, subpersistent; aments borne before the leaves begin to expand; staminate aments erect to ascending, dense, to 1.5 cm long, the flowers spirally arranged, each with 2 stamens, the filaments free, glabrous, the bract pale brown, pubescent, subpersistent, the gland 1; pistillate aments erect to ascending, dense, to 5 cm long, the flowers spirally arranged, each with a pubescent ovary, the bract pale brown, pubescent, subpersistent, the gland 1, the style about 1 mm long; capsules ovoid-conic, 4–6 mm long, densely tomentose, the pedicel about twice as long as the gland. April–May.

Bogs.

IA, IL, IN, OH (OBL).

Hoary willow.

The hoary willow is a species of bogs and moist prairies, where it associates with such species as *Betula pumila, Larix laricina, Toxicodendron vernix*, and *Vaccinium macrocarpon*. It is distinguished by its densely white-tomentose, revolute leaves.

7. **Salix caroliniana** Michx. Fl. Bor. Am. 2:226. 1803. Fig. 238.
Salix longipes Shuttlw. ex Anders. Öfv. Svensk. Vetensk. Acad. Förh. 15:114. 1858.
Salix nigra Marsh. var. *wardi* Bebb ex Ward, Bull, U.S. Nat. Mus. 22:114. 1881.
Salix wardi (Bebb) Bebb, Gard. & For. 8:363. 1895.
Salix longipes Shuttlw. var. *wardi* (Bebb) Schneider, Bot. Gaz. 65:22. 1918.

Trees to 20 m tall, with a trunk diameter up to 0.5 m, the bark rough and scaly, usually gray or dark reddish brown; twigs brownish, yellowish, reddish, or gray, pubescent, brittle near the base; leaves linear-lanceolate to oblong-lanceolate, acute to acuminate at the apex, rounded to subcuneate at the base, straight or falcate, glandular-serrulate, to 10 cm long, to 3 cm wide, more or less pubescent throughout, rarely glabrous, whitened on the lower surface, the petiole to 1 cm long, pubescent, usually with glands at the upper end; stipules (at least on the sprouts) glandular-serrulate, more or less auriculate, persistent; aments borne before the leaves begin to expand; staminate aments erect to ascending, slender, to 10 cm long, the flowers whorled, each with 4–8 stamens, the filaments free, pilose near the base, the bract obovate, yellow, villous, not persistent, the glands usually 2; pistillate aments erect to ascending, slender, rather densely flowered, to 10 cm long, the flowers whorled, each with a glabrous ovary, the bract oblong, yellow, villous, usually not persistent, the gland one, the style more or less absent; capsules conic to ovoid-conic, 4–6 mm long, granular, on pedicels up to 2 mm long, several times longer than the gland. April–May.

Usually in gravel beds of streams.

IL, IN, KS, KY, MO, OH (OBL).

Carolina willow.

Salix caroliniana differs primarily from *S. nigra* by its whitened lower leaf surface and from *S. amygdaloides* by its usually narrower leaves and its persistent stipules.

238. *Salix caroliniana* (Carolina willow).

a. Vegetative branch.
b. Leaf.
c. Staminate flower.
d. Pistillate flower and capsule.
e. Fruiting raceme.

8. **Salix discolor** Muhl. Neue Schrift. Ges. Nat. Fr. Berlin 4:234. 1803. Fig. 239.
Salix prinoides Pursh, Fl. Am. Sept. 613. 1814.
Salix sensitiva Barratt, Salic. Am. no. 8. 1840.
Salix discolor Muhl. var. *sensitiva* (Barratt) Bebb, Trans. Ill. State Agr. Soc. 3:587. 1859.
Salix discolor Muhl. var. *latifolia* Anderss. Sv. Vet. Akad. Handl. 6:84. 1867.
Salix imponens Gandoger, Fl. Europ. 21:167. 1890.

Shrubs or small trees to 6 m tall; twigs brown, pubescent, sometimes becoming glabrous by the second year; leaves alternate, simple, lanceolate to elliptic to obovate, acute at the apex, rounded to subcuneate at the base, serrate to crenate, to 10 cm long, to 3 cm broad, ferruginous-pilose when young, usually becoming glabrous or nearly so, dark green above, glaucous below, the petiole to 1.5 cm long, usually pubescent, without glands at the upper end; stipules prominent on sprouts; aments borne before the leaves begin to expand, sessile or nearly so; staminate aments erect to ascending, to 5 cm long, the flowers spirally arrranged, each with 2 stamens, the filaments free, glabrous or pilose near the base, the bract brown or black, conspicuously white-villous, persistent, the gland 1; pistillate aments erect to ascending, to 7 cm long, densely flowered, the flowers spirally arranged, each with a densely sericeous ovary, the bract brown or black, white-villous, subpersistent, the lowest ones tawny and usually not subtending a flower, the gland 1, the style up to 1 mm long; capsules oblongoid-conic, 5–12 mm long, puberulent, the pedicel about twice as long as the gland. March–May.

239. *Salix discolor* (Pussy willow). Vegetative branch (center). Staminate flower (left center). Pistillate inflorescence (upper center). Pistillate flower (lower left). Fruiting raceme (far right).

Marshes, swamps, fens.

IA, IL, IN, KY, MO, OH (FACW).

Pussy willow.

Salix discolor is variable in the degree and persistence of the pubescence of twigs and leaves. Leaves that remain permanently pubescent have been designated var. *latifolia* Anderss., but Argus (1964) speculates that these latter plants may be hybrids between *Salix discolor* Muhl. and *Salix humilis* Marsh. At any rate, the pubescent forms are not very well defined and are not recognized as distinct in this work.

The pussy willow sold by most florists is not this species but *S. caprea* L., the goat willow.

9. **Salix eriocephala** Michx. Fl. Bor. Am. 2:225. 1803. Fig. 240.
Salix rigida Muhl. var. *vestita* Anderss. Mon. Sal. 159. 1867.
Salix missouriensis Bebb, Gard. & For. 8:373. 1895.

Trees to 15 m tall, with a trunk diameter to 0.5 m, the bark black; twigs brown to gray-black, slender, tomentose; leaves alternate, simple, narrowly lanceolate to ovate-oblong, long-attenuate at the apex, cuneate or subcuneate at the base, serrulate, to 15 cm long, to 4 cm wide, pubescent beneath at maturity, at least on the nerves, dull green above, glaucous below, the petioles to 1 cm long, pubescent, without glands at the upper end; stipules reniform, to 1 cm long, persistent; aments borne before the leaves expand; staminate aments erect to ascending, to 5 cm long, the spirally arranged, each with 2 stamens, the filaments free, pilose at the base, the bract oblong, dark brown to black, pubescent, subpersistent, the gland 1; pistillate aments erect to ascending, to 10 cm long, the flowers spirally arranged, each with a glabrous ovary, the bract oblong, dark brown to black, pubescent, subpersistent, the gland 1, the style about 1 mm long; capsules conical-subulate, to 1 cm long, glabrous, on pedicels much longer than the gland. April.

Alluvial soil along streams.

IA, IL, IN, MO, OH (FACW), KS, NE (not listed).

Missouri wllow.

This is the same species that Bebb described in 1895 as *S. missouriensis*, but Michaux's *S. eriocephala* is clearly identical and predates Bebb's binomial.

Some botanists consider *S. eriocephala* to be a variety of *S. rigida*.

10. **Salix fragilis** L. Sp. Pl. 1017. 1753. Fig. 241.

Trees to 25 m tall, with a trunk diameter up to 1.5 m, the bark thick, rough, gray; twigs yellow, brown, or gray, slender, glabrous or pubescent, shiny, very brittle at the base; leaves alternate, simple, lanceolate, long-acuminate at the apex, cuneate at the base, straight, coarsely serrate, to 15 cm long, to 2.5 cm wide (sometimes much larger on sprouts), glabrous, green above, green or whitened below, the petiole to 1.5 cm long, glabrous, with glands at the upper end; stipules minute, caducous; aments borne as the leaves begin to expand; staminate aments erect to ascending, slender, to 6 cm long, the flowers spirally arranged, each with 2 (–4) stamens, the filaments free, pilose at the base, the bract oblong, yellow, sparsely pubescent, caducous, the glands 2; pistillate aments erect to ascending, slender, to 7 cm long, the flowers spirally arranged, each with a glabrous ovary, the bract oblong, yellow, sparsely pubescent, caducous, the glands 2, the style up to 1 mm long;

capsules conical-subulate, 4–5 mm long, glabrous, on pedicels about twice as long as the uppermost gland. April–May.

Along streams; near swamps.

IA, IL, IN, KY, MO, OH (FAC+), KS, NE (FAC).

Crack willow; brittle willow.

The branchlets are very brittle and break off readily. These broken branches easily take root in moist soil so that the species may spread rather rapidly.

240. *Salix eriocephala* (Missouri willow). a. Vegetative branch. b. Staminate flower. c. Pistillate flower.

The crack willow, a European native, is very similar to the weeping willow in most characters, except for its more coarsely serrate leaves. While most specimens of the crack willow lack decidedly pendulous branches, a few do possess this "weeping" character, making distinction from the weeping willow sometimes rather difficult.

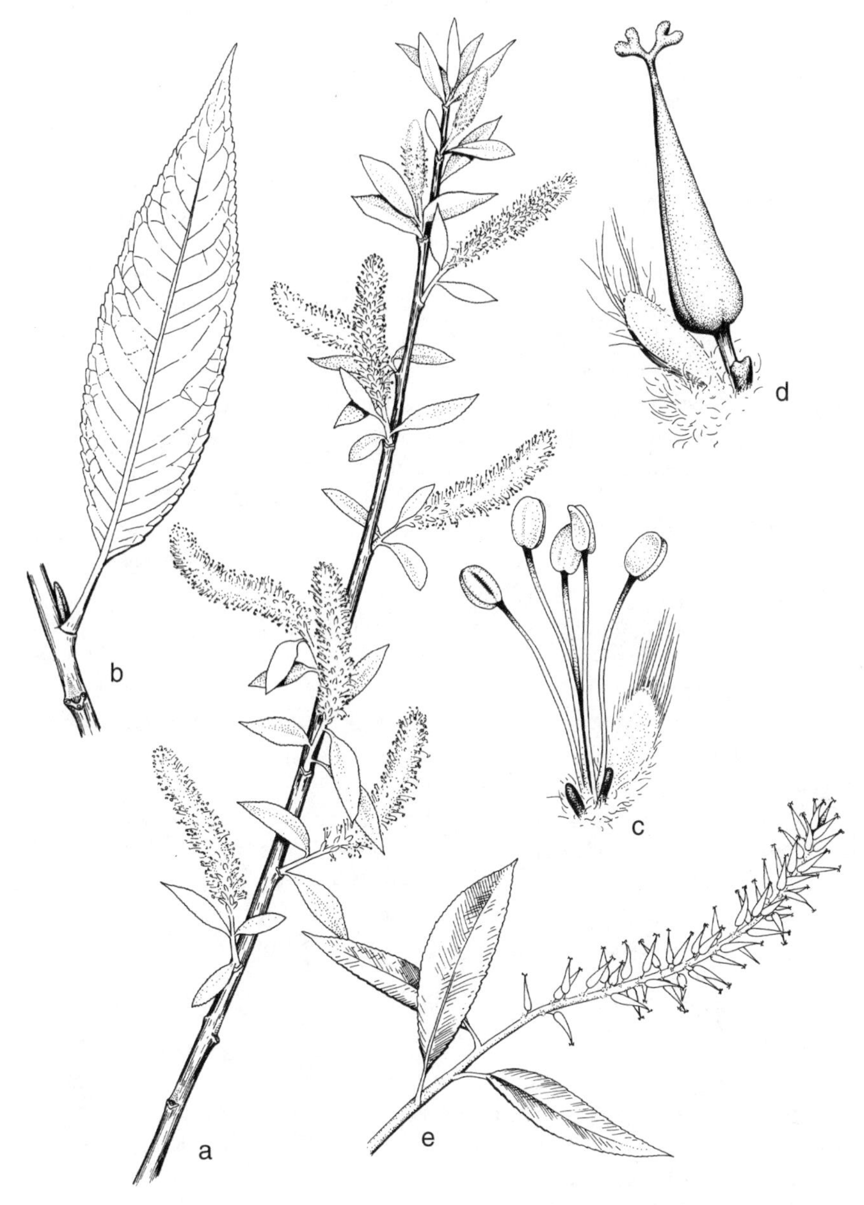

241. *Salix fragilis* (Crack willow).

a. Flowering branch.
b. Leaf.
c. Staminate flower.
d. Pistillate flower.
e. Immature fruiting branchlet.

11. **Salix glaucophylloides** Fern. var. **glaucophylla** (Bebb) Schneider, Journ. Arn. Arb. 1:157. 1920. Fig. 242.
Salix cordata Muhl. var. *glaucophylla* Bebb ex Babcock, The Lens 2:249. 1873.
Salix glaucophylla (Bebb) Bebb, Rep. Nat. Hist. Northwest. Univ. 1889:23. 1889, non Bess. (1822).
Salix glaucophylla (Bebb) Bebb var. *latifolia* Bebb, Rep. Nat. Hist. Northwest. Univ. 1889:23. 1889.
Salix glaucophylla (Bebb) Bebb var. *angustifolia* Bebb, Rep. Nat. Hist. Northwest. Univ. 1889:23. 1889.

Shrubs to 3 m tall; twigs yellowish or brown, slender, usually glabrous, shiny; leaves alternate, simple, lanceolate to oblong to narrowly ovate, acute or short-acuminate at the apex, rounded to subcordate at the base, serrate or serrate-crenate, to 12 cm long, to 4.5 cm wide, somewhat purbescent when young, becoming glabrous or nearly so at maturity except for the puberulent midrib beneath, green above, glaucous beneath, the petiole to 1.0 (–1.2) cm long, mostly pubescent, without glands at the upper end; aments borne as the leaves begin to expand; stipules ovate, glandular-serrate, to 1 cm long, glaucous, persistent; staminate aments erect to ascending, to 5 cm long, the flowers spirally arranged, each with 2 stamens, the filaments free or nearly so, glabrous, the bract oblong, dark brown to black, villous, subpersistent, the gland 1; pistillate aments erect to ascending, to 10 cm long, loosely flowered, the flowers spirally arranged, each with a glabrous ovary, the bract oblong, dark brown to black, villous, subpersistent, the gland 1, the style 1.0–1.3 mm long; capsules conical-subulate, 4–9 mm long, glabrous, on pedicels much longer than the gland and usually longer than the bract. May.

Open sand, shores; marshes.

IL, IN, OH (FACW), as *S. cordata.*

Blue-leaf willow.

I am following Schneider and Fernald in considering our material as a variety of the more northern *S. glaucophylloides.* It differs from var. *glaucophylloides* by its longer pedicels and loosely flowered pistillate aments.

Swink (1974) reports this taxon as an occupant of open sand near Lake Michigan, along with *Cornus stolonifera* and *Salix syrticola.* He also reports it as occurring on calcareous pond shores, old sand pits, and on marsh borders.

A narrow-leaved variant, described as var. *angustifolia* by Bebb, was first collected by H. H. Babcock from Cook County, Illinois.

12. **Salix X glatfelteri** Schneid. Journ. Arn. Arb. 3:79. 1922. (Not illustrated).

Trees to 12 m tall, with brown to blackish bark, the branchlets flexible, yellow-brown to brown, usually pubescent at first; leaves alternate, simple, lance-ovate, acuminate at the apex, subcuneate or rounded at the base, more or less falcate, glandular-serrate, to 10 cm long, to 1.5 cm wide, usually pubescent when young, glabrous or nearly so at maturity except for the pubescent midrib beneath, green on both sides, the petiole to 10 mm long, puberulent, usually with glands at the upper end; stipules (at least on the sprouts) glandular-serrate, auriculate, sometimes persistent. April–May.

Wet habitats.

IL, IN, MO, OH (not listed).

Hybrid black willow.

This plant is the reputed hybrid between *S nigra* and *S. amygdaloides*. It has the broader and sometimes falcate leaves of *S. amygdaloides* but has the lower green leaf surface of *S. nigra*.

242. *Salix glaucophylloides* var. *glaucophylla* (Blue-leaf willow).

a. Vegetative branch.
b. Staminate flower.
c. Pistillate flower.
d. Fruiting raceme.

13. **Salix interior** Rowlee, Bull, Torrey Club 27:253. 1900. Fig. 243.
Salix longifolia Muhl. Neve Schrift. Ges. Nat. Fr. Berlin 4:238. 1803, non Lam. (1778).
Salix interior Rowlee var. *wheeleri* Rowlee, Bull. Torrey Club 27:253. 1900.
Salix wheeleri (Rowlee) Rydb. ex Britt. Man., ed. 2, 1061. 1905.
Salix longifolia Muhl. var. *wheeleri* (Rowlee) Schneid. Bot. Gaz. 67:342. 1919.
Salix interior Rowlee f. *wheeleri* (Rowlee) Rouleau, Nat. Can. 71:268. 244.

Trees to 8 m tall, with shoots arising from vegetative buds on the roots, thus often forming dense colonies, with a trunk diameter up to 30 cm, the bark gray, rough at maturity; twigs brown to reddish-brown, slender, sericeous to nearly glabrous; leaves alternate, simple, linear to linear-lanceolate, acuminate at the apex,

243. *Salix interior* (Sandbar willow). a. Fruiting branch. b. Staminate flower. c. Pistillate flowers.

cuneate at the base, straight, remotely glandular-denticulate, to 10 cm long, to 1 cm wide (sometimes somewhat larger on sprouts), sericeous when young, usually becoming sparsely pubescent at maturity, green on both sides or sometimes whitish below, the petiole to 7 mm long, glabrous, eglandular; stipules minute or absent; aments borne as the leaves begin to expand; staminate aments erect to ascending, slender, sometimes branched, to 3 cm long, the flowers spirally arranged, each with 2 stamens, the filaments free, pilose in the lower half, the bract oblong, yellow, more or less pubescent, caducous, the gland 1; pistillate aments erect to ascending, slender, sometimes branched, to 7.5 cm long, the flowers spirally arranged, each with a sparsely sericeous or glabrous ovary, the bract narrowly oblong, yellow, pubescent, not persistent, with 1 gland, the style minute or absent; capsules narrowly conic-ovoid, 5–9 mm long, sparsely sericeous to glabrous, on pedicels twice as long as the gland. April–May.

Sandbars, banks or streams, wet fields, sandy beaches.

IA, IL, IN, KS, KY, MO, NE, OH (OBL).

Sandbar willow.

This species is distinguished by its narrow leaves with teeth farther apart than in any other willow. It sometimes grows in dense colonies.

The U.S. Fish and Wildlife Service calls this plant *S. exigua*, but that binomial should apply to a different species from the Rocky Mountains westward.

Plants with densely sericeous leaves may be called var. *wheeleri*.

Branched aments sometimes may occur.

14. **Salix lucida** Muhl. Neue Schrift. Ges. Nat. Fr. Berlin 4:239. Fig. 244.
Salix lucida var. *intonsa* Fern. Rhodora 6:2. 1904.

Shrubs or trees to 7 m tall, with a trunk diameter up to 30 cm; twigs reddish brown or yellowish, slender, glabrous or nearly so, shiny; leaves alternate, simple, lanceolate to ovate-lanceolate, acute to acuminate to long-attenuate at the apex, rounded to subcuneate at the base, straight, glandular-serrulate, to 15 cm long, to 4 cm wide, sometimes pubescent when young, glabrous at maturity, green and shiny above, green or pale beneath, the petiole to 1.5 cm long, more or less glabrous, with glands at the upper end; stipules (at least of the sprouts) rounded, glandular, up to 6 mm long; aments borne as the leaves begin to expand; staminate aments erect to ascending, slender to thickish, to 4 cm long, the flowers spirally arranged, each with 3–6 stamens, the filaments free, pilose near the base, the bract oblong, yellow, pubescent, persistent, the glands 2; pistillate aments erect to ascending, rather slender, to 5 cm long, the flowers spirally arranged, each with a glabrous ovary, the bract oblong, yellow, pubescent, rather persistent, the glands 2, the style less than 1 mm long; capsules narrowly ovoid, 5–7 mm long, glabrous, on pedicels at least four times as long as the gland. May.

Bogs, moist disturbed areas.

IA, IL, IN (FACW+), KS, NE, OH (FACW).

Shining willow.

The shining willow is most nearly related to *S. serissima* and the introduced *S. pentandra*. It differs from *S. serissima* by its larger stipules, long-attenuate leaves, and short capsules. It differs from *S. pentandra* by its larger stipules and long-attenuate leaves.

244. *Salix lucida* (Shining willow).

a. Vegetative branch.
b. Staminate flower.
c. Pistillate flower.
d. Fruiting branchlet.

15. **Salix nigra** Marsh. Arb. Am. 139. 1785. Fig. 245.
Salix falcata Pursh, Fl. Am. Sept. 2:614. 1814.
Salix nigra Marsh. var. *falcata* (Pursh) Torr. Fl. N. Y. 2:209. 1843.

Trees to 30 m tall, with a trunk diameter up to 1.5 m, the bark rough and scaly, dark brown; twigs brownish or brownish yellow, slender, pubescent at first, at length glabrous, brittle near the base; leaves alternate, simple, linear to linear-lanceolate, acute to acuminate at the apex, rounded to subcuneate at the base, straight, glandular-serrulate, to 10 cm long, to 1.5 cm wide, often pubescent when

245. *Salix nigra* (Black willow).
a. Flowering branch.
b. Leaf.
c. Staminate flower.
d. Pistillate flower.
e. Fruiting raceme.
f. Seed.

young, glabrous or nearly so at maturity except for the puberulent midrib beneath, green on both sides, the petiole to 8 mm long, puberulent, usually with glands at the upper end; stipules (at least on the sprouts) glandular-serrate, auriculate, more or less persistent; aments borne before the leaves begin to expand; staminate aments erect to ascending, slender, to 10 cm long, the flowers whorled, each with 3–6 stamens, the filaments free, pilose near the base, the bract obovate, yellow, pubescent, persistent, the glands 2 or more; pistillate aments erect to ascending, slender, loosely flowered, to 6 cm long, the flowers whorled, each with a glabrous ovary, the bract oblong, yellow, pubescent, the gland one, the style minute or absent; capsules ovoid to ovoid-conical, 3–4 mm long, glabrous, on pedicels less than 1 mm long but over twice as long as the gland. April–May.

Along streams, particularly in deep alluvial soil.

IA, IL, IN, KS, MO, NE (OBL), KY, OH FACW+).

Black willow.

The black willow attains a greater size than any other species of *Salix* in the central Midwest. It frequently grows with several trunks.

Although the twigs are brittle near the base, they are tough and flexible above. The soft, light wood makes the black willow useful in the making of baskets, packing cases, and toys.

Salix nigra is perhaps confused most often with *S. rigida* or *S. caroliniana. Salix rigida* usually has some or all its leaves at least 2 cm wide, and lacks any glands on the petioles. *Salix caroliniana* looks very much like *S. nigra* but is whitened on the lower surface of the leaves.

Salix nigra hybridizes with *S. amygdaloides* to form *S. X glatfelteri.*

16. **Salix pedicellaris** Pursh, Fl. Am. Sept. 611. 1814. Fig. 246.
Salix pedicellaris Pursh var. *hypoglauca* Fern. Rhodora 11:161. 1909.

Shrubs to 0.75 m tall, sometimes decumbent and rooting at the nodes; twigs yellow to reddish brown to gray, glabrous; leaves alternate, simple, oblong to oblong-obovate, obtuse to acute at the apex, rounded to subcuneate at the base, entire and somewhat revolute, to 7.5 cm long, to 2.5 cm wide, glabrous, dark green above, mostly glaucous beneath, the petiole to 6 mm long, glabrous, without glands at the upper end; stipules absent; aments borne as the leaves begin to expand; staminate aments erect to ascending, to 2 cm long, the flowers spirally arranged, each with 2 stamens, the filaments free or nearly so, glabrous, the bract pale yellow, pubescent, subpersistent, the gland 1; pistillate aments erect to ascending, to 3 cm long, the flowers spirally arranged, each with a glabrous ovary, with the bract pale yellow, pubescent, subpersistent, the gland 1; capsule oblongoid-conic, 4–8 mm long, glabrous, on pedicels much longer than the glands. April–May.

Bogs.

IA, IL, IN, OH (OBL).

Bog willow.

Salix pedicellaris is the only willow in the central Midwest with entire, revolute leaves completely glabrous beneath.

All specimens have leaves that are somewhat glaucous on the lower surface and have been segregated by some botanists as var. *hypoglauca* Fern.

246. *Salix pedicellaris* (Bog willow).
a. Habit, in fruit.
b. Staminate inflorescence.
c. Staminate flower.
d. Pistillate flower.

17. **Salix pentandra** L. Sp. Pl. 1016. 1753. Fig. 247.

Trees to 7 m tall, with a trunk diameter up to 30 cm; twigs brown to reddish brown, slender, glabrous, shiny; leaves alternate, simple, lanceolate to elliptic-oblong, acute to acuminate at the apex, rounded or subcuneate at the base, straight, glandular-serrulate, to 10 cm long, to 3 cm wide, glabrous, green above, green or a little pale below, the petiole to 1 cm long, glabrous, with glands at the upper end; stipules usually minute, not persistent; aments borne as the leaves begin to expand; staminate aments erect to ascending, slender, to 6 cm long, the flowers spirally arranged, each with 5 stamens, the filaments free, pilose near the base, the bract obovate, yellow, more or less pubescent, the glands 2, often united; pistillate aments erect to ascending, slender, densely flowered, to 6 cm long, the flowers

247. *Salix pentandra* (Bay-leaved willow). a. Fruiting branch. b. Staminate flower. c. Pistillate flower.

spirally arranged, each with a glabrous ovary, the bract oblong, yellow, pubescent, the glands 2, free, often cupular, the style about 1 mm long; capsules conical, 5–6 mm long, glabrous, on pedicels about twice as long as the gland. May–June.

Along creeks and streams.

Native of Europe; IA, IL, IN, KY, OH (not listed).

Bay-leaved willow; laurel willow.

The bay-leaved willow is seldom found in North America as an escape from cultivation.

The relatively short capsules distinguish this species from *S. serissima*, while the short fruiting pedicels distinguish it from *S. lucida*.

18. **Salix petiolaris** Sm. Trans. Linn. Soc. 6:122. 1802. Fig. 248.
Salix petiolaris Sm. var. *gracilis* Anderss. in DC. Prodr. 16:235. 1868.
Salix gracilis Anderss. var. *textoris* Fern. Rhodora 48:46. 1946.

Shrubs to 3 m tall; twigs dark brown to yellow-green, pubescent or nearly glabrous; leaves alternate, simple, linear to lanceolate, acuminate at the apex, cuneate at the base, serrate, denticulate, or subentire, to 12 cm long, to 2 cm wide, sericeous when young, becoming glabrous or sparsely sericeous at maturity, green above, glaucous beneath, the petiole to 1 cm long, pubescent, without glands at the upper end; stipules minute, caducous; aments borne as the leaves begin to expand; staminate aments erect to ascending, to 2 cm long, the flowers, the flowers spirally arranged, each with 2 stamens, the filaments free, glabrous or pilose at the base, the bract oblong, brown, pubescent, subpersistent, the gland 1; pistillate aments erect to ascending, 3–5 cm long, the flowers spirally arranged, each with a sericeous ovary, the bract oblong, brown, pubescent, subpersistent, the gland 1, the style minute or absent; capsule oblongoid-conical, 5–8 mm long, sericeous, the pedicels several times longer than the gland. April–June.

Low prairies, marshes, bogs.

IA, IL, IN (FACW+), KY, NE, OH (OBL).

Meadow willow.

The nomenclature regarding this taxon is confusing. Fernald (1946) calls it *S. gracilis* Anderss. var. *textoris* Fern., arguing that *S. petiolaris* Sm. is an entirely different European species. Ball (1948), on the other hand, believes that *S. petiolaris* Sm. was described from an American species cultivated in a European garden.

There is no doubt that *S. petiolaris* is related to *S. sericea*, with *S. petiolaris* differing primarily by its nearly glabrous leaves, pointed capsules, and later time for flowering.

The yellow-green twigs are distinctive.

19. **Salix purpurea** L. Sp. Pl. 1017. 1753. Fig. 249.

Shrubs to 4 m tall; twigs yellow-green, brown, or purple, slender, glabrous; leaves subopposite, simple, linear to spatulate, acute to acuminate at the apex, more or less rounded at the base, entire or serrulate, to 10 cm long, to 1.5 cm wide, glabrous, often purple-tinged, more or less glaucous, the petioles to 0.5 cm long, glabrous, without glands at the tip; stipules absent; aments flowering before the leaves begin to expand; staminate aments erect to ascending, slender, to 3 cm long, the flowers spirally arranged, each with 2 stamens, the filaments united, pubescent

248. *Salix petiolaris* (Meadow willow).

a. Vegetative branch.
b. Staminate flower.
c. Pistillate flower.
d. Fruiting branch.

near the base, the bract obovate, black-tipped, glabrous, subpersistent, the glands 2; pistillate aments erect to ascending, slender, to 3 cm long, the flowers spirally arranged, each with a pubescent ovary, the bract obovate, black-tipped, glabrous, subpersistent, the gland 1, the style minute; capsules broadly ovoid, 2–3 mm long, pubescent, sessile. April–May.

Wet ground, sometimes planted in mitigation sites.

Native to Europe; IA, IL, IN, MO (FACW), OH (NI).

Purple osier.

The purple osier differs from most other willows in several of its characteristics. Its leaves appear to be suboppositely arranged and often purplish, its two filaments are united as one, and its capsules are sessile.

The leaves often have a peculiar purplish but yet silvery-gray color. The young twigs frequently are purple.

249. *Salix purpurea* (Purple osier). a. Vegetative branch. b. Staminate flower. c. Pistillate flower.

20. **Salix rigida** Muhl. Neue Schrift. Ges. Nat. Fr. Berlin 4:236. 1803. Fig. 250.
Salix cordata Muhl. Neue Schrift. Ges. Nat. Fr. Berlin 4:236. May, 1803, non Michx. (March, 1803).
Salix angustata Pursh, Fl. Am. Sept. 613. 1814.
Salix cordata Muhl. var. *rigida* (Muhl.) Carey ex Gray, Man. Bot. 427. 1848.
Salix cordata β *angustata* (Pursh) Anderss. in DC. Prodr. 16:252. 1868.

Shrubs to 3 m tall; twigs reddish brown to yellowish, slender, pubescent when young, eventually becoming glabrous or nearly so; leaves alternate, simple, oblong-lanceolate, acute to acuminate at the apex, rounded or subcordate at the base, serrulate, to 10 cm long, to 4 cm wide, densely sericeous when young, usually becoming glabrous or nearly so at maturity, usually reddish purple when immature, the petioles to 1.5 cm long, pubescent, without glands at the upper end; stipules to 2 cm long, cordate, usually pubescent, persistent; aments borne just as the leaves begin to expand; staminate aments erect to ascending, to 3 cm long, the flowers spirally arranged, each with 2 stamens, the filaments free, glabrous, the bract obtuse, brown, pilose, not persistent, the gland 1; pistillate aments erect to ascending, to 5 cm long, the flowers spirally arranged, each with a glabrous ovary, the bract narrowly oblong, brown to black, pilose, subpersistent, the glands 2, the style less than 1 mm long, capsules narrowly ovoid, 4–5 mm long, glabrous, on pedicels usually less than twice as long as the gland. April–May.

Moist soil.

IA, IL, IN, KY, MO, OH (OBL), KS, NE (FACW+).

Heart-leaved willow.

Salix rigida, *S. glaucophylloides* var. *glaucophylla*, and *S. syrticola* are three taxa in which the leaves are usually cordate or subcordate at the base. *Salix rigida* differs from *S. syrticola* by its narrower leaves, its eglandular-serrulate leaves, and its shorter pistillate aments and capsules. It differs from *S. glaucophylloides* var. *glaucophylla* by its narrower, serrate to crenate leaves.

Salix rigida also resembles *S. nigra* vegetatively but differs by lacking petiolar glands and by never attaining the stature of a tree.

Much confusion has existed with respect to the nomenclature of this species. Muhlenberg described it in May of 1803 as *S. cordata*, but two months after Michaux had described a different North American species as *S. cordata.* Thus, with Muhlenberg's *S. cordata* unavailable for this species, the binomial *S. rigida* Muhl. becomes the earliest available binomial.

Pursh had described a narrow-leaved variant as *S. angustata* in 1814, which Andersson later reduced to a variety.

21. **Salix X rubens** Schrank, Baier. Fl. 1:226. 1789. (Not illustrated).

Trees to 25 m tall, with a trunk diameter up to 1.2 m, the bark thick, rough, gray; twigs yellow, brown, or gray, slender to rather stout, usually pubescent, shiny; leaves alternate, simple, lanceolate, acuminate at the apex, cuneate at the base, not falcate, with numerous blunt teeth, to 15 cm long, to 2.5 cm wide, green above, whitened beneath, glabrous except for pubescence on the veins beneath, the glands at the upper end, the petioles to 1.5 cm long; stipules minute, caducous; aments borne as the leaves begin to expand; staminate aments ascending, slender, to 6 cm long, the flowers with 2–4 stamens, the bract oblong, yellow, caducous, with 2 glands; pistillate aments ascending, slender, to 7.5 cm long, the bract oblong, yellow, caducous, with usually only 1 gland; capsules narrowly conical, 4–5 mm long, glabrous, the pedicels less than twice as long as the gland. April–May.

Wet ground.

Native of Europe and Asia; IA, IL, IN, KY, MO, NE, OH (not listed).

Hybrid crack willow.

This plant is reputed to be the hybrid between *S. alba* and *S. fragilis.* It is intermediate in pubescence between the two parents, and is usually more common. Older trees may have a trunk diameter up to 1.2 m.

250. *Salix rigida* (Heart-leaved willow). a. Vegetative branch. b. Staminate flower. c. Pistillate flower.

22. **Salix sericea** Marsh. Arb. Am. 140. 1785. Fig. 251.

Shrubs to 3 m tall; twigs brown, slender, glabrous or sparsely pubescent; leaves alternate, simple, narrowly lanceolate, acuminate at the apex, cuneate at the base, glandular-serrulate, to 10 cm long, to 2.5 cm wide, dark green and glabrous or pubescent above, with appressed silky hairs beneath, the petioles to 1 cm long, more or less pubescent, without glands at the upper end; stipules (on sprouts) lanceolate, caducous; aments borne before the leaves begin to expand; staminate aments erect to ascending, to 2.5 cm long, the flowers spirally arranged, each with 2 stamens, the filaments free, pilose near the base, the bract oblong, dark brown to black, pubescent, persistent, the gland 1; pistillate aments erect to ascending, to 2.5 cm

251. *Salix sericea* (Silky willow). a. Vegetative branch. b. Staminate flower. c. Pistillate flower. d. Fruiting branch.

long, the flowers spirally arranged, each with a sericeous ovary, the bract oblong, dark brown to black, pubescent, persistent, the gland 1, the style absent or minute; capsules ovoid-oblongoid, 3–5 mm long, sericeous, on pedicels several times longer than the gland. March–May.

Along streams, in bogs and fens.

IA, IL, IN, KY, MO, OH (OBL).

Silky willow.

The silky willow is a handsome shrub with its brown branches and silvery-silky leaves.

23. **Salix serissima** (Bailey) Fern. Rhodora 6:6. 1904. Fig. 252.
Salix lucida Muhl. var. *serissima* Bailey ex Arthur, Bailey, & Holway, Bull. Geol. & Nat. Hist. Surv. Minnesota 3:19. 1887.

Shrubs to 4 m tall; twigs reddish brown, slender, glabrous, shiny; leaves alternate, simple, lanceolate to oblong-lanceolate, short-acuminate at the apex,

252. *Salix serissima* (Autumn willow). a. Fruiting branch. b. Staminate flower. c. Pistillate flower. d. Capsule.

rounded to subcuneate at the base, straight, glandular-serrulate, to 10 cm long, to 3.5 cm wide, glabrous, green above, whitened below, the petiole to 1 cm long, glabrous, with glands at the upper end; stipules minute or absent; aments borne after the leaves have begun to expand; staminate aments erect to ascending, to 3 cm long, the flowers spirally arranged, each with 4–7 stamens, the filaments free, pilose near the base, the bract oblong, yellow, pubescent, persistent, the glands 2, more or less united; pistillate aments erect to ascending, to 4.5 cm long, the flowers spirally arranged, each with a glabrous ovary, the bract oblong, yellow, pubescent, not persistent, the glands 2, the style up to 1 mm long; capsules conical-subulate, 7–10 mm long, glabrous, on pedicels at least twice as long as the gland. June.

Bogs, fens.

IL, IN, OH (OBL).

Autumn willow.

This willow is closely related to *S. lucida* but differs by its larger capsules, its short-acuminate leaves, and its minute stipules. It is also similar to *S. petiolaris*, but *S. serissima* has reddish brown twigs.

The common name autumn willow refers to the late dehiscence of the capsule, the seeds being liberated from mid-July to late August.

24. **Salix X subsericea** (Anderss.) Schneid. Ill. Handb. Laubholzk. 1:65. 1904. Fig. 253.
Salix petiolaris Sm. var. *subsericea* Anderss. in DC. 16(2):234. 1864.

Shrubs to 3 m tall; twigs brown, slender, usually somewhat sericeous when young, usually becoming less pubescent by maturity; leaves alternate, simple, narrowly lanceolate, acuminate at the apex, cuneate at the base, entire to undulate to glandular-serrulate, to 12 cm long, to 3 cm wide, puberulent at least on the lower surface, the petioles up to 1 cm long, more or less pubescent; staminate aments erect to ascending, to 2 cm long, the flowers spirally arranged, each with 2 stamens the filaments free, usually pilose near the base, the bract oblong, brown with a black tip, pubescent, subpersistent, the gland 1; pistillate aments erect to ascending, to 3 cm long, the flowers spirally arranged, each with a sericeous ovary, the bract oblong, brown with a black tip, pubescent, subpersistent, the gland 1, the style minute or absent; capsule ovoid, up to 6 mm long, sericeous, on pedicels several times longer than the gland. April–May.

Wet soil.

IL, IN, OH (FACW).

Hybrid willow.

This taxon is a reputed hybrid between *Salix sericea* and *S. petiolaris* and it shares characteristics of both species. It is intermediate between these two species in degree of pubescence. It differs from both by having brown bracts with a black tip. It was originally described by Andersson as a variety of *S. petiolaris*.

253. *Salix X subsericea* (Hybrid willow).

a. Vegetative branch.
b. Leaf.

c. Staminate flower.
d. Pistillate flower.

107. SANTALACEAE—SANDALWOOD FAMILY

Herbs, shrubs, or trees, often root parasites; leaves simple, alternate or opposite; flowers actinomorphic, perfect or unisexual, borne in axillary or terminal inflorescences or solitary; calyx 4- or 5-parted, united; petals absent; disk present; stamens 4 or 5, usually attached to the base of the calyx lobes; ovary inferior, 1-locular; fruit a drupe or nut, 1-seeded.

The sandalwood family is composed of about thirty genera and more than four hundred species, most of them in the tropics of the Southern Hemisphere. Several of these tropical species have considerable economic importance.

1. **Comandra** Nutt.—Bastard Toadflax

Rhizomatous herbs, usually parasitic on the roots of other plants; leaves alternate, entire; flowers perfect, borne in cymes or panicles of small umbels; calyx 5-parted, united below; petals absent; disk 5-lobed; stamens 5, connected to the base of the calyx lobes by a tuft of hairs; ovary inferior, 1-locular; fruit a hard drupe (sometimes considered a nut) enclosed at the base by the persistent calyx, 1-seeded.

1. **Comandra umbellata** (L.) Nutt. Gen. N. Am. Pl. 1:157. 1818. Fig. 254.
Thesium umbellatum L. Sp. Pl. 208. 1753.
Comandra richardsiana Fern. Rhodora 7:48. 1905.

Herb with slender rhizomes at or just below the surface of the soil; aerial stems erect, slender, to 30 cm tall, glabrous; leaves alternate, simple, oblong to lance-ovate, obtuse to acute at the apex, cuneate at the base, entire, glabrous, pale green, to 4 cm long, to 2.5 cm wide, sessile or nearly so; flowers in small, usually terminal, flat-topped clusters; calyx 5-parted, the lobes 2–3 mm long, whitish; petals absent; disk shallowly lobed; stamens 5, opposite the lobes of the calyx; drupe dry, subglobose, to 6 mm in diameter. May–August.

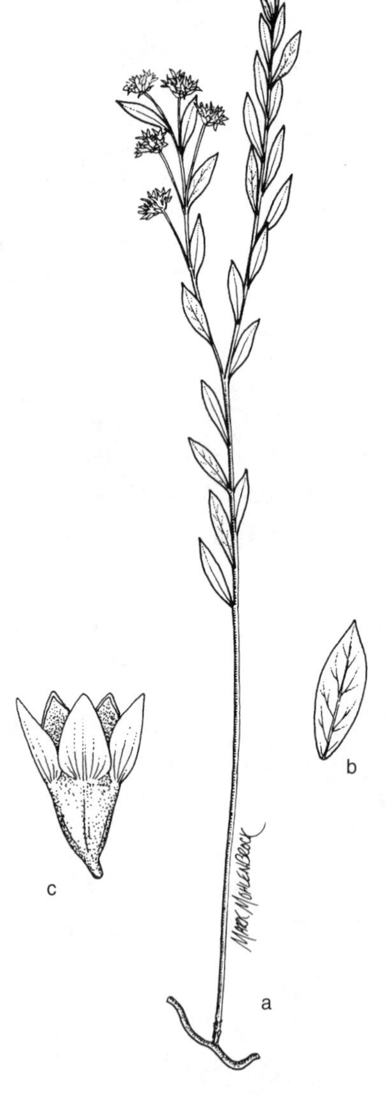

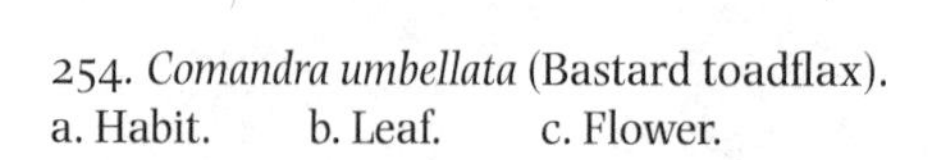
254. *Comandra umbellata* (Bastard toadflax).
a. Habit. b. Leaf. c. Flower.

Woods, wet prairies, fens.

IA, IL, IN, MO (FACU), KS, NE (UPL), KY, OH (FAC−).

Bastard toadflax.

An interesting feature about this species and most other species in the family is the association with other plants as root parasites. *Comandra umbellata* parasitizes the roots of several different kinds of trees.

Although usually found in dry habitats, it also occurs in wet prairies and fens.

This species has sometimes been called *C. richardsiana*.

108. SAPINDACEAE—SOAPBERRY FAMILY

Trees, shrubs, or herbaceous vines; leaves alternate, compound, without stipules; flowers in racemes or panicles, usually polygamous, actinomorphic or zygomorphic; sepals 4–5, free or united; petals 4–5, free, usually with a scale or appendage at the base of each; stamens usually 4–10, attached to a disk; ovary superior, usually 3-locular; fruit a berry or capsule, the seeds sometimes arillate.

Nearly 150 genera and more than 1,500 species comprise this family, most of them tropical or subtropical. In addition to the genera described below, *Koelreuteria paniculata*, the golden rain tree, is grown as an ornamental.

1. **Cardiospermum** L.—Balloon-vine

Herbaceous vines with tendrils; leaves alternate, ternately or biternately compound; flowers unisexual, in small clusters; sepals 4, in unequally sized pairs, green; petals 4, slightly unequal in size, each with a petaloid appendage at its base; stamens 8; disk present; ovary superior, 3-locular; fruit a bladderlike capsule, with 3 or 6 seeds.

Fourteen species comprise this genus, with most of them in tropical America.

1. **Cardiospermum halicacabum** L. Sp. Pl. 366. 1753. Fig. 255.

Annual vine from fibrous roots; stems much branched, bearing tendrils, usually glabrous; leaves alternate, ternately or biternately compound, the leaflets ovate, tapering to the base, coarsely toothed, to 4 cm long, to 3 cm wide, glabrous or nearly so; flowers in small clusters, each flower 4–5 mm across; sepals 4, in 2 unequal pairs, green; petals 4, free, white; stamens 8; disk present; ovary superior; capsules bladdery, globose or ovoid, 2.5–4.5 cm in diameter, with usually 6 seeds. July–September.

Along rivers and streams.

Native to tropical America; IL, IN, MO (FAC), KS (FAC−), KY, OH (FACU).

Balloon-vine.

This species is unique because of its unequally sized sepals, its four petals per flower, and its inflated, bladdery capsules. It has been introduced from tropical America. In the central Midwest, it grows in floodplains adjacent to large rivers.

109. SAPOTACEAE—SAPODILLA FAMILY

Trees or shrubs, usually with latex; leaves simple, alternate, entire, without stipules; flowers in axillary clusters, small, perfect, actinomorphic; sepals 5, free or united below, petals 5, united below; stamens usually 5, attached to the corolla tube, with staminodia often present; ovary superior, 4- to 12-locular, with one ovule per locule; fruit fleshy, indehiscent, with usually large seeds.

Approximately fifty genera and eleven hundred species, mostly tropical, are in this family.

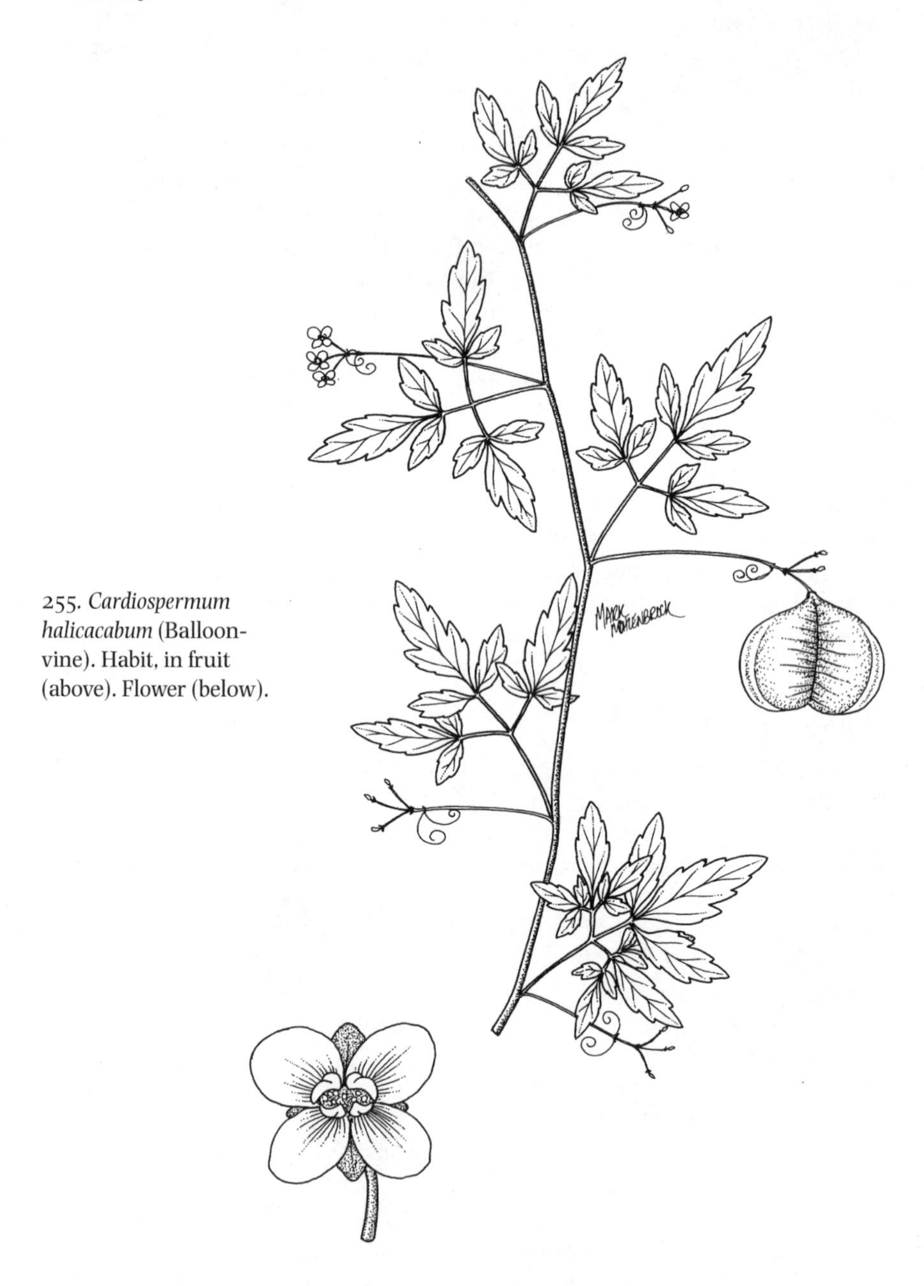

255. *Cardiospermum halicacabum* (Balloon-vine). Habit, in fruit (above). Flower (below).

1. **Sideroxylon** L.—Buckthorn

Trees or shrubs, usually with thorns; leaves alternate, simple, entire; flowers few to several in axillary clusters, small, actinomorphic, perfect, white; sepals 5, united at base; petals 5, united below, with a pair of appendages at each sinus; stamens 5, attached to the corolla tube, with 5 petaloid staminodia; ovary 5-locular; fruit fleshy, 1-seeded.

There are about twenty-five species, all in the New World, in this genus. For many years, our species were placed in *Bumelia*. In addition to the species described below, one other species of *Sideroxylon* occurs in upland woods in the southern part of the central Midwest.

1. **Sideroxylon lycioides** L. Sp. Pl. ed. 2, 279. 1762. Fig. 256.
Bumelia lycioides (L.) Pers. Syn. Pl. 1:237. 1805.

Tree to 10 m tall, with a trunk up to 20 cm in diameter; branches stout, flexible, usually with numerous curved thorns and short spur-like lateral branchlets,

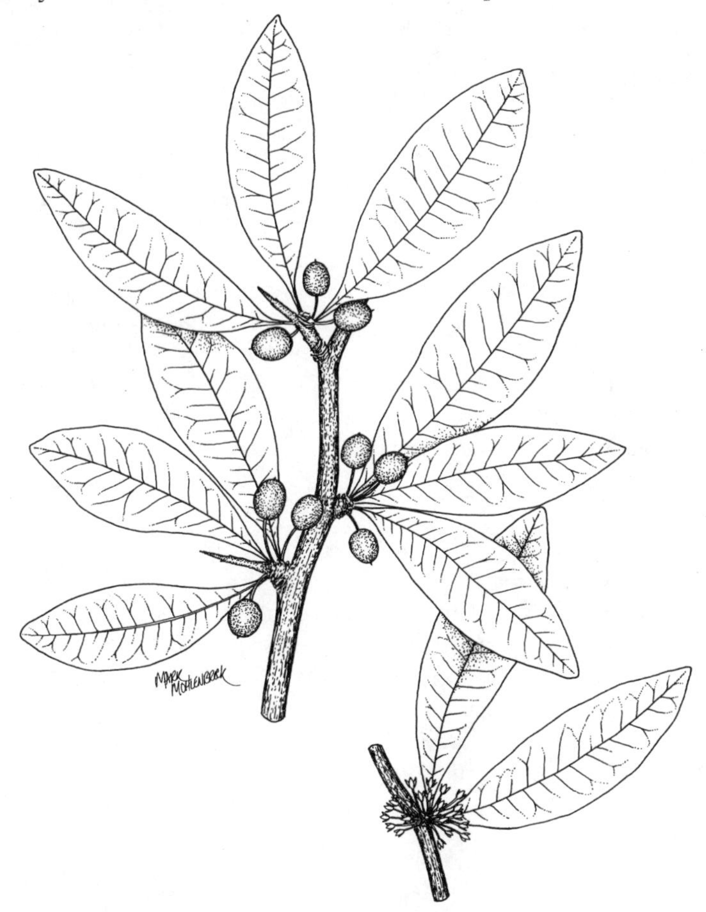

256. *Sideroxylon lycioides* (Smooth southern buckthorn). Fruiting branchlet (above). Flowering branchlet (below).

glabrous, red-brown, shiny, with numerous pale lenticels; leaves alternate, simple, elliptic to oblanceolate, more or less acute at the apex, tapering at the base, to 10 cm long, to 2.5 cm wide, entire, bright green and glabrous above, light green and pubescent below, the petioles slender, glabrous at maturity, up to 2 cm long; flowers many in axillary clusters, on glabrous pedicels up to 1 cm long; sepals 5, united below, green, glabrous; petals 5, united below, white, usually a little longer than the sepals; stamens 5, with 5 petaloid staminodia; ovary ovoid, pubescent at base; fruit fleshy, ovoid, red, up to 12 mm in diameter, with a single seed. June–July.

Along streams, swampy woods.

IL, IN, KY, MO (FACW).

Smooth southern buckthorn.

This small, spiny tree is recognized by its usually elliptic, entire leaves, its thorny branches, its small white flowers in axillary clusters, and its red, cherrylike fruits.

110. SARRACENIACEAE—PITCHER PLANT FAMILY

Perennial herbs; leaves basal, tubular or pitcher-shaped; flower solitary; sepals 4–5, free, persistent; petals 5, free, or absent; stamens numerous, free; pistil one, the ovary superior, 3- to 5-locular, with numerous ovules, the style peltate, usually lobed; fruit a loculicidal capsule, with numerous small seeds.

The Sarraceniaceae is composed of three genera and fifteen species of marsh herbs, all native to the New World.

Only the following genus is represented in the central Midwest.

1. **Sarracenia** L.—Pitcher Plant

Perennial herbs; leaves basal, pitcher-shaped or tubular; flower solitary, nodding, subtended by 3 or 4 bracts; sepals 5, free; petals 5, free; stamens numerous, free; pistil one, the ovary superior, 5-locular, with numerous ovules, the style peltate, umbrellalike, with 5 rays, each terminated by hooked stigmas; fruit a 5-locular capsule with many seeds.

Only the following species occurs in the central Midwest. Eight species comprise the genus.

1. **Sarracenia purpurea** L. Sp. Pl. 510. 1753. Fig. 257.

Perennial herbs; leaves basal, pitcher-shaped, with a wing on one side and an arching hood at the tip, curved, to 15 cm long, glabrous on the outside, densely clothed on the inside with stiff, reflexed bristles, the tube inflated and hollow, usually green with purple veins; flower solitary on a naked scape, up to 5 cm across; sepals 5, usually purplish, persistent; petals 5, free, obovate, purple, arched over the greenish-yellow style; style peltate, umbrellalike, 5-lobed; capsules granular, rugose, with many small seeds. May–June.

Bogs.

IL, IN, OH (OBL).

Pitcher plant.

The tubular pitcher is hollow and filled with liquid into which unwary organisms fall and drown. Reflexed bristles on the inner surface of the pitcher prevent captured insects from crawling out.

257. *Sarracenia purpurea* (Pitcher plant). Habit.

III. SAURURACEAE—LIZARD'S-TAIL FAMILY

Aromatic perennial herbs from rhizomes; leaves alternate, simple, entire, the stipules adnate to the petioles; inflorescence spicate or racemes of spikes, bracteate; flowers actinomorphic, perfect; sepals 0; petals 0; stamens 4–8; pistils 3–5, united, at least at the base, the ovaries superior or inferior, each with 1–many ovules; fruit a capsule or dry berry.

The family consists of four genera and six species. In addition of the two described below, *Houttynia* is sometimes grown as an ornamental.

1. Leaves, or most of them, basal, obtuse at the apex, truncate at the base; spikes to 3 cm long, stout, subtended by showy white bracts...1. *Anemopsis*
1. Leaves cauline, acuminate at the apex, cordate at the base; spikes more than 3 cm long, slender, not subtended by white bracts...2. *Saururus*

1. **Anemopsis** Hook. & Arn.—Yerba Mansa

Only the following species comprises the genus.

1. **Anemopsis californica** Hook. & Arn. Bot. & Beechey Voy. 390, pl. 92. 1840. Fig. 258.

Aromatic perennial from stout rhizomes, usually forming colonies; stems erect, glabrous, to 50 cm tall, usually unbranched; most leaves basal, oblong to elliptic, obtuse at the apex, truncate at the base, entire, glabrous, to 12 cm long, to 6 cm wide, on petioles to 30 cm long, the cauline leaves 1–2, alternate, to 4 cm long, acute at the apex, cuneate at the sessile base; spikes to 3 cm long, stout, pedunculate, resembling a large flower, but what appear to be petals are 4–8 white petallike bracts to 2 cm long and 1 cm wide; flowers actinomorphic, perfect, bracteolate, without a perianth; stamens 6; pistils 3, united, the ovaries sunken in the rachis, each with several ovules; fruit a compound capsule resembling a cone, to 3 cm long, dehiscent, ferruginous. June–July.

Wet meadows, along streams, around lakes, often in saline or alkaline areas.

KS, NE (OBL).

Yerba mansa.

This handsome species owes its attractiveness to the 4–8 white bracts that surround the small flowers in the short, stout spikes. It is particularly common in saline or alkaline areas.

2. **Saururus** L.—Lizard's-tail

Perennial herbs with rhizomes; leaves alternate, simple, entire, cordate, the petioles sheathing at the base, attached to the stipules; flowers perfect, small, in spikelike racemes, each subtended by bracteoles; perianth absent; stamens 4–8, free; pistils 3–5, united at the base, each with 1–2 ovules; fruit separating into 3–4 1-seeded segments.

Three species in Asia and ours comprise this genus.

1. **Saururus cernuus** L. Sp. Pl. 341. 1753. Fig. 259.

Perennial herb with aromatic rhizomes; stems erect, jointed, pubescent at first, usually glabrous at maturity, up to 1 m tall, branched or unbranched, usually

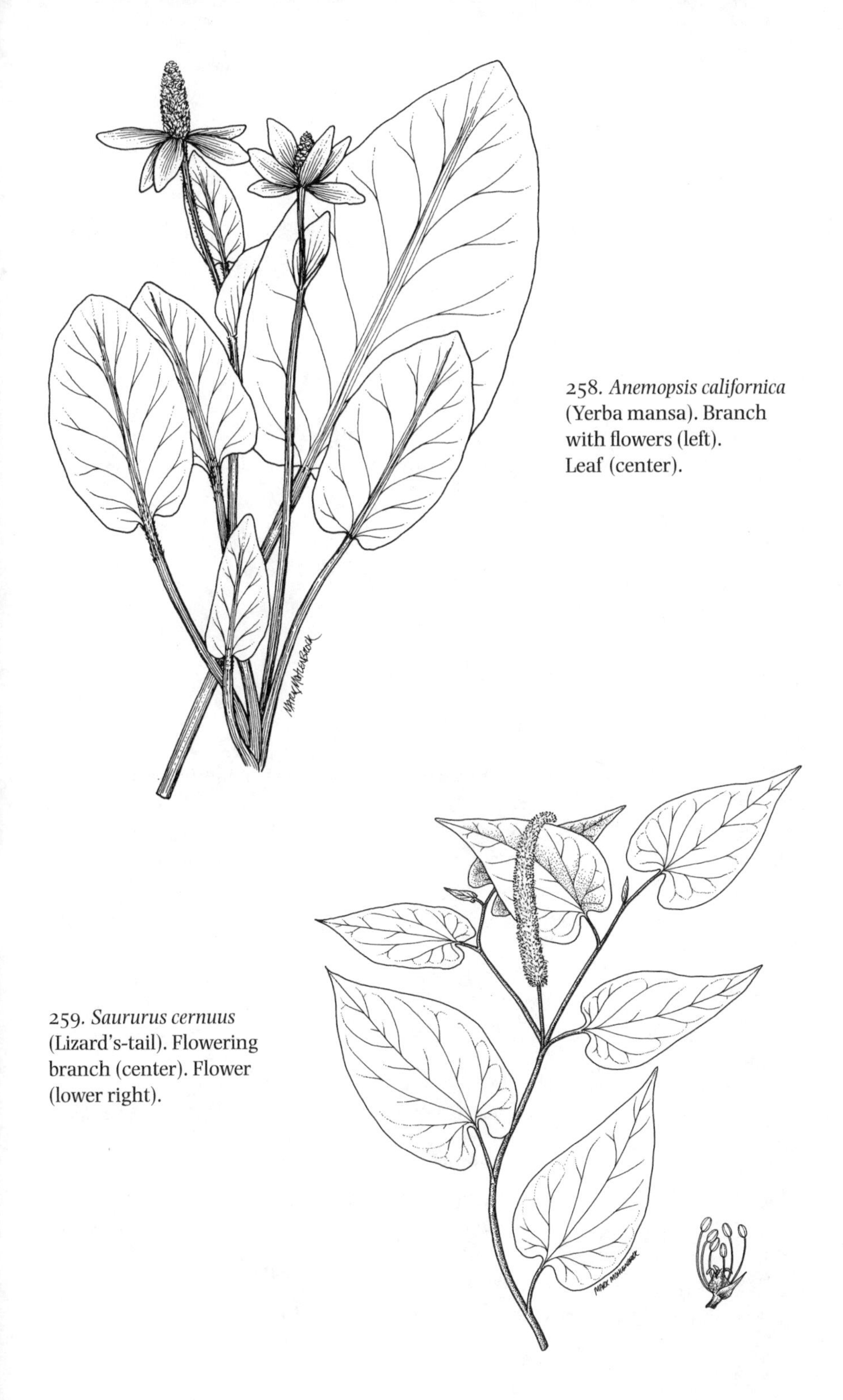

258. *Anemopsis californica* (Yerba mansa). Branch with flowers (left). Leaf (center).

259. *Saururus cernuus* (Lizard's-tail). Flowering branch (center). Flower (lower right).

zigzag at maturity; leaves alternate, simple, ovate, acute to acuminate at the apex, cordate at the base, to 15 cm long, to 8 cm wide, entire, pubescent when young, becoming glabrous, palmately 5- to 9-nerved, the petioles densely sheathing at the base; inflorescence a spikelike raceme, terminal but often overtopped by the uppermost leaves, 3–15 cm long, on peduncles up to 8 cm long, strongly curved or nodding at first, becoming straightened at maturity; flowers up to 300 per raceme, each subtended by a bracteole, actinomorphic, perfect, without a perianth, fragrant; stamens 4–8, white, up to 4 mm long, free; pistilis 3–5, united at base, up to 2 mm long, each with a recurved stigma; fruits separating at maturity, each segment up to 3 mm in diameter, strongly rugose. May–September.

Ditches, marshes, swamps, often in shallow water.

IA, IL, IN, KS, KY, MO, NE, OH (OBL).

Lizard's-tail.

The white color of the flowers is due to the stamens since there is no perianth. The flowers, densely crowded in spikelike racemes, have a mild, pleasant fragrance, as do the rhizomes. As the season progresses, the stems become strongly zigzag.

112. SAXIFRAGACEAE—SAXIFRAGE FAMILY

Herbaceous perennials; leaves simple or compound, basal or alternate, less commonly opposite, with or without stipules; flowers in various types of inflorescences, usually perfect, mostly actinomorphic; sepals (3–) 4–5 (–10); petals (3–) 4–5 (–10), free, or absent; hypanthium usually present; stamens various, sometimes reduced to staminodia; pistils 2–7, usually united at base, the ovary often inferior; fruits dry, dehiscent, with numerous seeds.

This family traditionally has included many diverse genera. In this work I am removing several of these genera from the Saxifragaceae and placing them in other families, *e.g.*, *Itea* in the Escalloniaceae, *Ribes* in the Grossulariaceae, and *Parnassia* in the Parnassiaceae. The Saxifragaceae as I recognize it has approximately thirty-five genera and six hundred species found in most parts of the world.

1. Leaves borne all along the stem.
  2. Leaves opposite .... 1. *Chrysosplenium*
  2. Leaves alternate.
    3. Leaves rounded at base, obtuse at apex, crenate; stamens 5 or 8 .... 1. *Chrysosplenium*
    3. Leaves tapering at base, acute at apex, serrate; stamens 10 .... 3. *Penthorum*
1. Leaves all basal, or rarely with one leaf on the stem.
  4. Leaves cordate, deeply crenate; petals fringed, white .... 2. *Mitella*
  4. Leaves tapering at base, entire or repand; petals not fringed, greenish .... 4. *Saxifraga*

1. **Chrysosplenium** L.—Golden Saxifrage

Small herbaceous annuals or perennials; leaves simple, cauline, opposite or alternate, usually crenate, somewhat succulent; flowers in cymes or solitary, perfect, actinomorphic; calyx lobes 4; petals absent; disk present; stamens 5 or 8; carpels 2, united at base, perigynous; fruit a 2-lobed, dehiscent capsule.

Forty species in cooler regions of the world comprise this genus.

1. Leaves, or some of them, opposite, often rounded at the base, obscurely toothed; flower usually solitary, greenish; stamens 8; stems decumbent .... 1. *C. americanum*
1. Leaves alternate, often cordate or rounded at the base, crenate; flowers in cymes, bright yellow; stamens 5 or 8; stems erect .... 2. *C. iowense*

1. **Chrysosplenium americanum** Schwein. in Hook. Fl. Bor. Am. 1:242. 1832. Fig. 260.
Chrysosplenium oppositifolium Walt. Fl. Carol. 140, 1788, *non* L. (1753).

260. *Chrysosplenium americanum* (Golden saxifrage). Habit. Flower (left). Fruit (lower right).

Annual herb with fibrous roots, often rooting at the nodes; stems decumbent, glabrous or nearly so, dichotomously branched, to 20 cm long; leaves mostly opposite, simple, cauline, broadly ovate, obtuse at the apex, rounded at the base, obscurely crenulate, glabrous or nearly so, to 1.8 cm long, to 1.5 cm wide, sessile or short-petiolate; flower usually solitary, actinomorphic, perfect, sessile, up to 3 mm across; calyx lobes 4, greenish or sometimes violet-tinged; petals absent; stamens 8; disk present; capsules 3–4 mm long, glabrous. March–June.

Mossy, wet areas, springheads.

IA, IN, KY, OH (OBL).

Golden saxifrage.

This northern species barely enters the northern part of our range. It differs from *C. iowense* by its mostly opposite cordate leaves and its solitary, greenish flowers.

2. **Chrysosplenium iowense** Rydb. in Britt. Man. 483. 1901. Fig. 261.

Annual herb with fibrous roots; stems erect, slender, branched, glabrous or somewhat pubescent, to 15 cm tall; leaves alternate, simple, cauline, suborbicular, rounded at the apex, cordate, deeply crenate, glabrous or somewhat pubescent, to 2.5 cm long, about as wide, petiolate; flowers in cymes, actinomorphic, perfect, 3.5–4.5 mm across; calyx lobes 4, bright yellow; petals absent; disk present; stamens 5 or 8; capsules 4–6 mm long. May–July.

Wet moss.

IA (OBL).

Iowa saxifrage.

This plant is distinguished by its bright yellow flowers and its alternate, suborbicular, petiolate leaves that are cordate at the base. In addition to Iowa, this species is known from Minnesota. It is sometimes considered to be a variety of the European *C. alternifolium*, a view not accepted in this work.

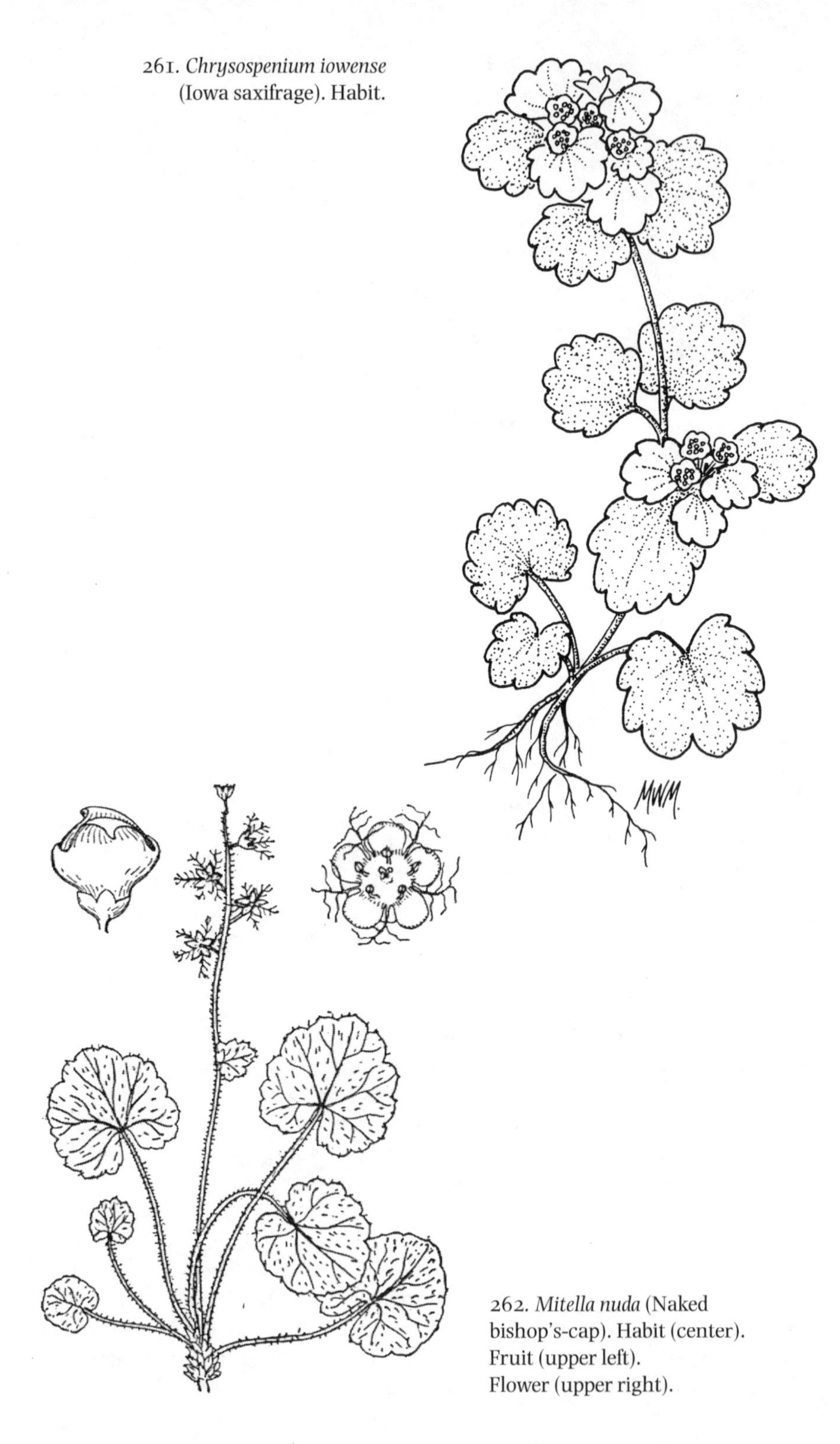

261. *Chrysospenium iowense* (Iowa saxifrage). Habit.

262. *Mitella nuda* (Naked bishop's-cap). Habit (center). Fruit (upper left). Flower (upper right).

2. **Mitella** L.—Bishop's-cap; Mitrewort

Herbs; leaves simple, palmately lobed, basal, sometimes with a solitary leaf or with a pair of leaves on the stem, the basal leaves long-petiolate; flowers several in spike-like racemes, actinomorphic, perfect; calyx 5-lobed, tubular below; petals 5, free, fringed; stamens 5 or 10; styles 2; capsules 2-beaked, 1-locular, with many seeds.

Mitella consists of about twleve species in North America and Asia.

In addition to the species listed below, *M. diphylla* occurs in rich woods in the central Midwest.

1. **Mitella nuda** L. Sp. Pl. 408. 1753. Fig. 262.

Perennial herb with slender rhizomes and stolons; flowering stems erect, slender, unbranched, pubescent, to 18 cm tall; leaves basal, or with one small, sessile leaf midway on the stem, orbicular, rounded at the apex, cordate at the base, strigose above, villous below, deeply crenate, 2.0–3.5 cm across; flowers several on a usually leafless scape, actinomorphic, perfect, on short, pubescent pedicels; calyx 5-lobed, tubular below, 1–2 mm long, green; petals 5, free, greenish yellow, fringed, 3–4 mm long; stamens usually 10; ovary glabrous, 1-loculr; styles 2; capsules 4–6 mm long. April–June.

Bogs, swampy woods.

IA (FACW).

Naked bishop's-cap.

This species with intricately fringed petals usually produces its flowers on leafless scapes, although sometimes one leaf may occur midway on the stem.

3. **Penthorum** L.—Ditch Stonecrop

Perennial herbs; leaves simple, alternate, serrulate, without stipules; flowers in spreading cymes, perfect, actinomorphic; sepals 5 (–7), united below; petals absent; stamens 10; pistils 5 (–7), united below; capsules 5-parted, each part beaked, with many seeds.

This genus consists of three species, the others Asian. In the past, some botanists have placed this genus in the Crassulaceae.

1. **Penthorum sedoides** L. Sp. Pl. 432. 1753. Fig. 263.

Rhizomatous perennial herb; stems erect or decumbent, glabrous, branched or unbranched, glabrous below but usually stipulate-glandular above; leaves simple, alternate, elliptic to lanceolate, acute to acuminate at the apex, tapering at the base, serrate, glabrous; cymes 2–4- (5-) branched, often spreading to form a star; flowers perfect, actinomorphic; sepals 5, united below, green or yellowish; petals absent; stamens 10; pistils 5, yellowish, united at base; fruit a 5-parted capsule, each part beaked, the beak falling away when the seeds are ready for dispersal. June–September.

Wet ground, marshes, sometimes in standing water.

IA, IL, IN, KS, KY, MO, NE, OH (OBL).

Ditch stonecrop.

The star-shaped arrangement of the cymes is distinctive as well as the 5-lobed, beaked fruits. The stem may become reddish, and there is often a dark reddish area on the stem immediately above the attachment of each leaf.

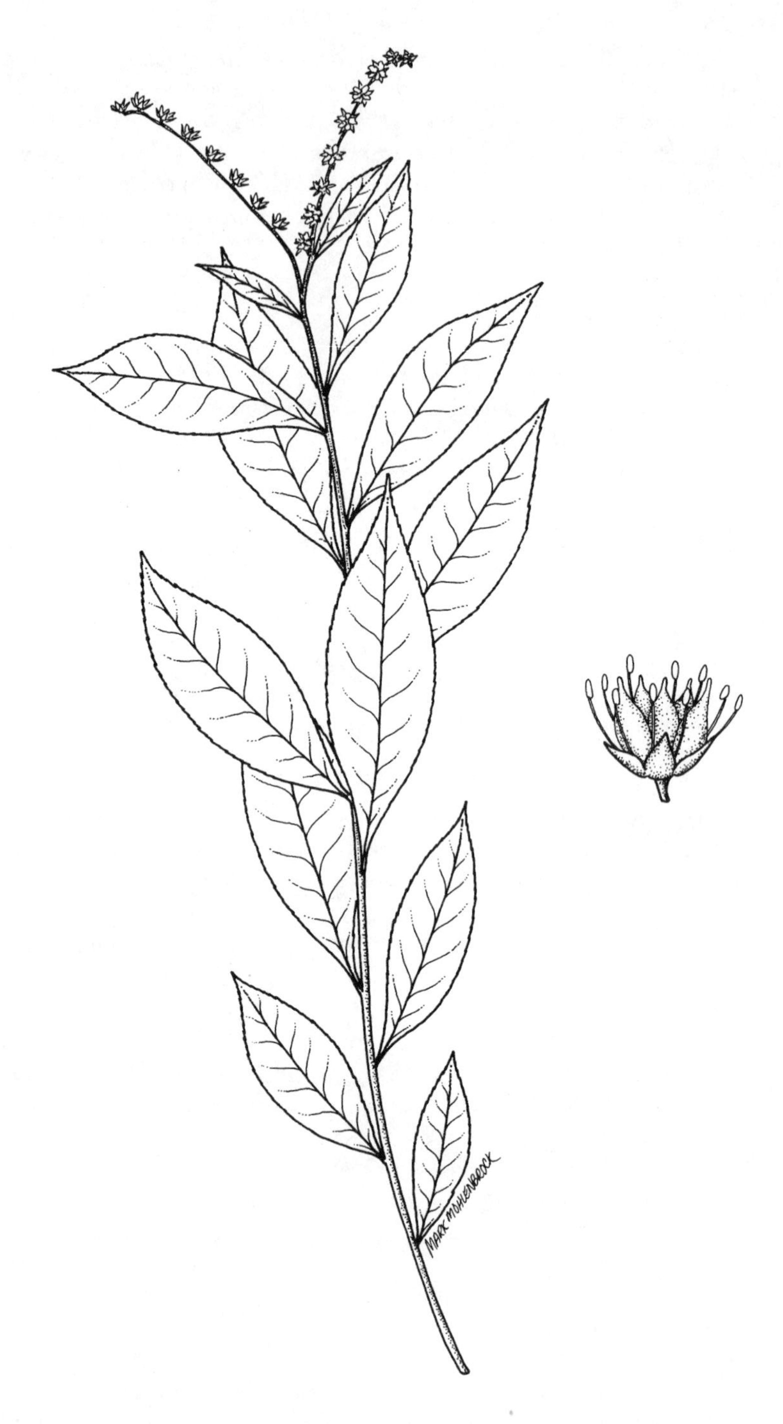

263. *Penthorum sedoides* (Ditch stonecrop). Habit (center). Flower (right).

4. **Saxifraga** L.—Saxifrage

Perennial herbs; leaves simple, mostly basal, occasionally with 1–2 or several alternate cauline leaves; inflorescence racemose; flowers perfect, actinomorphic; sepals 5, united at base; petals 5, free; hypanthium present; stamens 10; pistils 2, united at base; fruit a 2-lobed capsule or a pair of follicles.

There are about 350 species in the genus, found in the north temperate regions of the world and in the Andes Mountains.

1. **Saxifraga pensylvanica** L. Sp. Pl. 399. 1753. Fig. 264.

Perennial herb from thick rhizomes; leaves simple, forming a basal rosette, oblong to narrowly ovate, acute at the apex, tapering at the base, entire or glan-

264. *Saxifraga pensylvanica* (Swamp saxifrage). Habit (center). Fruit (upper left).

dular-denticulate, glabrous or nearly so on the lower surface, to 25 cm long, to 10 cm wide; flowers perfect, actinomorphic, borne in panicles at the tip of a glandular scape, the scape up to 1.5 m tall, the panicle up to 60 cm long, with linear to lanceolate bracts; sepals 5, united below, green, the tips reflexed; petals 5, free, linear to narrowly oblong, 2–3 mm long, greenish yellow to nearly white; stamens 10, with filiform filaments; capsules 3–5 mm long, divided nearly to the base. May–June.

Springs, bogs, marshes, wet meadows, swamps.

IA, IL, IN, KY, MO, OH (OBL).

Swamp saxifrage.

The large leaves that form a basal rosette and the tall scape that bears a terminal panicle of small, usually greenish yellow flowers are distinctive.

113. SCROPHULARIACEAE—FIGWORT FAMILY

Mostly herbaceous perennials or annuals; leaves usually simple, although sometimes deeply divided, basal or alternate or opposite, without stipules; inflorescence various; flowers perfect, actinomorphic or zygomorphic; sepals 4–5, usually united below; corolla 4- or 5-lobed, sometimes tubular, occasionally spurred at the base; stamens 2, 4, or 5, borne on the corolla tube, some of the stamens sometimes sterile, a staminodium sometimes present; ovary superior, 2-locular; fruit a capsule, with numerous seeds.

This worldwide family consists of nearly two hundred genera and more than four thousand species.

The following plants may occur in wetlands or in standing water in the central Midwest.

1. Leaves all basal .... 8. *Limosella*
1. At least some of the leaves cauline.
  2. Some or all of the leaves whorled .... 17. *Veronicastrum*
  2. Leaves all opposite or alternate.
    3. At least some of the leaves alternate.
      4. All leaves alternate .... 4. *Castilleja*
      4. Lower leaves opposite, upper leaves alternate (excluding bracteal leaves) ... 10. *Mazus*
    3. All leaves except the bracteal ones opposite.
      5. Anther-bearing stamens 2.
        6. Calyx and corolla lobes each 4; fruit heart-shaped .... 16. *Veronica*
        6. Calyx and corolla lobes each 5; fruit subacute or rounded at the tip, but not heart-shaped.
          7. Sterile stamens 2; bractlets absent .... 9. *Lindernia*
          7. Sterile stamens absent or minute; each flower subtended by 2 bractlets .... 6. *Gratiola*
      5. Anther-bearing stamens 4.
        8. One elongated sterile stamen present.
          9. Flowers sessile; leaves sharply and regularly serrate .... 5. *Chelone*
          9. Flowers pedicellate leaves irregularly and sometimes obscurely serrate .... 15. *Penstemon*
        8. Sterile stamens absent or represented only by a gland as broad as long.
          10. Some or all the leaves pinnatifid or pinnately lobed.
            11. Corolla lavender, to 5 mm long; plants up to 20 cm tall .... 7. *Leucospora*

11. Corolla creamy yellow, at least 15 mm long; plants more than 20 cm tall..... .... 14. *Pedicularis*

10. Leaves toothed or entire, not pinnatifid or pinnately lobed.

12. Flowers 1–2 in the axils of the cauline leaves.

13. Leaves entire.

14. Leaves ovate to orbicular .... 2. *Bacopa*

14. Leaves linear to elliptic.. 12. *Melampyrum*

13. Leaves toothed.

15. Calyx 5-parted nearly to the base; flowers up to 1 cm long, white with purple lines ....11. *Mecardonia*

15. Calyx tubular, with 5 lobes; flowers more than 1 cm long, violet or yellow or, if less than 1 cm long, yellow......................... 13. *Mimulus*

12. Flowers 3 or more in spikes, racemes, panicles, or corymbs.

16. Corolla campanulate; leaves usually entire ....1. *Agalinis*

16. Corolla salverform; leaves toothed ....3. *Buchnera*

1. **Agalinis** Raf.—Purple Gerardia; False Foxglove

Annual or perennial herbs; stems erect, branched; leaves opposite (except the bracteal ones), simple, entire, 1-nerved; inflorescence usually racemose, the flowers pink or purple, perfect, zygomorphic; calyx tubular, 5-lobed; corolla campanulate, 5-lobed; stamens 4, didynamous, attached to the corolla; ovary superior; fruit usually glabrous; seeds yellow, brown, or black.

The genus of sixty New World species used to be called *Gerardia*, but that name belongs to a genus in the Acanthaceae. Several species in the central Midwest are upland species.

1. Plants yellow-green, usually not turning black upon drying; calyx strongly reticulate-nerved; seeds yellow .... 4. *A. skinneriana*

1. Plants deep green, usually turning black upon drying; calyx weakly reticulate-nerved; seeds brown or black.

2. Pedicels more than twice as long as the calyx.

3. Calyx lobes up to 1 mm long; axillary fascicles of leaves rarely present..... 5. *A. tenuifolia*

3. Calyx lobes 1–2 mm long; axillary fascicles of leaves commonly present .... 1. *A. besseyana*

2. Pedicels less than twice as long as the calyx.

4. Calyx up to ½ as long as the capsule; style to 1 cm long; flowers up to 2 cm long........... .... 2. *A. paupercula*

4. Calyx ½ as long as the capsule or longer; style 1.5–2.0 cm long; flowers usually more than 2 cm long .... 3. *A. purpurea*

1. **Agalinis besseyana** (Britt.) Britt. in Britt. & Brown, Ill. Fl. N. U. S. ed. 2, 3:211. 1913. Fig. 265.
Agalinis tenuifolia (Vahl) Raf. var. *macrophylla* Benth. Comp. Bot. Mag. 1:209. 1836.
Gerardia besseyana Britt. Mem. Torrey Club 5:295. 1894.
Gerardia tenuifolia Vahl var. *macrophylla* Benth. in Hook. Comp. Bot. Mag. 1:174. 1836.

Annual herb with fibrous roots; stems erect, branched, glabrous, to 75 cm tall; leaves (except bracteal ones) opposite, simple, with axillary fascicles often present, acute at the apex, tapering at the sessile base, green but turning black upon drying, 2–4 cm long, 2–4 mm wide, scabrous at least on the upper surface; flowers on pedicels more than twice as long as the calyx; calyx 5-lobed, the lobes 1–2 mm long, green; corolla slightly zygomorphic, 5-lobed, purple, 10–15 mm long, glabrous within; stamens 4; capsules globose, glabrous, 3–4 mm in diameter; seeds brown or black. August–October.

Moist soil.

IL, KS, MO, NE (not listed).

Bessey's false foxglove.

This species is similar to *A. tenuifolia* and sometimes combined with it or considered to be a variety of it. It differs, however, from *A. tenuifolia* by its longer calyx and the usual presence of axillary fascicles of leaves. It occurs in the western part of our range.

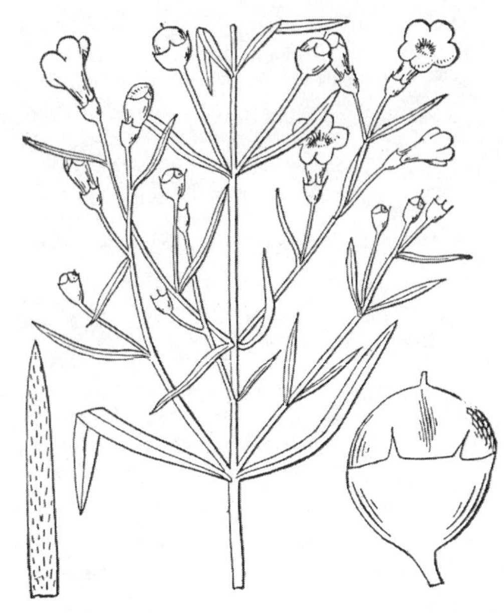

265. *Agalinis besseyana* (Bessey's false foxglove). Habit (center). Leaf (lower left). Fruit (lower right).

2. **Agalinis paupercula** (Gray) Britt. Mem. Torrey Club 5:295. 1894. Fig. 266.

Gerardia purpurea L. var. *paupercula* Gray, Syn. 2, part 1, 293. 1878.

Agalinis paupercula (Gray) Britt. var. *borealis* (Pennell) Pennell, Proc. Acad. Nat. Sci. Phila. 81:159. 1929.

Annual herb with fibrous roots; stems erect, branched, glabrous or nearly so, to 1.2 m tall; leaves (except the bracteal ones) opposite, simple, linear, acute at the apex, tapering at the sessile base, green but turning black upon drying, 1–2 cm long, up to 2 mm wide, sometimes scabrous; flowers on pedicels shorter than the calyx; calyx 5-lobed, the lobes lanceolate, about ½ as long as the tube; corolla slightly zygomorphic, 5-lobed, pale purple, 10–20 mm long, puberulent; stamens 4; styles 6–10 mm long; capsules a little longer than wide, up to 4.5 mm long, longer than the calyx; seeds brown or black. August–September.

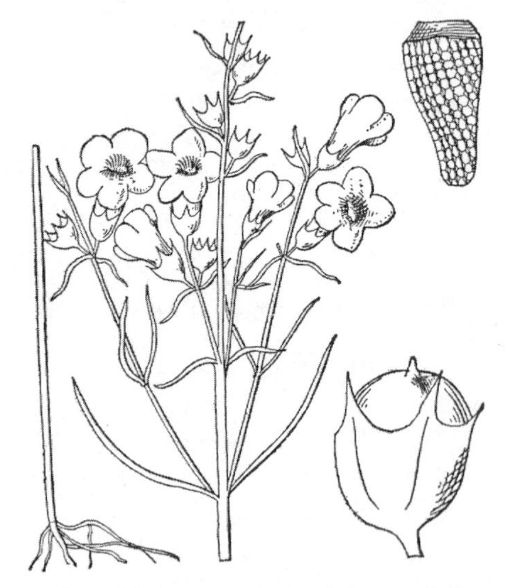

266. *Agalinis paupercula* (Small-flowered false foxglove). Habit (center). Fruit (lower right). Seed (upper right).

Moist soil, wet meadows.

IA, IL, IN (OBL), OH (FACW+).

Small-flowered false foxglove.

The very narrow leaves and smaller flowers distinguish this species from others in the genus. In the northern part of the central Midwest, plants with corollas 5–10 mm across and styles 6–8 mm long may be called var. *borealis*. In the southern part of our range, the corollas are 10–15 mm across and the styles 8–10 mm long.

3. **Agalinis purpurea** (L.) Pennell, Proc. Acad. Nat. Sci. Phila. 81:162. 1929. Fig. 267.
Gerardia purpurea L. Sp. Pl. 610. 1753.

267. *Agalinis purpurea* (Large purple false foxglove). Habit (center). Flower (lower left).

Annual herb with fibrous roots; stems erect, branched, glabrous to scabrous, to 85 cm tall; leaves opposite (except the bracteal ones), simple, linear, acute at the apex, tapering at the sessile base, green but turning black upon drying, 1.5–2.5 cm long, up to 2 mm wide, sometimes scabrous; flowers on pedicels shorter than the calyx; calyx 5-lobed, the lobes lanceolate to ovate, ⅓ to ½ as long as the tube; corolla slightly zygomorphic, 5-lobed, pale purple, 20–30 mm long, puberulent or sometimes nearly glabrous; stamens 4; styles 15–20 mm long; capsules globose, 4–6 mm in diameter, longer than the calyx; seeds brown or black. July–September.

Moist soil, borders of ponds, fens.

IA, IL, IN, KS, MO, NE (FACW), KY, OH (FACW−).

Large purple false foxglove.

This species, usually more common than *A. paupercula*, differs from this latter species by its larger flowers and longer styles. The short pedicels distinguish *A. purpurea* and *A. paupercula* from *A. besseyana* and *A. tenuifolia*.

4. **Agalinis skinneriana** (Wood) Britt. Ill. Fl. U. S., ed. 2, 3:212. 1913. Fig. 268.
Gerardia skinneriana Wood, Class-book 408. 1847.

Annual herb with fibrous roots; stems erect, 4-sided, branched or unbranched, scabrous, to 75 cm tall; leaves (except the bracteal ones) opposite, simple, yellow-green, not turning black upon drying, narrowly linear, acute at the apex, tapering at the sessile base, 12–20 mm long, 1.0–1.2 mm wide, usually scabrous; flowers on pedicels longer than the calyx and about as long as the corolla; calyx 2.5–3.0 mm long, with 5 minute teeth, the tube strongly reticulate; corolla slightly zygomorphic, 5-lobed, pale purple or pink, 10–15 mm long, the lobes ciliate; stamens 4; capsules oblongoid, 4–5 mm long, about twice as long as the calyx; seeds yellow. July–September.

Dry prairies, but occasionally in calcareous fens.

IL, IN, KS, MO, OH (not listed).

Pale false foxglove.

This species is readily recognized in the field by its yellow-green leaves. The leaves do not turn black upon drying. The similar appearing and more upland *A. gattingeri* has terete stems.

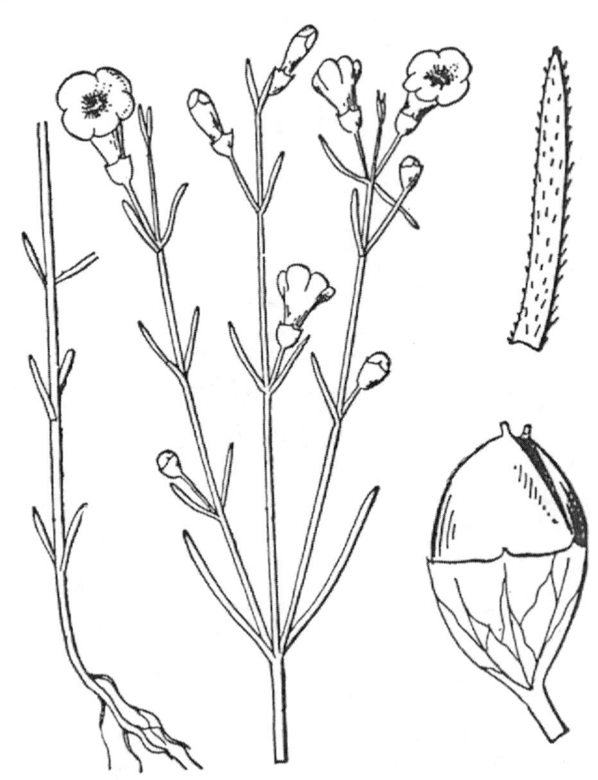

268. *Agalinis skinneriana* (Pale false foxglove). Upper part of plant (center). Lower part of plant (left). Leaf (upper right). Fruit (lower right).

5. **Agalinis tenuifolia** (Vahl) Raf. New Fl. N. Am. 2:64. 1836. Fig. 269.
Gerardia tenuifolia Vahl, Symb. Bot. 3:79. 1794.

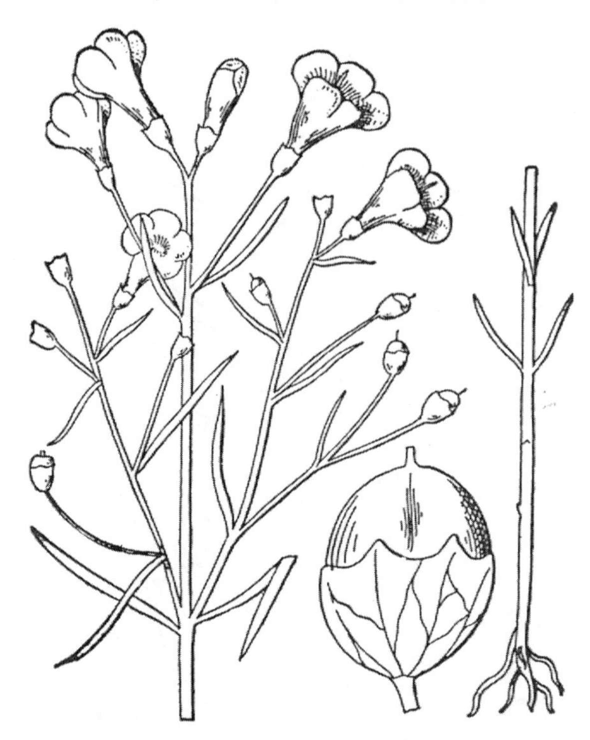

269. *Agalinis tenuifolia* (Narrow-leaved false foxglove). Upper part of plant (center). Lower part of plant (right). Fruit (bottom center).

Annual herb with fibrous roots; stems spreading to erect, branched, glabrous or scabrous, to nearly 1 m tall; leaves (except the bracteal ones) opposite, simple, green but turning black upon drying, narrowly linear, acute at the apex, tapering at the sessile base, 1.5–2.8 cm long, 1.0–1.4 mm wide, often scabrous; flowers on pedicels longer or equaling the length of the corolla; calyx to 4 mm long, 5-lobed, the lobes short-triangular, green; corolla slightly zygomorphic, 5-lobed, pink or pale purple, 10–15 mm long, glabrous or puberulent; stamens 4; capsules globose, 4–6 mm in diameter, glabrous, nearly twice as long as the calyx; seeds brown or black. August–October.

Moist areas, wet meadows, calcareous fens.
IA, IL, IN, KS, MO, NE (FACW), KY, OH (FAC).
Narrow-leaved false foxglove.

This rather common wetland species is distinguished by its narrowly linear leaves and it long-pedicellate flowers. It differs from the very similar *A. besseyana* by its slightly shorter calyx and by the absence of axillary fascicles of leaves.

2. **Bacopa** Aubl.—Water Hyssop

Annual or perennial herbs; stems usually creeping in mud or on floating mats in water; leaves opposite, simple, entire or shallowly toothed, sometimes succulent; flowers axillary, perfect, somewhat zygomorphic, sessile or pedicellate, not subtended by bractlets, or the bractlets minute; calyx deeply 5-parted, the lobes usually unequal; corolla 5-parted, the lobes 2-lipped to nearly equal; stamens 4, attached to the corolla, without staminodia; capsules with numerous seeds.

There are approximately one hundred species in this genus of tropical and warm temperate regions of the world. Only the following are known from the central Midwest, although others approach our range from the south.

1. Leaves ovate, crenate, lemon-scented; flowers blue, subtended by minute bractlets .... 1. *B. caroliniana*
1. Leaves orbicular or nearly so, entire, not lemon-scented; flowers white with a yellow center, not subtended by bractlets .... 2. *B. rotundifolia*

1. **Bacopa caroliniana** (Walt.) B.L. Robins. Rhodora 10:66. 1908. Fig. 270.
Obolaria caroliniana Walt. Fl. Carol. 166. 1788.

Creeping or floating perennial; stems glabrous below, lanate near the tip, to 20 cm long; leaves simple, opposite, ovate, obtuse at the apex, rounded or cordate at the clasping base, glabrous, lemon-scented, crenate, 5- to 9-nerved, glabrous or nearly so, up to 2.5 cm long; flower solitary in the leaf axils, on pedicels up to 1.5 cm long, subtended by minute bractlets; calyx 5-parted, the lobes unequal, up to 8 mm long, glandular-punctate; corolla 5-lobed, blue, puberulent within, up to 1 cm long; stamens 4; ovary superior, surrounded by a 10- to 12-lobed disk; capsules globose. April–September.

270. *Bacopa caroliniana* (Lemon bacopa). Habit (center). Flower (lower right).

Shallow water.

KY (OBL).

Lemon bacopa.

This species of the Coastal Plain has a disjunct location in Kentucky. Its lemon-scented herbage is distinctive.

2. **Bacopa rotundifolia** (Michx.) Wettst. Nat. Pflanzenf. IV. 3b:76. 1891. Fig. 271.
Monniera rotundifolia Michx. Fl. Bor. Am. 2:22. 1803.
Macuillamia rotundifolia (Michx.) Raf. Neogenyton 2. 1825.

Creeping perennial, often forming floating mats when in water; stems pubescent, up to 15 cm long; leaves opposite, simple, broadly ovate to suborbicular, rounded at the apex, clasping at the base, 5- to 13-nerved, entire, glabrous or nearly so, 2–3 cm long, nearly as wide; flower solitary in the leaf axils on pubescent pedicels up to 2 mm long; calyx deeply 5-parted, the lobes unequal in size, up to 8 mm long; corolla white with a yellow center, 5-parted, campanulate, the lobes somewhat unequal, up to 8 mm long; stamens 4; ovary superior; stigmas 2; capsules globose, glabrous, 6–8 mm in diameter, with numerous seeds. May–September.

Muddy shores, ponds, often in standing water.

IA, IL, IN, KS, MO, NE (OBL), KY, OH (NI).

Round-leaved water hyssop.

This species is recognized by its often orbicular, entire leaves that clasp the stem and its creeping or floating habit. Its white flowers have a yellow center.

271. *Bacopa rotundifolia* (Round-leaved water hyssop). Habit (left). Flower (right).

3. **Buchnera** L.—Blue Hearts

Biennial or perennial herbs, turning black upon drying; leaves opposite below, sometimes alternate above, simple; flowers actinomorphic, perfect, usually borne in spikes, bracteate; calyx tubular, 5-toothed; corolla salverform, deeply 5-lobed, the tube usually curved; stamens 4, didynamous; ovary superior; fruit a capsule with numerous seeds.

Nearly one hundred species are in this genus, most of them in the tropics.

Only the following occurs in the central Midwest and usually in upland habitats.

1. **Buchnera americana** L. Sp. Pl. 630. 1753. Fig. 272.

Perennial herbs, turning black upon drying; stems erect, usually unbranched, hispid, to 85 cm tall; leaves mostly opposite, simple, lanceolate to oblong-lanceolate, acute to acuminate at the apex, tapering at the sessile base, coarsely toothed except for the uppermost, rough-hispid, 5–10 cm long, 2.5–6.0 cm wide; flowers in spikes, bracteate, the spikes up to 12 cm long, the bracts shorter than the calyx; calyx 5-toothed, strigose, to 10 mm long; corolla salverform, 5-lobed, purple, the tube 10–15 mm long, the lobes 5–8 mm long; stamens 4, included; capsules slightly curved, to 8 mm long, strigose, with numerous seeds. July–September.

272. *Buchnera americana* (Blue hearts). Leaves (left). Inflorescence (center). Fruit (right).

Moist calcareous prairies.

IA, IL, IN, MO (FAC−), KS, KY, OH (FACU).

Blue hearts.

Although not usually a wetland species, this plant sometimes occurs in moist calcareous prairies. Its stems and leaves are harshly scabrous. The slightly curved fruits are distinctive.

4. **Castilleja** Mutis—Paintbrush

Partially parasitic herbs attached to the roots of other plants; leaves alternate, simple; flowers perfect, zygomorphic, in dense spikes, usually with colorful bracts longer than the flowers; calyx tubular, sometimes 2-cleft at the tip; corolla tubular, 2-lipped, the tube shorter than the calyx; stamens 4, didynamous; fruit a capsule, with many seeds.

There are approximately 150 species in this genus, most of them in the western United States. Only the following may sometimes occur in central Midwest wetlands.

1. **Castilleja coccinea** (L.) Spreng. Syst. 2:775. 1825. Fig. 273.
Bartsia coccinea L. Sp. Pl. 602. 1753.
Castilleja coccinea (L.) Spreng. f. *lutescens* Farw.Am. Midl. Nat. 8:276. 1923.

273. *Castilleja coccinea* (Scarlet paintbrush). Flowers and bracts (left). Bottom of plant (right). Capsule (lower left). Flower (upper right). Stamen (lower right).

Annual or biennial herb with thickened roots; stems erect, mostly unbranched, pubescent, to 85 cm tall; leaves alternate, simple, sessile, broadly linear to oblong, acute at the apex, tapering at the base, the cauline leaves deeply cleft, the basal leaves usually entire, pubescent, to 7 cm long; flowers sessile in spikes subtended by deeply cleft bracts, the bracts usually red or occasionally yellow; calyx tubular, 2-cleft at the top, 2–5 cm long; corolla 2-lipped, greenish yellow, to 2.5 cm long, the tube shorter than the calyx; stamens 4; capsules oblongoid, to 1.2 cm long. April–July.

Moist calcareous prairies, wet meadows, sand prairies, sand flats.

IA, IL, IN, KS, KY, MO, OH (FAC).

Scarlet paintbrush.

Although most of the plants of this showy species have red bracts, a few have yellow bracts. These latter plants may be known as f. *lutescens*.

The showy appearance of this species is due to its bracts and not to its corollas.

5. **Chelone** L.—Turtlehead

Perennial herbs; leaves opposite, simple, serrate, petiolate; flowers in terminal and axillary spikes, perfect, zygomorphic, bracteate; calyx united below, 5-lobed, green; corolla tubular, 2-lipped, the lower lip 3-lobed; fertile stamens 4, with 1 sterile stamen, the filaments woolly; fruit a capsule with numerous, compressed, winged seeds.

Four species, all in eastern North America, comprise this genus. Two of them occur in the central Midwest.

1. Corolla creamy white; leaves sessile or on petioles up to 5 mm long .... 1. *C. glabra*
1. Corolla rose-purple; leaves on petioles 5 mm long or longer .... 2. *C. obliqua*

1. **Chelone glabra** L. Sp. Pl. 611. 1753. Fig. 274.
Chlonanthes tomentosa Raf. New Fl. Am. 2:20. 1836.
Chelone glabra L. var. *linifolia* N. Coleman, Kent Sci. Inst. Misc. Publ. 2:27. 1874.
Chelone glabra L. f. *tomentosa* (Raf.) Pennell, Torreya 19:117. 1919.
Chelone glabra L. var. *linifolia* N. Coleman f. *velutina* Pennell & Wherry, Bartonia 10:23. 1929.

Perennial herb; stems erect, branched, glabrous, sometimes glaucous, to 1.2 m tall; leaves opposite, simple, linear-lanceolate to lanceolate to lance-ovate, acuminate at the apex, tapering or somewhat rounded at the base, sharply serrate, glabrous, rarely pubescent, sometimes whitened below, to 15 cm long, to 3 cm wide, sessile or on petioles up to 5 mm long; flowers in spikes, sessile, perfect, zygomorphic, subtended by 2–3 rather large, glabrous bracts; calyx united at base, 5-lobed, the lobes up to 4 mm long, green, glabrous; corolla 2-lipped, creamy white, to 3.5 cm long, the two lips barely open; fertile stamens 4, sterile stamen 1 and white; capsules ovoid, rounded at the top, glabrous, 6–8 mm long . July–October.

Bogs, fens, marshes.

IA, IL, IN, KS, KY, MO, NE, OH (OBL).

White turtlehead.

When in flower, this is one of the most beautiful species in wetlands. The corolla, which barely opens at maturity to permit pollinators to enter the flower, resembles the outline of a turtle's head in side view.

274. *Chelone glabra* (White turtlehead). Habit.

Most plants in our area have glabrous leaves more than 2 cm wide. Uncommon plants with tomentose leaves may be distinguished as f. *tomentosa*. Some plants have very narrow leaves less than 2 cm wide. If they are glabrous, the plants may be known as var. *linifolia*. If they are velvety-pubescent, they may be called var. *linifolia* f. *velutina*.

2 **Chelone obliqua** L. Syst., ed. 2, no. 4. 1767. Fig. 275.
Chelone obliqua L. var. *speciosa* Pennell & Wherry, Bartonia 10:19. 1929.

Perennial herb; stems erect, somewhat 4-sided, branched or unbranched, usually glabrous, to 90 cm tall; leaves opposite, simple, linear-lanceolate to lance-ovate, acuminate at the apex, tapering or somewhat rounded at the base, sharply serrate, usually glabrous, to 12 cm long, to 4 cm wide, on petioles 5–15 mm long; flowers in spikes, sessile, perfect, zygomorphic, subtended by 2–3 rather large, ciliate bracts; calyx united at base, 5-lobed, the lobes up to 5 mm long, green, glabrous; corolla 2-lipped, rose-purple, to 3.5 cm long; fertile stamens 4, sterile stamen 1 and greenish yellow; capsules ovoid, rounded at the top, glabrous, 6–8 mm long. August–October.

Low woods, swampy meadows, fens.

IA, IL, IN, MO (OBL).

Pink turtlehead.

This beautiful species difrers from *C. glabra* by its rose-purple flowers, greenish yellow sterile stamens, and longer petioles.

275. *Chelone obliqua* (Pink turtlehead). Habit, in flower (center). Flower (lower left).

6. **Gratiola** L.—Hedge Hyssop

Annual or perennial herbs; leaves simple, opposite, sessile; flowers in the leaf axils, subtended by a pair of bractlets; sepals 5, free or united at base, equal; corolla 5-lobed, bilabiate, tubular, campanulate, pubescent within at the base; anther-bearing stamens 2, with 2 small sterile stamens on 2 slender filaments; ovary superior, with 2 stigmas; fruit a capsule with many seeds.

Gratiola consists of about twenty species found in most of the world. The following occur in central Midwest wetlands.

1. Corolla bright yellow throughout or at least with a yellow tube.
   2. Corolla bright yellow throughout; leaves up to 6 mm wide; capsules 2–3 mm long............ ......1. *G. aurea*
   2. Corolla white, with a yellow tube; leaves more than 6 mm wide; capsules 3–5 mm long.... ......2. *G. neglecta*
1. Corolla white or creamy white, without a yellow tube.
   3. Corolla creamy white; pedicels stout, 3–10 mm long; sterile stamens minute; capsules 5–9 mm in diameter ...... 3. *G. virginiana*
   3. Corolla white; pedicels slender, up to 1.5 mm long; sterile stamens a pair of filiform filaments; capsules about 2 mm in diameter...... 4. *G. viscidula*

1. **Gratiola aurea** Muhl. Cat. 2. 1813. Fig. 276.

Perennial herb; stems prostrate to ascending, branched or unbranched, somewhat 4-angled, glandular-pubescent, to 30 cm long; leaves opposite, simple, lanceolate to narrowly oblong, acute at the apex, tapering at the sessile base, entire or remotely toothed, glandular-pubescent, to 2.5 cm long, to 6 mm wide; flowers in the leaf axils, on glandular-pubescent petioles up to 15 mm long, subtended by 2 green bractlets; calyx united at base, 5-parted, the lobes lanceolate, 4–5 mm long; corolla bright yellow, somewhat zygomorphic, to 15 mm long, tubular below; fertile stamens 2; sterile stamens 2; capsules more or less globose, 2–3 mm in diameter, shorter than the calyx. July–September.

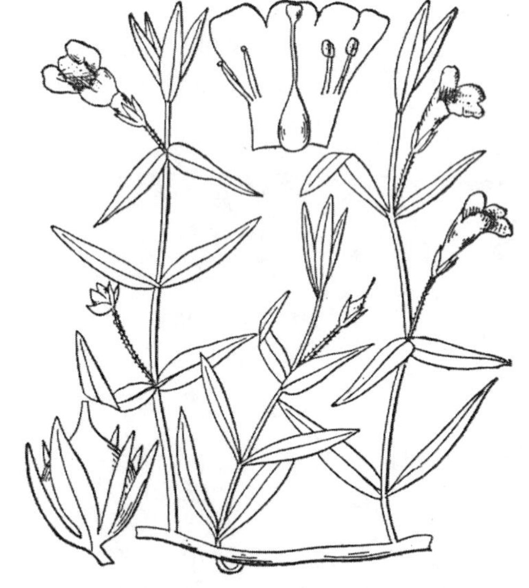

276. *Gratiola aurea* (Golden hedge hyssop). Habit (center). Fruit (lower left). Flower, cut open (upper center).

Wet soil.

IL (OBL).

Yellow hedge hyssop.

The bright yellow flowers are distinctive for this species. It also has entire or remotely toothed leaves.

2. **Gratiola neglecta** Torr. Cat. Pl. N. Y. 89. 1819. Fig. 277.
Gratiola missouriana Beck, Am. Journ. Sci. 10:258. 1826.

Annual herb with fibrous roots; stems branched or unbranched, erect, glandular-pubescent, to 30 cm tall; leaves opposite, simple, lanceolate to oblanceolate, acute at the apex, denticulate, glabrous or pubescent, to 5 cm long, to 2.5 cm wide; flowers axillary, on slender pedicels at least 1 cm long, subtended by a pair of foliaceous bractlets; sepals 5, green, free or united at base; corolla 5-lobed, the lobes creamy white, the tube yellow, to 12 mm long; sterile stamens minute; anther-bearing stamens 2; ovary superior; capsules globose to ovoid, 3–5 mm long. May–October.

Wet soil, sometimes in shallow water.

IA, IL, IN, KS, KY, MO, NE, OH (OBL).

Clammy hedge hyssop.

This species differs from other *Gratiola* species by its yellow corolla tube and its slender pedicels. Its stems are glandular-pubescent. It differs from species of *Lindernia* by its pair of foliaceous bractlets at the base of each flower.

277. *Gratiola neglecta* (Clammy hedge hyssop). Habit (center). Flower (lower right).

3. **Gratiola virginiana** L. Sp. Pl. 17. 1753. Fig. 278.
Gratiola sphaerocarpa Ell. Bot. S. C. & Ga. 1:14. 1816.

Annual or biennial herb with fibrous roots; stems usually unbranched, erect, more or less glabrous, usually somewhat fleshy, to 40 cm tall; leaves opposite, simple, lanceolate to elliptic, acute at the apex, serrulate, glabrous or nearly so, to 6 (–7) cm long, to 2.5 cm wide; flowers axillary, on stout pedicels usually less than 1 cm long, subtended by a pair of bractlets; sepals 5, green, free or nearly so; corolla 5-lobed, creamy white, rarely with a yellowish tube, 8–12 mm long; sterile stamens minute; anther-bearing stamens 2; ovary superior; capsules globose, 5–9 mm in diameter. May–October.

Wet soil, sometimes in standing water.

IA, IL, IN, KS, KY, MO, OH (OBL).

Round-fruited hedge hyssop.

This species is very similar to *G. neglecta* but is stouter in most aspects. In particular, the fruits are larger than those of *G. neglecta*.

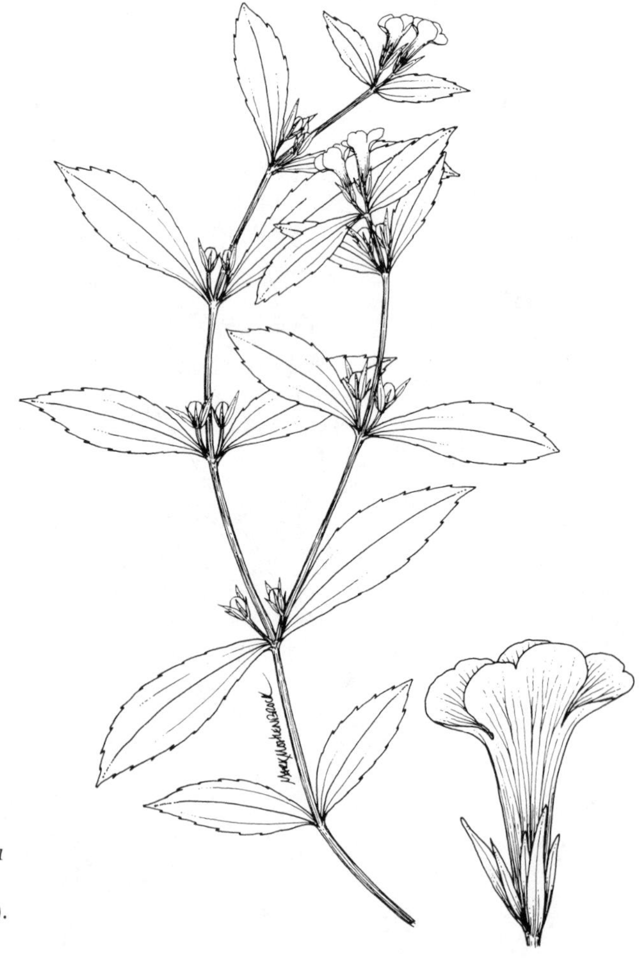

278. *Gratiola virginiana* (Round-fruited hedge hyssop). Habit (center). Flower (lower right).

4. **Gratiola viscidula** Pennell, Proc. Nat. Acad. Sci. Phila. 81:184. 1929. Fig. 279.

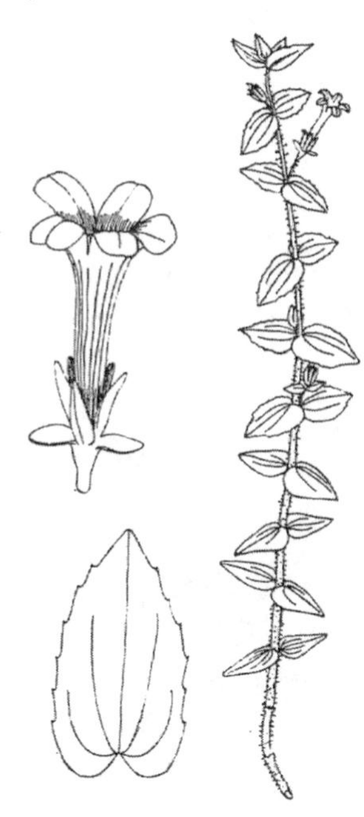

279. *Graiola viscidula* (Sticky hedge hyssop). Habit (right). Flower (upper left). Leaf (lower left).

Perennial herb from creeping rhizomes; stems erect, to 60 cm tall, usually glandular-pubescent and viscid; leaves opposite, simple, ovate to oblong, acute at the apex, denticulate, glabrous or pubescent, to 2.5 cm long, to 1.5 cm wide; flowers axillary, on slender pedicels to 1.5 mm long, subtended by a pair of foliaceous bracts; sepals 5, green, up to 7 mm long, free or united at the base; corolla lobes white, with blue lines and a yellow center, to 15 mm long; sterile stamens a pair of filiform filaments; anther-bearing stamens 2; ovary superior; capsules globose, about 2 mm in diameter. June–September.

Marshes, wet soil.

KY, MO, OH (OBL).

Sticky hedge hyssop.

The sticky pubescence on the stems and usually the leaves is distinctive for this species.

7. **Leucospora** Nutt.—Leucospora

Only the following species comprises the genus.

1. **Leucospora multifida** (Michx.) Nutt. Jour. Acad. Nat. Sci. Phila. 7:87. 1834. Fig. 280.
Capraria multifida Michx. Fl. Bor. Am. 2:22. 1803.
Conobea multifida (Michx.) Benth. in DC. Prodr. 10:391. 1846.

Annual herb from fibrous roots; stems erect or spreading or decumbent, much branched, pubescent, to 20 cm long; leaves opposite, pininately divided, the ultimate divisions to 3 cm long, pubescent, petiolate; flowers 1 or 2 in the axils of the leaves, on slender pedicels up to 1 cm long, not subtended by bracteoles; sepals 5, united at base, green, 3–4 mm long; corolla lavender, bilabiate, tubular, 3–5 mm long; stamens 4; ovary superior; capsules ovoid, 2–3 mm long, with numerous seeds. May–June.

Wet soil, shallow water at edge of lakes.

IA, IL, IN, MO (FACW+), KS, KY, NE, OH (OBL).

Leucospora.

This tiny species is readily distinguished by its opposite, pininately divided leaves and its small, lavender, bilabiate flowers.

8. **Limosella** L.—Mudwort

Dwarf annuals with creeping stolons; leaves basal, fleshy, entire, without stipules; peduncles 1-flowered; sepals 5, united below; corolla 5-lobed, campanulate, actinomorphic; stamens 4; ovary superior; fruit a capsule.

There are about fifteen species in this genus, found throughout the world.

280. *Leucospora multifida* (Leucospora). Habit (center). Flower (lower left).

1. **Limosella aquatica** L. Sp. Pl. 631. 1753. Fig. 281.

Annual herb with fibrous roots; leaves basal, tufted, somewhat fleshy, narrowly oblong to narrowly elliptic, entire, to 5 cm long, to 1 cm wide; flower solitary on a recurved pedicel; calyx 5-lobed, green; corolla campanulate, 5-lobed, white, 2.5–3.5 mm long; stamens 4; ovary superior; capsules globose, up to 3 mm in diameter. July–September.

Wet shores.

MO, NE (OBL).

Mudwort.

This dwarf annual is distinguished by its tufts of narrow, basal, rather fleshy leaves and its 1-flowered peduncles.

9. **Lindernia** All.—False Pimpernel

Annual or biennial herbs; leaves sessile, opposite, entire or toothed, without stipules; flower solitary in the axils of the leaves, not subtended by bracts or bracteoles; sepals 5, free or nearly so, equal; corolla bilabiate, 5-lobed; fertile stamens 2, with 2 sterile stamens; ovary superior; fruit a capsule.

281. *Limosella aquatica* (Mudwort). Habit in fruit (left). Flower (right).

There are about seventy species in this genus worldwide. *Lindernia* differs from *Gratiola* by the absence of small bracteoles at the base of each flower.

1. All pedicels longer than the subtending leaves; leaves rounded at the base .... 1. *L. anagallidea*
1. Lower or all pedicels shorter than the subtending leaves; leaves tapering at the base.. 2. *L. dubia*

1. **Lindernia anagallidea** (Michx.) Pennell, Acad. Nat. Sci. Phila. Monogr. 1:152. 1935. Fig. 282.
Gratiola anagallidea Michx. Fl. Bor. Am. 1:6. 1803.
Lindernia dilatata Muhl. ex Ell. Bot. S.C. & Ga. 1:16. 1816.

Annual herb from fibrous roots; stems slender, branched, erect, glabrous, to 20 cm tall; leaves opposite, simple, ovate to elliptic, acute at the apex, rounded at the base, entire or slightly toothed, glabrous, to 2 cm long, to 1.5 cm wide; flower solitary in the axils of the leaves, on slender pedicels longer than the leaves; sepals 5, green, free; corolla bilabiate, 5-lobed, white to lavender, to 10 mm long; fertile stamens 2, with 2 sterile stamens; ovary superior; capsules slightly longer than the calyx. May–October.

Wet ground, sandbars, occasionally in shallow water.

IA, IL, IN, KS, KY, MO, NE, OH (OBL).

Slender false pimpernel.

This species is very similar to *L. dubia*, differing by its pedicels longer than the subtending leaves.

282. *Lindernia anagallidea* (Slender false pimpernel). Habit (center). Flower (lower left).

2. **Lindernia dubia** (L.) Pennell, Acad. Nat. Sci. Phila. Monogr. 1:141. 1935. Fig. 283.
Gratiola dubia L. Sp. Pl. 17. 1753.
Ilysanthes riparia Raf. Ann. Nat. 13. 1820.
Lindernia dubia (L.) Pennell var. *riparia* (Raf.) Fern. Rhodora 44:444. 1942.

Annual herb from fibrous roots; stems erect, branched, glabrous, to 35 cm tall; leaves opposite, simple, oblong to elliptic, acute at the apex, tapering at the base, entire or sparingly toothed, glabrous, to 3 cm long, to 1.5 cm wide; flower solitary in the axils of the leaves, on slender pedicels shorter than the leaves, or sometimes the lowermost flower on a pedicel longer than the leaves; sepals 5, free, green; corolla 5-lobed, bilabiate, white to lavender, to 10 mm long; fertile stamens 2, with 2 sterile stamens; ovary superior; capsules usually slightly shorter than the calyx. May–October.

Wet ground, mudflats, sometimes in shallow water.

IA, IL, IN, KS, KY, MO, NE, OH (OBL).

False pimpernel.

Some plants have all the pedicels shorter than the leaves, and other plants have the lowest flower on a pedicel longer than the leaves. These latter plants have been called var. *riparia*, and these are intermediate with *L. anagallidea*.

283. *Lindernia dubia* (False pimpernel). Habit (center). Flower (lower right).

10. **Mazus** Lour.

Herbs; leaves basal and cauline and opposite or alternate; flowers zygomorphic, perfect, in terminal racemes, subtended by minute bracteoles; calyx 5-lobed, campanulate, green; corolla bilabiate, 5-lobed, with small projections in the throat; fertile stamens 4; ovary superior, with 2 stigmas; fruit a capsule.

Twenty species, native to Asia and Australia, comprise this genus. It is becoming popular as a diminutive ornamental called a steppable.

1. **Mazus pumilus** (Burm. f.) van Steenis, Nova Guinea n.s. 9:31. 1958. Fig. 284.
Lobelia pumila Burm. f. Fl. Ind. 186. 1768.

Annual herb with fibrous roots; stems slender, erect, branched, pubescent, to 15 cm tall; basal leaves spatulate to oblong, obtuse to acute at the apex, more or less rounded at the base, to 5 cm long, to 2.5 cm wide, pubescent, coarsely but irregularly dentate, the cauline leaves similar but smaller, the lower opposite, the upper usually alternate; flowers up to 8 (–10) in terminal racemes, zygomorphic, perfect; calyx 5-lobed, campanulate, green, 4–5 mm long, elongating in fruit; corolla bilabiate, bluish with yellow and white markings, 7–10 mm long; stamens 4; ovary superior; capsules globose, 3–4 mm in diameter, a little shorter than the calyx. May–October.

Wet soil, bottomland forests.

IL, MO (UPL), KY, OH (FACU−), as *M. japonicus.*

Mazus.

This small annual differs from *Mecardonia* by its campanulate calyx and from *Lindernia, Gratiola,* and *Bacopa* by having four fertile stamens. It is native to Asia.

This species used to be called *M. japonicus*, but that binomial belongs to a different species.

284. *Mazus pumilus* (Mazus). Habit (center). Flower (lower left).

11. **Mecardonia** Ruiz & Pavon

Perennial herbs; leaves opposite, simple, serrate; flower solitary in the axil of the leaves, zygomorphic, perfect, subtended by a pair of bracteoles; sepals 5, free or nearly so, green; corolla bilabiate, 5-lobed, the tube pubescent within; fertile stamens 4; ovary superior, with 2 stigmas; fruit a capsule.

There are fifteen species in this genus, all in North and South America.

1. **Mecardonia acuminata** (Walt.) Small, Fl. S.E. U.S. 1065. 1903. Fig. 285.
Gratiola acuminata Walt. Fl. Carol. 61. 1788.
Bacopa acuminata (Walt.) B. L. Robins. Rhodora 10:66. 1908.
Pagesia acuminata (Walt.) Pennell, Proc. Nat. Acad. Sci. U.S.A. 1:65. 1935.

Perennial herb from a thickened rootstock; stems erect, slender, branched, 4-angled, glabrous, to 65 cm tall; leaves opposite, simple, oblong to lanceolate, acute at the apex, tapering to the sessile base, glabrous, serrate, to 4 cm long, to 1.5 cm wide; flower solitary in the axil of the leaves, with a pair of bracteoles at the base, the pedicels slender, up to 3 cm long; sepals 5, unequal, free or nearly so, green, 6–9 mm long; corolla bilabiate, white with purple striations, to 10 mm long; fertile stamens 4; ovary superior; capsules oblongoid, glabrous, to 10 mm long, about as long as the calyx. June–September.

Wet soil, mud flats.

IL, KY, MO (OBL), KS (NI).

Purple hedge hyssop.

This species, although similar in appearance to *Gratiola, Lindernia,* and *Bacopa*, differs by its four fertile stamens. It is also similar to *Mazus* but differs by all its leaves opposite and its nearly free sepals.

285. *Mecardonia acuminata* (Purple hedge hyssop). Habit (center). Flower (right).

12. **Melampyrum** L.—Cow-wheat

Hemiparasitic annual herbs; leaves simple, opposite, lobed or toothed, without stipules; flower solitary in the upper leaf axils, perfect; calyx 4- to 5-lobed, campanulate; corolla 5-lobed, tubular below; fertile stamens 4; ovary superior; fruit a 1- to 4-seeded capsule.

Melampyrum consists of fifteen species in the Northern Hemisphere.

1. **Melampyrum lineare** Desr. in Lam. Encycl. 4:22. 1796. Fig. 286.
Melampyrum americanum Michx. Fl. Bor. Am. 2:16. 1803.
Melampyrum lineare Desr. var. *americanum* (Michx.) Beauv. Mem. Soc. Phys. & Hist. Nat. Geneve 38:474. 1806.
Melampyrum lineare Desr. var. *latifolium* Bart. Comp. Pl. Phila. 2:50. 1818.
Melampyrum lineare Desr. var. *pectinatum* (Pennell) Fern. Rhodora 44:463. 1942.

Annual herb from fibrous roots; stems unbranched to very bushy-branched, to 50 cm tall, glabrous; leaves opposite, simple, linear to elliptic, acute at the apex, tapering at the base, entire except for the bracteal leaves, to 6 cm long, sessile or on short petioles; flowers in the axils of sharply toothed bracteal leaves; sepals 5-lobed, campanulate, green; corolla 5-parted, white, usually with yellow tips, up to 12 mm long; stamens 4; ovary superior; capsules flat, ovate, with white seeds often with brown tips. July–August.

Bogs, marshes, sand dunes.

IL, IN (FAC−), KY, OH (FACU).

Cow-wheat.

Typical var. *lineare*, with entire bracteal leaves, does not occur in the central Midwest. Sparsely toothed bracteal leaves in our area are known as var. *americanum*. Extremely pectinate bracteal leaves on plants from northwestern Indiana are known as var. *pectinatum*. Plants with elliptic leaves may be called var. *latifolium*.

286. *Melampyrum lineare* (Cow-wheat). Habit (center). Lower part of plant (left). Flower (upper left). Fruit (upper right). Seed (lower right).

13. **Mimulus** L.—Monkey-flower

Perennial herbs; leaves simple, opposite, without stipules; flower solitary in the leaf axils, showy; calyx 5-toothed, 5-angled, the teeth unequal; corolla bilabiate, 5-lobed; stamens 4, attached to the corolla tube; ovary superior, bilocular; fruit a capsule with numerous seeds.

Mimulus consists of about 120 species in many parts of the world.

1. Plants erect; corolla purple.
  2. Leaves petiolate; stems slightly winged...1. *M. alatus*
  2. Leaves sessile or even clasping; stems wingless.......................................4. *M. ringens*
1. Plants spreading to ascending; corolla yellow.
  3. Leaves oval to orbicular; calyx in flower 5–8 mm long, in fruit 10–12 mm long; throat of corolla open ....2. *M. glabratus*
  3. Leaves ovate to broadly lanceolate; calyx in flower 8–17 mm long, in fruit 10–25 mm long; throat of corolla nearly closed...3. *M. guttatus*

1. **Mimulus alatus** Sol. in Ait. Hort. Kew. 2:361. 1789. Fig. 287.

Perennial herb from slender rhizomes; stems erect, 4-angled, the angles usually slightly winged, glabrous, to 1.2 m tall; leaves opposite, simple, lanceolate to ovate, acute at the apex, tapering or rounded at the base, serrate, glabrous, up to 10 cm long, up to 4 cm wide, on somewhat winged petioles up to 2 cm long; flower solitary in the upper axils, on pedicels up to 1.5 cm long; calyx strongly 5-angled, with 5 teeth 1.0–1.5 mm long, green; corolla bilabiate, 2.0–2.5 cm long, blue to violet; stamens 4, attached to the corolla tube; ovary superior; capsules cylindric, shorter than and closely enclosed by the calyx. June–September.

Wet ground, swamps, wet woods, marshes.

IA, IL, IN, KS, KY, MO, NE, OH (OBL).

Winged monkey-flower.

This species is very similar to *M. ringens*, differing by its petiolate leaves and wing-angled stems.

2. **Mimulus glabratus** HBK.Nov. Gen. Sp. 2:370. 1817. Fig. 288.
Mimulus geyeri Torr. in Nicollet, Rep. Hydrogr. Upper Miss. 157. 1843.
Mimulus jamesii Torr. & Gray var. *fremontii* Benth. ex DC. Prodr. 10:371. 1846.
Mimulus glabratus HBK. var. *fremontii* (Benth.) A.L. Grant, Ann. Mo. Bot. Gard. 11: 190. 1925.

Perennial herb; stems prostrate to ascending, sometimes rooting at the lower nodes, glabrous or minutely pubescent; leaves opposite, simple, oval to suborbicular, subacute to obtuse at the apex, rounded or cordate at the base, entire or denticulate, glabrous or minutely pubescent, to 1.5 cm long, nearly as wide, the upper sessile, the lower petiolate; flower solitary in the leaf axils, on pedicels shorter than the leaves; calyx campanulate, 5-toothed, 5–8 mm long in flower, 10–12 mm long in fruit, green; corolla bilabiate, up to 2.2 cm long, yellow, the throat open; stamens 4; ovary superior; capsules cylindrical. June–October.

Springs, fens, often in shallow standing water.

IA, IL, IN, KS, MO, NE (OBL).

287. *Mimulus alatus* (Winged monkey-flower). Flowering branch (above). Leaves (below).

288. *Mimulus glabratus* (Yellow monkey-flower). Habit.

Yellow monkey-flower.

This handsome species has yellow flowers and oval to suborbicular opposite leaves.

Typical var. *glabratus* is native to the southwestern United States, Mexico, and South America. Our plants are var. *fremontii*, and they differ from var. *glabratus* by their shorter calyces.

3. **Mimulus guttatus** Fisch. ex DC. Cat. Pl. Horti Monsp. 127. 1813. Fig. 289.

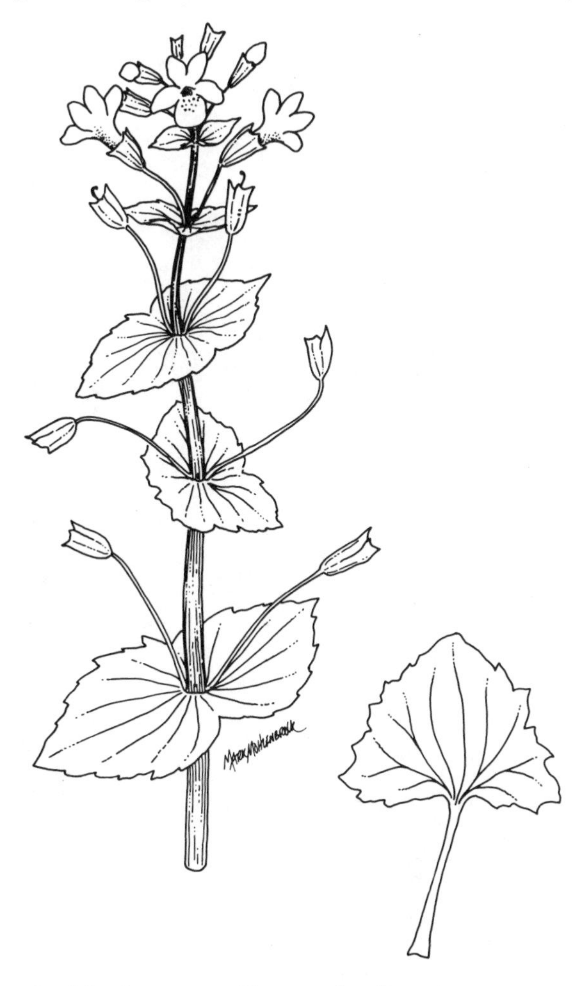

289. *Mimulus guttatus* (Seep monkey-flower). Habit (center). Lower leaf (lower right).

Perennial herb; stems ascending to decumbent, glabrous, often rooting at the nodes, to 50 cm long; leaves opposite, simple, broadly ovate to broadly lanceolate, acute at the apex, rounded or tapering at the base, glabrous, coarsely dentate, longitudinally veined, to 1.5 cm long, to 1.2 cm wide, the lower on long petioles, the upper sessile; flowers in terminal racemes, remote, actinomorphic, perfect, on pedicels up to 2 cm long; calyx campanulate, 5-toothed, to 17 mm long in flower, to 25 mm long in fruit, green, the teeth obtuse, the uppermost tooth the longest; corolla bilabiate, to 4 cm long, yellow with red spots in the throat, the throat nearly closed; stamens 4, attached to the corolla tube; ovary superior; capsules cylindrical. June–October.

Wet soil.

NE (OBL).

Seep monkey-flower.

This species of the western United States barely enters our range in Nebraska.

The yellow flowers with red spots and the nearly closed throat of the corolla are distinctive.

The calyx is larger than in the yellow-flowered *M. glabratus*.

4. **Mimulus ringens** L. Sp. Pl. 634. 1753. Fig. 290.
Mimulus minthodes Greene, Leaflets Bot. 2:1. 1909.
Mimulus ringens L. var. *minthodes* (Greene) A. L. Grant, Am. Mo. Bot. Gard. 11:31. 1925.

Perennial herb; stems erect, 4-angled, the angles not winged, glabrous, to 1.2 m tall; leaves opposite, simple, elliptic to lanceolate to oblanceolate, acute to acuminate at the apex, rounded or narrowed at the sessile and sometimes clasping base, serrate to crenate, glabrous, up to 10 cm long, up to 4 cm wide; flower solitary in the upper axils, actinomorphic, perfect, on stout pedicels up to 5 cm long; calyx strongly 5-angled, green, with 5 linear-tipped lobes, to 15 mm long; corolla blue to violet, bilabiate, to 3 cm long; stamens 4; ovary superior; capsules cylindrical, shorter than the persistent calyx. June–September.

290. *Mimulus ringens* (Sessile monkey-flower). Upper part of plant (center). Fruit (lower right).

Wet ground in woods, swamps, marshes.

IA, IL, IN, KS, KY, MO, NE, OH (OBL).

Sessile monkey-flower.

Typical var. *ringens* has leaves that are rounded at the base and often clasping, while var. *minthodes* has leaves that taper to the sessile base.

This species is similar in appearance to *M. alatus* but is readily distinguished by its sessile leaves.

14. **Pedicularis** L.—Lousewort; Betony

Mostly hemiparasitic perennial herbs; leaves simple, opposite or alternate, pinnatifid to bipinnatifid, without stipules; flowers in spikes or racemes, actinomorphic, perfect, subtended by leafy bracts; calyx 5-lobed, campanulate to tubular; corolla 5-lobed, bilabiate; stamens 4, concealed by the upper lip of the corolla; ovary superior, bilocular; capsules more or less falcate, asymmetrical.

There are more than five hundred species of *Pedicularis*, most of them in boreal and north temperate regions of the world and in the Rocky Mountains.

In addition to the species listed below, *P. canadensis* occurs in upland habitats in the central Midwest.

1. **Pedicularis lanceolata** Michx. Fl. Bor. Am. 2:18. 1803. Fig. 291.

291. *Pedicularis lanceolata* (Swamp betony). Upper part of plant (center). Leaf (lower right). Flower, cut open (upper right). Capsule (lower left).

Perennial herb from short rhizomes; stems erect, usually unbranched, glabrous or sparsely pubescent, up to 75 cm tall; leaves simple, opposite, oblong to lanceolate, pinnatifid with crenate lobes, to 10 cm long, glabrous or nearly so, sessile or short-petiolate; flowers actinomorphic, perfect, crowded in terminal spikes from the upper leaf axils; calyx green, to 1 cm long, 2-lobed, the lobes unequal; corolla pale yellow, incurved, to 2.5 cm long; stamens 4; ovary superior; capsules ovoid, slightly longer than the calyx. August–October.

Fens, bogs, wet meadows, swamps.

IA, IL, IN, MO, NE (FACW+), KY, OH (FACW).

Swamp betony; swamp lousewort.

This species is readily distinguished by its pinnatifid opposite leaves and pale yellow flowers in terminal spikes.

15. **Penstemon** Mitchell—Beardstongue

Perennial herbs; leaves opposite, simple; flowers zygomorphic, perfect, in racemes or panicles; calyx 5-parted; corolla bilabiate, tubular; fertile stamens 4, the fifth stamen reduced to a glabrous or bearded filament; ovary superior; fruit a capsule, with many seeds.

Penstemon consists of about 280 species, mostly in the western United States. Many species have beautiful flowers and are popular ornamentals.

1. **Penstemon alluviorum** Pennell, Man. SE. Fl. 1203. 1933. Fig. 292.

Perennial from slender rhizomes; stems erect, unbranched, puberulent, to 1 m tall; leaves opposite, simple, lanceolate to elliptic, acute to acuminate (at least the lowermost) at the apex, tapering or somewhat rounded at the base, sharply serrate, pubescent on both surfaces, to 10 cm long, to 6 cm wide; inflorescence an erect to ascending panicle, the branches very sparsely glandular; calyx 5-parted, green, 3–5 mm long, with scarious margins; corolla bilabiate, white, 17–23 mm long; stamens 4, the sterile filament sparsely pubescent; ovary superior; capsules 8–9 mm long, with many seeds. May–June.

Bottomland forests.

IL, IN, MO (FACW+), KY, OH (FACW).

Lowland beardstongue.

This species is similar to *P. digitalis* and often considered to be synonymous with it. It differs, however, by its puberulent stems and leaves, the sparsely glandular axis of the inflorescence, and its smaller calyx, corolla, and capsules.

I have only observed this species in wet forests, particularly in floodplains.

292. *Penstemon alluviorum* (Lowland beardstongue). Habit.

16. **Veronica** L.—Speedwell

Annual or perennial herbs; leaves simple, opposite or alternate, without stipules; flowers axillary or in terminal racemes, perfect; calyx 4-lobed; corolla rotate, 4-lobed, the lobes slightly unequal; stamens 2; ovary superior; capsules usually obcordate, flattened.

Approximately three hundred species comprise this genus, mostly in north temperate regions.

1. Flower solitary in the axils of the cauline leaves .... 4. *V. peregrina*
1. Flowers in racemes.
   2. Leaves linear to linear-lanceolate; axis of raceme zigzag............................... 5. *V. scutellata*
   2. Leaves lanceolate to oblong to oval; axis of raceme straight.
      3. Leaves petiolate; pedicels eglandular.. 1. *V. americana*
      3. Leaves sessile, sometimes clasping; pedicels glandular-pubescent.
         4. Fruiting pedicels ascending; corolla bluish, 4–5 mm across; capsules barely notched at summit .... 2. *V. anagallis-aquatica*
         4. Some of the fruiting pedicels horizontally spreading; corolla white to pale rose, 3–4 mm across; capsules deeply notched at summit....................................... 3. *V. comosa*

1. **Veronica americana** (Raf.) Schwein. ex Benth. in DC. Prodr. 10:468. 1846. Fig. 293.
Veronica beccabunga L. var. *americana* Raf. Med. Fl. 109. 1830.

Somewhat succulent perennial with rhizomes; stems erect or sometimes creeping, glabrous, to nearly 1 m long; leaves simple, opposite, lanceolate to narrowly ovate, acute at the apex, tapering at the base, serrate or nearly entire, up to 7 cm long, up to 3 cm wide, twice or more times longer than wide, with short petioles; racemes terminal, arching, 4- to 15-flowered; flowers perfect, on glabrous pedicels; sepals 4-parted, united at base, about 1.5 mm long; corolla 4-lobed, 5–10 mm across, pale blue; stamens 2; ovary superior; capsules flat, orbicular, notched at the summit, about 3 mm in diameter. June–August.

Shallow water of springs; swamps.

IA, IL, IN, KS, KY, MO, NE, OH (OBL).

American brooklime.

This is the only wetland species of *Veronica* in the central Midwest that has racemose inflorescences and petiolate leaves.

2. **Veronica anagallis-aquatica** L. Sp. Pl. 12. 1753. Fig. 294.

Annual or perennial herbs with fibrous roots; stems creeping, with ascending branches, glabrous or minutely glandular-pubescent, to 1 m long; leaves simple, opposite, elliptic to narrowly ovate, acute or subacute at the apex, tapering or cordate at the sessile or clasping base, serrate to nearly entire, glabrous or minutely glandular-pubescent, to 10 cm long, to 5 cm wide; racemes terminal, more or less straight, 20- to 50-flowered; flowers perfect, on glandular-pubescent pedicels; sepals 4, united at base, green, about 3 mm long; corolla 4-lobed, 4–5 mm across, pale blue; stamens 2; ovary superior; capsules ovate to suborbicular, flat, slightly notched at the summit, about 3 mm long, on ascending pedicels. June–August.

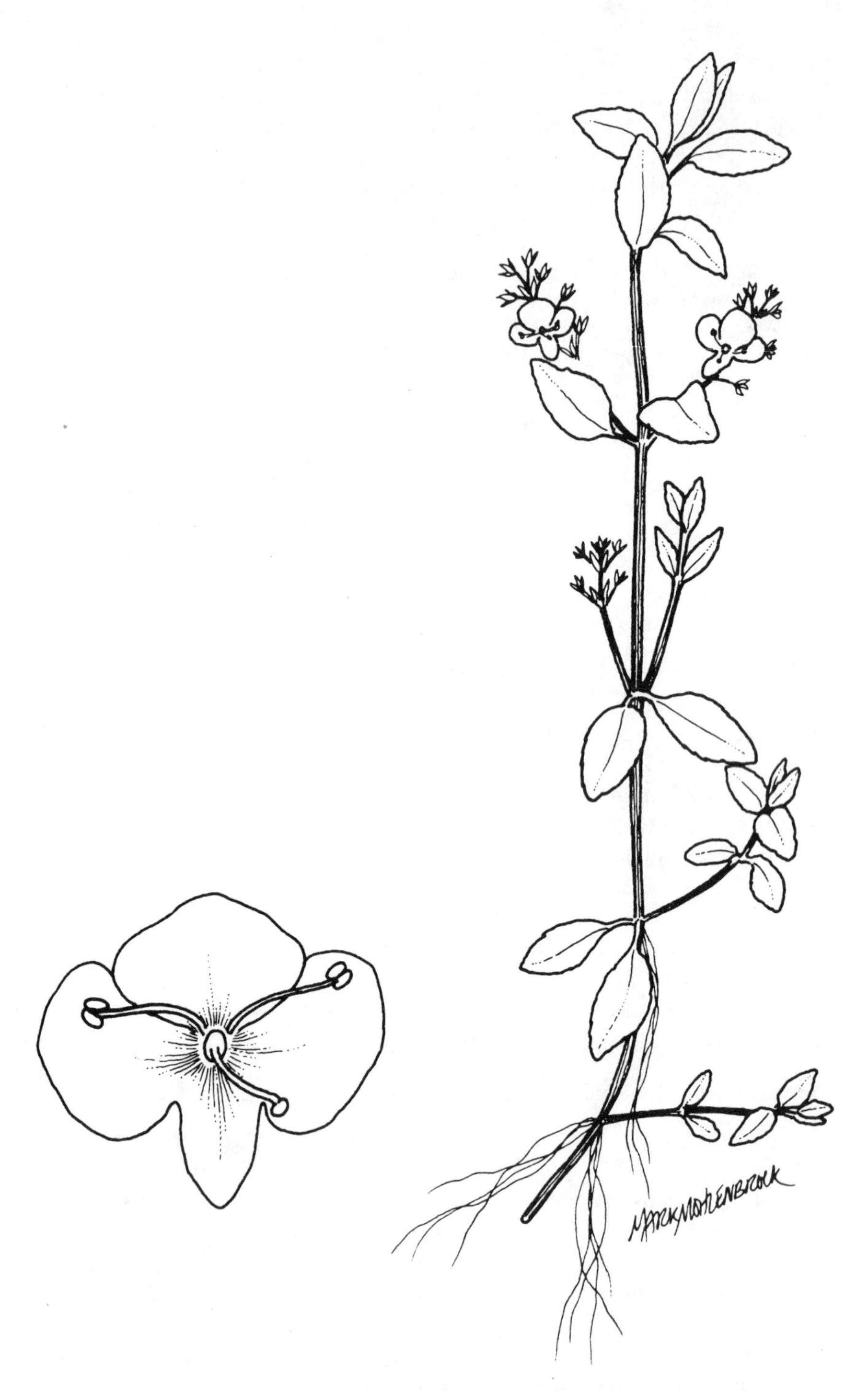

293. *Veronica americana* (American brooklime). Habit (right). Flower (lower left).

Shallow water of springs and streams; swamps.

IA, IL, IN, KS, KY, MO, NE, OH (OBL).

Water speedwell; brook pimpernel.

Similar to *V. comosa, V. anagallis-aquatica* differs by itrs slightly larger flowers, ascending fruiting pedicels, and a scarcely notched capsule at the summit.

3. **Veronica comosa** Richter, Pl. Eur. 276. 1903. Fig. 295.
Veronica catenata Pennell, Rhodora 23:37. 1921.

Annual or perennial herb with fibrous roots; stems creeping, with ascending branches, glabrous or minutely glandular-pubescent, up to 1 m long; leaves simple, opposite, lanceolate to narrowly oblong, acute at the apex, tapering or rounded at the sessile or sometimes clasping base, glabrous or minutely glandular-pubescent,

294. *Veronica anagallis-aquatica* (Water speedwell). Habit (left). Flower (lower right).

usually entire, to 6 cm long, to 1.5 cm wide; racemes ascending, 5- to 35-flowered, the flowers perfect, on glandular-pubescent pedicels; sepals 4, united at base, green, 2.5–3.0 mm long; corolla 4-lobed, 3–4 mm across, white to pale rose; stamens 2; capsules broadly ovate, flat, deeply notched at the summit, about 3 mm long, the lowest on horizontally spreading pedicels. June–September.

Shallow water of springs, streams, sloughs, wet ditches.

IA, IL, IN, KS, KY, MO, NE, OH (OBL), as *V. catenata*.

Water speedwell.

This species differs from the similar *V. anagallis-aquatica* by it paler and slightly smaller flowers, its deeply notched capsules, and its horizontally spreading lower fruiting capsules.

If *V. comosa* and *V. catenata* apply to two different species, then our plants should be called *V. catenata*.

295. *Veronica comosa* (Water speedwell). Habit, with flowers (center). Leaves (right). Flower and fruit (lower left).

4. **Veronica peregrina** L. Sp. Pl. 14. 1753. Fig. 296.
Veronica xalapensis HBK. Nov. Gen. Sp. 2:389. 1818.
Veronica peregrina L. var. *xalapensis* (HBK.) Pennell, Torreya 19:167. 1919.

Annual herb with fibrous roots; stems erect or ascending, branched or unbranched, glabrous, to 40 cm tall; leaves simple, opposite, or those subtending the flowers sometimes alternate, linear to narrowly oblong, obtuse to subacute at the apex, tapering at the sessile base, entire or with a few low teeth, glabrous, to 3 cm long, to 6 (–7) mm wide; flower solitary in the axil of the upper leaves, perfect, sessile; calyx lobes 4, free except for the very base, 3–5 mm long; corolla 4-lobed, white, 2–3 mm across; stamens 2; ovary superior; capsules obcordate, notched at the summit, flat, 3–4 mm long, glabrous or glandular-pubescent. April–August.

Moist soil, often weedy, occasionally in shallow water.

IA, IL, IN, MO (FACW+), KS, NE (OBL), KY, OH (FACU−).

Purslane speedwell.

This is an extremely common plant, often growing in disturbed areas, but sometimes in undisturbed wetlands. The tiny white flowers are solitary in the axils of the upper leaves. Plants with glandular-pubescent fruits may be known as var. *xalapensis*.

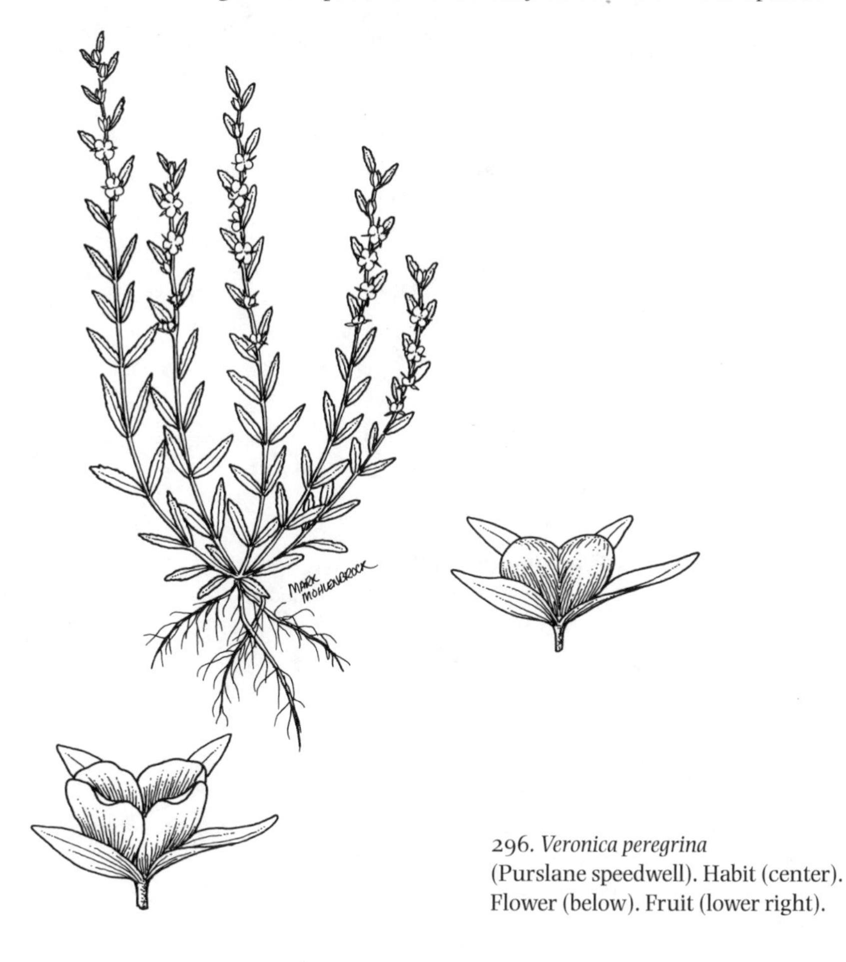

296. *Veronica peregrina* (Purslane speedwell). Habit (center). Flower (below). Fruit (lower right).

5. **Veronica scutellata** L. Sp. Pl. 12. 1753. Fig. 297.

Perennial herb with slender rhizomes and stolons; stems decumbent to ascending, branched or unbranched, glabrous or sometimes slightly pubescent, to 75 cm long; leaves simple, opposite, linear to linear-lanceolate, acute to acuminate at the apex, tapering at the sessile base, entire or with minute glandular teeth, glabrous, to 8 cm long, to 1.5 cm wide; racemes several, with a zigzag axis, 5- to 20-flowered, the flowers on spreading or reflexed slender pedicels; calyx lobes 4, united at base, green, 2–3 mm long; corolla 4-lobed, pale blue, 6–10 mm across; stamens 2; ovary superior; capsules flattened, broader than high, notched at the summit, 2.5–3.5 mm long, glabrous or nearly so. April–July.

297. *Veronica scutellata* (Marsh speedwell). Upper part of plant (center). Flower (lower left). Fruit (lower right).

Bogs, swamps, around ponds.

IA, IL, IN, OH (OBL), NE (not listed).

Marsh speedwell.

This speedwell differs from all others by its linear to linear-lanceolate leaves.

17. **Veronicastrum** Fabr.—Culver's-root

Perennial herbs; leaves whorled, simple; inflorescence spicate, branched, appearing candelabra-like; flowers zygomorphic, perfect; calyx 4- to 5-parted, zygomorphic; corolla tubular, 4- to 5-parted, nearly actinomorphic, perfect; stamens 2, attached to the corolla tube, long-exserted; ovary superior; fruit a capsule.

The species below and one in eastern Asia comprise this genus.

1. **Veronicastrum virginicum** L. Sp. Pl. 9. 1753. Fig. 298.
Leptandra virginica (L.) Nutt. Gen. 1:7. 1818.

Perennial herb; stems erect usually unbranched, pubescent or glabrous, to 2 m tall; leaves simple, in whorls of 3–6, lanceolate to narrowly ovate, acuminate at the apex, tapering at the base, sharply serrate, glabrous or pubescent, to 15 cm long, to 10 cm wide, the petioles to 1 cm long; inflorescence a cluster of terminal spikes to

15 cm long, resembling a candelabra; flowers more or less actinomorphic except for the calyx, perfect; calyx 4- to 5-parted, green, to 2.5 mm long; corolla 4- to 5-lobed, tubular below, pink or white, 7–9 mm long; stamens 2, attached to the corolla tube, long-exserted; ovary superior; capsule ovoid, 4–5 mm long, glabrous. June–August.

Wet meadows, prairies, thickets, woods.

IA, IL, IN, KS, MO, NE (FAC), KY, OH (FACU).

Culver's-root; candelabra plant.

This species is readily recognized by its whorls of 3–6 sharply serrate leaves and its candelabra-shaped inflorescence. It may occur in both wet and dry habitats. Plants may be glabrous or pubescent.

298. *Veronicastrum virginicum* (Culver's-root). Habit (right). Flower (lower left).

114. SOLANACEAE—NIGHTSHADE FAMILY

Herbs or less frequently somewhat woody, rarely climbing; leaves usually alternate, simple or compound; inflorescence various; flowers perfect, actinomorphic; sepals 5, united, often persistent on the fruit; petals 5, united into a rotate, campanulate, or funnelform corolla; stamens usually 5, attached to the corolla tube; ovary superior, 2- to 6-locular, with numerous ovules on axile placentae; fruit a capsule or berry.

The Solanaceae is a family found in both temperate and tropical regions of the Old and New World. Botanists recognize about eighty genera and more than three thousand species, including such important plants as the potato (*Solanum tuberosum*), eggplant (*S. melongena*), tomato (*Lycopersicum esculentum*), strawberry tomato (*Physalis* spp.), and red pepper (*Capsicum* spp.), and several ornamentals—butterfly plant (*Schizanthus* spp.), salpiglossis (*Salpiglossis* spp.), petunia (*Petunia* spp.), nierembergia (*Nierembergia* spp.), browallia (*Browallia* spp.), brunfelsia (*Brunfelsia* spp.), and the Chinese lantern (*Physalis alkekengi*). In addition, several species in the family have poisonous properties, while others are drug plants—belladonna (*Atropa belladonna*), henbane (*Hyoscyamus* spp.), and stramonium (*Datura* spp.). Tobacco (*Nicotiana* spp.) also belongs to this family.

1. **Solanum** L.—Nightshade

Herbs or sometimes vines (in our area); leaves alternate, simple, but sometimes lobed; inflorescence variable; flowers perfect, actinomorphic; sepals 5, green, united below; corolla rotate, usually 5-parted; stamens 5; ovary superior; fruit a berry with many seeds.

There are about two thousand species of *Solanum* found in many parts of the world.

1. **Solanum dulcamara** L. Sp. Pl. 185. 1753. Fig. 299.
Solanum dulcamara L. var. *villosissimum* Desv. Obs. Pl. Angers 112. 1818.
Solanum dulcamara L. f. *albiflorum* House, Bull, N.Y. State Mus. 254:613. 1824.

Woody perennial; stems climbing, branched, glabrous or villous, to nearly 3 m long; leaves ovate, acuminate at the apex, rounded at the base, often 3-lobed or 3-parted with 2 small basal lobes, glabrous or villous, to 10 cm long, on slender petioles; inflorescence cymose, the flowers perfect, to 1.5 cm across, on spreading or pendulous, articulated pedicels; calyx rotate, green, 5-lobed, the lobes oblong, obtuse, glabrous or pubescent; corolla 5-lobed nearly to the base, the lobes lanceolate, acute to acuminate, violet to white; stamens 5, connivent; berry globose to ovoid, glabrous, red, to 1.5 cm in diameter, subtended for a while by the subpersistent calyx. June–October.

Woods, thickets, near ponds and lakes.

IA, IL, IN, KS, MO, NE (FAC), KY, OH (FAC−).

Bittersweet nightshade.

This species, sometimes planted as a garden ornamental, can be found in a wide variety of habitats, from disturbed areas to more "natural" localities, including low woods. There is variation in the degree of lobing of the leaves, in flower color ranging from violet to white, and in degree of pubescence of the stems and leaves. White-flowered plants are sometimes designated as f. *albiflorum*. Plants with villous stems and leaves are sometimes called var. *villosissimum*.

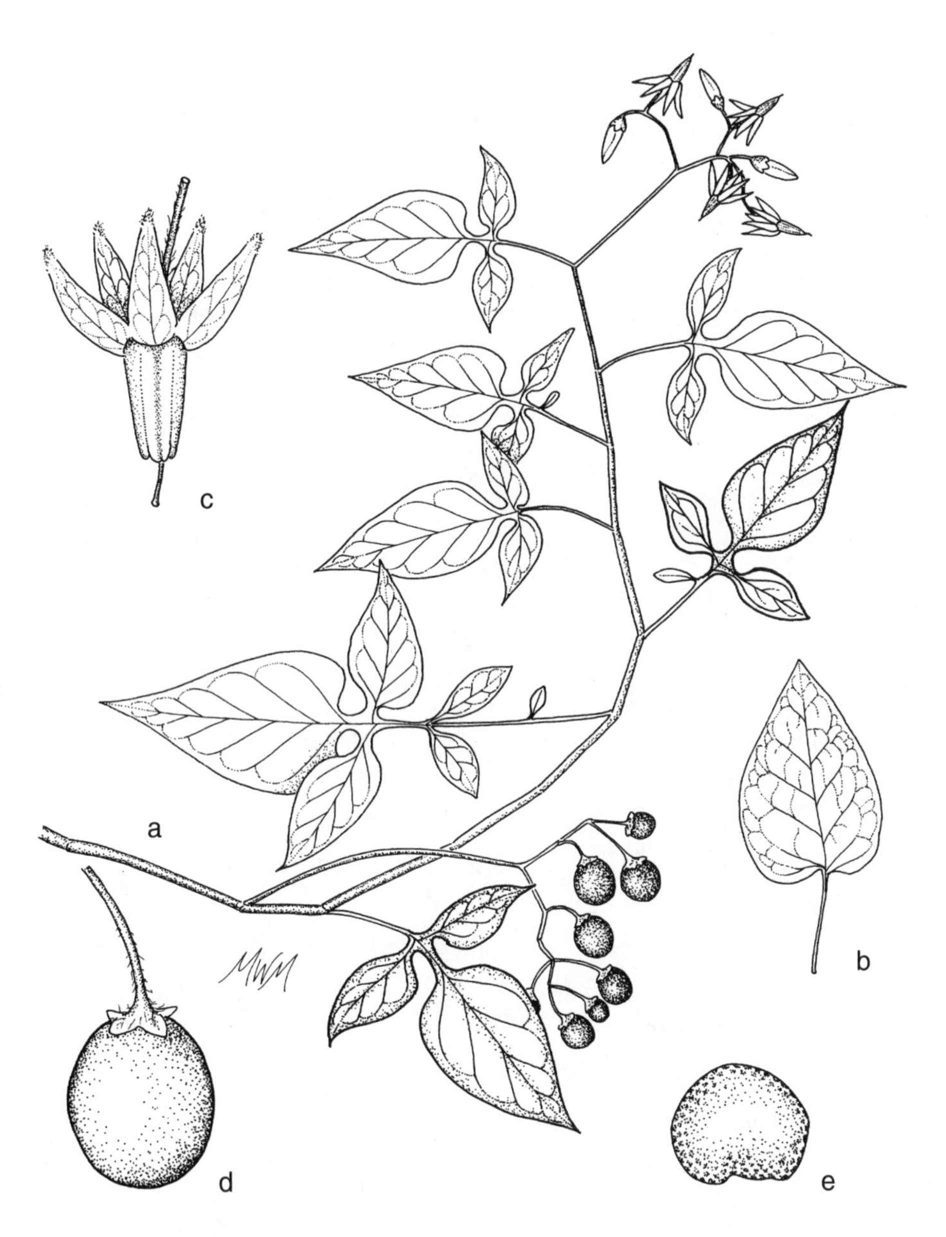

299. *Solanum dulcamara* (Bittersweet nightshade).

a. Habit.
b. Unlobed leaf.
c. Flower.
d. Fruit.
e. Seed.

115. STYRACACEAE—STORAX FAMILY

Shrubs or trees; leaves alternate, simple, with stellate pubescence, without stipules; flowers often showy, actinomorphic, perfect; calyx 4- to 8-parted, adherent to the ovary; petals 4–8, free or united at the base; stamens the same number as the petals or twice as many as the petals; ovary superior or inferior, 3- to 6-locular; fruit a drupe or capsule.

Approximately ten genera and 150 species are in the family that is found in temperate and subtropical regions of the world.

1. **Styrax** L.—Storax

Shrubs or small trees; leaves alternate, simple, with stellate pubescence; flowers axillary in short racemes, showy, actinomorphic, perfect; calyx 5-toothed, united below, adherent to the base of the ovary; petals 5, tubular below, the lobes somewhat recurved, white; stamens 10, attached to the base of the corolla tube; ovary 3-locular, at least below; fruit 1-seeded, dry, surrounded at the base by the persistent calyx.

This genus, found in many parts of the world, consists of approximately 120 species.

1. **Styrax americanus** Lam. Encycl. 1:82, 1783. Fig. 300.

Shrub to 4 m tall, with many branches; leaves alternate, simple, elliptic to oblong, acute to acuminate at the apex, tapering or somewhat rounded at the base, irregularly obscurely toothed, to 10 cm long, to 3.5 cm wide, green and glabrous above, paler and sometimes with sparse stellate hairs below, on petioles to 1.5 cm long; flowers 3–4 in racemes, perfect, actinomorphic; calyx 5-toothed, united below, 3–4 mm long, green; petals 5, tubular below, the lobes recurved, white, 10–12 mm long; stamens 10; fruit subglobose, 6–8 mm in diameter. May–June.

Swampy woods.

IL, IN, KY, MO, OH (OBL)

American snowbell.

This shrub of swampy woods has beautiful white flowers during late spring. The obscure teeth of the alternate leaves are distinctive.

116. TILIACEAE—BASSWOOD FAMILY

Trees or shrubs, rarely herbs; leaves alternate, simple, with stipules; flowers often in cymes, actinomorphic, usually perfect; sepals usually 5, free; petals usually 5, free; stamens numerous; ovary superior, 2- to many-locular; fruit a capsule or dry and indehiscent.

There are fifty genera and nearly five hundred species in this family, found mostly in the tropics and subtropics.

1. **Tilia** L.—Basswood

Trees; leaves alternate, simple, stipulate, with palmate venation; flowers in pedunculate cymes, the peduncle attached near the middle of a foliaceous bract; sepals 5, free, green; petals 5, free, white or creamy, fragrant; stamens numerous with occasional petaloid staminodia; ovary superior, 5-locular; fruit a hard, dry, 1-seeded nut.

Nearly fifty species comprise this genus, most all of them in the Northern Hemisphere.

1. **Tilia heterophylla** Vent. Mem. Acad. Paris 4:16. 1802. Fig. 301.
Tilia americana L. var. *heterophylla* (Vent.) Loud. Arb. Frut. Brit. 1:375. 1838.

Tree to 20 m tall, with trunk diameter up to 0.7 m, the crown broadly spreading; bark brown, scaly, deeply furrowed; twigs slender, red-brown, glabrous, the buds ovoid, red, usually glabrous, up to 6 mm long; leaves alternate, simple, ovate, acute to acuminate at the apex, cordate at the asymmetrical base, serrate, green and glabrous or with scattered pubescence above, silvery-white and stellate-tomentose below, up to 20 cm long, up to 10 cm wide, the petioles glabrous or sparsely pubes-

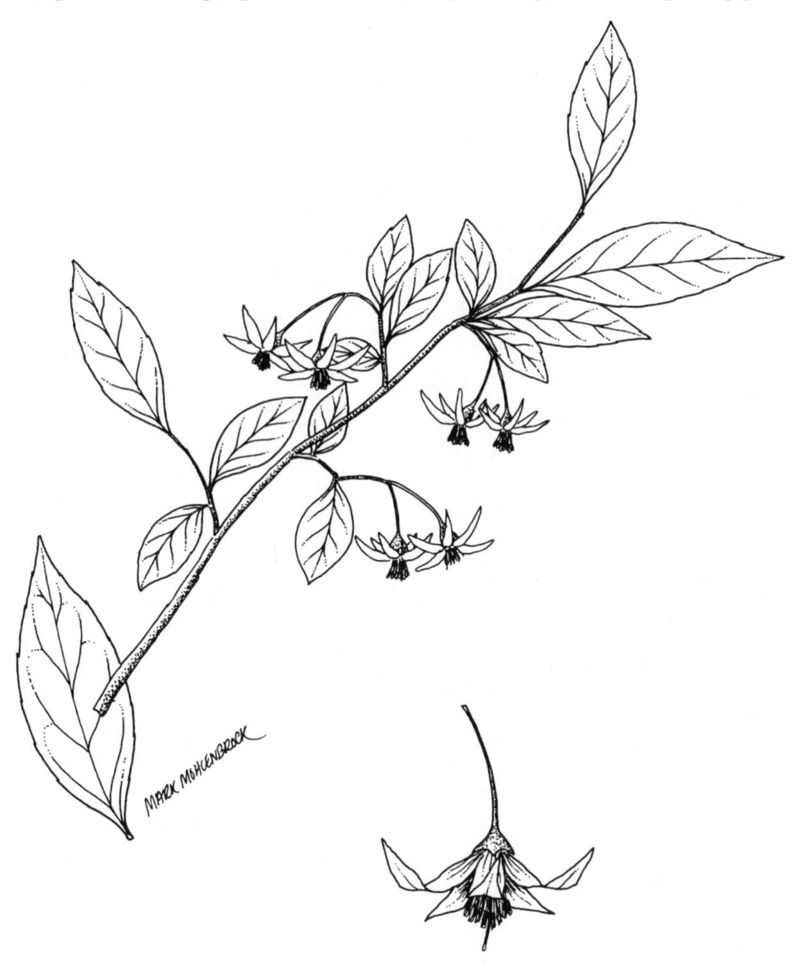

300. *Styrax americanus* (American snowbell). Branchlet, with flowers (above). Leaf (lower left). Flower (lower right).

cent, up to 4.5 cm long; flowers several in a cyme, the cyme on a glabrous peduncle attached near the middle of a foliaceous bract, fragrant; sepals 5, free, green; petals 5, free, 8–12 mm long, yellow-green, pubescent; stamens numerous; ovary superior, 5-locular; fruit globose, hard, dry, indehiscent, 6–8 mm in diameter. June–July.

Bottomland forests, swampy woods.

IA, IL, IN, KY, MO, OH (U. S. Fish and Wildlife Service does not distinguish this species from *T. americana*, which they list as FACU); however, *T. heterophylla* occurs in wetter situations than *T. americana*.)

White basswood.

Although the flowers and fruits of this species are similar to those of *T. americana*, the leaves and the habitat convince me that these are two different species.

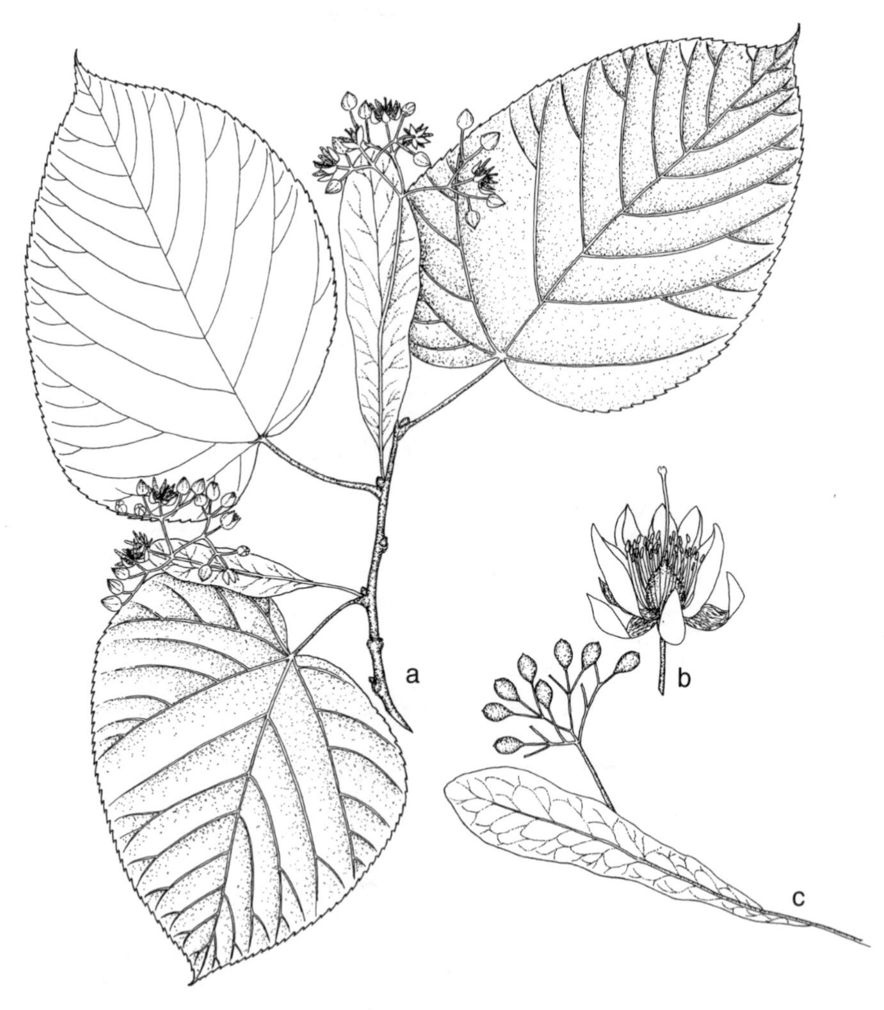

301. *Tilia heterophylla* (White basswood). a. Flowering branch. b. Flower. c. Bract, with fruits.

117. ULMACEAE—ELM FAMILY

Monoecious trees or shrubs; leaves simple, alternate, with caducous stipules; flowers perfect or unisexual, zygomorphic, solitary or variously arranged; sepals 4–8, free or united at the base; petals absent; stamens 4–8, free, arising from the hypanthium; pistil one, the ovary superior, bicarpellate, unilocular, with one pendulous ovule per locule, the styles 2; fruit a samara, nut, or drupe.

The Ulmaceae is composed of fifteen genera and about 175 species. Most of these are in the Northern Hemisphere.

The family is divided into two tribes. Tribe Ulmeae, which includes *Ulmus* and *Planera*, has united sepals, solid pith, and a samara or nutlike fruit. Tribe Celtideae, including *Celtis*, has free sepals, chambered pith, and a drupe.

1. Leaves with strong lateral veins arising from the main vein at a distance above the very base of the blade; pith of branches solid; sepals united; fruit a samara or nut.
    2. Fruit a samara; all flowers perfect, appearing before the leaves; leaves mostly doubly toothed............ 3. *Ulmus*
    2. Fruit a wingless nut; at least some of the flowers unisexual, appearing with the leaves; leaves mostly singly toothed............ 2. *Planera*
1. Leaves with a pair of strong lateral veins arising from the main vein at the very base of the blade; pith of branches chambered; sepals free, or nearly so; fruit a drupe............ 1. *Celtis*

1. **Celtis** L.—Hackberry

Trees or shrubs, usually with warty bark; twigs with chambered pith; bud scales 2-ranked; leaves alternate, simple, serrate, the stipules caducous; flowers polygamous, solitary or fascicled; calyx deeply 4- to 5-lobed, or the sepals free; petals absent; stamens 5–6, free, exserted; pistil one, the ovary superior, sessile, 1-locular, 1-ovulate, the styles deeply 2-lobed; fruit a 1-seeded drupe.

Celtis is a genus of about seventy species found primarily in the Northern Hemisphere.

Most species of the genus are attacked by a gall insect that causes the production of gnarled broomlike clusters of branchlets, a disfiguration known as witch's-broom.

The taxa of *Celtis* are extremely variable and seemingly intergrade in an almost hopeless manner. Usually exact identification can only be made if mature drupes are present.

1. Drupes orange, red, or brownish, 5–8 mm in diameter; seeds 4–7 mm long; leaves most often lanceolate, entire or sparingly serrate, smooth or somewhat scabrous on the upper surface. .... 1. *C. laevigata*
1. Drupes purple, black, or orange-red, 8–11 mm in diameter; seeds 7–9 mm long; leaves most often ovate, sharply serrate, harshly scabrous on the upper surface............ 2. *C. occidentalis*

1. **Celtis laevigata** Willd. Berl. Baumz. ed. 2, 81. 1811. Fig. 302.
Celtis mississippiensis Bosc, Encycl. Met. Agr. 7:577. 1822.
Celtis occidentalis L. var. *mississippiensis* (Bosc) Schneck. Ann. Rep. Geol. Surv. Ind. 7:559. 1876.
Celtis smallii Beadle in Small, Fl. SE. U.S. 365. 1903.
Celtis laevigata Willd. var. *smallii* (Beadle) Sarg. Bot. Gaz. 67:223. 1919.

Trees to 30 m tall, the oblong crown rather open, the trunk diameter up to 0.75 m; bark gray, smooth except for numerous wartlike outgrowths; twigs slender, gray, glabrous, the buds ovoid, acute, brown, glabrous or nearly so, appressed against the twigs, to 3 mm long; leaf scars crescent-shaped, slightly elevated, with 3 bundle scars: leaves lanceolate to lance-oblong, less commonly ovate, acute to acuminate at the apex, cuneate to rounded to subcordate at the usually asymmetrical base, entire or sparsely toothed, less commonly regularly toothed, smooth or rarely harshly scabrous on the upper surface, glabrous or pubescent at least along the veins on the lower surface, to 10 cm long, to 4 cm wide, coriaceous to membranous, the petioles to 1 cm long, glabrous or pubescent; staminate flowers several in axillary fascicles; pistillate flowers 1–3 in the leaf axils; sepals 4–5, free or nearly so, ovate-lanceolate, greenish yellow, not persistent in fruit; drupes ellipsoid to globose, orange, red, or brownish at maturity, 5–8 mm long, the seed 4–7 mm long. April–May.

Low woods, floodplains, wet forests.

IL, IN, KS, KY, MO (FACW).

Sugarberry.

302. *Celtis laevigata* (Sugarberry). a. Leafy twig. b. Leaf variation. c. Flower.

Two varieties occur in wetlands. Typical var. *laevigata* has entire or sparsely toothed leaves, while var. *smallii* has regularly toothed leaves.

A third variety, var. *texana*, with harshly scabrous leaves, occurs on cliffs and in dry woods.

2. **Celtis occidentalis** L. Sp. Pl. 1044. 1753. Fig. 303.
Celtis crassifolia Lam. Encycl. 4:138. 1796.
Celtis occidentalis L. var. *crassifolia* (Lam.) Gray, Man. Bot., ed. 2, 297. 1856.
Celtis occidentalis L. var. *canina* (Raf.) Sarg. Bot. Gaz. 67:217. 1919.

Trees to 30 m tall, the crown broad and irregular, the trunk diameter up to 1.5 m; bark dark brown or gray, smooth except for numerous wartlike outgrowths; twigs slender, gray, glabrous, the buds slenderly ovoid, acute, brown, pubescent,

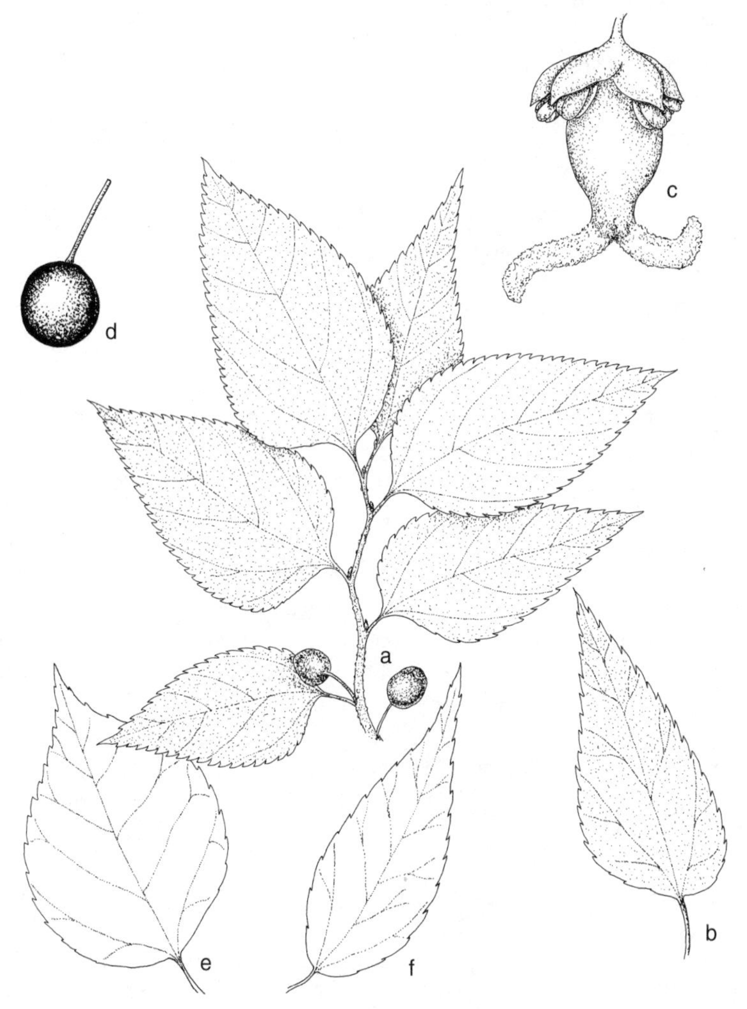

303. *Celtis occidentalis* (Hackberry). a. Leafy branch, with fruits. b, e, f. Leaf variations. c. Flower. d. Fruit.

appressed against the twigs, to 6 mm long; leaf scars crescent-shaped, slightly elevated, with 3 bundle scars; leaves ovate to oblong-ovate, acuminate at the apex, cuneate to truncate to rounded at the asymmetrical base, serrate, smooth or usually harshly scabrous on the upper surface, glabrous or pilose along the veins on the lower surface, to 12 cm long, to 9 cm wide, coriaceous to membranous, the petiole glabrous or pubescent, to 1.5 cm long; staminate flowers several in axillary fascicles; pistillate flowers 1–3 in the leaf axils; sepals 4–5, free or nearly so, linear-oblong, greenish yellow, not persistent in fruit; drupes globose to ovoid, purple, black, or orange-red at maturity, 8–11 mm long, the seed 7–9 mm long. April–May

Low woods, bottomland forests.

IA, IL, IN, KY, MO, OH (FACU), KS, NE (FAC−).

Hackberry.

Three varieties may be recognized in the central Midwest. Leaves that are harshly scabrous on the upper surface and with orange-red drupes are var. *occidentalis*. Plants with leaves more or less smooth on the upper surface and with purple or black drupes also occur. If most of the leaves are more than half as broad as long, they may be called var. *pumila*. If most of the leaves are less than half as long as broad, they may be called var. *canina*.

2. **Planera** J.F. Gmel.—Water Elm; Planer Tree

Only the following species comprises this genus.

1. **Planera aquatica** (Walt.) J.F. Gmel. Syst. Nat., ed. 13, 2:150. 1791. Fig. 304.
Anonymos aquatica Walt. Fl. Carol. 230. 1788.
Planera gmelinii Michx. Fl. Bor. Am. 2:248. 1803.

Tree to 10 m tall, the crown broad and irregular, the trunk diameter up to 0.4 m; bark brown or gray, with large scales; twigs slender, puberulent, becoming scaly, brown or gray, the buds subglobose, acute, brown or reddish brown, puberulent; leaf scars suborbicular, slightly elevated, with several bundle scars; leaves oblong-ovate, acute at the apex, rounded or subcordate at the slightly asymmetrical base, singly crenate-serrate, pubescent when young, becoming glabrous or nearly so and scabrellous at maturity, to 6 cm long, to 2.5 cm wide, the petioles to 6 mm long, puberulent, the stipules ovate, red, caducous; staminate flowers fasciculate from the outer bud scales on twigs of the previous year, on very short pedicels; perfect flowers in clusters of 1–3 from the axils of the current leaves, on elongated pedicels; calyx campanulate, deeply 4- to 5-lobed, greenish yellow; stamens few, exserted drupes oblongoid, stipitate, subtended by the persistent calyx, tipped by the persistent styles, pale brown, tuberculate, to 8 mm long, with one dark brown, ovoid seed. March–April.

Swampy woods.

IL, KY, MO (OBL).

Water elm; planer tree.

The small, pointed buds, scaly bark, single-toothed leaves, and drupes distinguish the water elm from the genus *Ulmus*. Beneath the scaly outer bark is a reddish brown inner bark.

The tree is too rare, and the wood is too light and soft, to be of any commercial value.

3. **Ulmus** L.—Elm

Trees with furrowed bark; twigs often flexuous, sometimes winged, with small, solid pith; bud scales 2-ranked; leaves alternate, simple, usually doubly serrate, the stipules linear to lanceolate, caducous; flowers mostly perfect, clustered in fascicles, cymes, or racemes, borne on twigs of the preceding year; calyx campanulate, 4- to 9-lobed; petals absent; stamens 3–9, free, exserted beyond the calyx; pistil one, the ovary superior, sessile or stipitate, 1- to 2-locular, with 1 ovule per locule, the styles deeply 2-lobed; fruit a 1-seeded samara.

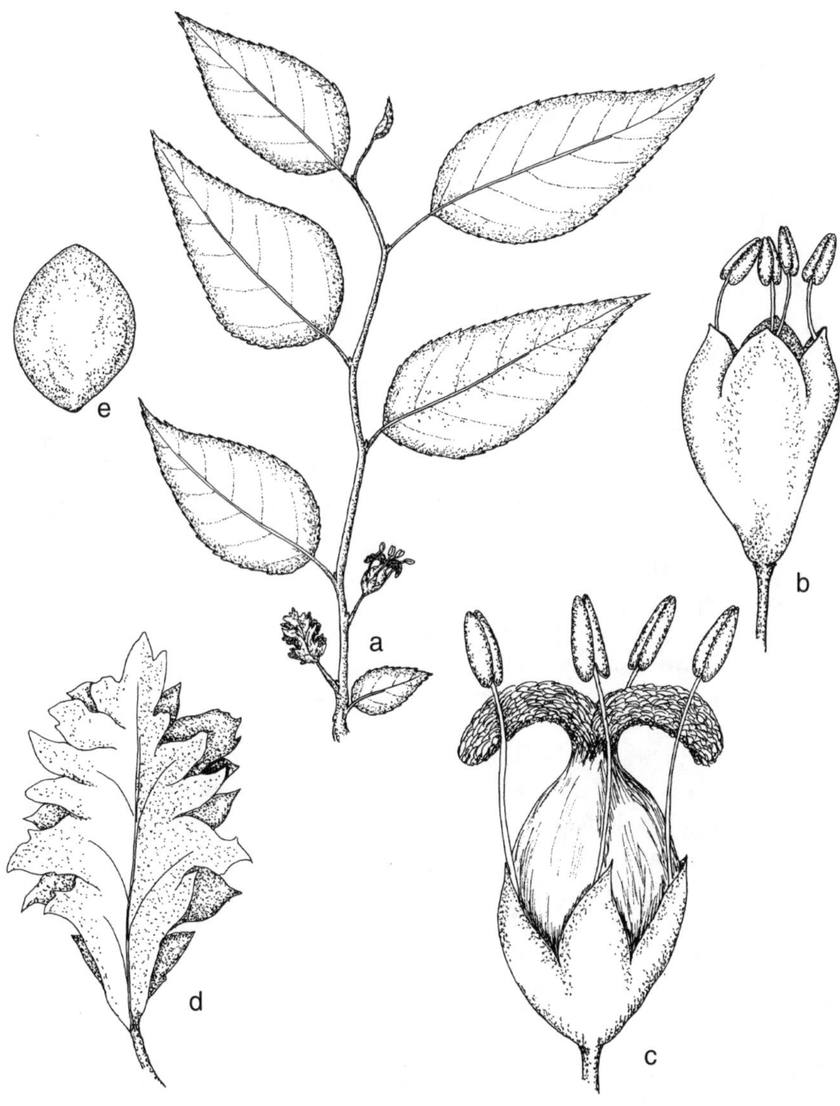

304. *Planera aquatica* (Water elm).

a. Leafy branch, with flowers.
b. Staminate flower.
c. Perfect flower.
d. Fruit.
e. Seed.

There are approximately eighteen species of *Ulmus* native to the Northern Hemisphere.

In addition to the species enumerated below, *Ulmus parviflora* (Chinese elm), *U. glabra* (Scotch elm), and *U. pumila* (Siberian elm) are sometimes grown as ornamentals.

1. None of the branches corky-winged .... 2. *U. americana*
1. Some of the branches corky-winged ....1. *U. alata*

1. **Ulmus alata** Michx. Fl. Bor. Am. 1:173. 1803. Fig. 305.

Trees to 20 m tall, usually much smaller, the crown mostly oblong, the trunk diameter up to 0.7 m; bark light reddish brown, with flat ridges separated by shallow fissures; twigs slender, light brown to gray, glabrous or nearly so, usually winged with two thin, corky wings, the buds ovoid, acute, dark brown, glabrous or puberulent, to 4 mm long; leaf scars crescent-shaped, slightly elevated, with 3 bundle scars; leaves oblong-lanceolate to oblong-ovate, acute or acuminate at the apex, cuneate or rounded at the slightly asymmetrical base, doubly serrate, pubescent at first but becoming glabrous or nearly so at maturity on the upper surface, softly pubescent below, even at maturity, to 9 cm long, to 4 cm wide, the petioles pubescent, to 8 mm long, the stipules narrow; flowers few in a fascicle, the pedicels slender and pendulous; calyx campanulate, to 3 mm long, 5- to 9-lobed, the lobes ovate, glabrous; stamens 5–9, exserted; samaras oblong, to 8 mm long, to 6 mm wide, subtended by the persistent calyx, the apex bifid, the base stipitate, the wings long-ciliate. February–April.

Rocky upland woods, bluff-tops, less commonly in low ravines and swampy forests.

IL, IN, KS, KY, MO (FACU).

Winged elm.

The winged elm is a conspicuous species with its broad, corky wings. In specimens where the wings are greatly developed, the wings may measure as much as 2.5 cm across. A few specimens observed may be nearly destitute of wings.

Winged elm is one of the characteristic trees of exposed sandstone bluff-tops, where it is nearly always associated with *Quercus marilandica*, *Q. stellata*, *Juniperus virginiana*, and *Vaccinium arboreum*. Under such xeric conditions, the winged elm is a tree of small, gnarled stature.

Occasionally winged elm grows in wooded ravines and swampy woods, where it is associated with *Celtis laevigata*. In these conditions, it may reach a height of 20 meters.

The leaves turn a dull yellow in the autumn.

The wood, which is hard and heavy but difficult to split, was used in the making of tool handles.

2. **Ulmus americana** L. Sp. Pl. 226. 1753. Fig. 306.

Trees to 30 m tall, the crown broadly rounded, the trunk diameter up to 1.5 m (rarely larger); bark dark gray, furrowed; twigs slender, flexuous, glabrous or puberulent, the buds to 6 mm long, ovoid, acute, light brown, glabrous; leaf scars crescent-shaped, slightly elevated, with 3 bundle scars; leaves obovate to oval-oblong, acuminate at the apex, the asymmetrical base cuneate on one side, rounded on the other, doubly serrate, glabrous or scabrous on the upper surface, glabrous to soft-pubescent on the lower surface, to 15 cm long, up to half as wide, becoming bright yellow in the autumn, the petioles to 6 mm long, glabrous or nearly so, the caducous stipules glabrous or pubescent; flowers fascicled, the pedicels slender, glabrous or puberulent, 1.2–2.5 cm long, jointed near the base; calyx campanulate,

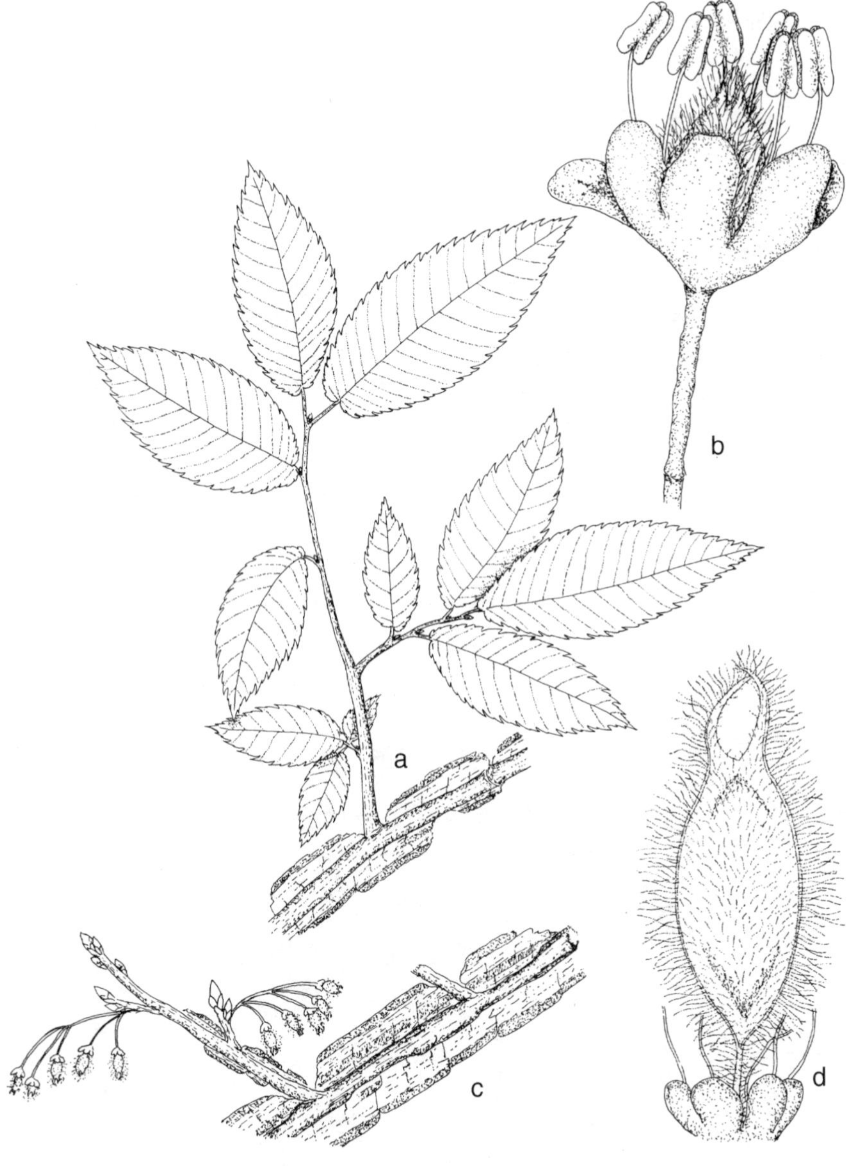

305. *Ulmus alata* (Winged elm).

a. Leafy branch.
b. Flower.

c. Twig, with fruits.
d. Fruit.

to 4 mm long, 5- to 9-lobed, the lobes oblong, obtuse, green, puberulent; stamens 5–9, exserted, the filaments stramineous to light brown, the anthers bright red; samaras ovoid to oblongoid, to 1.2 cm long subtended by the persistent calyx, the apex deeply notched, the base stipitate, the wings reticulate-nerved, ciliate on the margins. February–April.

Floodplain woods, bottomland forests.

IA, IL, IN, KY, MO, OH (FACW−), KS, NE (FAC).

American elm.

The American elm is a popular shade tree because of its beautiful widely spreading crown. It is becoming uncommon primarily because of the Dutch elm disease.

The leaves of this species may be glabrous or moderately pubescent on the upper surface, but never as rough-pubescent as *U. rubra*. A piece of outer bark of the American elm, when broken to show a cross-section, has alternating layers of light and dark inner bark.

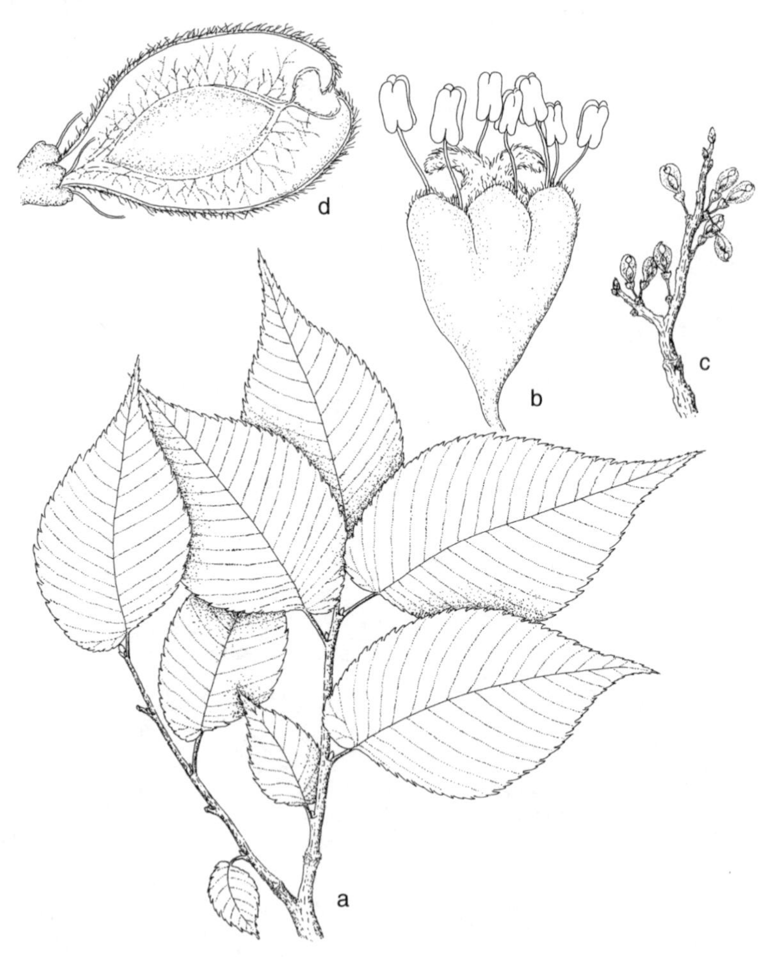

306. *Ulmus americana* (American elm). a. Leafy branch. b. Flower. c. Twig, with fruits. d. Fruit.

118. URTICACEAE—NETTLE FAMILY

Herbs (in our area), shrubs, or trees; leaves alternate or opposite, simple, with stipules; flowers unisexual or perfect, monoecious or dioecious, variously arranged; calyx 2- to 5-lobed, or the sepals free; petals absent; stamens 2–5, free; ovary superior, 1-locular, with one basal ovule; style 1; fruit an achene.

The Urticaceae contains about forty genera and six hundred species distributed over much of the world.

Probably the most economically important member of the family is *Boehmeria nivea* of China, Japan, and Malaysia, the source of ramie fibers.

1. Leaves opposite.
  2. Leaves and usually the stems pubescent, at least sparsely so.
    3. Stinging hairs present; pistillate flowers with free sepals .... 4. *Urtica*
    3. Stinging hairs absent; pistillate flowers with united sepals .... 1. *Boehmeria*
  2. Leaves and stems completely glabrous..3. *Pilea*
1. Leaves alternate...2. *Laportea*

1. **Boehmeria** Jacq.—Stingless Nettle

Perennial herbs without stinging hairs; leaves mostly opposite, simple, toothed, stipulate; flowers unisexual, monoecious or dioecious; staminate flowers in racemes or spikes, the calyx deeply 4-lobed, the stamens 4, free; pistillate flowers in racemes or spikes, the calyx tubular, 2- to 4-lobed or entire, enclosing the ovary; fruit an achene enclosed by the persistent calyx.

Boehmeria is a genus of about fifty species native primarily to the tropics. If differs from *Urtica* by the absence of stinging hairs and the tubular pistillate calyx.

1. **Boehmeria cylindrica** (L.) Sw. Prodr. 34. 1788. Fig. 307.
Urtica cylindrica L. Sp. Pl. 984. 1753.
Urtica capitata L. Sp. Pl. 985. 1753.
Boehmeria drummondiana Wedd. Ann. Sci. Nat. Ser. IV, 1:201. 1854.
Boehmeria cylindrica (L.) Sw. var. *drummondiana* (Wedd.) Wedd. in DC. Prodr. 16:202. 1869.

Perennial herbs; stems erect, simple or branched, glabrous to rough-pubescent, to 80 cm tall; leaves ovate-lanceolate to ovate, acute to acuminate at the apex, cuneate to rounded at the base, coarsely dentate, smooth to scabrous above, glabrous or pubescent beneath, 3-nerved from the base, to 7 cm long, up to half as wide, the petiole to 9 cm long, glabrous or pubescent, the stipules lance-subulate; flowers monoecious or dioecious, the staminate in interrupted axillary spikes, the pistillate in usually uninterrupted axillary spikes; staminate calyx deeply 4-lobed, pubescent; pistillate calyx tubular, entire or very shallowly 4-toothed; achene ellipsoid, to 2 mm long, enclosed by the persistent calyx, glabrous or beset with hooked hairs. July–October.

Moist woods, wet woods, marshes.

IA, IL, IN, KS, MO, NE (OBL), KY, OH (FACW+).

Stingless nettle.

Plants with leaves harshly scabrous on the upper surface may be known as var. *drummondiana*.

307. *Boehmeria cylindrica* (Stingless nettle).
a. Leafy branch, with pistillate flowers.
b. Staminate flower.
c. var. *drummondiana*. Fruit.
d. Seed.
e. Leafy branch, with fruits.

2. **Laportea** Gaud.—Wood Nettle

Perennial herbs with stinging hairs; leaves alternate, simple, toothed, stipulate; flowers unisexual, monoecious or dioecious, arranged in compound cymes; staminate flowers with 5 sepals and 5 stamens; pistillate flowers with 4 sepals in 2 unequal pairs and a superior ovary with a single style; achene reflexed.

Laportea is a genus of about two dozen species, primarily native to the topics. This is the only genus of Urticaceae with stinging hairs and alternate leaves in the central Midwest.

1. **Laportea canadensis** (L.) Gaud. Bot. Voy. Freyc. 498. 1826. Fig. 308.
Urtica canadensis L. Sp. Pl. 985. 1753.
Urtica divaricata L. Sp. Pl. 985. 1753.
Urticastrum divaricatum (L.) Kuntze, Rev. Gen. Pl. 635. 1891.

Perennial herbs; stems erect, simple or branched, beset with stinging hairs, to 2 m tall; leaves ovate, acute to acuminate at the apex, cuneate at the base, sharply serrate, 3-nerved from the base, glabrous or bristly-hairy, to 15 cm long, to 10 cm wide, the petioles to 10 cm long, setose; flowers unisexual, borne in large compound cymes, the staminate with 5 nearly free sepals and 5 stamens, the pistillate with 4 unequal sepals and a slender superior ovary; achenes ovate, compressed, glabrous, to 3 mm long, reflexed at maturity. June–September.

308. *Laportea canadensis* (Woodland nettle).

a. Leafy branch, with inflorescences.
b. Staminate flower.
c. Pistillate flowers.
d. Fruit.

Moist soil, wet woods.

IA, IL, IN, KS, KY, MO, NE, OH (FACW).

Wood nettle.

This species bears numerous, coarse, stinging hairs, which break off when touched. The substance contained in the hairs causes a very uncomfortable sensation when contacting the skin. Since the wood nettle grows in dense thickets in low woods, it causes a formidable barrier to anyone wishing to cross through an area where this species grows.

3. **Pilea** Lindl.—Clearweed

Annual or perennial herbs without stinging hairs; stems translucent; leaves opposite, simple, toothed, with united stipules; flowers unisexual, monoecious or dioecious, arranged in cymes or glomerules; staminate flowers with the calyx deeply 4-lobed and with 4 free stamens; pistillate flowers with the calyx 3-lobed, with 3 scalelike staminodia and one superior ovary without a style; fruit an achene partly enclosed by the persistent calyx.

There are about two hundred species of *Pilea* primarily found in tropical America.

1. Achenes black, about 1.5 mm long.. 1. *P. fontana*
1. Achenes green, often with purple speckles, about 1 mm long.............................. 2. *P. pumila*

1. **Pilea fontana** (Lunell) Rydb. Brittonia 1:87. 1931. Fig. 309.
Adicea fontana Lunell, Am. Midl. Nat. 3:7. 1913.
Adicea opaca Lunell, Am. Midl. Nat. 3:8. 1913.
Pilea opaca (Lunell) Rydb. Brittonia 1:87. 1931.

Annual herbs from fibrous roots; stems pellucid, erect or decumbent, simple or branched, glabrous, to 50 cm long; leaves ovate, acute or acuminate at the apex, rounded or nearly so at the base, coarsely crenate, 3-nerved from the base, not translucent, glabrous or sparsely pubescent, to 10 cm long, the petiole to 4 cm long, glabrous; flowers unisexual, the staminate with a 4-lobed calyx, the pistillate with a 3-lobed calyx; achenes ovate, flat, black, averaging 1.5 mm long and broad, partly enclosed by the usually purplish, persistent calyx. July–September.

Moist soil.

IA, IL, IN, MO (FACW), KS, NE (OBL), KY, OH (FACW+).

Black-seeded clearweed.

Pilea opaca, another black-seeded plant, has been reported from the entire Midwest, but the characters of *P. opaca* are not distinct enough from those of *P. fontana* to merit recognition of *P. opaca*.

2. **Pilea pumila** (L.) Gray, Man. Bot. ed. 1, 437. 1848. Fig. 310.
Urtica pumila L. Sp. Pl. 984. 1753.
Adicea pumila (L.) Raf. ex Torr. Fl. N. Y. 2:223. 1843.
Adicea deamii Lunell, Am. Midl. Nat. 3:10. 1913.
Pilea pumila (L.) Gray var. *deamii* (Lunell) Fern. Rhodora 38:169. 1936.

Annual herbs from fibrous roots; stems pellucid, erect or decumbent, simple or branched, glabrous, to 70 cm long; leaves ovate, acute or acuminate at the apex, cuneate or rounded at the base, coarsely crenate, 3-nerved from the base, translucent, sparsely pubescent, to 15 cm long, the petiole to 5 cm long, glabrous; flowers unisexual, the staminate with a 4-lobed calyx, the pistillate with a 3-lobed calyx, the lobes lanceolate; achene ovate, flat, green, with purple speckles, averaging 1 mm wide and about as long, partly enclosed by the whitish or greenish persistent calyx. July–September.

Moist soil.

IA, IL, IN, KY, MO, OH (FACW), KS, NE (FAC).

Clearweed.

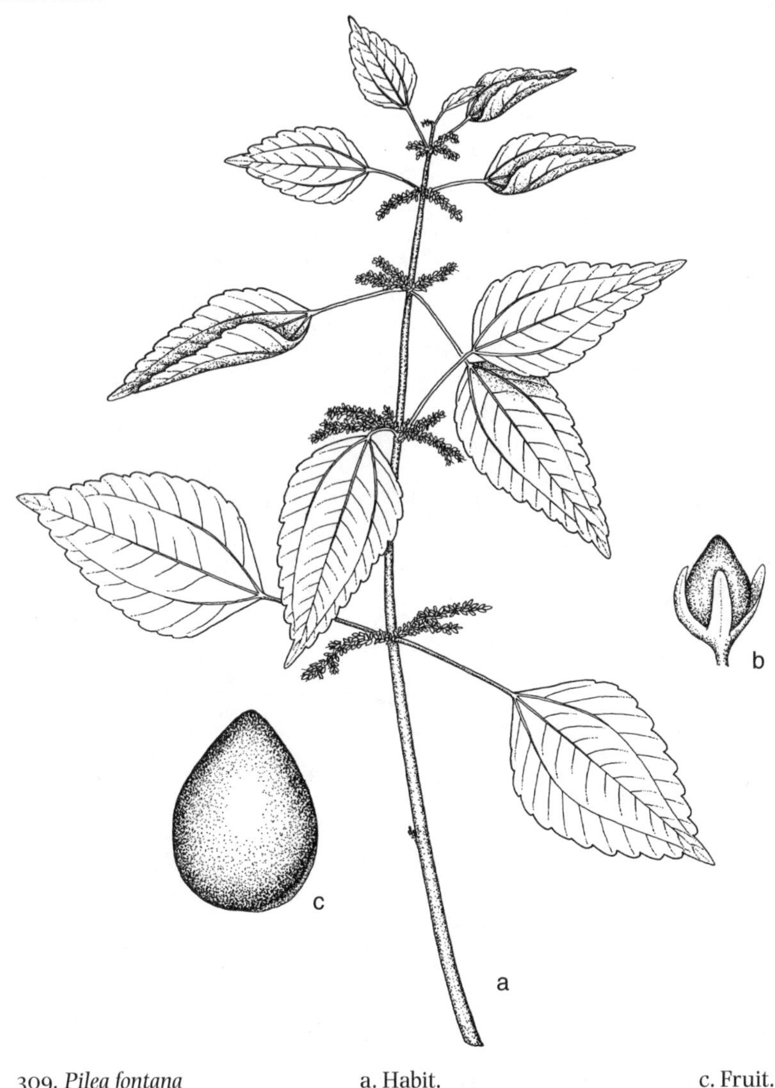

309. *Pilea fontana* (Black-seeded clearweed).
a. Habit.
b. Fruit, with persistent calyx.
c. Fruit.

Pilea pumila often grows in extensive numbers in low, moist, shaded soil. It is frequent also on rotted logs that have fallen in swamps.

The green seeds speckled with purple distinguish this species from *P. fontana*. Although var. *deamii* is sometimes recognized as distinct because of its rounded leaf bases and more numerous teeth on the leaves, it does not seem to me to be worthy of recognition.

The clear stem, to which the plant owes its common name, is an interesting phenomenon.

4. **Urtica** L.—Nettle

Annual or perennial herbs, often with stinging hairs; leaves opposite, simple, toothed, stipulate; flowers unisexual, monoecious or dioecious; staminate flowers

310. *Pilea pumila* (Clearweed).
a. Habit.
b. Fruit, with persistent calyx.
c. Fruit.

in racemes or spikes, the calyx deeply 4-lobed, the stamens 4, free; pistillate flowers in racemes or spikes, the calyx unequally 4-lobed; fruit an achene enclosed by the persistent calyx.

Urtica is a genus of about thirty species found throughout much of the World. The genus differs from other genera of Urticaceae by its stinging hairs and opposite leaves.

1. Leaves ovate, cordate at the base; plants to 75 cm tall .... 1. *U chamaedryoides*
1. Leaves lanceolate to ovate-lanceolate, rounded at the base; plants to 2 m tall .... 2. *U. dioica*

1. **Urtica chamaedryoides** Pursh, Fl. Am. Sept. 113. 1814. Fig. 311.

Annual herbs; stems ascending, simple or branched, sparsely bristly, to 75 cm tall; leaves ovate, obtuse to acute at the apex, mostly cordate at the base, coarsely crenate, mostly pubescent, 3- to 5-nerved from the base, to 5 cm long, to 3.5 cm wide, becoming abruptly smaller toward the apex, the petioles to 3 cm long, glabrous or puberulent, the stipules very narrow, to 5 mm long; flowers monoecious, androgynous, in nearly globose spikes from the axils and also terminal; calyx deeply 4-parted, glabrous or puberulent; stamens 4, free; achenes oblong-ovoid, sparsely pubescent, twice as long as the subpersistent calyx. April–June.

Swampy woods, river banks.

IL, KS, KY, MO, OH (FACU).

Nettle.

The flowers bloom from April to June.

The crenate and cordate leaves are distinctive for this species.

2. **Urtica dioica** L. Sp. Pl. 984. 1753. Fig. 312.
Urtica gracilis Ait. Hort. Kew. 3:341. 1789.
Urtica procera Muhl. ex Willd. Sp. Pl. 4:353. 1805.
Urtica dioica L. var. *procera* (Muhl.) Wedd. Mon. Fam. Urt. 78. 1856.

Perennial monoecious or dioecious herbs; stems erect, simple or branched, cinereous-pubescent to setose, sometimes glabrous or nearly so below, to 2 m tall; leaves lanceolate to ovate, acute to acuminate at the apex, cordate to rounded at the base, coarsely serrate, 3- to 5-nerved from the base, glabrous, pubescent, or setulose on both surfaces, to 15 cm long, up to half as wide, the petioles to 5 cm long, glabrous, pubescent, or setulose, the stipules lanceolate, usually pubescent, to 1.5 cm long; flowers monoecious or dioecious, in cymose-paniculate inflorescences; calyx deeply 4-lobed, pubescent; stamens 4, free; achene ovoid, pubescent, about twice as long as the subpersistent calyx. June–September.

Rich woods, moist disturbed areas.

IA, IL, IN, MO (FAC+), KS, NE (FACW), KY, OH (FACU).

Stinging nettle.

I am combining *U. gracilis* and *U. procera* with *U. dioica* because the differences listed by Fernald (1950) to distinguish these three entities seem very questionable.

Most of the central Midwest material had cinereous-pubescent stems and leaves, with only a small number of stinging bristles. Both monoecious and dioecious plants have been seen in the central Midwest.

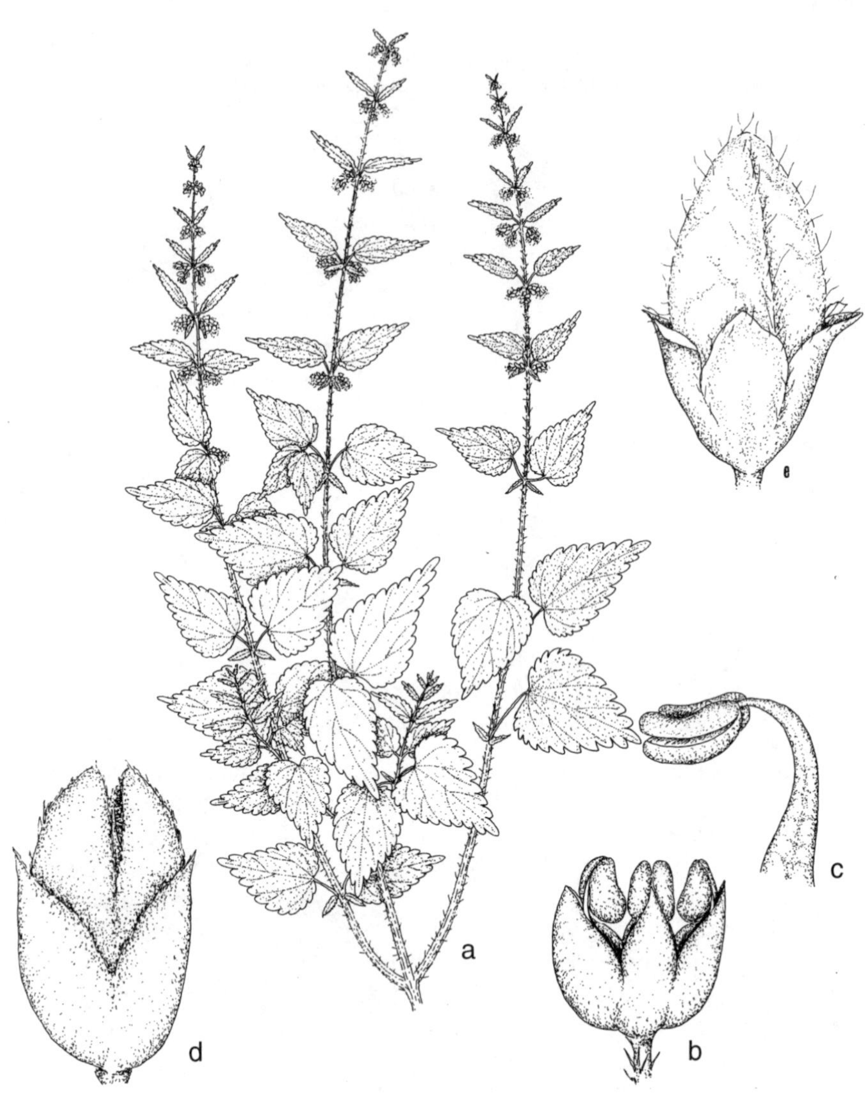

311. *Urtica chamaedryoides* (Nettle).

a. Leafy branches.
b. Staminate flower.
c. Stamen.
d. Pistillate flower.
e. Fruit.

312. *Urtica dioica* (Stinging nettle).

a. Flowering branch.
b. Leaf variation.
c. Staminate flower.
d. Pistillate flower.
e. Fruit.

119. VALERIANACEAE—VALERIAN FAMILY

Herbs; leaves opposite, simple or compound, without stipules; flowers perfect or unisexual, actinomorphic or zygomorphic; calyx absent or resembling pappus; corolla 5-lobed, usually tubular, often bilabiate, sometimes with a spur; stamens usually 3, attached to the corolla tube; ovary inferior, 3-locular; fruit dry and indehiscent.

Approximately thirteen genera and three hundred species comprise this family.

1. Cauline leaves, or some of them, pinnatifid or pinately compound; calyx divided into 5–15 setaceous lobes .... 1. *Valeriana*
1. Cauline leaves entire or toothed, not pinnately compound; calyx lobes reduced to short teeth or absent .... 2. *Valerianella*

1. **Valeriana** L.—Valerian

Perennial herbs, usually with thickened roots or rhizomes; leaves opposite, simple or pinnately compound, without stipules; flowers perfect or unisexual, only slightly zygomorphic; calyx divided into 5–15 setiform lobes; corolla 5-parted, united below, sometimes gibbous; stamens 3; ovary inferior; fruit an achene.

Nearly two hundred species, found worldwide, are in this genus. *Valeriana pauciflora*, a species of rich woods, also occurs in the central Midwest.

1. Leaves parallel-veined, densely ciliate; cauline leaves pinnately divided .... 1. *V. ciliata*
1. Leaves pinnately veined, glabrous; cauline leaves pinnatifid .... 2. *V. uliginosa*

1. **Valeriana ciliata** Torr. & Gray. Fl. N. Am. 2:49. 1841. Fig. 313.
Valeriana edulis Nutt. ssp. *ciliata* (Torr. & Gray) F.G. Mey. Ann. Mo. Bot. Gard. 38:428. 1951.
Valeriana edulis Nutt. var. *ciliata* (Torr. & Gray) Boivin, Nat. Can. 93:1062. 1966.

313. *Valeriana ciliata* (Valerian). Lower part of plant (left). Upper part of plant (right). Fruit (lower left). Flower (upper left).

Perennial herb from large, tuberous roots; stems erect, stout, short-hairy, to 75 cm tall; leaves thick, parallel-veined, densely ciliate, entire, the basal leaves linear to spatulate, obtuse to acute at the apex, tapering to the base, simple and usually entire to less commonly pinnately divided, to 30 cm long, to 5 cm wide, with a stout, winged petiole, the cauline leaves opposite, usually pinnately divided, sometimes undivided, smaller than the basal leaves; flowers in an elongated thyrse or panicle, crowded at first but becoming less crowded at maturity, usually unisexual; calyx reduced to 5–15 setiform bristles; corolla white, in staminate and perfect flowers, 2.5–3.5 mm long, in pistillate flowers 0.8–1.2 mm long; stamens 3; ovary inferior; achenes glabrous, ovoid to oblongoid, 3–4 mm long. May–June.

Wet prairies, calcareous fens.

IA, IL, IN (OBL), OH (FACW).

Valerian.

This rather robust plant is readily distinguished by its parallel-veined leaves, the basal ones usually undivided, the cauline ones usually pinnately divided. The tuberous roots are often large and deep in the ground.

The similar *V. edulis* from the Pacific coast has pubescent achenes and glabrous leaves and stems.

2. **Valeriana uliginosa** (Torr. & Gray) Rydb. ex Britt. Man. Fl. N. States 878. 1901. Fig. 314.

Valeriana sylvatica Sol. var. *uliginosa* Torr. & Gray. Fl. N. Am. 2:47. 1841.

Valeriana sitchensis Bong. ssp. *uliginosa* (Torr. & Gray) F.G. Mey. Ann. Mo. Bot. Gard. 38:399. 1951.

Valeriana septentrionalis Rydb. var. *uliginosa* (Torr. & Gray) Gl. Phytologia 4:25. 1952.

314. *Valeriana uliginosa* (Bog valerian). Habit. Fruit (upper left).

Perennial herb from thickened rhizomes and fibrous roots; stems erect, rather stout, glabrous or nearly so, to 1 m tall; leaves thin, pinnately veined, glabrous or nearly so, the basal leaves in rosettes, broadly elliptic to obovate, simple or pinnately divided, to 15 cm long, to 8 cm wide, on long petioles, the cauline leaves opposite, pinnatifid, to 15 cm long, to 10 cm wide, glabrous or nearly so except for marginal cilia; flowers in a corymb up to 15 cm across, perfect, bracteolate, the bractlets to 1 cm long, ciliate at first, becoming nearly glabrous; calyx reduced to 5–15 setiform bristles; corolla white, 5-lobed, 5–7 mm long; stamens 3; ovary inferior; achenes glabrous, lanceoloid to oblongoid, 3.5–4.5 mm long. June–July.

Calcareous fens, swamps, wet woods, bogs, wet meadows.

IL, IN, OH (OBL).

Bog valerian, marsh valerian.

This species, rare in the central Midwest, lacks the tuberous roots and parallel-veined leaves of *V. ciliata.*

Our plants are sometimes considered to be a variety of either *V. septentrionalis* or *V. sitchensis.*

2. **Valerianella** Mill.—Corn Salad

Annual or biennial herbs from fibrous roots; stems usually dichotomously branched; leaves opposite, simple, without stipules; flowers in cymes, actinomorphic, perfect; calyx absent or reduced to minute teeth; corolla 5-parted, united below; stamens 3; ovary inferior; fruit 3-locular, with only one locule bearing a seed.

Approximately sixty species are in this genus, all in the Northern Hemisphere. Most of the species are difficult to identify, usually requiring fruits to be present.

1. Corolla 1.5–2.0 mm long; bracts and bractlets spinulose-ciliate .... 4. *V. radiata*
1. Corolla 3–5 mm long; bracts and bractlets ciliate except sometimes at the tip.
    2. Fruits ovoid to ovoid-ellipsoid, longer than broad.
        3. Fruits 3–4 mm long, 2.5 mm broad or broader .... 1. *V. chenopodiifolia*
        3. Fruits up to 2.5 mm long, 0.7–1.2 mm broad... 2. *V. intermedia*
    2. Fruits orbicular to globose, as broad as long.
        4. Fruits flattened, 3.0–3.5 mm long, about 3 mm broad............................. 3. *V. paterllaria*
        4. Fruits not flattened, 2.0–2.5 mm long, about 2 mm broad .... 5. *V. umbilicata*

1. **Valerianella chenopodiifolia** (Pursh) DC. Prodr. 4:627. 1830. Fig. 315.
Fedia chenopodifolia Pursh, Fl. Am. Sept. 2:727. 1814.

Annual with fibrous roots; stems dichotomously branched, erect, angular, more or less pubescent, to 45 cm tall; basal leaves spatulate to obovate, obtuse at the apex, tapering or slightly rounded at the base, the cauline leaves opposite, sessile, lanceolate, acute at the apex, tapering to the base, glabrous, to 3 cm long, to 1.5 cm wide, usually serrulate; flowers in cymules, actinomorphic, perfect, subtended by acute, eciliate bracts; calyx obsolete; corolla 5-lobed, white, 3–5 mm long; stamens 3; ovary inferior; fruits 3–4 mm long, 2.5–2.8 mm wide. May–June.

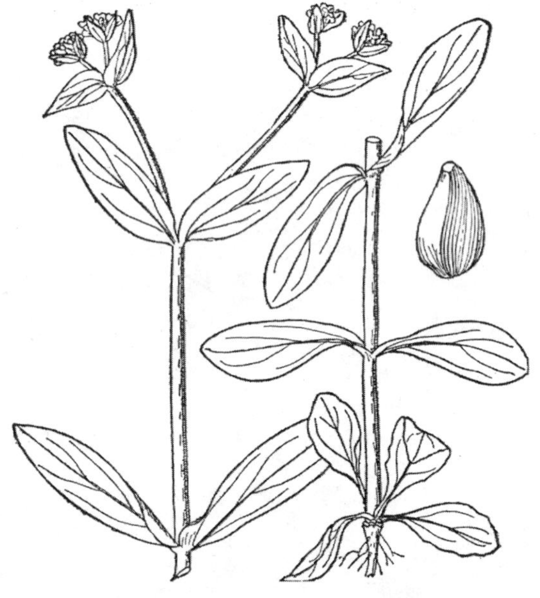

315. *Valerianella chenopodiifolia* (Great Lakes corn salad). Upper part of plant (left). Lower part of plant (right). Seed (right).

Wet meadows, floodplain forests.

IL, IN, KY, OH (not listed)

Great Lakes corn salad, goosefoot corn salad.

This species is distinguished by the combination of acute, eciliate bracts and fruits 3–4 mm long and 2.5–2.8 mm wide.

2. **Valerianella intermedia** Dyal, Rhodora 40:202–204. 1940. Not illustrated.
Valerianella radiata (L.) Dufr. var. *intermedia* (Dyer) Gl. Phytologia 4:25. 1952.

Annual with fibrous roots; stems dichotomously branched, erect, angular, usually pubescent on the angles, to 65 cm tall; basal leaves spatulate to oblong, obtuse at the apex, tapering or rounded at the base, the cauline leaves opposite, sessile, lanceolate, acute at the apex, tapering to the base, glabrous, to 3 cm long, to 1.5 cm wide, serrulate; flowers in cymules, actinomorphic, perfect, subtended by acute, eciliate bracts; calyx obsolete; corolla 5-lobed, white, 3–5 mm long; stamens 3; ovary inferior; fruits ellipsoid to oblongoid, 2.0–2.5 mm long, 1.0–1.2 mm wide. May–June.

Wet meadows, along streams.

IL, IN, KY, OH (considered a variation of *V. umbilicata*).

Intermediate corn salad.

This species is sometimes combined with *V. radiata* and sometimes with *V. patellaria*. It differs from both by its larger corollas and its very narrow fruits.

3. **Valerianella patellaria** (Sull. ex Gray) Porter, Am. Midl. Nat. 6:386. 1872. Not illustrated.
Fedia patellaria Sull. ex Gray, Man. Bot. N. U.S. 183. 1848.

Annual herb with fibrous roots; stems dichotomously branched, erect, usually pubescent on the angles, to 1 m tall; basal leaves spatulate to oblong, obtuse at the apex, tapering or rounded at the base, the cauline leaves opposite, sessile, lanceolate to oblong, obtuse to subacute at the apex, sessile, glabrous, to 3 cm long, to 1.5 cm wide, serrulate; flowers in cymules, actinomorphic, perfect, subtended by acute, eciliate bracts; calyx obsolete; corolla 5-lobed, white, 3–5 mm long; stamens 3; ovary inferior; fruits orbicular, flattened, 3.0–3.5 mm long, about 3 mm wide. May–June.

Wet meadows, moist soil.

IL, IN, OH (as a variety of *V. umbilicata*).

Corn salad.

This species differs from others in the genus by its flattened, orbicular fruits. It is often considered a variety of *V. umbilicata*.

4. **Valerianella radiata** (L.) Dufr. Hist. Val. 57. 1811. Fig. 316.
Fedia locusta L. var. *radiata* L. Sp. Pl. 1:34. 1753.
Valerianella radiata (L.) Dufr. var. *missouriensis* Dyal, Rhodora 40:206–207. 1938.
Valerianella radiata (L.) Dufr. var. *fernaldii* Dyal, Rhodora 40:207. 1938.

Annual herb with fibrous roots; stems dichotomously branched, erect, usually pubescent on the angles, to 65 cm tall; basal leaves oblong to spatulate, obtuse at the apex, usually rounded at the base, the cauline leaves opposite, simple, obtuse at the apex, rounded at the sessile base, glabrous or pubescent, entire or toothed only at the base; flowers in cymules, actinomorphic, perfect, subtended by lanceolate, acute, and sometimes spinulose-ciliate bracts; calyx obsolete; corolla 5-lobed, white, 1.5–2.0 mm long; stamens 3; ovary inferior; fruits flattened, ellipsoid to ovoid, 1.4–1.8 mm long, up to 1.2 mm wide, glabrous or puberulent. April–May.

Moist woods, wet meadows, old fields, roadsides.

IL, IN, KS, MO (FAC), KY, OH (FAC+).

Corn salad.

This species, with usually perfect, dichotomous branching, is the only *Valerianella* in central Midwestern wetlands with flowers only 1.5–2.0 mm long.

Minor variations occur in fruit size, and these have been given varietal names by Dyal, but they do not seem to me to be worthy of recognition.

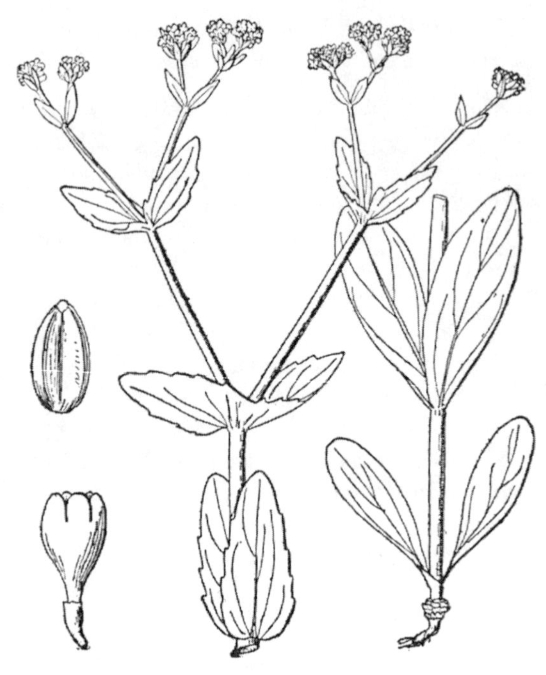

316. *Valerianella radiata* (Corn salad). Upper part of plant (center). Lower part of plant (right). Seed (center left). Flower (lower left).

5. **Valerianella umbilicata** (Sull.) Porter, Am. Midl. Nat. 6:387. 1872. Fig. 317.
Fedia umbilicata Sull. Am. Journ. Sci. 42:50. 1842.

Annual herb with fibrous roots; stems dichotomously branched, erect, usually puberulent on the angles, to 75 cm tall; basal leaves spatulate to oblong, obtuse at the apex, tapering or rounded at the base, the cauline leaves opposite, simple, lanceolate to oblong, obtuse to subacute at the apex, sessile, usually glabrous, to 3 cm long, to 1.5 cm wide, serrulate; flowers in cymules, actinomorphic, perfect, subtended by acute, eciliate bracts; calyx obsolete; corolla 5-lobed, white, 3–5 mm long; stamens 3; ovary inferior; fruits subglobose, not flattened, 2.0–2.5 mm in diameter. May–June.

Wet meadows.

Il, IN (FAC), KY, OH FACW).

317. *Valerianella umbilicata* (Corn salad). Upper part of plant (left). Lower part of plant (right). Section of flower (center left).

Corn salad.

This species differs from the very similar *V. patellaria* by its subglobose fruits 2.0–2.5 mm in diameter.

120. VERBENACEAE—VERBENA FAMILY

Herbaceous or woody plants; stems usually 4-angled; leaves opposite, simple or compound, without stipules; inflorescence various; calyx mostly 4- or 5-parted; corolla 4- or 5-lobed, united below, usually somewhat zygomorphic; stamens usually 4; ovary superior, often 4-lobed; fruit 4 nutlets or fleshy.

There are about one hundred genera in this family with approximately twenty-six hundred species. Many of them, particularly the woody ones, occur in the tropics.

Two genera occur in central Midwest wetlands.

1. Plants trailing; calyx strongly 2-lipped; calyx teeth 4; flowers in a dense, rounded head .... 1. *Phyla*
1. Plants erect; calyx nearly actinomorphic; calyx lobes 5; flowers in terminal spikes .... 2. *Verbena*

1. **Phyla** Lour.—Fog-fruit; frog-fruit

Trailing perennial herbs; leaves simple, opposite, toothed; flowers crowded into small heads or spikelike racemes on long peduncles, perfect; calyx bilabiate, 5-lobed, tubular below; corolla nearly actinomorphic, united below; stamens 4; ovary superior, 1-locular; fruit an achene.

Ten species native to North America are in the genus. Our species are sometimes placed in the genus *Lippia*.

1. Leaves acute to acuminate at the apex, with 5–11 teeth on each margin, some of the teeth extending below the middle of the leaf .... 2. *P. lanceolata*
1. Leaves more or less rounded at the apex, with 1–7 teeth on each margin, all the teeth above the middle.
    2. Leaves with 3–7 teeth on each margin .... 3. *P. nodiflora*
    2. Leaves with 1–2 teeth on each margin .... 1. *P. cuneifolia*

1. **Phyla cuneifolia** (Torr.) Greene, Pittonia 4:47. 1899. Fig. 318.
Zappania cuneifolia Torr. Ann. Lyc. Nat. Hist. N.Y. 2:234. 1827.

Trailing perennial herb, rooting at the nodes; stems 4-angled, glabrous or puberulent; leaves simple, opposite, very stiff, oblanceolate, rounded at the apex, strongly cuneate at the base, with 1–2 teeth on each side near the summit, glabrous or nearly so, to 2 cm long, to 8 mm wide; flowers many, crowded into heads, the heads 1–2 cm long, up to 1 cm thick, on long peduncles up to 5 cm long, each flower subtended by acute, cinereous bracts; calyx bilabiate; corolla white to purple, the four lobes 2.0–4.5 mm long, the tube 4–5 mm long; stamens 4; achenes oblongoid. May–September.

Wet areas.

IL, MO (NI), KS, NE (FAC).

Hoary fog-fruit; wedge-leaf fog-fruit.

This species, mostly in the Great Plains, is recognized by its stiff, narrow leaves that have only 1–2 teeth on each side near the summit.

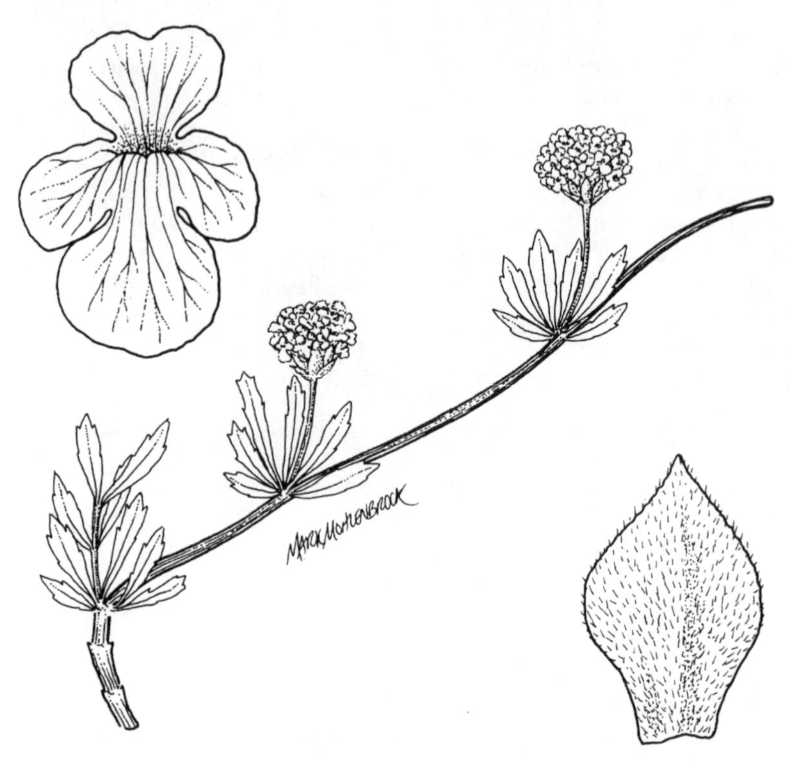

318. *Phyla cuneifolia* (Hoary fog-fruit). Habit (center). Petals (upper left). Bract (lower right).

2. **Phyla lanceolata** (Michx.) Greene, Pittonia 4:47. 1899. Fig. 319.
Lippia lanceolata Michx. Fl. Bor. Am. 2:15. 1803.

Trailing perennial herb, rooting at the nodes, with ascending flowering branches to 15 cm tall; stems 4-angled, glabrous or nearly so; leaves simple, opposite, not stiff, lanceolate to narrowly elliptic to ovate, acute to acuminate at the apex, tapering to the base, with 5–11 teeth on each margin that extend below the middle of the leaf, glabrous or nearly so, to 6 cm long, to 4 cm wide, short-petiolate; flowers crowded into heads on peduncles arising from the upper leaf axils, the heads 1–2 cm long, the peduncles slender, up to 9 cm long; each flower perfect, subtended by bracts 2.0–3.5 mm long and longer than the calyx; calyx bilabiate, 2–3 mm long; corolla pink to white, the four lobes up to 2 mm long, the tube 1.5–2.2 mm long; stamens 4; achenes oblongoid. May–September.

Wet ground, occasionally in standing water.

IA, IL, IN, KS, KY, MO, NE, OH (OBL).

Fog-fruit; frog-fruit.

This is the common species of *Phyla* in the central Midwest and the only one with acute to acuminate leaves that bear teeth to below the middle of the leaf. The flowers open in rings from the bottom of the head to the top so that the number of purple buds per head gets smaller and smaller during the flowering season.

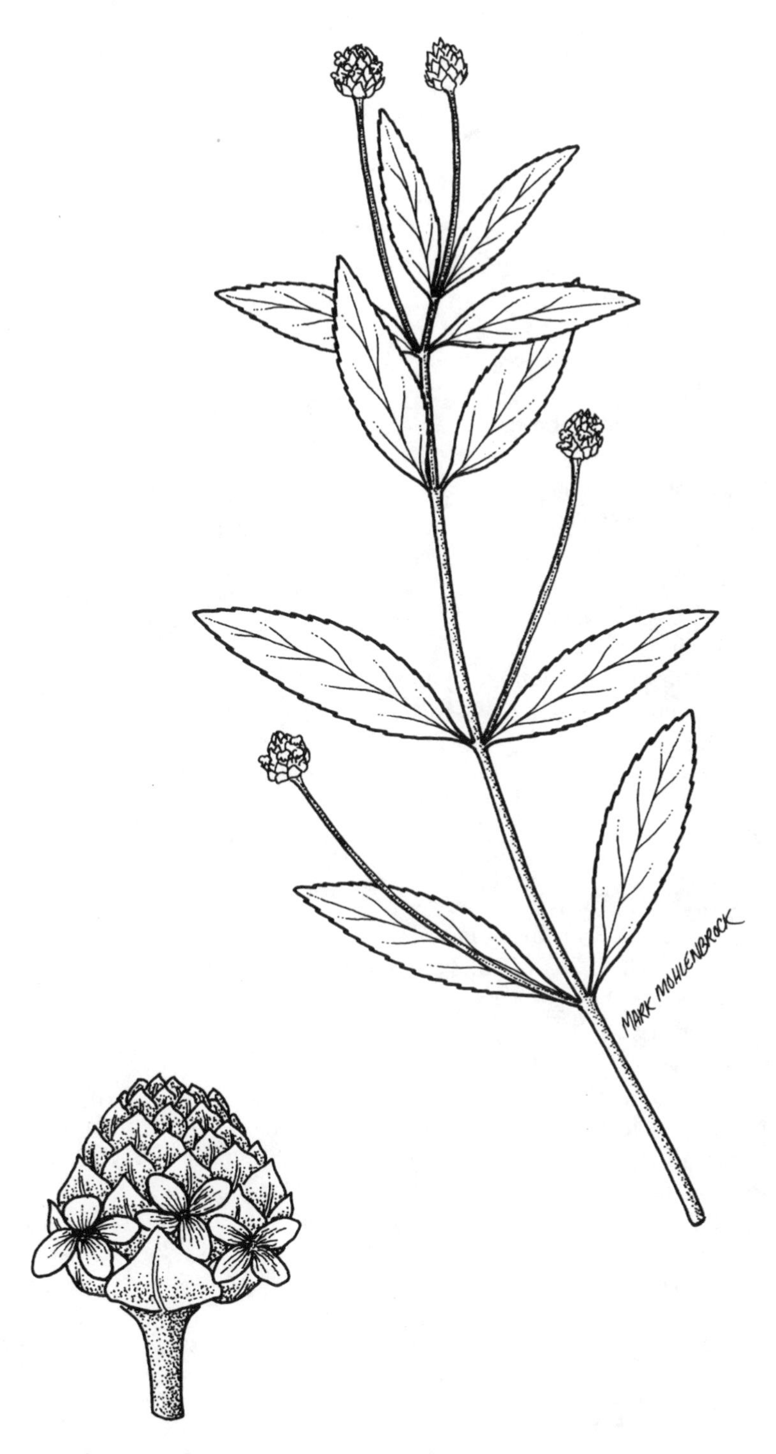

319. *Phyla lanceolata* (Fog-fruit). Habit (right). Flowering head (lower left).

3. **Phyla nodiflora** (L.) Greene, Pittonia 4:46. 1899. Fig. 320.
Verbena nodiflora L. Sp. Pl. 1:20. 1753.
Lippia nodiflora (L.) Michx. Fl. Bor. Am. 2:15. 1803.

Trailing perennial herb, rooting at the nodes, with ascending flowering branches to 15 cm tall; stems 4-angled, glabrous or nearly so; leaves simple, opposite, spatulate to obovate, rounded at the apex, tapering to the base, with 3–7 teeth on either margin above the middle of the leaf, glabrous or nearly so, to 4 cm long, to 2 cm wide, sessile or very short-petiolate; flowers crowded into cylindric spikes on peduncles arising from the upper leaf axils, the spikes up to 2.5 cm long, the peduncles up to 5 cm long; calyx bilabiate, 2–3 mm long; corolla pink to white, the four lobes up to 2 mm long, the tube 1.5–2.0 mm long; stamens 4; achenes oblongoid. June–October.

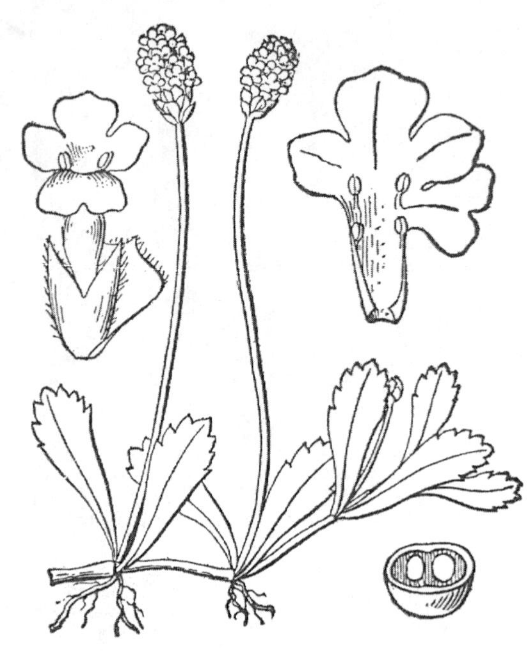

320. *Phyla nodiflora* (Nodding fog-fruit). Habit (center). Flower (upper left). Flower, cut open (upper right). Cross-section of ovary (lower right).

Wet ground.

KS (NI), KY, MO (FACW).

Nodding fog-fruit; nodding frog-fruit.

This species, which occurs in the very southern part of the central Midwest, has elongated flowering spikes and round-tipped leaves with 3–7 teeth on each margin above the middle.

2. **Verbena** L.—Vervain

Mostly perennial herbs; stems 4-angled; leaves opposite, serrate or deeply divided; inflorescence spicate; flowers perfect, slightly zygomorphic; calyx 5-lobed, tubular, persistent on the fruit; corolla 5-lobed, the lobes slightly unequal, united below into a tube; stamens 4, included; ovary superior, 4-lobed; fruit a group of 4 nutlets.

Nearly 250 species comprise this genus. Most of them are in the New World.

Only the following occur in wetlands in the central Midwest.

1. Some or all of the flowers overlapping, the calyx of one reaching beyond the base of the calyx of the flower above, blue or purple .... 1. *V. hastata*
1. None or few of the flowers overlapping, the calyx of one at most reaching only to the base of the calyx of the flower above, white .... 2. *V. urticifolia*

1. **Verbena hastata** L. Sp. Pl. 1:20. 1753. Fig. 321.

Perennial herbs with rhizomes; stems erect, 4-angled, with short spreading to reflexed hairs, to 1.5 m tall, unbranched below, branched above, often purple; leaves simple, opposite, rugose, lanceolate to narrowly ovate, acuminate at the apex, rounded or tapering to the base, coarsely serrate and occasionally hastately lobed near the base, harshly pubescent, to 18 cm long, to 8 cm wide, petiolate; flowers crowded into 3–15 very slender, terminal, erect spikes, most of the flowers overlapping, perfect, slightly zygomorphic, up to 4.5 mm long, subtended by narrow bracts; calyx 5-lobed, the lobes curving inward over the fruit; corolla 5-lobed, purple or blue, the lobes 2.5–4.5 mm long, the tube 2–4 mm long; stamens 4, included; ovary shallowly 4-lobed; nutlets 4 in a cluster, linear, 1.5–2.0 mm long, smooth or faintly striate. June–October.

321. *Verbena hastata* (Blue vervain). Habit (left). Leaf variations (right).

Wet soil, in prairies, woods, marshes, meadows, swamps.

IA, IL, IN, MO, KY, OH (FACW+), KS, NE (FACW).

Blue vervain; blue verbena; purple vervain; purple verbena.

This is the common vervain in wetlands in the central Midwest. It may grow in shallow water, particularly in marshes. It is distinguished by its very slender, terminal spikes bearing small, blue or purple, crowded flowers, its coarsely toothed, rugose, opposite leaves, and its usually purple stems.

2. **Verbena urticifolia** L. Sp. Pl. 1:20. 1753. Fig. 322.
Verbena urticifolia L. var. *leiocarpa* L. M. Perry & Fern. Rhodora 38:441–442. 1936.

Perennial herb with rhizomes; stems erect, 4-angled, the stems pubescent, often with stiff hairs, to 1.75 m tall, unbranched below, branches above, green; leaves simple, opposite, rugose, ovate-lanceolate to ovate, acuminate at the apex, tapering to the base, coarsely serrate, occasionally hastately lobed near the base, harshly pubescent or sometimes velutinous on the lower surface, to 20 cm long, to 10 cm wide, petiolate; flowers strongly interrupted in terminal slender spikes to 15 cm long, the flowers perfect, slightly zygomorphic, up to 4 mm long, subtended by narrow bracts; calyx 5-lobed, pubescent, the lobes more or less equal; corolla 5-lobed, white, the lobes 2–4 mm long; stamens 4; ovary shallowly 4-lobed; nutlets 4 in a cluster, linear, 1.5–2.2 mm long, glabrous or corrugated. June–October.

322. *Verbena urticifolia* (White verbena). Stem and leaves (center). Inflorescence (left). Flowers (upper right). Capsule (lower right). Seed (next to lower right).

Wet or dry fields, thickets, woods.

IA, IL, IN, MO (FAC+), KS, NE (UPL), KY, OH (FACU).

White verbena; white vervain.

This species is recognized by its slender spikes with strongly interrupted white flowers. Plants with leaves velutinous on the lower surface and with smooth nutlets only 1.5–1.7 mm long may be called var. *leiocarpa*.

Hybrids between this species and *V. hastata* are called *V. engelmanniana* Moldenke.

121. VIOLACEAE—VIOLET FAMILY

Herbs; leaves alternate or basal, stipulate; flowers axillary, perfectly irregular; sepals 5, persistent; petals 5, the lower one usually spurred; stamens 5, the anthers connivent or connate; ovary superior, 1-locular, with 3 parietal placentae; valves of capsule folding lengthwise at maturity, expelling the seeds.

There are about fifteen genera and over eight hundred species in the Violaceae. There are representatives of this family in most parts of the world, with the greatest concentration of genera being in South America. Although most species are herbaceous, the genera *Rinorea, Decorsella, Corynostylis, Hybanthus*, and *Paypayrola* contain small tree forms.

1. **Viola** L.—Violet

Herbaceous annuals or perennials, with leafy stems or acaulescent; leaves basal or cauline, simple, toothed or lobed, stipulate; flowers usually of two types; petaliferous flowers with 5 auriculate sepals, 5 petals somewhat unequal, the lowermost spurred, 5 stamens somewhat coherent but not connate, the two lowest ones with appendages projecting into the spurred petal, a superior ovary, 1-locular; cleistogamous flowers with 5 small sepals, 5 minute greenish petals, 5 stamens (only 2 or 3 producing pollen), and a single 1-locular ovary; capsule dehiscent by means of 3 valves, several-seeded.

The genus *Viola* is composed of about five hundred species, most of them occurring in north temperate regions of the world, although a great number live in the Andes Mountains of South America.

The cleistogamous flowers, found in most of our species, are primarily responsible for seed production in the genus. These flowers, which are produced at the end of short horizontal stalks, are usually concealed beneath the soil and the leaf mold until the seeds have ripened. Then the capsules are lifted into the air so that the seeds may be expelled. The cleistogamous flowers are produced later in the season, after the showy petaliferous flowers have terminated blooming. These flowers never open and actually look like buds.

The petaliferous flowers, which bloom early in the season, rarely set seed. Although the arrangement of the pollen-bearing anthers in a ring around the pistil seems ideal for the process of pollination, there are various mechanisms in the flower that prevent pollination from taking place.

Ripened capsules split into three boat-shaped valves, each containing several seeds. As the thin sides of the valves become dry, they contract, eventually expelling the seeds forcefully to a distance reportedly up to nine feet.

1. Plants acaulescent, the leaves and peduncles seeming to arise from out of the ground.
  2. Flowers basically blue or white with deep purple veins.
    3. Some of the leaves hastately or palmately lobed .... 12. *V. sagittata*
    3. None of the leaves lobed.
      4. Leaves less than ½ as long as broad .... 12. *V. sagittata*
      4. Leaves more than ½ as long as broad.
        5. Spurred petal beardless within or nearly so.
          6. Petioles and leaves uniformly and densely villous .... 13. *V. sororia*
          6. Petioles and leaves glabrous or only sporadically and sparsely pubescent.

7. Lateral petals with hairs swollen at the tip; leaf blades somewhat pubescent on the upper surface .... 5. *V. cucullata*
7. Lateral petals with hairs slender throughout; leaf blades usually glabrous.
8. Leaves elongate-triangular, tapering to an elongated tip ... 8. *V. missouriensis*
8. Leaves cordate-ovate, rounded to short-pointed at the tip .... 10. *V. pratincola*
5. Spurred petal bearded within.
9. Sepals obtuse; peduncles longer than petioles .... 9. *V. nephrophylla*
9. Sepals acute; peduncles about equaling petioles .... 1. *V. affinis*
2. Flowers basically white.
10. Lateral petals bearded .... 2. *V. blanda*
10. Lateral petals beardless.
11. Leaves cordate at base .... 7. *V. macloskeyi*
11. Leaves tapering to base.
12. Leaves linear, lanceolate, or oblanceolate .... 6. *V. lanceolata*
12. Leaves oblong to ovate .... 11. *V. primulifolia*
1. Plants with stems bearing alternate leaves and axillary flowers.
13. Flowers blue .... 4. *V. conspersa*
13. Flowers white.
14. Stipules entire .... 3. *V. canadensis*
14. Stipules toothed .... 14. *V. striata*

1. **Viola affinis** LeConte, Ann. Lyc. N. Y. 2:138. 1826. Fig. 323.
Viola illinoensis Greene, Pittonia 4:293. 1901.

Perennial from stout rhizomes; leaves ascending, on long, usually glabrous, petioles; blades glabrous or with short stiff hairs on upper surface of basal lobes, ovate to lance-ovate, acute at the apex, cordate at the base, crenate-serrate, to 12 cm long, to 6 cm wide (sometimes even larger after anthesis); petaliferous flowers blue or purple, sometimes white at base, on glabrous peduncles rarely longer than the petioles; sepals lanceolate, acute, green, occasionally with a hyaline margin, ciliate; lateral petals bearded; spurred petal bearded; cleistogamous flowers ovoid, borne on prostrate or arched-ascending peduncles; capsules ellipsoid, 5–9 mm long, glabrous or puberulent, green with occasional purple speckles; seeds pale or dark brown, to 2 mm long. April–June.

Moist woods; wet woods.

IL, IN, KY, MO, OH (FACW).

Blue violet.

This blue violet is an inhabitant of moist to wet woodlands. It resembles both *V. pratincola* (*V. papilionacea*) and *V. missouriensis*, but these latter violets do not have a bearded spurred petal.

Russell (1965) uses as important diagnostic characters for *V. affinis* the presence of stiff hairs on the upper surface of the basal lobes of the leaves and the prostrate peduncles of the cleistogamous flowers. Although these characters seem to hold up on most specimens of this species, some material, otherwise referable to *V. affinis*, lack this leaf pubescence, while still other specimens have arched-ascending peduncles of the cleistogamous flowers.

Near the turn of the century, E. L. Greene found and named a violet that he collected along the Sangamon River near Monticello, Illinois as *V. illinoensis*. I have examined the type and find it identical to *V. affinis*.

323. *Viola affinis* (Blue violet).

a. Habit.
b. Capsule.
c. Seed.

2. **Viola blanda** Willd. Hort. Berol. pl. 24. 1806. Fig. 324.
Viola incognita Brainerd, Rhodora 7:84. 1905.
Viola incognita Brainerd var. *forbesii* Brainerd, Bull. Torrey Club 38:8. 1918.

Perennial from very slender rootstocks and vigorous stolons; leaves ascending, on villous petioles; blades softly hairy when young, the upper surface becoming nearly glabrous at maturity, ovate to reniform, abruptly short-pointed at the apex, cordate at the base, crenate, rugose, to 12 cm long, to 10 cm wide; petaliferous flowers white, with purple veins, borne on pubescent peduncles rarely longer than the petioles; sepals ovate-lanceolate, usually glabrous; lateral petals bearded; spurred petal beardless; cleistogamous flowers borne on prostrate or low-arching, usually pubescent peduncles; capsules ovoid, to 10 mm long, glabrous, usually purplish; seeds obovoid, brown to olivaceous, to 2 mm long. May–June.

Woods, sometimes wet; swamps; bogs.

IN (FACW−), KY, OH (FACW).

Large-leaved white violet.

324. *Viola blanda* (Large-leaved white violet).

a. Habit.
b. Leaves and flower.

The large-leaved white violet differs from other white-flowered stemless violets by its bearded lateral petals and its cordate leaves. The leaves and petioles always bear some pubescence. During the summer, this species produces vigorous vegetative stolons, which sometimes densely carpet a wooded hillside.

This plant is sometimes called *V. incognita*, but I believe *V. incognita* and *V. blanda* should be considered the same.

Typical *V. blanda* is said to have both surfaces of the leaf pubescent, while var. *forbesii* is pubescent only on the upper surface. There seems to be no geographical or ecological correlation to this.

3. **Viola canadensis** L. var. **rugulosa** (Greene) C. L. Hitchcock, Vasc. Pl. Pac. N.W. 3:442. 1961. Fig. 325.
Viola rugulosa Greene, Pittonia 5:26. 1902.

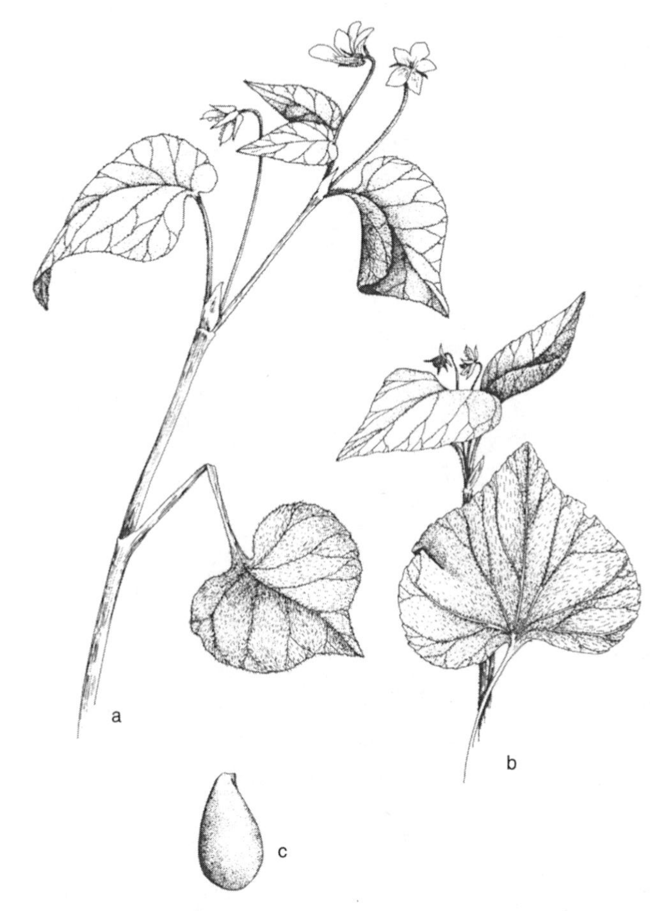

325. *Viola canadensis* var. *rugulosa* (Wrinkled violet). a, b. Habit. c. Seed.

Perennial from long, cordlike rhizomes; stems rather stout, glabrous or pubescent, to 60 cm tall; basal leaves ascending, on usually pubescent petioles, the blades pubescent, ovate to reniform, acute at the apex, cordate, crenate, to 10 cm long and wide; cauline leaves similar but mostly ovate and progressively smaller toward the apex; stipules lanceolate, usually pubescent, entire; petaliferous flowers axillary, white, the outside usually tinged with violet; sepals subulate, glabrous; lateral petals bearded; spurred petal beardless, yellow at the base, with purple veins; cleistogamous flowers borne later from axillary peduncles; capsules ovoid to subglobose, to 12 mm long, usually pubescent; seeds brown, to 2.2 mm long. April–September.

Wet woods.

IA, IL, NE (not listed by the U.S. Fish and Wildlife Service).

Wrinkled violet.

Some botanists maintain this taxon as a distinct species, but I am following Russell (1965) and others in considering it a variety of the more eastern *V. canadensis.*

4. **Viola conspersa** Reichenb. Plantae Criticae 1:44, pl. 52, fig. 108. 1823. Fig. 326.
Viola muhlenbergii Torr. Fl. N. & Mid. U. S. 1:256. 1824.
Viola canina L. var. *muhlenbergii* (Torr.) Trautv. Act. Hort. Petrop. 5:28. 1877.

Perennial herb from branched rhizomes; stems several, tufted, ascending, to 20 cm tall, glabrous; basal leaves nearly orbicular, obtuse at the apex, cordate at the

326. *Viola conspersa* (American dog violet). a. Habit. b. Capsule.

base, glabrous or nearly so, crenate-serrate, up to about 3 cm in diameter, the upper cauline leaves similar but tending to the somewhat smaller, more reniform to ovate, subacute at the apex; stipules lance-ovate; petaliferous flowers axillary, pale violet, on glabrous peduncles to 8 cm long, the peduncles at length usually overtopping the leaves; sepals lanceolate, glabrous; lateral petals bearded; spur 4–7 mm long; style bent downward, puberulent; cleistogamous flowers borne later in the same axils or in axils of later leaves; capsules ellipsoid, 4–5 mm long; seeds light brown, narrowly obovoid, to 2 mm long, about one-half as thick. May–June.

Wet woods; perched swamps on moraines.

IL, IN, KY, OH (FAC).

American dog violet.

The American dog violet belongs to that group of violets with leafy stems and pale violet flowers. No other violet in the central Midwest has this particular combination of characters. The similar but cream-flowered *V. striata* is further distinguished by its ciliate sepals.

5. **Viola cucullata** Ait. Hort. Kew. 3:228. 1789. Fig. 327.
Viola palmata L. var. *cucullata* (Ait.) Gray, Bot. Gaz. 11:254. 1886.

Perennial herb from stout rhizomes; leaves ascending, on long, usually glabrous, petioles; blades glabrous or nearly so, ovate to nearly reniform, acute or subacute at the apex, cordate at the base, crenate-serrate, to 10 cm long, to 6 cm wide (sometimes even larger after anthesis); petaliferous flowers blue or purple, borne on glabrous peduncles longer than the petioles; sepals lanceolate, acute, green, ciliate, with auricles to 4 mm long; lateral petals bearded with knob-tipped hairs; spurred petal beardless; cleistogamous flowers sagittate-lanceolate, borne on long, erect or ascending, peduncles; capsules ovoid-cylindric, to 15 mm long, glabrous, green, seeds dark brown to black, to 1.5 mm long. April–June.

Bogs, marshes, wet woods.

IA, IL, IN, MO (OBL), KY, OH (FACW+).

Marsh blue violet.

The diagnostic feature most frequently used to distinguish this species is the presence of knob-tipped hairs on the lateral petals. However, it is not unusual to find flowers of *V. pratincola* or *V. sororia* with hairs at least swollen in the upper half.

Perhaps an equally good character to distinguish petaliferous forms of *V. cucullata* is that the spurred petal is shorter than the lateral petals, rather than being subequal as in *V. pratincola* and *V. sororia.*

In general, the peduncles of the cleistogamous flowers in *V. cucullata* are erect or nearly so, while those of *V. pratincola* and *V. sororia* are generally prostrate or low-arching. The cleistogamous flowers are long and narrow in *V. cucullata*, rather than ovoid in the other similar species.

6. **Viola lanceolata** L. Sp. Pl. 934. 1753. Fig. 328.
Viola vittata Greene, Pittonia 3:258. 1898.
Viola lanceolata L. var. *vittata* (Greene) Weatherby & Grisc. Rhodora 36:48, 1934.
Viola lanceolata L. ssp. *vittata* (Greene) Russell, Am. Midl. Nat. 54:484. 1955.

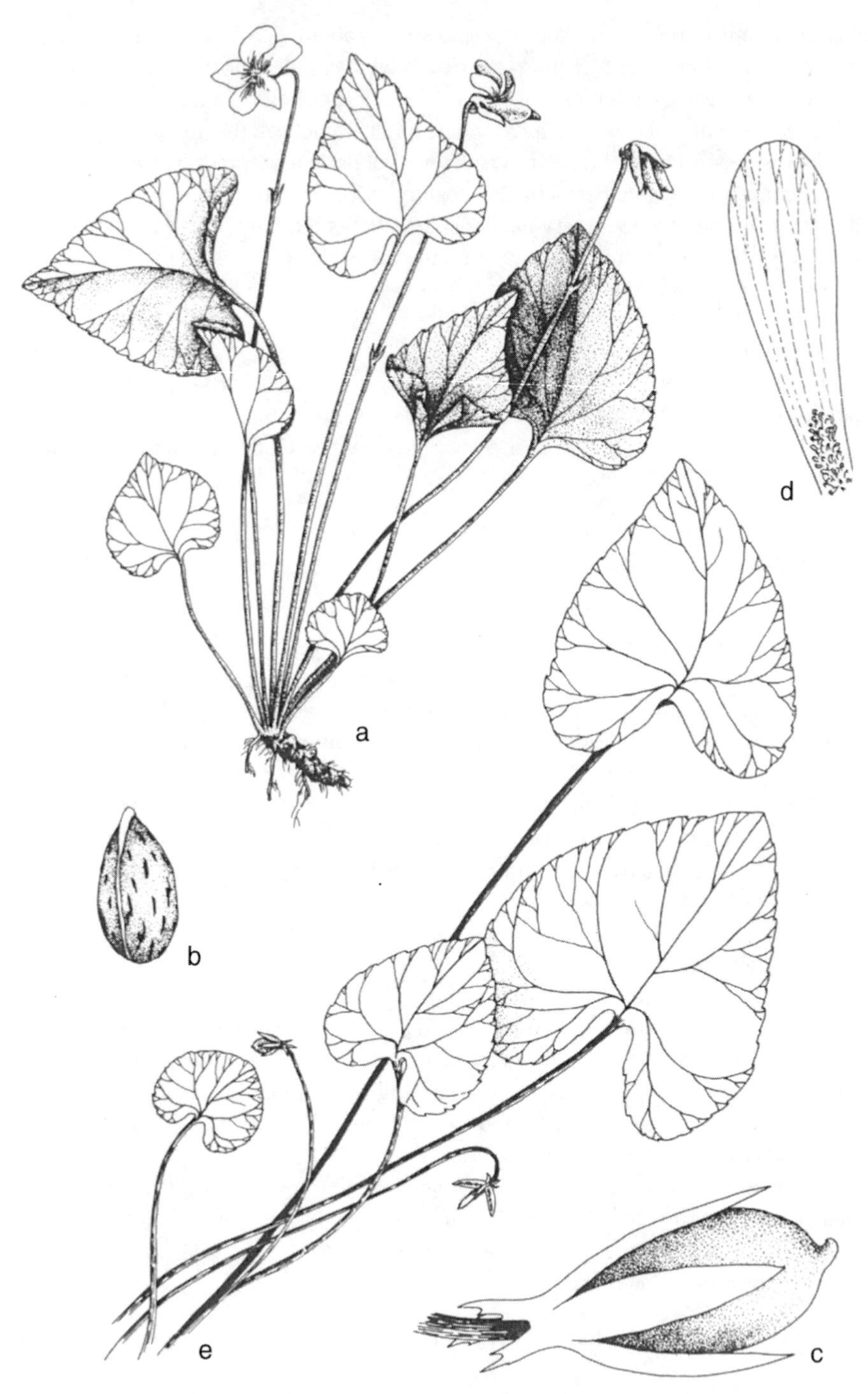

327. *Viola cucullata* (Marsh blue violet).

a. Habit.
b. Seed.
c. Capsule.
d. Petal.
e. Leaves, with fruits.

Perennial herb from slender rootstocks and numerous stolons; leaves ascending, on glabrous, often reddish, petioles; blades glabrous or sometimes sparsely pubescent beneath, linear-lanceolate to lanceolate to oblanceolate, obtuse to subacute at the apex, cuneate at the base, crenulate, to 12 cm long, to 2.5 cm wide; petaliferous flowers white, with purple veins, borne on glabrous peduncles usually longer than the petioles; sepals lanceolate, glabrous; lateral petals beardless; spurred petal beardless; cleistogamous flowers borne on erect, glabrous peduncles; capsules ellipsoid, to 12 mm long, glabrous, green, seeds obovoid, brown to olivaceous, to 1.5 mm long. May–June.

328. *Viola lanceolata* (Lance-leaved violet). a. Habit. b. Flower. c. Leaves and capsules. d, f. Habit. var. *vittata.* e. Flower. var. *vittata.*

Wet meadows; wet sands.
IA, IL, IN, KY, MO, OH (OBL), KS, NE (not listed).
Lance-leaved violet.
Typical var. *lanceolata* has leaves that are three to five times longer than broad, while var. *vittata* has leaves six to fifteen times longer than broad.

7. **Viola macloskeyi** Lloyd ssp. **pallens** (Banks) M. S. Baker, Madroño 12:60. 1953. Fig. 329.
Viola rotundifolia Michx. var. *pallens* Banks ex DC. Prodr. 1:295. 1824.
Viola pallens (Banks) Brainerd, Rhodora 7:247. 1905.

Small perennial herb from slender rootstocks and filiform stolons; leaves ascending, on pubescent petioles; blades glabrous, ovate to reniform, obtuse at the apex, cordate at the base, to 8 cm long, to 8 cm wide; petaliferous flowers white, with purple veins, fragrant, borne on usually pubescent peduncles longer than the petioles; sepals linear-lanceolate, glabrous; lateral petals sparsely pubescent; spurred petal beardless; cleistogamous flowers borne on ascending, usually pubescent, peduncles; capsules ellipsoid, to 12 mm long, glabrous, green; seeds black, to 1.5 mm long. April–May.
Bogs, marshes.
IA, IL, IN, KY, MO, OH (OBL).
Northern white violet.
This tiny violet differs from other violets in the central Midwest by its filiform stolons, white flowers, and sparsely pubescent lateral petals.
The similar *V. blanda*, which has pubescence on the upper surface of the leaf, is sometimes confused with *V. macloskeyi.*

8. **Viola missouriensis** Greene, Pittonia 4:141. 1900. Fig. 330.

Perennial herb from stout rhizomes; leaves ascending, on long, usually glabrous, petioles; blades glabrous or nearly so, ovate-deltoid, tapering straight from the base to the attenuate apex, or the margins sometimes slightly concave, cordate at the base, coarsely crenate-serrate, with the serrations relatively few along the upper one-third of the margins, to 12 cm long, to 6 cm wide; petaliferous flowers often pale violet with a darkened area surrounding a whitish center, sometimes uniformly violet throughout, on glabrous peduncles shorter or longer than the petioles; sepals lance-oblong, obtuse to subacute, green with narrow whitened margins, somewhat ciliate, the auricles up to 2 mm long; lateral petals bearded; spurred petal beardless; cleistogamous flowers ovoid, borne on prostrate or low-arching peduncles; capsules ellipsoid, 9–13 mm long, glabrous, green and occasionally speckled with brown; seeds light brown, to 2 mm long. April–May.
Wet woods, bottomland forests.
IA, IL, IN, MO (FACW), KS, NE (FAC), KY, OH (NI).
Missouri violet.
This glabrous blue-flowered violet may usually be distinguished by the deltoid leaf shape and the few serrations along the upper margins of the leaves.
Viola affinis, which may have a similar leaf-shape and glabrous blades, has a bearded spurred petal. *Viola pratincola* and nearly glabrous forms of *V. sororia* do not

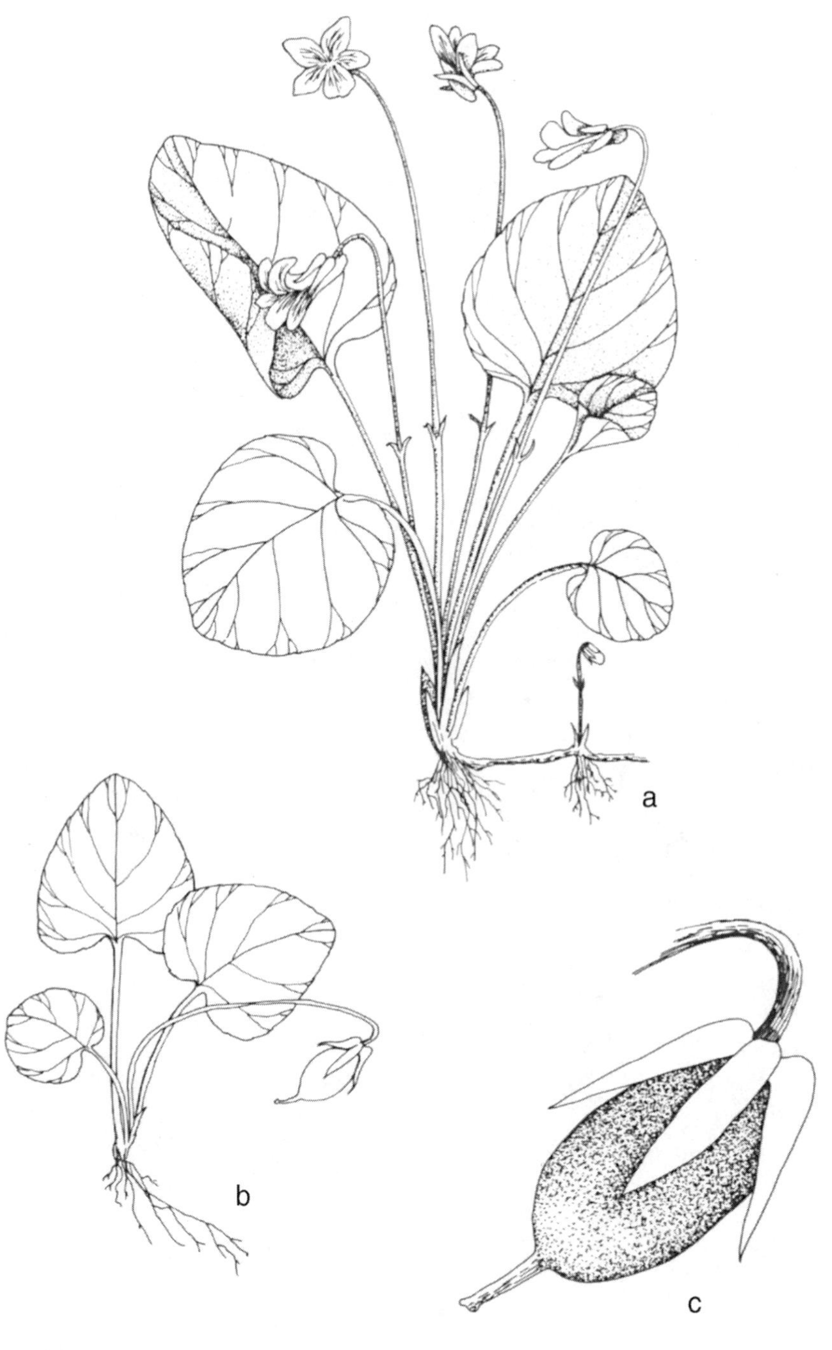

329. *Viola macloskeyi* ssp. *pallens* (Northern white violet).

a, b. Habit.
c. Capsule.

have tapering leaves. In addition, the margins of the leaves of these latter species are regularly serrate to the apex.

Most petaliferous flowers of *Viola missouriensis* are pale violet with a darkened band around a whitish center.

Viola missouriensis generally grows in lower, wetter woods than do *V. pratincola* and *V. sororia*.

330. *Viola missouriensis* (Missousri violet). a. Habit, in flower. b. Habit, in fruit. c. Leaves.

9. **Viola nephrophylla** Greene, Pittonia 3:144. 1896. Fig. 331.

Perennial herb from stout rhizomes; leaves ascending, on long, usually glabrous, petioles; blades glabrous or minutely pubescent on the upper surfaces of the basal lobes, the earliest nearly orbicular and purplish beneath, the later leaves broadly ovate to nearly reniform, subacute to rounded at the apex, cordate at the base, crenate-serrate, to 8 cm long, to 6 cm wide; petaliferous flowers violet, on glabrous peduncles as long as or longer than the petioles; sepals ovate to lanceolate, obtuse, green, glabrous; lateral petals bearded; spurred petal pubescent; cleistogamous flowers ovoid, on erect or arching peduncles; capsule ellipsoid, to 10 mm long, glabrous, green; seeds olive-brown, to 2 mm long. May–June.

331. *Viola nephrophylla* (Northern bog violet). a. Habit. b. Flower. c. Capsule. d. Seed.

Fens, calcareous marshes.

IA, IL, IN, MO (FACW+), KS, NE, OH (FACW).

Northern bog violet.

Although *V. nephrophylla* resembles *V. pratincola* (=*V. papilionacea* in part), it differs by its more rounded leaves and by the pubescence of the spurred petal. *Viola affinis*, which is similar and which also has a bearded spurred petal, has a more elongated leaf.

The earliest leaves of *V. nephrophylla* are usually purplish on the undersurface.

10. **Viola pratincola** Greene, Pittonia 4:64. 1899. Fig. 332.

Perennial herb from stout rhizomes; leaves ascending, on long, glabrous petioles; blades glabrous throughout, ovate to nearly reniform, subacute to acute at the apex, cordate at the base, crenate-serrate, to 12 cm long, to 12 cm wide; petaliferous flowers violet, on glabrous peduncles usually longer than the petioles; sepals lanceolate, green, glabrous; lateral petals bearded; spurred petal beardless; cleistogamous flowers ovoid, borne on prostrate to low-arching peduncles; capsules ellipsoid, to 15 mm long, green, glabrous; seeds dark brown, to 2 mm long. April–May.

332. *Viola pratincola* (Smooth blue violet). a. Habit. b. Immature capsule.

Woods, sometimes in floodplains.

IA, IL, IN, KS, MO, NE (FAC), KY, OH (NI).

Smooth blue violet.

All the specimens referred to *V. pratincola* have been identified previously as *V. papilionacea*. Russell (1965) discusses the authenticity of Pursh's *V. papilionacea*, concluding that there is no such species. He further postulates that Pursh's binomial actually should be referred to one of the less pubescent races of *V. sororia*.

Viola pratincola is characterized as a completely glabrous, stemless blue violet with a beardless spurred petal and unlobed leaves. *Viola sororia* (including nearly glabrous forms previously referred to as *V. papilionacea*) shows some pubescence on the blades and/or the petioles.

11. **Viola primulifolia** L. Sp. Pl. 934. 1753. Fig. 333.

Perennial herb from slender rootstocks and filiform stolons; leaves ascending, on glabrous or pubescent petioles; blades glabrous or villous on one or both sides, oblong to ovate, subacute at the apex, subcordate, truncate, or somewhat cuneate at the base, crenulate, to 12 cm long, to 8 cm wide; petaliferous flowers white, with purple veins, borne on glabrous or pubescent peduncles longer than the petioles; sepals lanceolate, glabrous; lateral petals sparsely pubescent or glabrous; spurred petal glabrous; cleistogamous flowers borne on erect, glabrous or sparsely pubescent, peduncles; capsules ellipsoid, to 12 mm long, glabrous, green; seeds obovoid, reddish-brown, to 1.7 mm long. April–May.

Wet meadows.

IA, IL, IN (FACW+), KY, OH (FAC+).

Primrose-leaved violet.

This violet is similar to *V. lanceolata* in most characters of the flowers and capsules but differs markedly in the shape of the leaves. Throughout its overall range, *V. primulifolia* exhibits a great variation in pubescence of the leaves, petioles, and peduncles.

This violet is the hybrid between *V. lanceolata* and *V. macloskeyi*.

12. **Viola sagittata** Ait. Hort. Kew. 3:287. 1789. Fig. 334.
Viola ovata Nutt. Gen. Am. 1:148. 1818.
Viola sagittata Ait. var. *ovata* (Nutt.) Torr. & Gray, Fl. N. Am. 1:138. 1838.

Perennial herb from stout rhizomes; leaves ascending, on glabrous or pubescent petioles usually longer than the blades; blades pubescent throughout to nearly glabrous, lanceolate to oblong-lanceolate, obtuse to acute to acuminate at the apex, cordate to truncate at the base, coarsely serrate, at least some of them hastately lobed, to 10 cm long, less than half as wide; petaliferous flowers violet or blue, borne on glabrous or pubescent peduncles usually longer than the petioles; sepals lanceolate, glabrous; lateral petals bearded; spurred petal pubescent; cleistogamous flowers borne on erect, usually pubescent peduncles; capsules ovoid, to 15 mm long, green; seeds brown, to 2 mm long. March–May.

Fields, prairies, woods, swampy meadows.

IA, IL, IN, MO (FACW−), KS (FAC), KY, OH (FACW).

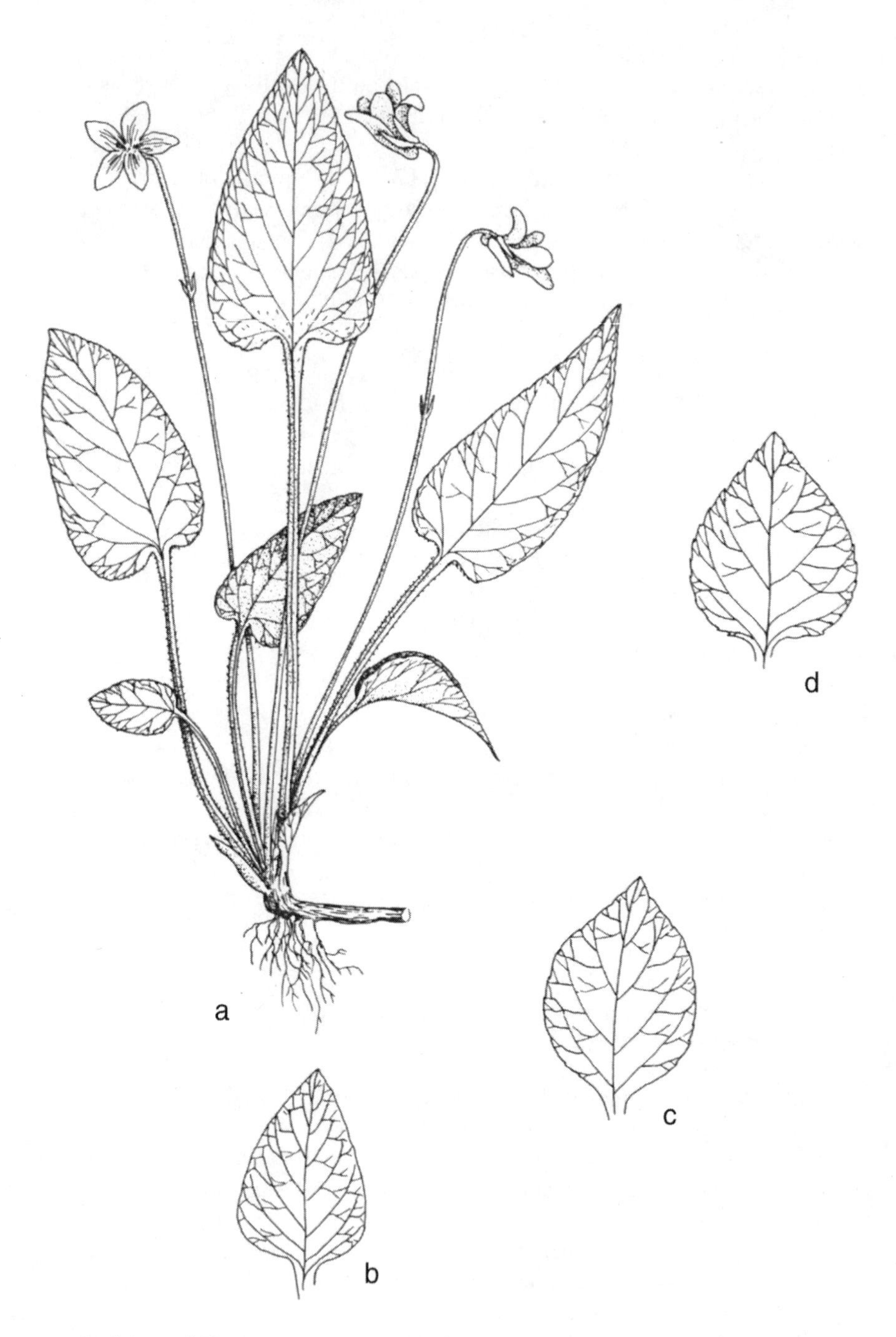

333. *Viola primulifolia* (Primrose-leaved violet).

a. Habit.
b, c, d. Leaves

Arrow-leaved violet.

The arrow-leaved violet is distinguished by its elongated leaves, which are sometimes hastately lobed.

Viola sagittata grows in a variety of habitats, ranging from moist or dry open woods to fields to prairies to swampy meadows.

334. *Viola sagittata* (Arrow-leaved violet).

a. Habit, in flower.
b. Habit, in fruit.

13. **Viola sororia** Willd. Enum. Hort. Berol. 1:72. 1809. Fig. 335.
Viola papilionacea Pursh, Fl. Am. Sept. 1:173. 1814, in part.
Viola priceana Pollard, Proc. Biol. Soc. Wash. 16:127. 1903.

Perennial herb from stout rhizomes; leaves ascending on long pubescent (rarely glabrous) petioles; blades densely or sparsely pubescent on both surfaces, ovate, obtuse to acute, cordate, crenate-serrate, to 10 cm long, to 10 cm wide; petaliferous flowers violet or blue (or occasionally blue-gray), on usually pubescent peduncles shorter than to longer than the petioles; sepals ovate to oblong, ciliate;

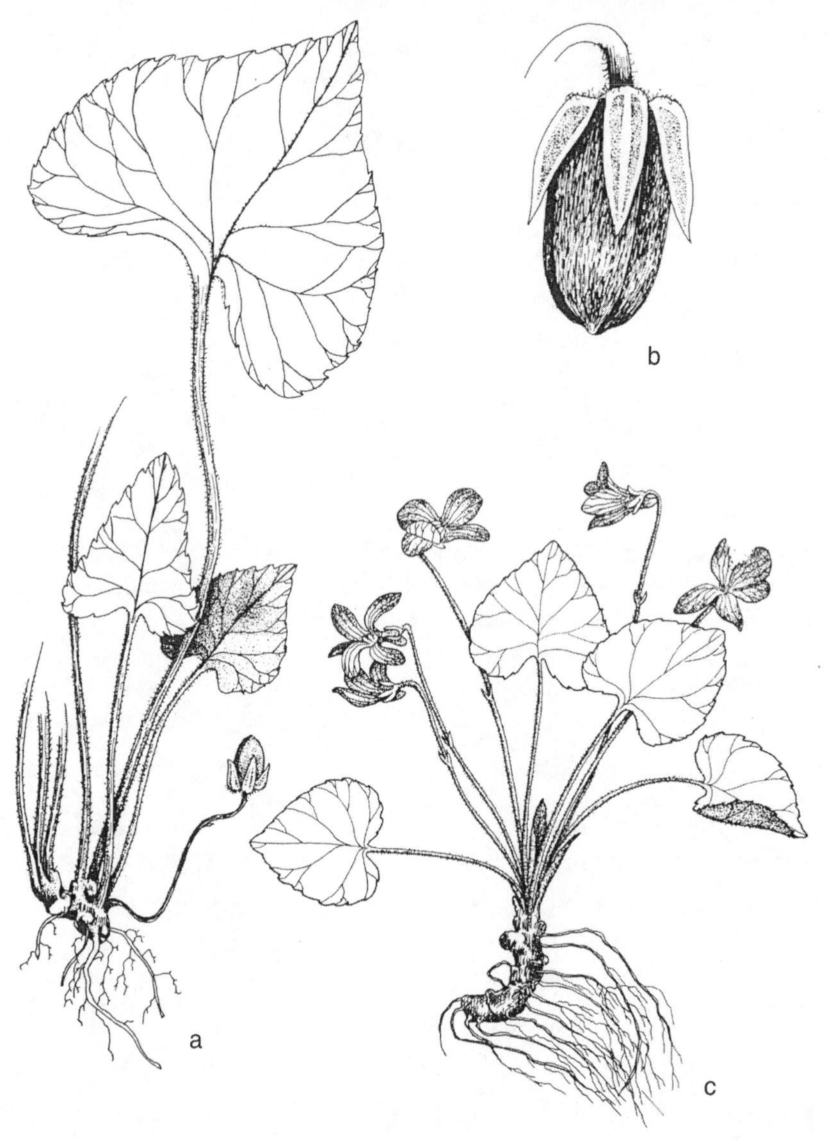

335. *Viola sororia* (Woolly blue violet). a. Habit, in fruit. b. Capsule. c. Habit, in flower.

lateral petals bearded; spurred petal usually bearded; cleistogamous flowers ovoid, on short, prostrate, usually pubescent peduncles; capsule ovoid, glabrous, green or purple, to 15 mm long; seeds dark brown, to 2.5 mm long. March–May.

Woods, occasionally in floodplains.

IA, IL, IN, KY, MO, OH (FAC−), KS, NE (FAC).

Woolly blue violet.

As recognized in this work, and following Russell (1965), *Viola sororia* is characterized as a stemless blue violet with at least some pubescence uniformly distributed on both surfaces of the leaf and on the petioles. In addition, the large ovoid capsules are borne on short, prostrate peduncles. Of the other stemless blue violets with unlobed leaves, *Viola pratincola* is completely glabrous, *V. nephrophylla* has the upper leaf surfaces glabrous or sparsely pubescent and the lower leaf surfaces hirsutulous, *V. cucullata* has club-shaped hairs on the lateral petals, *V. nephrophylla* has glabrous peduncles and eciliate sepals, *V. missouriensis* has a beardless spurred petal and a more deltoid leaf shape, and *V. affinis* has glabrous petioles and peduncles and often nearly glabrous leaves.

Variation in *V. sororia* occurs in the degree of pubescence on the leaves and petioles. Nearly glabrate forms approach *V. pratincola.*

14. **Viola striata** Ait. Hort. Kew. 3:290. 1789. Fig. 336.

Perennial herb from short rhizomes; stems several, usually tufted, reclining to ascending, angular, to 30 cm long at anthesis, becoming up to 60 cm long later, glabrous; basal leaves orbicular to broadly ovate, obtuse at the apex, cordate at the base, glabrous or nearly so, crenulate, rugulose, up to 5 cm in diameter, soon withering, the upper cauline leaves similar but acute at the apex, membranaceous, persistent until autumn; stipules lance-oblong, fimbriate the upper three-fourths of their length; petaliferous flowers axillary, creamy-white with purple lines, on glabrous peduncles usually longer than the subtending leaves; sepals linear-lanceolate, ciliate, the auricles up to 4 mm long; lateral petals bearded; spur 4–5 mm long; style bent downward, puberulent below the tip; cleistogamous flowers borne later in the same axils or in axils of later leaves; capsules ovoid, to 6.5 mm long, more than half as thick; seeds light brown, narrowly obovoid, to 2.5 mm long. April–June.

Low woods; moist meadows; wet ditches.

IA, IL, IN, KY, MO, NE, OH (FACW).

Common white violet; cream violet.

The common white violet is easy to recognize because of its cream-colored flowers borne from the axils of cauline leaves and its large, fimbriate stipules. *Viola conspersa*, with similar stipules, has pale violet flowers.

In ideal situations, the common white violet grows in great abundance. In early spring, this species forms dense basal rosettes of leaves, but these usually wither when the cauline leaves begin to develop.

336. *Viola striata* (Common white violet).
a. Habit.
b. Capsule.
c. Seed.
d. Stipule.

122. VITACEAE—GRAPE FAMILY

Woody vines with tendrils; leaves alternate, simple or compound, with stipules; flowers actinomorphic, perfect or unisexual, in racemes or cymes opposite the leaves; calyx obsolete or minute and 5-parted; petals 4–5, free, small; stamens 4–5; ovary superior, 2-locular; fruit a berry.

Eleven genera and about eleven hundred species comprise this family.

1. Leaves compound.
   2. Leaves palmately compound, with 5 (7) leaflets ....2. *Parthenocissus*
   2. Leaves termately or pinnately compound, with 11 or more leaflets .... 1. *Ampelopsis*
1. Leaves simple.
   3. Pith white.. 1. *Ampelopsis*
   3. Pith brown .... 3. *Vitis*

1. **Ampelopsis** Michx.

Woody vines with tendrils; pith white; leaves alternate, simple or pinnately compound, with stipules; flowers actinomorphic, perfect or unisexual, in cymes; sepals shallowly 5-lobed; petals 5, free, caducous; disk present; stamens 5, with only staminodia in pistillate flowers; ovary superior, 2-locular; fruit a berry, with 1–4 seeds.

About twenty species in North America and Asia comprise this genus.

1. Leaves compound ....1. *A. arborea*
1. Leaves simple ....2. *A. cordata*

1. **Ampelopsis arborea** (L.) Rusby, Mem. Torrey Club 5:221. 1894. Fig. 337.
Vitis arborea L. Sp. Pl. 203. 1753.

Woody vines with tendrils, sometimes shrublike, the bark with lenticels; leaves alternate, bi- or tri-pinnately compound, up to 3.5 m long, the petioles up to 3 cm long, the leaflets mostly ovate, acute at the apex, rounded at the base, dark green, shiny, glabrous, sharply toothed, to 4 cm long, to 3 cm wide, sessile or on short petiolules; flowers in cymes, actinomorphic, usually perfect; calyx shallowly toothed; petals 5, greenish, to 5 mm long; disk thick; stamens 5; ovary superior; berry globose, dark purple to black, to 5 mm in diameter, with a peppery taste, with 1–3 seeds. June–July.

Swampy woods, along rivers.

IL, IN, KY, MO, OH (FACW).

Peppervine.

This is the only vine in central Midwest wetlands with alternate leaves that are bi- or tri-pinnately compound. The berries have a peppery taste.

2. **Ampelopsis cordata** Michx. Fl. Bor. Am. 1:159. 1803. Fig. 338.

Woody vines, usually with tendrils, the twigs usually somewhat pubescent, with lenticels, to 25 m long; leaves alternate, simple, broadly ovate, acute at the apex, usually truncate at the base, serrate, glabrous except for pubescence on the veins below, to 10 cm long, to 6 cm wide, the petioles glabrous, up to 2.5 cm long; flowers in cymes, actinomorphic, perfect; calyx shallowly toothed; petals 5, greenish, to 5

337. *Ampelopsis arborea* (Peppervine). Habit, in fruit (center). Flower (right).

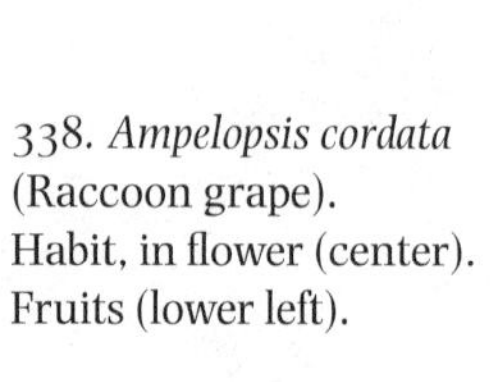

338. *Ampelopsis cordata* (Raccoon grape). Habit, in flower (center). Fruits (lower left).

mm long; disk cup-shaped; stamens 5; ovary superior; berries blue-black, globose, 4–5 mm in diameter, inedible, with 1–2 seeds. May–June.

Swampy woods, bottomland forests.

IA, IL, IN, KY, MO, OH (FAC+), KS, NE (UPL).

Raccoon grape.

This species is distinguished from species of *Vitis* by its usually truncate leaf base and its cymose flowers and fruits. Although a few leaves may have a cordate base, the specific epithet *cordata* seems inappropriate.

2. **Parthenocissus** Planch.—Woodbine

Woody vines, either climbing or trailing, with tendrils; leaves alternate, palmately compound, with stipules, the leaflets 5; flowers perfect or unisexual, actinomorphic, in cymes or short panicles; sepals obsolete; petals 5, free; disk attached to ovary; stamens 5; ovary superior, 2-locular; fruit a berry.

Parthenocissus consists of fifteen species in North America and Asia.

1. Tendrils without adhesive disks; leaves shiny on the upper surface; berries 8–10 mm in diameter .... 1. *P. inserta*
1. Most or all the tendrils ending in an adhesive disk; leaves dull on the upper surface; berries 5–7 mm in diameter .... 2. *P. quinquefolia*

1. **Parthenocissus inserta** (Kerner) K. Fritsch, Excur. Oester. 321. 1922. Fig. 339.
Parthenocissus vitacea Hitchc. Spring Fl. Manhattan 26. 1894, misapplied.

Woody creeper with or without tendrils, the tendrils not terminated by adhesive disks, to 5 m long; leaves alternate, palmately compound with 5 leaflets, the leaflets obovate to elliptic, acute to acuminate at the apex, tapering to the base, sharply toothed at least above the middle, to 20 cm long, to 8 cm wide, green and shiny on the upper surface, paler on the lower surface, glabrous or somewhat pubescent, at least on the veins on the lower surface, on petiolules up to 1.5 cm long; flowers in branched cymes, actinomorphic, perfect; calyx obsolete; petals 5, free, greenish white, to 4 mm long; stamens 5; ovary superior; berries blue, globose, 8–10 mm in diameter, with 3–4 seeds. June–July.

Woods, floodplain forests.

IA, IL, IN, KY, MO, OH (FACU), KS, NE (FAC).

Woodbine.

Many persons are unaware of another species with leaves that resemble Virginia creeper (*P. quinquefolia*). *Parthenoscissus inserta* either lacks tendrils or does not have adhesive disks attached to the tendrils so that this species does not climb. In addition, the leaflets of *P. inserta* may sometimes be much larger than those of *P. quinquefolia* and are usually shiny.

2. **Parthenocissus quinquefolia** (L.) Planch. in DC. Mon. Phan. 5, part 2, 448. 1887. Fig. 340.
Hedera quinquefolia L. Syn. Pl. 202. 1753.

Woody vines with tendrils, the tendrils with adhesive disks at the tip, the stems glabrous or pubescent, to 20 m long; leaves alternate, palmately compound, with 5

339. *Parthenocissus inserta* (Woodbine). Habit.

340. *Parthenocissus quinquefolia* (Virginia creeper). Habit, in flower (center). Flower (lower right).

leaflets, the leaflets elliptic to oblanceolate, acute to acuminate at the apex, rounded or tapering at the base, sharply toothed, at least above the middle, to 15 cm long, to 6 cm wide, green and dull on the upper surface, paler on the lower surface, glabrous or pubescent, at least on the lower surface, sessile or nearly so; flowers in branched cymes, actinomorphic, perfect; calyx obsolete; petals 5, free, greenish white, to 4 mm long; stamens 5, ovary superior; berries blue, globose, 5–7 mm in diameter, with 1–3 seeds. June–July.

Woods, fields, floodplain forests.

IA, IL, IN, MO (FAC−), KS, NE (FAC), KY, OH (FACU).

Virginia creeper.

This vine is always distinguished from poison ivy (*Toxicodendron radicans*) by having 5 leaflets per leaf. The leaves usually turn red in the autumn.

3. **Vitis** L.—Grape

Woody vines with tendrils, the brown pith usually interrupted; bark without lenticels (except for *V. rotundifolia*); leaves alternate, simple, coarsely toothed or sometimes lobed, usually cordate at the base, with stipules; flowers perfect or unisexual, actinomorphic, in racemes, the plants sometimes dioecious; calyx obsolete; petals 5, free, caducous; stamens 5; disk 5-lobed; ovary superior; fruit a berry, with 1–4 seeds.

About sixty species comprise the genus, all in the Northern Hemisphere.

1. Bark with lenticels; pith continuous; racemes up to 4 cm long; berries up to 20 mm in diameter .... 5. *V. rotundifolia*
1. Bark without lenticels; pith interrupted; racemes more than 4 cm long; berries up to 12 mm in diameter.
    2. Leaves with cobwebby hairs on the lower surface or merely silvery-glaucous.
        3. Seeds 5–8 mm long; branches not angular .... 1. *V. aestivalis*
        3. Seeds 4–5 mm long; branches angular .... 2. *V. cinerea*
    2. Leaves glabrous on the lower surface, green, or with hairs merely in the vein axils.
        4. Leaves unlobed or shallowly 3-lobed .... 6. *V. vulpina*
        4. Leaves, or at least those on the fertile branches, sharply 3- or 5-lobed.
            5. Berries black; partitions in pith at the nodes 4–5 mm thick; leaves scarcely or not at all ciliate on the margins, very deeply lobed .... 3. *V. palmata*
            5. Berries bluish, glaucous; partitions in pith at the nodes up to 2 mm thick; leaves ciliate on the margins, usually lobed less than halfway to the middle .... 4. *V. riparia*

1. **Vitis aestivalis** Michx, Fl. Bor. Am. 2:230. 1803. Fig. 341.
Vitis argentifolia Munson ex Bailey, Gentes Herb. 3:194. 1934.
Vitis aestivalis Michx. var. *argentifolia* (Munson) Fern. Rhodora 38:428. 1936.

High-climbing woody vines, with tendrils; branches terete, without lenticels, pubescent, at least when young; bark loose, shredding; leaves alternate, simple, broadly ovate, acute at the apex, cordate at the base, coarsely dentate, sometimes 3- or 5-lobed, white-tomentose and usually glaucous beneath, to 20 cm long, nearly as wide, the petioles glabrous or sparsely pubescent, to 5 cm long; flowers in racemes to 18 cm long, actinomorphic, perfect or unisexual; calyx obsolete; petals 5, free, caducous; stamens 5; disk present; ovary superior; berries globose, to 10 mm in diameter, black and usually somewhat glaucous. May–July.

Dry woods, wet woods, thickets.
IA, IL, IN, KY, MO, OH (FACU), KS, NE (UPL).
Summer grape.

This species occurs mostly in dry woods, but it may also grow on occasion in wet woods. It is similar in appearance to *V. cinerea* because of the whitened lower leaf surface, but this latter species has more pubescent leaves, pubescent petioles, angular twigs, and smaller berries.

Plants that are strongly glaucous on the lower surface may be called *V. argentifolia* or *V. aestivalis* var. *argentifolia.*

The berries are sweet.

2. **Vitis cinerea** (Engelm.) Engelm. Am. Nat. 2:321. 1868. Fig. 342.
Vitis aestivalis Michx. var. *cinerea* Engelm. in Gray, Man. ed. 5, 679. 1867.

High-climbing woody vines, with tendrils; branches angular, without lenticels, permanently gray- or white-pubescent; bark loose, shredding; leaves alternate,

341. *Vitis aestivalis* (Summer grape). Habit.

simple, ovate, acute at the apex, cordate at the base, coarsely dentate, sometimes 3- or 5-lobed, permanently gray- or white-tomentose on the lower surface, not glaucous, to 20 cm long, to 18 cm wide, the petioles with persistent pubescence, to 5 cm long; flowers in racemes to 16 cm long, actinomorphic, perfect or unisexual; calyx obsolete; petals 5, free, caducous; stamens 5; disk present; ovary superior; berries globose, to 8 mm in diameter, black and sometimes slightly glaucous. June–July.

Wet woods, floodplain forests, along rivers and streams.

IA, IL, IN, KS, MO, NE (FACW−), KY, OH (FACW).

Winter grape.

This vine may have stem diameters up to 12 centimeters. The berries are sweet.

Vitis cinerea is distinguished from *V. aestivalis* by its angular twigs, pubescent petioles, and smaller berries.

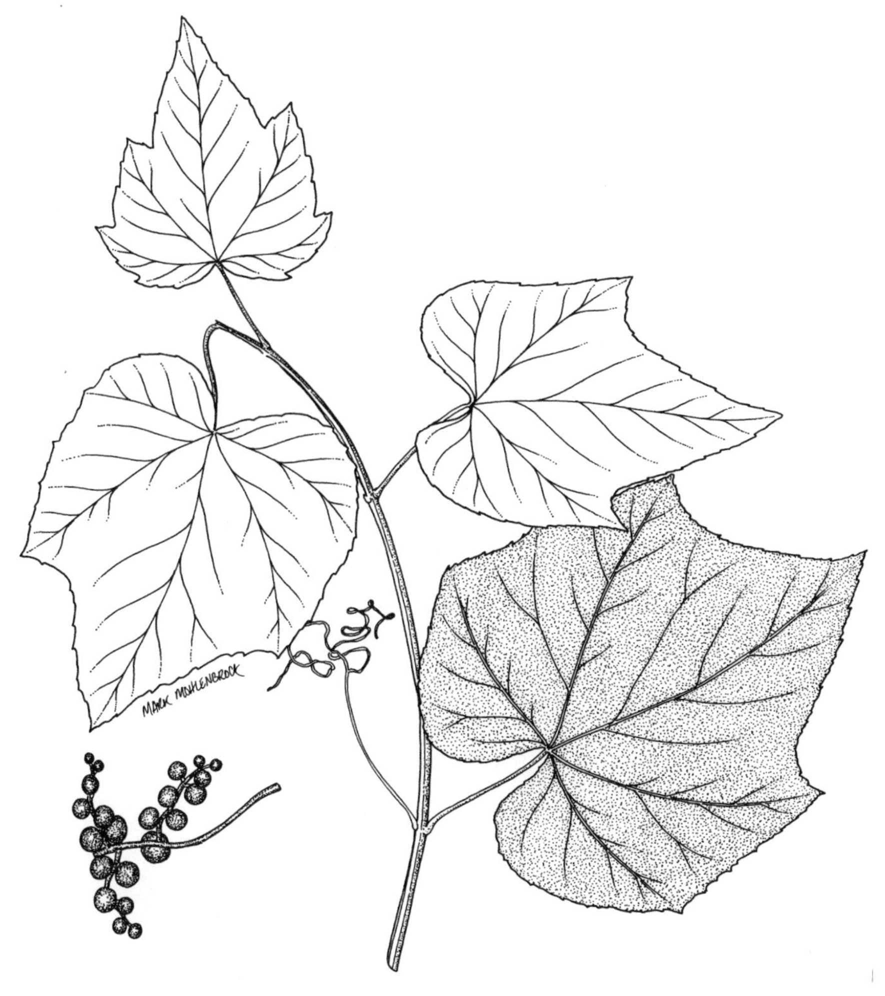

342. *Vitis cinerea* (Winter grape). Leafy branch (center). Fruits (lower left).

3. **Vitis palmata** Vahl, Symb. Bot. 3:42. 1794. Fig. 343.

High-climbing woody vines, with tendrils; branches terete, slender, without lenticels, red to purple-red, glabrous or nearly so; bark usually loose, shredding; leaves alternate, simple, ovate, acuminate at the apex, cordate at the base, serrate, usually deeply 3- or 5-lobed, to 10 cm long, nearly as wide, glabrous or pubescent on the veins beneath, the petioles to 7 cm long, glabrous or nearly so; flowers in racemes to 15 cm long, actinomorphic, perfect or unisexual; calyx obsolete; petals 5, free, caducous; stamens 5; disk present; ovary superior; berries globose, to 8 mm in diameter, black. June–July.

Swampy woods, wet woods.

IL, IN, MO (OBL), KY (NI).

Catbird grape.

The distinguishing features of this grape are its red branches, deeply lobed, acuminate leaves, and black berries that are never glaucous. The berries are sweet.

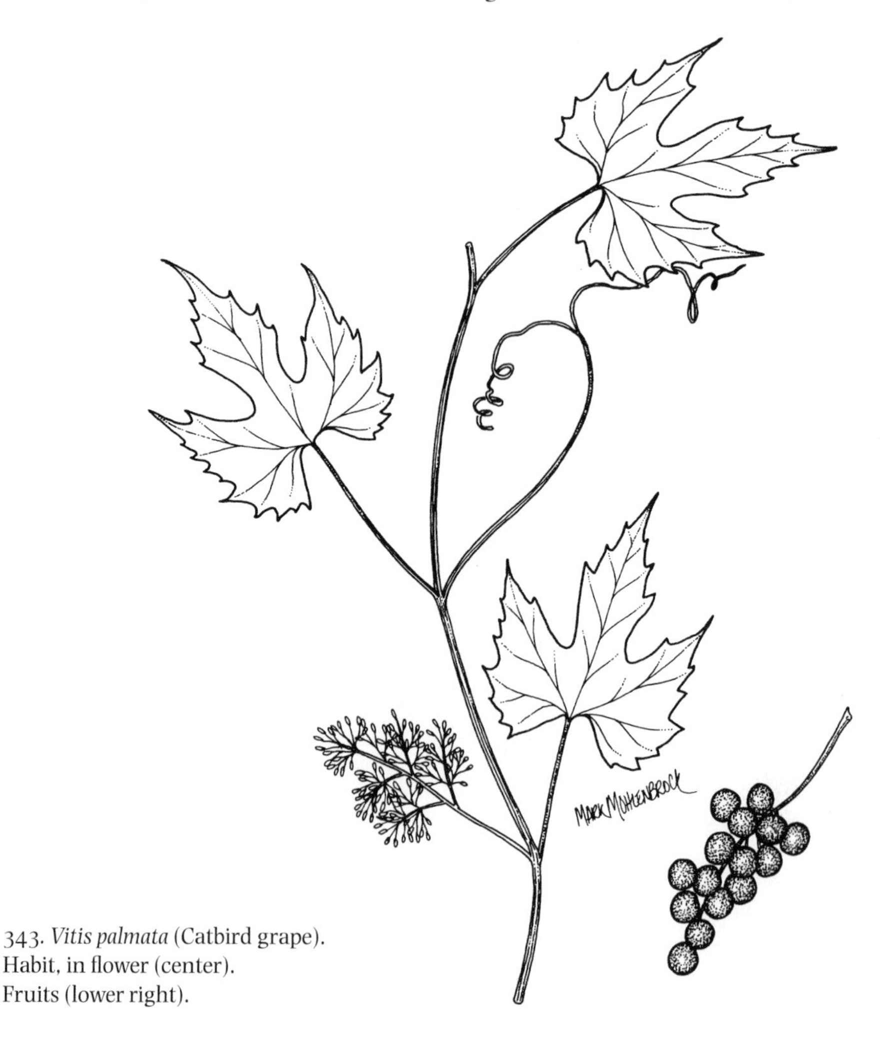

343. *Vitis palmata* (Catbird grape).
Habit, in flower (center).
Fruits (lower right).

4. **Vitis riparia** Michx. Fl. Bor. Am. 2:231. 1803. Fig. 344.
Vitis cordifolia Michx. var. *riparia* (Michx.) Gray, Man. ed. 2, 78. 1856.
Vitis riparia Michx. var. *praecox* Engelm. ex Bailey, Amer. Gard. 14:353. 1893.

High-climbing woody vines, with tendrils; branches terete, without lenticels, glabrous or pubescent; bark usually loose, shredding; leaves alternate, simple, ovate, acuminate at the apex, cordate at the base, coarsely dentate, often 3- or 5-lobed, green on the lower surface, usually with pubescence on the veins on the lower surface, the petioles to 6 cm long, glabrous or occasionally pubescent; flowers in racemes to 15 cm long, actinomorphic, perfect or unisexual; calyx obsolete; petals 5, free, caducous; stamens 5; disk present; ovary superior; berries globose, up to 12 mm in diameter, glaucous. May–July.

Wet woods, moist woods, thickets.

IA, IL, IN, MO (FACW−), KS, NE (FAC), KY, OH (FACW).

Riverbank grape; frost grape.

This species is distinguished by its lower leaf surface green, the presence of 3 or 5 lobes or at least three teeth larger than the others, and its strongly glaucous berries. The berries are sweet.

Plants with berries only 6–7 mm in diameter and with racemes only up to 6 cm long have been called var. *praecox*.

344. *Vitis riparia* (Riverbank grape). Habit, in fruit.

5. **Vitis rotundifolia** Michx. Fl. Bor. Am. 2:231. 1803. Fig. 345.
Muscadinia rotundifolia (Michx.) Small, Fl. Se. U.S. 757. 1903.

High-climbing woody vines, with tendrils; branches terete, with conspicuous lenticels, the pith not interrupted by diaphragms, glabrous; bark tight, not shredding; leaves alternate, simple, orbicular in outline, acute at the tip, cordate at the base, uniformly dentate, glabrous on both surfaces or occasionally with tufts of hairs in the axils of the veins beneath, to 10 cm long, about as wide, the petioles up to 5 cm long, glabrous; flowers in racemes up to 4 cm long, actinomorphic, perfect or unisexual; calyx obsolete; petals 5, free, caducous; stamens 5; disk present; ovary superior; berries globose, up to 2 cm in diameter, purple, very sweet. May–June.

Swampy woods.

KY (FAC−), MO (FACW).

Muscadine.

The berries are so large that they resemble plums, and they make superb jelly.

Because of the large, purple berries, the presence of lenticels on the twigs, the continuous pith, and the very short racemes, this species could well deserve being placed in its own genus, a concept proposed by John K. Small in 1903.

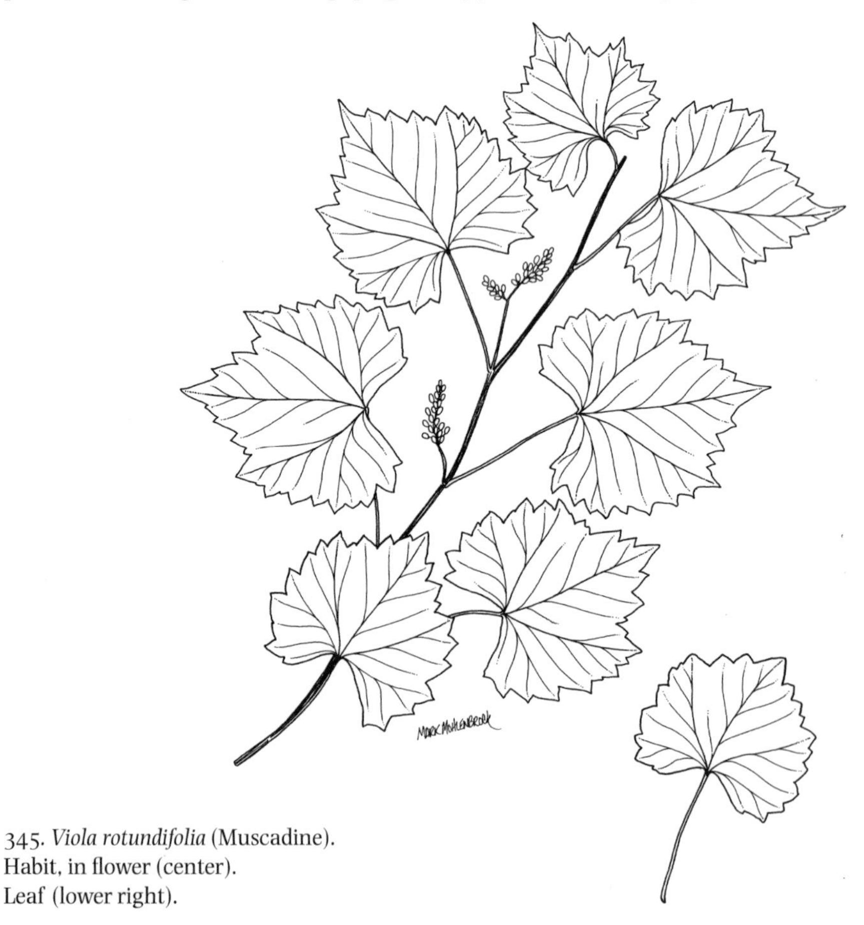

345. *Viola rotundifolia* (Muscadine).
Habit, in flower (center).
Leaf (lower right).

6. **Vitis vulpina** L. Sp. Pl. 203. 1753. Fig. 346.
Vitis cordifolia Michx. Fl. Bor. Am. 2:231. 1803.
Vitis cordifolia Michx. var. *vulpina* (L.) Eaton, Man. Bot., ed. 2, 497. 1818.

High-climbing woody vines, with tendrils; branches terete, without lenticels, glabrous or nearly so; bark usually loose, shredding; leaves alternate, simple, ovate, acuminate the apex, cordate at the base, coarsely and uniformly dentate, rarely lobed, green on the lower surface, glabrous except for occasional hairs in the leaf axils on the lower surface, the petioles to 6 cm long, glabrous; flowers in racemes to 12 cm long, actinomorphic, perfect or unisexual; calyx obsolete; petals 5, free, caducous; stamens 5; disk present; ovary superior; berries globose, to 8 mm in diameter, black, shiny, not glaucous. May–June.

Bottomland forests, wet woods, thickets.

IA, IL, IN, MO (FACW−), KS, KY, NE, OH (FAC).

Fox grape.

Because of the green lower leaf surface, this species resembles *V. riparia* but differs by its unlobed leaves, its more sparse pubescence, and its smaller, black, shiny berries.

The berries are edible.

346. *Vitis vulpina* (Fox grape). Leafy branch (below). Fruits (above).

Glossary

Acaulescent. Seemingly without aerial stems.
Achene. A type of one-seeded, dry, indehiscent fruit with the seed coat not attached to the mature ovary wall.
Actinomorphic. Having radial symmetry; regular, in reference to a flower.
Acuminate. Gradually tapering to a point.
Acute. Sharply tapering to a point.
Adaxial. Toward the axis; when referring to a leaf, the upper surface.
Ament. A spike of unisexual, apetalous flowers; a catkin.
Anther. The terminal part of a stamen which bears pollen.
Anthesis. Flowering time.
Antrorse. Project forward.
Apical. At the apex.
Apiculate. Abruptly short-pointed at the tip.
Appressed. Lying flat against the surface.
Areole. A small area between leaf veins.
Aristate. Bearing an awn.
Articulated. Jointed.
Attenuate. Gradually becoming narrowed.
Auricle. An earlike lobe.
Auriculate. Bearing an earlike process.
Awn. A bristle usually terminating a structure.
Axillary. Borne from an axil.

Beard. A tuft of stiff hairs.
Berry. A type of fruit where the seeds are surrounded only by fleshy material.
Bidentate. Having two teeth.
Bifid. Two-cleft.
Bifoliolate. Bearing two leaflets.
Bifurcate. Forked.
Biglandular. Bearing two glands.
Bilabiate. Two-lipped.
Bipinnate. Divided once into distinct segments, with each segment in turn divided into distinct segments.
Bipinnatifid. Divided partway to the center, with each lobe again divided partway to the center.
Bisexual. Referring to a flower that contains both stamens and pistils.
Biternate. Divided into three segments two times.
Bract. An accessory structure at the base of many flowers, usually appearing leaflike.
Bracteate. Bearing one or more bracts.
Bracteole. A secondary bract.
Bractlet. A small bract.
Bristle. A stiff hair or hairlike growth; a seta.
Bulblet. A small bulb.

Callosity. Any hardened thickening.
Calyx. The outermost segments of the perianth of a flower, composed of sepals.
Campanulate. Bell-shaped.
Canescent. Grayish hairy.
Capillary. Threadlike.
Capitate. Forming a head.
Capsule. A dry, dehiscent fruit composed of more than one carpel.
Carpel. A simple pistil, or one member of a compound pistil.
Cartilaginous. Firm but flexible.
Catkin. A spike of unisexual, apetalous flowers; an ament.
Caudate. With a tail-like appendage.
Caudex. The woody base of a perennial plant.
Cauline. Belonging to a stem.
Cespitose. Growing in tufts.
Chaffy. Covered with scales.
Chartaceous. Papery.
Cilia. Marginal hairs.
Ciliate. Bearing cilia.
Ciliolate. Bearing small cilia.
Circumscissile. Usually referring to a fruit which dehisces by a horizontal, circular line.
Claw. A narrow, basal stalk, particularly of a petal.
Coherent. The growing together of like parts.
Coma. A tuft of hairs at the end of a seed.
Compressed. Flattened.
Concave. Curved on the inner surface; opposed to convex.
Connate. Union of like parts.
Convex. Curved on the outer surface.

Convolute. Rolled lengthwise.
Cordate. Heart-shaped.
Coriaceous. Leathery.
Corm. An underground, vertical stem with scaly leaves, differing from a bulb by lacking fleshy leaves.
Corolla. The segments of a flower just within the calyx, composed of petals.
Corona. A crown of petal-like structures.
Corrugated. Folded or wrinkled.
Corymb. A type of inflorescence where the pedicellate flowers are arranged along an elongated axis but with the flowers all attaining about the same height.
Corymbiform. Shaped like a corymb.
Cotyledon. A seed leaf..
Crateriform. Cone-shaped but sunken in the center at the top.
Crenate. With round teeth.
Crenulate. With small, round teeth.
Crest. A ridge.
Crisped. Curled.
Cucullate. Hood-shaped.
Culm. The stem that terminates in an inflorescence.
Cuneate. Wedge-shaped; tapering to the base.
Cupular. Shaped like a small cup.
Cuspidate. Terminating in a very short point.
Cyathium. A cuplike structure enclosing flowers.
Cyme. A type of broad and flattened inflorescence in which the central flowers bloom first.
Cymose. Bearing a cyme.

Deciduous. Falling away.
Decumbent. Lying flat, but with the tip ascending.
Decurrent. Adnate to the petiole or stem and then extending beyond the point of attachment.
Deflexed. Turned downward.
Dehiscent. Splitting at maturity.
Deltoid. Triangular.
Dentate. With sharp teeth, the tips of which project outward.
Denticulate. With small, sharp teeth, the tips of which project outward.
Diffuse. Loosely spreading.
Digitate. Radiating from a common point, like the fingers from a hand.
Dilated. Swollen; expanded.
Dimorphic. Having two forms.
Dioecious. With staminate flowers on one plant, pistillate flowers on another.
Disarticulate. To come apart; to become disjointed.
Disk. An enlarged outgrowth of the receptacle.
Divergent. Spreading apart.
Drupe. A type of fruit in which the seed is surrounded by a hard, dry covering which, in turn, is surrounded by fleshy material.

Ebracteate. Without bracts.
Echinate. Spiny.
Eciliate. Without cilia.
Eglandular. Without glands.
Ellipsoid. Referring to a solid object that is broadest at the middle, gradually tapering to both ends.
Elliptic. Broadest at the middle, gradually tapering to both ends.
Emarginate. Having a shallow notch at the extremity.
Epidermis. The outermost layer of cells.
Epunctate. Without dots.
Erose. With an irregularly notched margin.
Exudate. Secreted material.

Falcate. Sickle-shaped.
Fascicle. Cluster.
Ferruginous. Rust-colored.
Fetid. Foul-smelling.
Fibrous. Referring to roots borne in tufts.
Filament. That part of the stamen supporting the anther.
Filiform. Threadlike.
Fimbriate. Fringed.
Flaccid. Weak; flabby.
Flexible. Readily or easily bent.
Flexuous. Zigzag.
Follicle. A type of dry, dehiscent fruit that splits along one side at maturity.
Friable. Breaking easily into small particles.
Funnelform. Shaped like a funnel.

Geniculate. Bent.
Gibbous. Swollen on one side.
Glabrate. Becoming smooth.
Glabrous. Without pubescence or hairs.
Gland. An enlarged, spherical body functioning as a secretory organ.

Glandular. Bearing glands.
Glaucous. With a whitish covering that can be rubbed off.
Globose. Round; globular.
Glomerule. A small compact cluster.
Glutinous. Covered with a sticky secretion.

Hastate. Spear-shaped; said of a leaf that is triangular with spreading basal lobes.
Hirsute. With stiff hairs.
Hirsutulous. With minute stiff hairs.
Hirtellous. Finely hirsute.
Hispid. With rigid hairs.
Hispidulous. With minute rigid hairs.
Hoary. Grayish white, usually referring to pubescence.
Hood. That part of an organ, usually of a flower, that is strongly concave and arching.
Horn. An accessory structure found in certain flowers.
Hyaline. Transparent.
Hypanthium. A development of the receptacle beneath the calyx.

Imbricate. Overlapping
Indehiscent. Not splitting open at maturity.
Indurate. Hard.
Inferior. Referring to the position of the ovary when it is surrounded by the adnate portion of the floral tube or is embedded in the receptacle.
Inflexed. Turned inward.
Inflorescence. A cluster of flowers.
Internode. The area between two adjacent nodes.
Involucel. A cluster of bracteoles that subtends a secondary flower cluster.
Involucre. A circle of bracts that subtends a flower cluster.
Involute. Rolled inward.

Keel. A ridgelike process.

Laciniate. Divided into narrow, pointed divisions.
Lanceolate. Lance-shaped; broadest near base, gradually tapering to the narrower apex.
Lanceoloid. Referring to a solid object that is broadest near base, gradually tapering to the narrower apex.
Latex. Milky juice.
Leaflet. An individual unit of a compound leaf.
Legume. A dry fruit usually dehiscing along two sides at maturity.
Lenticel. Corky openings on bark of twigs and branches.
Lenticular. Lens-shaped.
Lepidote. Scaly.
Linear. Elongated and uniform in width throughout.
Lobulate. With small lobes.
Locular. Referring to the locule, or cavity of the ovary or the anther.
Loculicidal. Said of a capsule that splits down the dorsal suture of each cell.
Lunate. Crescent-shaped.
Lustrous. Shiny.
Lyrate. Pinnatifid, with the terminal lobe much larger than the lower ones.

Moniliform. Constricted at regular intervals to resemble a string of beads.
Monoecious. Bearing both sexes in separate flowers on the same plant.
Monomorphic. Having but one form.
Mucilaginous. Slimy.
Mucro. A short, abrupt tip.
Mucronate. Possessing a short, abrupt tip.
Mucronulate. Possessing a very short, abrupt tip.
Muricate. Minutely spiny.

Nectariferous. Producing nectar.
Nigrescent. Blackish.
Node. That place on the stem from which leaves and branchlets arise.
Nutlet. A small nut.

Oblanceolate. Reverse lance-shaped; broadest at apex, gradually tapering to narrow base.
Oblong. Broadest at the middle, and tapering to both ends, but broader than elliptic.
Oblongoid. Referring to a solid object that, in side view, is nearly the same width throughout.
Obovoid. Referring to a solid object that is broadly rounded at the apex, becoming narrowed below.
Obpyramidal. Referring to an upside-down pyramid.

Obtuse. Rounded at the apex.
Ocrea. A sheathing stipule, often tubular.
Ocreola. A secondary, usually tubular, sheath.
Opaque. Referring to an object that cannot be seen through.
Orbicular. Round.
Oval. Broadly elliptic.
Ovary. The lower swollen part of the pistil that produces the ovules.
Ovoid. Referring to a solid object that is broadly rounded at the base, becoming narrowed above.

Palmate. Divided radiately, like the fingers of a hand.
Pandurate. Fiddle-shaped.
Panduriforn. Fiddle-shaped.
Panicle. A type of inflorescence composed of several racemes.
Papilla. A small wart.
Papillate. Bearing small warts, or papillae.
Papillose. Bearing pimplelike processes.
Pappus. The modified calyx in the Asteraceae; a similar structure in the Valerianaceae.
Papule. A pimplelike projection.
Pectinate. Pinnatifid into close, narrow segments; comblike.
Pedicel. The stalk of a flower of an inflorescence.
Pedicellate. Bearing a pedicel.
Peduncle. The stalk of an inflorescence.
Peduncled. Provided with a peduncle.
Peltate. Attached away from the margin, in reference to a leaf.
Perennial. Living more than two years.
Perfect. Bearing both stamens and pistils in the same flower.
Perfoliate. Referring to a leaf that appears to have the stem pass through it.
Perianth. Those parts of the flower including both the calyx and corolla.
Pericarp. The ripened ovary wall.
Petal. One segment of the corolla.
Petaloid. Resembling a petal in texture and appearance.
Petiolate. Bearing a petiole, or leafstalk.
Petiole. The stalk of a leaf.
Petiolulate. Bearing a petiolule, or leafletstalk.
Petiolule. The stalk of a leaflet.
Pilose. Bearing soft hairs.
Pinnate. Divided once into distinct segments.
Pinnatifid. Said of a simple leaf or leaf-part that is cleft or lobed only partway to its axis.
Pistil. The ovule-producing organ of a flower normally composed of an ovary, a style, and a stigma.
Pistillate. Bearing pistils but not stamens.
Plicate. Folded.
Plumose. Bearing fine hairs, like the plume of a feather.
Procumbent. Lying on the ground.
Pruinose. Having a waxy covering.
Puberulent. With minute hairs.
Pubescent. Bearing some kind of hairs.
Punctate. Dotted.
Pyramidal. Shaped like a pyramid.
Pyriform. Pear-shaped.

Quadrate. Four-sided.

Raceme. A type of inflorescence where pedicellate flowers are arranged along an elongated axis.
Rachis. A primary axis.
Receptacle. That part of the flower to which the perianth, stamens, and pistils are usually attached.
Reflexed. Turned downward.
Reniform. Kidney-shaped.
Repand. Wavy along the margin.
Repent. Creeping.
Resinous. Producing a sticky secretion, or resin.
Reticulate. Resembling a network.
Reticulum. A network.
Retrorse. Pointing downward.
Retuse. Shallowly notched at a rounded apex.
Revolute. Rolled under from the margin.
Rhizome. An underground horizontal stem, bearing nodes, buds, and roots.
Rhombic. Becoming quadrangular.
Rosette. A cluster of leaves in a circular arrangement at the base of a plant.
Rotate. Flat and circular.
Rufescent. Reddish brown.
Rufous. Red-brown.
Rugose. Wrinkled.
Rugulose. With small wrinkles.

Saccate. Sac-shaped.
Sagittate. Shaped like an arrowhead.

Salverform. Referring to a tubular corolla that abruptly expands into a flat limb.
Samara. An indehiscent winged fruit.
Scaberulous. Slightly rough to the touch.
Scabrous. Rough to the touch.
Scape. A leafless stalk bearing a flower or inflorescence.
Scarious. Thin and membranous.
Scurfy. Bearing scaly particles.
Secund. Borne on one side.
Sepaloid. Resembling a sepal in texture.
Septate. With dividing walls.
Septicidal. Said of a capsule that splits between the locules.
Sericeous. Silky; bearing soft, appressed hairs.
Serrate. With teeth that project forward.
Serrulate. With very small teeth that project forward.
Sessile. Without a stalk.
Seta. Bristle.
Setaceous. Bearing bristles, or setae.
Setiform. Bristle-shaped.
Setose. Bearing setae.
Sinuate. Wavy along the margins.
Sinus. The cleft between two lobes or teeth.
Spatulate. Oblong, but with the basal end elongated.
Spicate. Bearing a spike.
Spike. A type of inflorescence where sessile flowers are arranged along an elongated axis.
Spikelet. A small spike.
Spinescent. Becoming spiny.
Spinose. Bearing spines.
Spinule. A small spine.
Spinulose. Bearing small spines.
Spur. A saclike extension of the flower.
Stamen. The pollen-producing organ of a flower composed of a filament and an anther.
Staminate. Bearing stamens but not pistils.
Staminodium. A sterile stamen.
Stellate. Star-shaped.
Stipe. A stalk.
Stipel. A small stipe.
Stipitate. Possessing a stipe.
Stipular. Pertaining to a stipule.
Stipule. A leaflike or scaly structure found at the point of attachment of a leaf to the stem.
Stolon. A slender, horizontal stem on the surface of the ground.
Stoloniferous. Bearing stolons.
Stramineous. Straw-colored.
Striate. Marked with grooves.
Strigillose. With short, appressed, straight hairs.
Strigose. With appressed, straight hairs.
Strigulose. With short, appressed, straight hairs.
Style. That elongated part of the pistil between the ovary and the stigma.
Subcuneate. Nearly wedge-shaped.
Suborbicular. Nearly spherical.
Subulate. With a very short, narrow point.
Succulent. Fleshy.
Suffused. Spread throughout; flushed.
Superior. Referring to the position of the ovary when the free floral parts arise below the ovary.

Tendril. A spiraling, coiling structure that enables a climbing plant to attach itself to a supporting body.
Terete. Round, in cross-section.
Ternate. Divided three times.
Tomentose. Pubescent with matted wool.
Tomentulose. Finely pubescent with matted wool.
Tomentum. Woolly hair.
Torulose. With small contractions.
Translucent. Partly transparent.
Trifoliolate. Divided into three leaflets.
Trigonous. Triangular in cross-section.
Truncate. Abruptly cut across.
Tuber. An underground fleshy stem formed as a storage organ at the end of a rhizome.
Tubercle. A small, wartlike process.
Tuberculate. Warty.
Turgid. Swollen to the point of bursting.
Turion. A swollen asexual structure.

Umbel. A type of inflorescence in which the flower stalks arise from the same level.
Undulate. Wavy.
Unisexual. Bearing either stamens or pistils in one flower, but not both.
Urceolate. Urn-shaped.
Utricle. A small, one-seeded, indehiscent fruit with a thin covering.

Verrucose. Warty.
Verticil. A whorl.
Verticillate. Whorled.

Villous. With long, soft, slender, unmatted hairs.
Virgate. Wandlike.
Viscid. Sticky.
Whorl. An arrangement of three or more structures at a point on the stem.

Zygomorphic. Bilaterally symmetrical.

Illustration Credits

Illustrations 6, 7, 8, 9, 10, 11, 12, 13, 16, 22, 23, 28, 29, 31, 36, 40, 43, 44, 45, 46, 47, 88, 90, 93, 96, 100, 128, 130, 131, 132,133, 137, 145, 156, 157, 175, 180, 183, 185, 186, 188, 189, 195, 196, 197, 198, 200, 202, 210, 211, 213, 214, 218, 220, 225, 228, 254, 255, 256, 259, 261, 263, 264, 267, 270, 271, 274, 277, 280, 282, 283, 285, 287, 292, 296, 298, 299, 300, 302, 303, 304, 305, 306, 307, 308, 309, 310, 311, 312, 319, 321, 337, 338, 340, 341, 342, 343, 344, 345, and 346 were prepared by Mark W. Mohlenbrock.

Illustrations 103, 104, 110, and 114 were prepared by Miriam Wysong Meyer.

Illustrations 41, 50, 53, 54, 55, 56, 57, 58, 59, 60, 61, 62, 63, 64, 65, 66, 67, 68, 69, 70, 71, 76, 77, 79, 80, 81, 82, 83, 150, 152, 154, 167, 170, 173, 229, 231, 232, 233, 234, 235, 236, 239, 240, 241, 242, 244, 245, 246, 247, 248, 249, 250, 251, 252, and 253 were prepared by Paul W. Nelson.

Illustration 129 was prepared by Mark W. Mohlenbrock and is reprinted from *Flowering Plants: Basswoods to Spruges* by Robert H. Mohlenbrock, from the Illustrated Flora of Illinois series, Southern Illinois University Press.

Illustrations 84, 85, 86, 87, 89, 92, 94, 95, 97, 323, 324, 325, 326, 327, 328, 329, 330, 331, 332, 333, 334, 335, and 336 were prepared by Fredda Burton and are reprinted from *Flowering Plants: Hollies to Loasa* by Robert H. Mohlenbrock, from the Illustrated Flora of Illinois series, Southern Illinois University Press.

Illustrations 1, 2, 3, 4, 5, 98, 99, 101, 105, 106, 107, 108, 109, 111, 115, 116, 117, 118, 119, 120, 122, 123, 124, 126, 127, and 257 were prepared by Miriam Wysong Meyer and are reprinted from *Flowering Plants: Magnolias to Pitcher Plants* by Robert H. Mohlenbrock, from the Illustrated Flora of Illinois series, Southern Illinois University Press.

Illustrations 51 and 52 were prepared by Paul Nelson and are reprinted from *Flowering Plants: Smartweeds to Hazelnuts* by Robert H. Mohlenbrock, from the Illustrated Flora of Illinois series, Southern Illinois University Press.

Illustrations 230, 237, 238, and 243 were prepared by Paul Nelson and are reprinted from *Flowering Plants: Willows to Mustards* by Robert H. Mohlenbrock, from the Illustrated Flora of Illinois series, Southern Illinois University Press.

Illustrations 14, 19, 34, 38, 49, 72, 181, 182, 187, 190, 206, 209, 215, 275, and 278 were prepared by Mark W. Mohlenbrock and are reprinted from *Northeast Wetland Flora: Field Office Guide to Plant Species* by Robert H. Mohlenbrock.

Illustration 201 was prepared by Mark W. Mohlenbrock and is reprinted from *Southern Wetland Flora: Field Office Guide to Wetland Plants of the South* by Robert H. Mohlenbrock for the United States Department of Agriculture.

Illustrations 78, 121, 139, 140, 216, 258, 281, 289, 293, 294, 318, and 339 were prepared by Mark W. Mohlenbrock and are reprinted from *Western Wetland Flora: Field Office Guide to Plant Species* by Robert H. Mohlenbrock.

Illustration 219 was prepared by Mark W. Mohlenbrock and is reprinted from *Midwestern Wetland Flora: Field Office Guide to Plant Species* by Robert H. Mohlenbrock for the United States Department of Agriculture.

Illustrations 17, 18, 20, 21, 24, 25, 30, 32, 39, 42, 48, 102, 112, 113, 134, 135, 138, 147, 165, 169, 172, 176, 177, 178, 179, 191, 192, 193, 194, 203, 204, 207, 208, 260, and 262 are reprinted from *An Illustrated Flora of the Northern United States and Canada*, vol. 2, by Nathaniel Britton and Addison Brown. Charles Scribner's Sons, 1913; reprinted 1970 by Dover Publications, New York.

Illustrations 212, 217, 221, 222, 223, 224, 227, 265, 266, 268, 269, 272, 273, 276, 286, 290, 291, 297, 313, 314, 315, 316, 317, 320, and 322 are reprinted from *An Illustrated Flora of the Northern United States and Canada*, vol. 3, by Nathaniel Britton and Addison Brown. Charles Scribner's Sons, 1913; reprinted 1970 by Dover Publications, New York.

Illustrations 15 and 205 are reprinted from *A Great Lakes Wetland Flora: A Complete, Illustrated Guide to the Aquatic and Wetland Plants of the Upper Midwest* by Steve Chadde. Pocketflora Press, Calumet, Michigan, 1998.

Illustration 136 is reprinted from *The Complete Trees of North America: Field Guide and Natural History* by Thomas S. Elias. Copyright Thomas S. Elias.

Illustrations 26 and 91 are reprinted from *Aquatic and Wetland Plants of the Southeastern United States: Dicotyledons* by Robert K. Godfrey and Jean W. Wooten. Athens: University of Georgia Press. Copyright University of Georgia Press.

Illustrations 27, 33, 37, 151, 158, 226, 279, 284, 288 and 295 are reprinted from *Flora of Missouri*, vol. 2, by Julian A Steyermark, 1963, rev. 2006, George Yatskievych. St. Louis: Missouri Botanical Garden Press. With permission of Missouri Botanical Garden Press.

Illustration 199 is reprinted from *Manual of the Vascular Flora of the Carolinas* by Albert E. Radford, Harry E. Ahles, and C. Ritchie Bell. Copyright 1964, 1968 by the University of North Carolina Press.

Illustrations 141, 142, 143, 144, 146, 148, 149, 153, 155, 159, 160, 161,162, 163, 164, 166, 168, 171, 174, and 184 are reprinted from *Manual of the Trees of North America (Exclusive of Mexico) in Two Volumes* by Charles Sprague Sargent. Copyright 1905, 1922 by Charles Sprague Sargent.

Illustrations 35 and 125 are reprinted from *Plant Life of Kentucky: An Illustrated Guide to the Vascular Flora* by Ronald L. Jones. Copyright 2005 by University Press of Kentucky.

Index to Genera and Species

Names in roman type are accepted names, while those in italics are synonyms and are not considered valid. Page numbers in bold refer to pages that have illustrations.

Index to Common Names

Robert H. Mohlenbrock is in his sixty-third year of plant study. After receiving his doctorate from Washington University in St. Louis, Mohlenbrock taught botany and plant taxonomy at Southern Illinois University Carbondale for thirty-four years, sixteen of which he served as chairman of the department. During his career at SIU, he was major professor for ninety graduate students and carried out a vigorous program of research. To date, his research has resulted in the publication of sixty-one books and more than five hundred other publications. Since 1984, he has written "This Land," a monthly column for *Natural History* magazine, published by the American Museum of Natural History in New York. Since retiring from SIU in 1990, Mohlenbrock has been a senior scientist for Biotic Consultants, where he teaches weeklong wetland plant identification classes for government employees and consultants throughout the country.